Contemporary Research Trends in Agricultural Engineering 2017

BIO Web of Conferences Volume 10 (2018)

Krakow, Poland
25 – 27 September 2017

Editor:

Anna Szelag-Sikora

ISBN: 978-1-5108-6006-3

Printed from e-media with permission by:

Curran Associates, Inc.
57 Morehouse Lane
Red Hook, NY 12571

proceedings.com

Some format issues inherent in the e-media version may also appear in this print version.

Printed by Curran Associates, Inc. (2018)

For additional information, please contact EDP Sciences – Web of Conferences at the address below.

EDP Sciences – Web of Conferences
17, Avenue du Hoggar
Parc d'Activité de Courtabœuf
BP 112
F-91944 Les Ulis Cedex A
France

Phone: +33 (0) 1 69 18 75 75
Fax: +33 (0) 1 69 28 84 91

contact-edps@webofconferences.org

Additional copies of this publication are available from:

Curran Associates, Inc.
57 Morehouse Lane
Red Hook, NY 12571 USA
Phone: 845-758-0400
Fax: 845-758-2633
Email: curran@proceedings.com
Web: www.proceedings.com

TABLE OF CONTENTS

Statement of Peer review

In submitting conference proceedings to BIO Web of Conferences, I certify to the Publisher that I adhere to the **Policy on Publishing Integrity** of the journal in order to safeguard good scientific practice in publishing.

1. All articles have been subjected to peer review administered by the proceedings editors.
2. Reviews have been conducted by expert referees, who have been requested to provide unbiased and constructive comments aimed, whenever possible, at improving the work.
3. Proceedings editors have taken all reasonable steps to ensure the quality of the materials they publish and their decision to accept or reject a paper for publication has been based only on the merits of the work and the relevance to the journal.

Title, date and place of the conference

CONTEMPORARY RESERCH TRENDS IN AGRICULTURAL ENGINEERING
(KRAKOW, POLAND, SEMPTEMBER 25-27, 2017)

Proceedings editor(s): ANNA SZELĄG-SIKORA

Date and editor's signature 02.02.2018

17, avenue du Hoggar - PA de Courtabœuf – BP 112 - 91944 Les Ulis Cedex A (France)
Tél. : 33 (0)1 69 18 75 75 - Fax : 33(0)1 69 07 45 17 – www.edpsciences.org

Investigation for additional head loss in helically coiled flexible pipe of a mobile fertigation system

Aliaksandr Basareuski[1*]

[1]Republican Unitary Enterprise «Scientific and practical centre of the National Academy of sciences of Belarus of agriculture mechanization», Belarus

Abstract. One of the major parameters of a hydraulic machine is its energy intensity, i.e. its power/output ratio. This article presents the results of theoretical studies of the effects of additional head loss of transporting fluid through helically coiled flexible pipe on specific energy consumption of fertigation machinery and pipeline transport. Computational methods have been used to determine that pressure loss will be 15-20% higher on the average than the loss at a straight section, depending on the average radius of coil helix/design pipe diameter ratio. Correlations have been obtained that allow for justification of major mobile fertilizer system parameters based on the need to minimize pressure loss.

1 Problem Statement

In agricultural production, machinery with flexible coiled pipes is finding ever-widening applications. Thus, mobile hose-reel irrigation machines (HRIM) are used worldwide. Flexible pipes coiled on reels allow for using advanced fertigation patterns, increase the distance between closed irrigation system pipelines and irrigation valves, completely automate the irrigation process, and reduce its labor intensity and energy intensity. In animal husbandry, such transport systems are used in milking plants, for fodder preparation and distribution, and for manure water wash. Pipeline transport is efficient; it combines loading and unloading operations with transportation and provides high sanitary benefits. At the same time, the major limitation of such systems is the increased pressure loss in the water supply system. To reduce energy consumption per production unit, the water supply system must deliver fluid with the lowest pressure loss. As required methods for flexible pipe dimensioning are missing, pressure loss reduction in fluid transport and, thus, energy intensity reduction in the described systems still remain a challenging issue.

1.1 Past Studies and Publications

Investigations of researchers in Germany, Switzerland, Italy, France, England, USA, etc. established that plastic pipes are hydraulically smooth [1]. Therefore, a number of authors have obtained semi-empirical equations for the Darcy factor, which do not take roughness into account.

It should be noted that there have been some studies of hydraulic resistance in pipes helically coiled on a reel [2, 3, 4-7]. However, up to the present time, this issue has been little studied and is therefore very relevant.

It should be noted that the fluid behavior in a helically coiled pipe is somewhat different from that in a straight pipe. Thus, the transition to turbulence occurs at higher *Re* values than in a straight pipe [2, 3]. In the above articles, the common ground is that all the authors determine the friction factor based on the ratio of the coil diameter to the flexible pipe outside diameter (D_c/d_o). However, the correlations obtained in articles [2, 3, 4] only cover the area up to $Re \leq 10^5$ and are therefore not applicable to dimensioning of flexible pipe systems (machinery) with high-turbulence fluid flow.

Particular attention should be given to article [7] that states that head loss in flexible pipe coiled on a reel is 10...20% higher than at a straight section, depending on the radius of coil helix. The limitation of this article is that the experiments only cover small-diameter pipes, and extrapolation to larger diameters is hardly acceptable.

The aim of our research is to evaluate the effects of additional resistance in helically coiled flexible pipe on energy intensity of mobile fertigation equipment and pipeline transport.

2 Research Results

One of the major parameters of a hydraulic machine is its energy intensity, i.e. its power/output ratio. For pipeline transport, energy consumption will be in proportion to flow and head. Therefore, energy intensity is determined by head. The higher head loss (hydraulic resistance) is, the higher energy consumption is for pumping fluid through a water supply system.

For machinery with helically coiled flexible pipes, energy intensity, in addition, will depend on certain

*Corresponding author: labmkr@yandex.ru

design limitations. According to the recommendations in [8], the inside coil diameter must be at least 20 times larger than the flexible pipe outside diameter. This limitation is mainly due to creation of high pipe ovality (over 3%) when coiling flexible pipe on a reel. However, as experiments have shown, if HDPE 63 or higher grade is used, in order to reduce material consumption, this ratio can be decreased to 13.6 without increasing ovality [9]. A further decrease in D_c/d_o< 13.6 will lead to a significantly increased ovality. Therefore,

$$\frac{D_c}{d_o} = 13.6 \div 20, \quad (1)$$

where D_c is the flexible pipe coil diameter [m];
d_o is the flexible pipe outer diameter [m].

The ratio (1) expresses the design limitation of the parameters of coiling flexible pipe onto a reel.

We will explore hydraulic parameters of helically coiled flexible pipe on the example of HRIMs. Depending on capacity, their energy intensity ranges from 55 to 64 kWh/ha [10]. Fig. 1 shows a 3D model and a general view of OGD-50 fertilizer injector prototype *(a)* operated together with a mobile HRIM and the hydraulic parameter dynamics chart *(b)* during the irrigation cycle of this fertilizer system.

a)

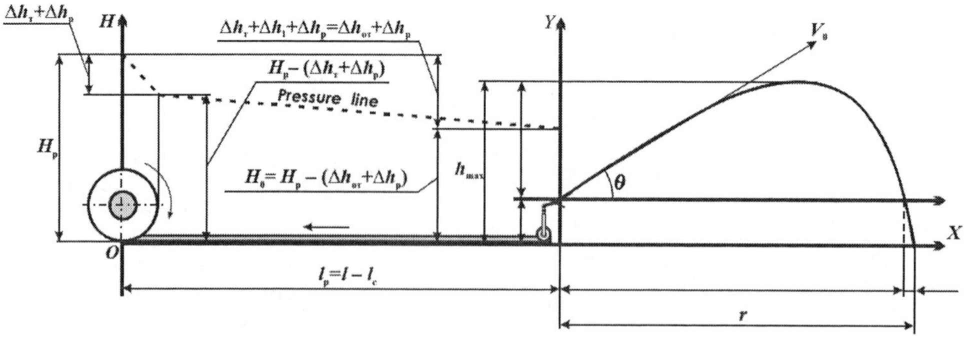

b)

Fig. 1. Mobile fertilizer system: *a)* 3D model of the equipment and general view of the system; *b)* chart for determining hydraulic parameters; H_p – working head [m]; H_0 – sprinkler head pressure [m]; Δh_p – head loss in drive [m]; h_{max}– maximum altitude of jet flight [m]; r – jet range[m]; θ – tilt angle of sprinkler [rad]; V_0 – initial jet speed [m/s].

The radius of coil helix for the i-th flexible pipe layer can be determined using the following equation:

$$R_{ci} = R_r + \frac{k_1 d_o}{2} + (i-1)k_1 d_o, \quad (2)$$

where R_r is the reel radius [m];
k_1 is the flexible pipe radial ovality factor equal to 0.99 [8].

Then the average radius of coil helix will be as follows:

$$R_c = \frac{\sum_{i=1}^{m}\left(R_r + \frac{k_1 d_o}{2} + (i-1)k_1 d_o\right)}{m} = \frac{k_1 d_o m^2 + 2mR_r}{2m} = R_r + \frac{k_1 d_o m}{2}, \quad (3)$$

where m is the number of flexible pipe layers.

Taking into account equation (3), the Weisbach equation for a smooth pipe curve [11] for coiled pipeline with n coils and m layers will be as follows:

$$\xi_{\text{bend}} = 4n\left(0.131 + 1.844\left(\frac{d_{\text{d}}}{2R_r + k_1 d_o m}\right)^{3,5}\right). \quad (4)$$

where ξ_{bend} is the local hydraulic resistance factor;
d_{d} is the flexible pipe design diameter [m];
δ is the flexible pipe curvature [rad].

Aronov's article [6] is especially noteworthy; the author established that in the high turbulence area, the curves $\lambda = f(Re)$ for curved and straight pipes run in parallel, and the total hydraulic resistance factor in curved pipes is a sum of factors taking into account the main flow friction and transverse circulation. The resistance factor ratio in curved and straight pipes is as follows:

$$C = \frac{\lambda_{\text{curv}}}{\lambda_{\text{str}}} = \frac{\lambda_{\text{str}} + \lambda_{\text{bend}}}{\lambda_{\text{str}}} > 1, \quad (5)$$

where λ_{curv} is the total hydraulic resistance factor in a curved pipe;
λ_{str} is the hydraulic resistance factor in a straight pipe;
λ_{bend} is the hydraulic resistance factor corresponding to an additional loss.

Based on his studies and depending on the ratio of the average radius of coil helix to the flexible pipe design diameter (R_c/d_d), Aronov recommends the values of C [6] that we have approximated using the following correlation:

$$C = \frac{1.816}{\left(\frac{R_c}{d_d}\right)^{0.156}}. \quad (6)$$

Then head loss in helically coiled flexible pipe calculated using the Darcy-Weisbach equation and taking into account condition (5) and Konakov equation [12] is as follows:

$$\Delta h_{\text{т}} = \frac{0.081 l_c Q^2 C}{d_d^5 g\left(9.489 + 0.782\ln\frac{Q}{d_d}\right)^2}, \quad (7)$$

where l_c the length of flexible pipe coiled on the reel [m];
Q is the flow rate (throughput) [m^3/s];
g is the gravity acceleration [m/s^2],

and taking into account equations (3) and (5) we obtain the following:

$$\Delta h_{\text{т}} = \frac{1.472 l_c Q^2}{d_d^5 g\left(9.489 + 0.782\ln\frac{Q}{d_d}\right)^2 \cdot \left(\frac{R_{\text{т}} + \frac{1}{2}k_1 d_o m}{d_d}\right)^{0.156}}. \quad (8)$$

The table shows pressure loss calculated using equations (4) and (8) in helically coiled pipe.

Table 1. Analysis of equations for hydraulic dimensioning of helically coiled pipe

Hydraulic Resistance Factor	Head Loss	Head Loss Increase [%]
$\xi_{\text{bend}} = 4n\left(0.131 + 1.844\left(\frac{d_d}{2R_r + k_1 d_o m}\right)^{3,5}\right)$	Borda equation: $\Delta h_{\text{м}} = \frac{32Q^2 n}{\pi^2 d_d^4 g}\left(0{,}131 + 1{,}844\left(\frac{d_d}{2R_r + k_1 d_o m}\right)^{3,5}\right)$, $\Delta h_{\text{т}} = \Delta h_1 + \Delta h_r$	27.4
$\lambda_{\text{curv}} = \frac{C}{(1.8\lg\text{Re} - 1.5)^2} = \frac{1.812}{\left(\frac{R_r + \frac{1}{2}k_1 d_o m}{d_d}\right)^{0.156} \cdot (1.5 - 0.782\ln\text{Re})^2}$	Darcy and Weisbach equation: $\Delta h_{\text{т}} = \frac{1.472 l_c Q^2}{d_d^5 g\left(9.489 + 0.782\ln\frac{Q}{d_d}\right)^2 \cdot \left(\frac{R_r + \frac{1}{2}k_1 d_o m}{d_d}\right)^{0{,}156}}$	18.7

Notes.
1. Calculations were made for $d_d = 60.65 \cdot 10^{-3}$ and pipeline length l_c =100 m at $Q = 10.91 \cdot 10^{-3}$ m^3/s; R_c =0.75 m; n = 13; m = 4.7.
2. At water temperature 15°C, $\nu = 1 \cdot 10^{-6}$ m^2/s (taken into account in numerical factors).

It can be seen from the table that the increase in head loss in helically coiled pipe, as compared with a straight section, is 27.4% when calculated according to the first equation and 18.7% when calculated according to the second equation. In the second case, the result is close to the experimental data, according to which loss increases by 10...20% depending on the R_c/d_d ratio [13]. In the first case, the equation gives a large error and is only suitable for estimations.

Fig. 2 shows, based on (8) and Konakov equation [12], a chart is given for $\Delta h_т = f(Q)$ correlation per 100 m of pipeline and $d_d = 60.65 \cdot 10^{-3}$ m for straight pipe and helically coiled pipe at different R_c/d_d ratios.

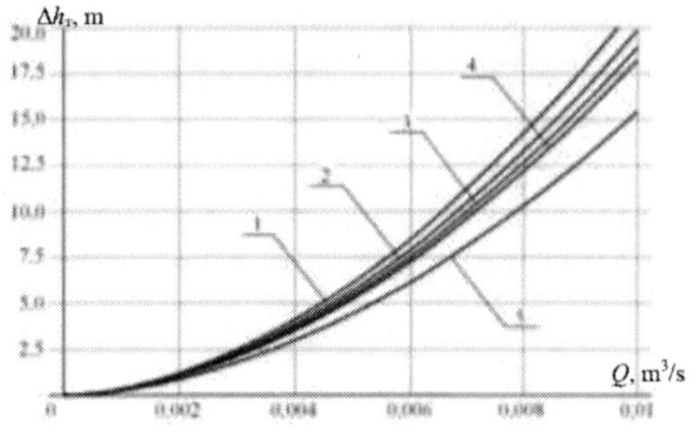

Fig. 2. Chart for head loss/flow correlation: 1 – $\frac{R_c}{d_d} = 5.77$; 2 – $\frac{R_c}{d_d} = 9.07$; 3 – $\frac{R_c}{d_d} = 12.37$; 4 – $\frac{R_c}{d_d} = 15.66$; 5– straight pipeline.

It is clear from the chart that as the ratio R_c/d_d increases, loss $\Delta h_т$ will decrease and approach loss Δh_1. Thus, at $Q = 0.008$ m³/s, loss at a straight pipeline section will be $\Delta h_т = 10.3$ m, while in coiled pipeline with $R_c/d_d = 5.77$, loss will be $\Delta h_т = 14.2$ m, and with $R_c/d_d = 15.66$, loss will be $\Delta h_т = 12.2$ m, which is 37.9 and 18.4% higher than at a straight section, respectively.

However, the resulting chart (Fig. 2) demonstrates pressure loss with fully coiled flexible pipe (at the end of the irrigation cycle). Actual total pressure loss will consist of the coiled pipe loss and the straight section loss:

$$\Delta h_{общ} = \Delta h_т + \Delta h_1 . \tag{9}$$

where Δh_1 is head loss along flexible pipe length [m].

After placing equation (8) and Konakov equation [12] in (9) and necessary transformations, we obtain the following:

$$\Delta h_{от} = \frac{Q^2\left(1.472 l_c + 0.727 l_d \left(\frac{2R_r + k_1 d_o m}{d_d}\right)^{0.156}\right)}{d_d^5 g\left(9.489 + 0.782 \ln\frac{Q}{d_d}\right)^2 \cdot \left(\frac{R_r + \frac{1}{2}k_1 d_o m}{d_d}\right)^{0.156}} . \tag{10}$$

Fig. 3 shows a chart of the correlation obtained for coiling onto a reel at $Q = 10.91 \cdot 10^{-3}$ m³/s, $R_r = 0.75$ m and $l = 370$ m.

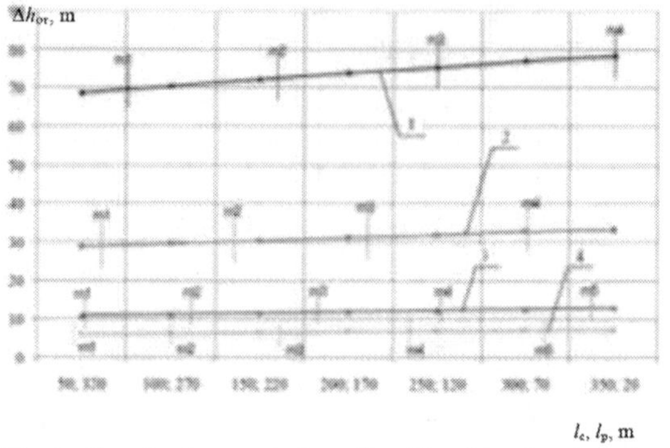

Fig. 3. Chart for the correlation of head loss/pipe length during its coiling onto the reel: 1 – $d_o = 0.075$ m; 2 – $d_o = 0.090$ m; 3 – $d_o = 0.110$ m; 4 – $d_o = 0.125$ m; m1, m2…mn – number of flexible pipe layers.

Thus, for a pipeline with $d_o = 0.075$ m, the loss difference in the coiled and uncoiled condition is 11.7 m, while for $d_o = 0.090$ m, this difference will be 5.6 m and for $d_o = 0.110$ m and $d_o = 0.125$ m, the differences are 2.4 m and 1.4 m, respectively. To reduce loss, R_r must be increased in accordance with (1). This is mostly feasible for pipelines with $d_o = 0.110$ m or higher, as with smaller d_o, specific material consumption will increase with the increase of R_r. It can also be seen from the figure that with identical structural parameters, the number of layers m increases with the increase of d_o.

Further, we will see how head loss varies depending on the inside reel diameter R_r at similar parameters (Fig. 4).

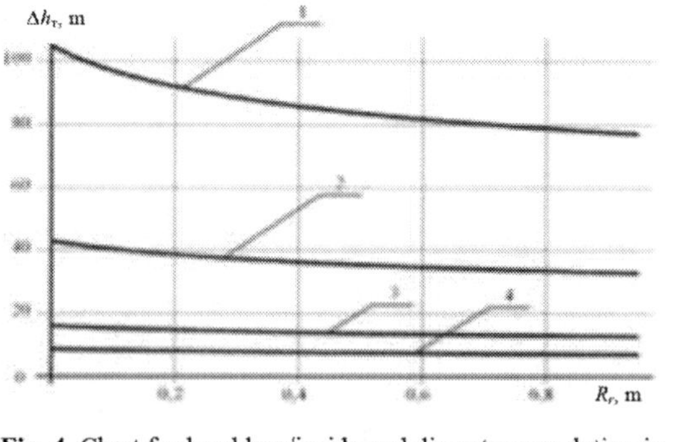

Fig. 4. Chart for head loss/inside reel diameter correlation in flexible pipe: 1 – $d_o = 0.075$ m and $m = 4.1$; 2 – $d_o = 0.09$ m and $m = 4.7$; 3 – $d_o = 0.110$ m and $m = 5.3$; 4 – $d_o = 0.125$ m and $m = 5.6$.

As we see, head loss increases as the inside reel diameter R_r and flexible pipe diameter d_o decrease. Thus, with $d_o = 0.075\ m$ and $R_r = 0.6\ m$, head loss is $\Delta h_т = 81.8\ m$. At the same time, with $R_r = 0.85\ m$ (41.7 % increase), head loss is $\Delta h_т = 78.2\ m$. Thus, with the same flexible pipe length, head loss decreased by 3.6 m (4.4%). Similarly, for the indicated parameters with $d_o = 0.09\ m$, head loss reduction is 1.4 m (4%) and, accordingly, with $d_o = 0.110\ m$ and $d_o = 0.125\ m$, head loss reduction is 0.5 m (3.8%) and 0.3 m (3.7%).

Thus, the above energy consumption of HRIM can be reduced to an average of 45–50 kWh/ha by justifying efficient design and operating system parameters based on the need to minimize pressure loss.

3 Conclusion

To justify the design parameters and operating modes of machines and systems with helically coiled flexible pipes, it is necessary to accurately determine pressure loss in the water supply system.

It is established that the ratio of the coil diameter to the flexible pipe outer diameter must be within the range of D_c/d_o= 13.6–20. Computational methods have been used to determine that pressure loss, depending on the average radius of coil helix/ pipe diameter ratio R_c/d_o, will be 15–20% higher on the average than the loss at a straight section (equation 10).

The analysis of equation (10) showed that, in order to reduce head loss (energy consumption), it is necessary to increase R_r and d_o in the above range of D_c/d_o ratio (equation 1). This is mostly feasible for pipelines with d_o = 0.110 *m* or higher, as with smaller d_o, specific material consumption will increase with the increase of R_r. With identical structural parameters, the number of layers *m* increases with the increase of d_o. Due to the reduced pressure loss, energy consumption for transporting fluid through the water supply system of the fertilizer equipment and the irrigation system can be reduced by 15–18%.

The correlations obtained will allow accurately determine and minimize pressure loss. This will help to reduce energy consumption with the same capacity and determine required design parameters for fertigation systems.

References

1. M.Ya. Ruzin. Gidravlicheskiy raschet plastmassovykh truboprovodov. Vodosnabzh. i sanit. tekhnika № 9, 1–5 (1962)
2. V.G Fastovskiy, A.E. Rovinskiy Issledovanie teplootdachi v spiral'nom kanale. Teploenergetika, № 1, 39–41(1957)
3. Yu.V. Kvitkovskiy. Gidravlicheskoe soprotivlenie plavno izognutykh trub. Sb. nauch. tr. Moskovskiy in-t inzh. zheleznod. transp. M., Vyp. 176, 61–63 (1963)
4. I.E. Idel'chik. Spravochnik po gidravlicheskim soprotivleniyam. M.: Mashinostroenie (1975)
5. I.Z. Aronov. O dvizhenii zhidkosti v izognutykh trubakh-zmeevikakh. Izvest. vyssh. uchebn. zavedeniy. Energetika, № 3, 65–74 (1961)
6. I.Z. Aronov. O gidrodinamicheskom podobii pri dvizhenii zhidkosti v izognutykh trubakh-zmeevikakh. Izvest. vyssh. uchebn. zavedeniy. Energetika, № 4, 130–132 (1962)
7. S.S. Savushkin, S.B. Khrol', T.M. Nekrasova. K voprosu o dopolnitel'nom soprotivlenii, voznikayushchem pri dvizhenii zhidkosti v polietilenovom shlange, navitom na baraban. Sb. nauch. tr. VNIIGiM, 134–140 (1983)
8. Truby napornye iz polietilena. Tekhnicheskie usloviya: GOST 18599–2001. Minsk: Mezhgos. sovet po standartizatsii, metrologii i sertifikatsii: Belorus. gos. in-t standartizatsii i sertifikatsii (2003)
9. E.Ts. Peterson. Ekspluatatsionnye svoystva orositel'nykh napornykh truboprovodov iz polietilena. Elgava (1987)
10. V.N. Dashkov, A.N. Basareuski. Sovremennaya energosberegayushchaya tekhnika dlya poverkhnostnogo poliva. Energoobespechenie i energosberezhenie v sel'skom khozyaystve. M.: VIESKh, Ch. 2, 168–173 (2008)
11. B.O. Botuk. Gidravlika. M.: Vysshaya shkola (1962)
12. F.A. Shevelev. Issledovanie osnovnykh gidravlicheskikh zakonomernostey turbulentnogo dvizheniya v trubakh. M.: VODGEO (1953)
13. V.N. Dashkov, N.F. Kapustin, A.N. Basareuski. Opredelenie gidravlicheskikh poter' v gibkom truboprovode barabanno-shlangovoy dozhdeval'noy ustanovki. Energosberegayushchie tekhnologii i tekhnicheskie sredstva v sel'skokhozyaystvennom proizvodstve. Minsk: BGATU, Ch. 1. – S. 72–78 (2008)

Comparison of price tendencies at domestic agro-food market of Ukraine and at global food market

Oleksandr Bitter[1], *Lyudmyla* Petryshyn[2]

[1]Lviv National Ivan Franko University, Ukraine
[2]Lviv National Agrarian University in Lviv, Ukraine

Abstract. The article describes price tendencies at the global and domestic agro-food market of Ukraine. It is proved that FAO Food Price Index, prices for food at the domestic market, as well as prices of food products sell by agricultural enterprises coincide. The research presents peculiarities of the tendencies at the markets of certain kinds of products, defines their statistical characteristics and outlines their impact on economic behavior of agricultural enterprises.

1 Introduction

Under current conditions, cereals and grain legumes are the main products at agrarian enterprises of Ukraine. Comparing to other kinds of agricultural products, their production has considerable advantages. A high level of its production profitability is the first one. A high liquidity is the second advantage, i.e. cereals can be sold easier than, for example, vegetables or meat. Third, modern technologies of cereals production can be easily mastered, and do not require specific knowledge and skills of personnel. Forth, there is a developed system of infrastructure to master the technologies, particularly systems of supply of the necessary material and technical resources. Fifth, natural and climatic conditions of the country are favorable for growing of cereals and grain legumes. Thus, potential of domestic agrarian enterprises is substantial in the field. Domestic economists and experts consider it is quite possible to increase amount of grain growing in Ukraine up to 100 million ton, as compared to the achieved level of 50-60 million ton.

. Technologically, it is quite real to reach such indicators. However, under conditions of market economy, it is not so important to use capabilities of production of a certain kind of products, but supply its sale. Both national and global market demonstrate tendencies, making negative impact on opportunities of grain sale and on prices for it. Particularly, social-political and financial situation in the countries, traditionally buying Ukrainian grain, i.e. Middle Eastern countries, first Egypt and Libya, is much deteriorated. There are proofs that similar problems can happen in Saudi Arabia, because of considerable fall of prices for oil. Markets of cereals in the Western Europe and other developed countries of the world are oversupplied. Main difference of the global grain market is that supply exceeds demand. Thus, globally prices for grain have fallen during the last five years.

Governments of countries in the world pay much attention to direct and indirect regulation of prices for food products, because it helps solution of the problems of food safety, income of rural producers, living standards of population, etc. However, under conditions of open economy, opportunities of the governments are limited in the field. Prices at global and national markets are closely connected.

The issue of pricing for agricultural products is studied in the works of numerous Ukrainian scientists. Among them, one should mark works of M. Artus [1], V. Hudak [2], V. Olefir [3], V. Oniehina [4], V. Okhrimenko [5], Kym Anderson [6]. The researchers gives complex description of different aspects of the problem. However, issue of relations between prices at the global food market and domestic agro-food market is poor studied. The given research aims made to fill the gap.

2 Materials and methods

Global statistics of prices for food products, as well as Ukrainian state statistics, contains numerous researches, devoted to comparison of prices dynamics at the global food and domestic agro-food market of Ukraine. Food and Agriculture Organization (FAO) monthly publishes FAO Food Price Index on its website concerning six groups: food products generally, meat, dairy products, cereals, oilseed and sugar [7]. Value of the six indexes are measured by comparison of their level at a certain period with an average level in 2002-2004. Data of the State Office of Statistics of Ukraine help to make correct determination of the values of similar indexes at the domestic agro-food market [8].

*Corresponding author: abitter@ukr.net
*Corresponding author: lydkar@ukr.net

Certain difficulties of methodic character occur because FAO Food Price Index is published in the form of basic index. However, in the process of the research of price tendencies it is necessary to use chain indexes. Considering regularities of relations between basic and chain indexes, the scientist overcomes the obstacle.

Methodologically, it is of great importance to bring indexes of prices at the domestic and global markets to a comparative form. Prices at the global market are determined in dollar equivalent and thus, prices at the domestic market should be defined in the same currency. Thus, it is necessary to transfer prices at the domestic market, revealed in hryvnia, into USD by the way of their division by the average exchange rate of the national currency in the studied period (a year or a month). Corresponding information is presented on the website of the National Bank of Ukraine [9].

3 Results and discussion

Prices for all main groups of agricultural products, sold by agrarian enterprises, increased in 2000-2015. However, rates of their raise were compared to the rates of increase of FAO global prices. The lowest rates were demonstrated by USD sell prices for the groups of cereals and grain legumes. In the studied period, they increased from 81,58 to 133,34 USD/ton, i.e. by 63,4 %. Approximately the same rates of growth were observed in prices for sugar beets. Much higher rates of price increase were seen for the products of animal breeding. In 2000-2015, the highest rates of USD price increase were demonstrated by oilseeds (3,57 times) and for fruits and berries (3,72 times). Such dynamics of prices makes base for economic behavior of agrarian enterprises.

Relatively long-lasting and stable accelerating growth of prices for oilseeds, including sunflower, rape and soybean, has caused a substantial expansion of the area under the crops. The similar conclusion is made concerning rapid development of horticulture and berry growing at agrarian enterprises of Ukraine.

The research has defined a considerable agreement of the rates of price changes for the three groups of agro-food products at the global and national markets. However, to obtain more accurate estimates it is necessary to carry out a more detailed comparison of the corresponding lines of dynamics. Two lines of dynamics of FAO Price Index for food products, i.e. chain and basic, are taken as a fundamental. Domestic dynamic lines are brought to a comparative form.

It is confirmed that integrated basic index of dollar prices for food products constituted 82,4 % at the domestic market. Its peak was reached in 2008 and constituted 161,9 %. Next increase was observed in 2010-2011, but it did not reach the level of 2008. Since 2008, in Ukraine, prices for food products have almost twice fallen in dollar equivalent. Under such conditions, import of food products is getting less important in the structure of their consumption, because it is not profitable to bring most of the products from abroad.

The current price situation supplies favorable conditions for export of the most kinds of food products from the country. Increase of dollar prices is particular for almost all kinds of products, produced by agrarian enterprises. Only dollar prices for vegetables were lower in 2015 than the average level in 2002-2004. At the same time, prices for cereals and grain legumes increased by 64,1 %, for oilseeds – by 91,7 %, fruits and berries – by 156,4 %, cattle and poultry (in live weight) – by 31,7 %, milk and dairy products – by 53,6 %, eggs – by 62,9 %. Thus, in the field of the products sell by agricultural enterprises, price tendencies are contrary to the tendencies at the domestic food market. It confirms improvements of economic situation of agrarian enterprises in cooperation with processing enterprises of AIC, on one hand, and growth of their export potential, on the other hand.

It is determined that in 2000-2015, integrated index of dollar prices for sale of agricultural products by agrarian enterprises made 239,2 %, including products of crop production – 249,7 %, and products of animal breeding – 258,9 %. However, degree of growth of prices for agricultural products was higher in favorable years, as compared to the degree of growth of prices for food products, and degree of fall was lower in unfavorable years. Fig 1. demonstrates values of the price indexes (indicator of the current year as compared to the indicator of the previous year).

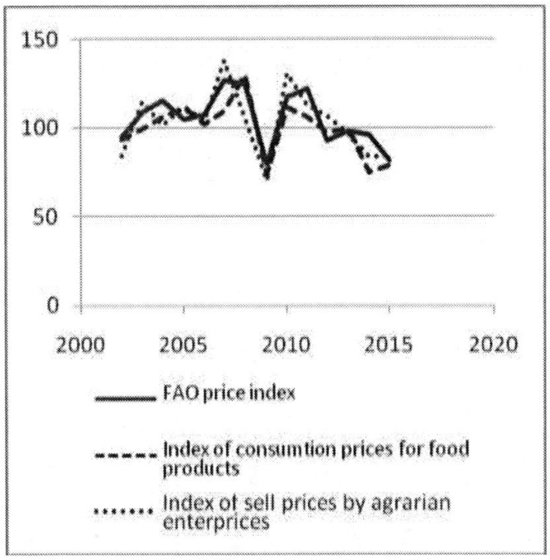

Fig. 1. Chain (indicator of the current year as compared to the indicator of the previous year) index of FAO global prices, consumption price index at food market of Ukraine) and index of sell prices for agricultural products by agrarian enterprises.

Basing on the comparison of the three lines of dynamics, one can make conclusion about a high level of their instability in the whole period of the study. However, the three indexes clearly demonstrate mutual dependence. In case the key index among them, i.e. FAO price index of a certain year, increases or decreases, as compared to the previous year, one observes increase or decrease of the two other indexes.

It can happen not in the same year, but with one-year lag.

Nevertheless, until 2012 particularly great differentiation of the indicators was observed concerning dynamics of sell prices for products of agrarian enterprises. In case they increase in a certain year, the raise is much higher than FAO prices or prices at the domestic food market. In case they decrease, the fall is also at rather high rate. However, since 2013, degree of their fluctuation has been average between the degree of FAO price fluctuation and prices at the domestic food market.

Comparison of dynamics of the same three indexes according to the basic method of their calculation demonstrates some other correlation. In 2002-2006, sell prices for products of agrarian enterprises increased much slower, comparing to the rates of growth of FAO global prices for food products and prices for food products at the national agro-food market (fig. 2).

In 2008-2010, they either exceeded them or were lower. Since 2009-2010, one has observed a clear tendency, i.e. rates of increase of sell prices for agricultural products, produced by agrarian enterprises, have been approximately the average of the rates of growth of FAO global prices and prices for food products at the national agro-food market.

The tendency is permanent in the last five years. Responding to the changes, agrarian enterprises grow their export potential, because they expect that the mentioned tendency will be long-lasting and benefits from export of agricultural products will prevail over opportunities to gain profit from their sale at the domestic market. Such economic behavior of agrarian enterprises is quite logical, in spite of the remarks of numerous domestic scientists about excessive export focus of production at large agrarian enterprises.

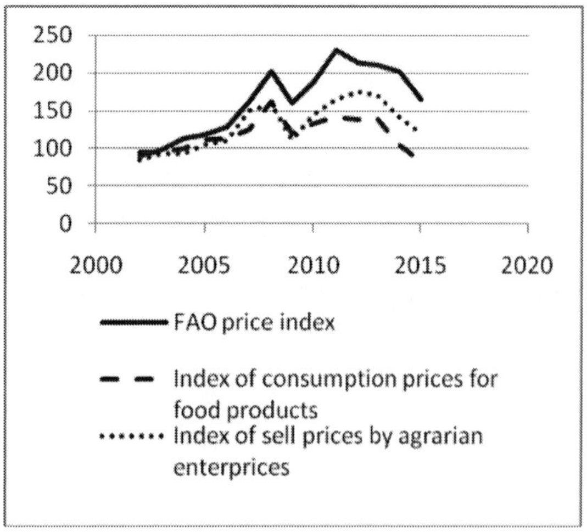

Fig.2. Basic (indicator of the current year to the average indicator in 2002-2004) FAO global price index, consumption price index at the food market of Ukraine and index of sell prices for agricultural products by agrarian enterprises

For small agrarian enterprises, growth of export potential is an extra complicated task, because it is difficult for them to meet all the requirements to the quality of exported commodities and control for them. Thus, one can see the situation, when supply of agricultural products at the domestic market is the business of small agrarian enterprises, and large enterprises are engaged in export of the products.

The authors of the article use statistical indicators of variation and correlation for a deeper analysis of price environment of agrarian enterprises development. Coefficient of variation, as a ratio of the standard deviation to the value of arithmetical average, revealed in percent, demonstrates a level of likeness of a certain complex of statistical indicators. Table 1 presents values of variation coefficients of chain indexes of prices for the whole complex of agro-food products and for separate kinds of them, defined by FAO in 2002-2015.

Table 1. Coefficients of variation of FAO chain price index, consumption prices at the domestic food market and sell prices for the products, sold by agrarian enterprises in 2002-2015.

Kinds of products	FAO price index	Index of consumption prices at the domestic food market	Index of sell prices for the products, sold by agrarian enterprises
Agro-food products	14,0	14,9	17,4
Bread and cereal products, cereals and grain legumes	18,6	15,2	28,0
Vegetable. oils, oilseeds	21,7	18,8	24,8
Sugar, sugar beet	24,0	21,3	16,8
Meat and meat products, cattle and poultry	10,7	18,9	23,8
Milk and dairy products	25,7	17,9	27,9

Coefficients of variation, presented in the table 1, characterize degree of likeness of prices for the determined kinds of products in a period, i.e. degree of their stability in the studied period. The carried analysis proves a rather high level of instability of FAO food prices, coefficient of correlation of the price index for the whole complex constitutes 14,0 %. The value of the coefficient could be much higher, but prices for meat and meat products demonstrated relatively stable character and they occupied a substantial share in the structure of the global agro-food market. Variation coefficient of all other kinds of products was some higher or lower than 20%. Among the above-mentioned kinds of products, prices for milk and dairy products were the least stable in the period of 2002-2015. Degree of variability of the global price indexes for sugar was not much lower.

In the process of correlation analysis of the lines of indexes of FAO food prices, consumption prices index at the domestic food market and sell prices for the products, sold by agrarian enterprises, the researchers obtain data about degree of relations between the lines.

Corresponding values of correlation coefficient are demonstrated in the table 2.

Coefficient of correlation between FAO price indexes and indexes of consumption prices at the domestic food market constituted 0,843 in the period of 2002-2015. Thus, price situation in AIC of the country heavily depends on its dynamics at the global agro-food market.

Table 2. Coefficient of correlation between FAO chain price indexes, consumption price indexes at the domestic food market and sell prices for the products, sold by agrarian enterprises in 2002-2015.

Kinds of products	FAO price index: indexes of consumption prices at the domestic food market	FAO price index: indexes of sell prices for the products, sold by agrarian enterprises	Indexes of consumption prices at the domestic food market: indexes of sell prices for the products, sold by agrarian enterprises
Agro-food products	0,843	0,810	0,728
Bread and cereal products, cereals and grain legumes	0,801	0,468	0,620
Vegetable. oils, oilseeds	0,624	0,708	0,275
Sugar, sugar beet	0,497	0,817	0,640
Meat and meat products, cattle and poultry	0,808	0,790	0,939
Milk and dairy products	0,644	0,894	0,851

One should stress that at the markets of certain kinds of agro-food products, the defined degree of mutual dependence substantially differed in 2002-2015. Particularly, time dynamics of consumption prices at the global and national markets of bread and cereal products was closely connected (coefficient of correlation constituted 0,801).

In 2002-2015, ratio between price tendencies at the markets of vegetable oils and oilseeds were even more various. Coefficient of correlation between the indexes of prices for vegetable oils at the global and national markets constituted 0,624. At the same time, the same indicator in dynamics of prices both at the global market and for products of agrarian enterprises was much higher and constituted 0,708. However, degree of mutual dependence between dynamics of prices for vegetable oils at the domestic market and dynamics of prices for oilseeds at agrarian enterprises was so low that it could be neglected. Dynamics of sell prices for sunflower, rape and soybean, sold by agrarian enterprises, was determined not by dynamics of prices for products of their processing at the domestic market, but by dynamics of prices at the global market. Similar situation was observed at the markets of sugar in the studied period.

The highest indicators of mutual dependence between price tendencies at the global and national markets were observed at the markets of meat and meat products. Contrary to the other kinds of products, dynamics of sell prices for cattle and poultry, sold by agrarian enterprises, was determined by dynamics of prices for meat and meat products at the domestic consumption market, which demonstrated high indicators of dependence on dynamics of prices at the global market. Almost the same situation concerning price correlations occurred at the markets of milk and dairy products in 2002-2015. However, national market of the products was considerably independent from the global market in the studied period.

4 Conclusions

Statistical processing of the data about dynamics of prices for food products generally and their separate kinds, with application of dispersive and correlation analysis, results in conclusions that price environment of agrarian enterprises development is mainly defined by price tendencies at the global market, not at the national market. Such situation proves a weak position of the national government in opportunities to activate its impact on price situation. However, it does not mean that the tendency should be supported in future. Obviously, the government, which cares about income of domestic commodity producers in agricultural field, as well as about living standards of population, should influence price tendencies, because it is one of its tasks.

References

1. M.M. Artus, State support and financial mechanism of pricing for agricultural products: abstract of PhD thesis, Kyiv, Ukraine (2011).
2. V.V. Hudak, IE, **7**, 221-225 (2011)
3. V.K. Olefir, EF, **1** (2012), 143-150.
4. V.M. Oniehina,*State regulation of prices andincome of agricultural producers*: monograph, NNTs "IAE", 558 (2007)
5. V.I. Okhrimenko, EAIC, **7** (2011), 116-122.
6. Kym Anderson Agricultural price distortions: trends and volatility, past, and prospective Agricultural Economics Volume 44, Issue s1.November (2013), 163–171.
7. FAO Food Price Index, available at:www.fao.org/worldfoodsituation/foodpricesindex/en/.
8. State Statistics Service of Ukraine, available at: http://www.ukrstat.gov.ua.
9. National Bank of Ukraine, available at:https://bank.gov.ua/control/uk/index.

Investment Management of Agro-Ecological Farms

Victoria Borisova[1,*], *Iryna* Samoshkina[1]

[1] Sumy National Agrarian University, Department of Finance, Sumy National Agrarian University, G. Kondratyeva Str., bld. 160 Sumy, 40021, Ukraine

Abstract. The effective development of farms requires the environmental and economic investing, which provides for investments in improving the soil quality that can be enhanced through a fresh look at the problem of degradation, given that agricultural land is considered to be produced, not primary resource, similar to all other kinds of resources accounted in agriculture. The degradation and restoration of land productivity are the problems of the analysis of investment updates, modified with the aim of establishing the relationship of the soil quality and relevant investments. The effective investing of additional capital is crucial for the successful investment activities of agro-ecological farms, while investment resources are subject to management. The stirring up of investment activities is possible on the basis of the environmental and economic planning of farms as the important link in the structure of the organizational and economic mechanism of farm operation, the efficient investment management. The article focuses on the directions of attracting investments in agro-ecological farms and agricultural enterprises to improve their business performance. The investment attractiveness of environmental agribusiness is justified. The features of management and trends in the development of the investment activities of agro-ecological farms are identified.

1 Introduction

Agroecology is the study of ecological processes applied to the design, development, and management of agriculture. Agro-ecological is a specialist asset manager focused on farmland as an asset class managed ecologically.

Management of all aspects of the investment activity of enterprise occupies the central place in the organizational and economic mechanism of intensification of the investment process in agribusiness. It includes management of investment projects, economic and financial levers.

During the period of 1991–2016 in Ukraine crop acreage decreased by 16.8%; forage acreage decreased by 4.6 times, that negatively affected the development of fodder supply of the livestock industry. During that period planting of lucrative export-oriented crops significantly increased: sunflower by 2.8 times, rape by more than 10 times that violates the crop rotation system and results in soil depletion. Acreage under potatoes and vegetables and melon crops has decreased by 5.0%, now these crops are concentrated mainly in households [1].

Motivation to attract investment in agro-ecological production depends on financial stability, flexibility and stimulating character of the tax policy, completion of the reform of ownership. Creation of favorable investment climate in the agrarian sector and promotion of investment activity of all economic entities, considering the current financial and material-technical provision of agricultural enterprises and agro-ecological farms need solution of the issue of quality changes in approaches to the management of agro-industrial production as a whole, every industry and business entity. Therefore, it is essential to form an effective mechanism of the ecological and economic investment management in the agrarian sector, adequate to the conditions of current economic status, taking into account the specifics of agricultural production, that will be able to create profitable conditions for stimulating savings of agricultural capital, and increasing the efficiency of its use, that is possible due to the development of ecological agribusiness.

2 Research goals, tasks, objects and methods

Goal of the research is to draw up scientifically grounded proposals for the development of ecological and economic investments of agricultural enterprises and agro-ecological farms, promotion of investments in the agribusiness to increase its economic efficiency and to expand the production of environmentally friendly farm products. Implementation of this goal is associated with the solution to the following tasks:

• to study the features and trends of investment in the economy of Ukraine;

• to justify the essence of investment attractiveness of agro-ecological entrepreneurship; make classification of organizational, economic and ecological measures of agricultural land protection;

* Corresponding author: vikibor@ukr.net

• to analyze the features of investment management of agro-ecological farms.

Object of the research is ecological and economic investments of agricultural enterprises and agro-ecological farms in Ukraine, practice of its implementation as a tool of investment management.

The following methods were used in the research: abstract-logical (to clarify the conceptual provisions of the development of investment management of agro-ecological farms, to make the economic assessment of the agricultural technologies of deficit-free humus balance), analytical (to justify the information-analytical support for the process of management of attracting investments in agro-ecological farms), statistical (to analyze capital investments in Ukraine by types of economic activities, investment activities of agro-ecological farms; to make the calculation and economic assessment of the agricultural technologies of deficit-free humus balance in growing winter wheat), normative (to justify the value of humus depending on the cost of raw materials for its formation), etc.

3 Environmental and economic investment agro-ecological farms

3.1 Features of attracting investments in the economy of Ukraine

The experience of the advanced economies shows that the agribusiness sector is attractive for investments, as there is always a steady demand for agrarian production, which has no tendency towards reduction, and the agricultural products (in contrast to their production technologies) get never out of date "technically and technologically", that is confirmed by the evolutionary process of economic development.

In Ukraine in 2016 economic entities from all sources of financing in agriculture, forestry, fishery invested 1.95 billion USD in fixed capital, that is 49.65% more than in 2015. The share of investment in these economic activities is 14.05% of the national volume of investments in fixed capital (in 2015 – 11.04%) [2]. Capital investments by types of economic activity are given in Table 1.

Table 1. Capital investments by types of economic activity for 2014-2016 (million USD)*.

Economic branch	2014	2015	2016
Total	14,667	11,798	13,875
Agriculture, forestry and fisheries, including:	1,256	1,303	1,950
• agriculture, hunting and provision of the related services	1,229	1,266	1,918
• forestry and timber logging	25	34	30
• fisheries	2	2	2
Industry	5,765	3,786	4,548
Construction	2,410	1,878	1,717
Other services	3,980	3,529	3,710
Total (%)	100.00	100.00	100.00
Agriculture, forestry and fisheries (%), including:	8.56	11.04	14.05
• agriculture, hunting and provision of the related services (%)	8.38	10.73	13.82
• forestry and timber logging (%)	0.17	0.29	0.22
• fisheries (%)	0.01	0.02	0.01
Industry (%)	39.31	32.09	32.78
Construction (%)	16.43	15.92	12.37
Other services (%)	27.14	29.91	26.75

* Calculated by the author according to the Office for National Statistics of Ukraine, excluding the temporarily occupied territory of the Autonomous Republic of Crimea and Sevastopol; without part of the antiterrorism operation zone [2]

The capital investments in the amount of 0.42 billion USD, accounting for 16.8% of their total volume in Ukraine [2, 3], were made in the development of the enterprises of agriculture, forestry and fisheries during January–March 2017. The most important direction of investments in the reproduction of natural resources used in the agrarian sector is the financing of measures aimed at improving soil fertility, as soil is irreplaceable and geographically limited means of production.

The unsatisfactory ecological state of Ukrainian farmland, primarily black soil, is of particular concern. 11 million tons of humus, 0.4 million tons of phosphorus and 7 million tons of potassium are annually taken out of soil that is 2.3 times more than the quantity thereof put in with fertilizers [4]. Over the past decade, the process of soil dehumification has accelerated. The trend towards the soil quality deterioration during the production of monocultures has been detected even before the mechanization and chemicalization which have triggered it by affecting adversely the soil bioproductivity through the compaction, accumulation of pesticides and other harmful substances. The irrational use of mineral fertilizers and chemical means of plant protection leads to an increase in the scale of soil contamination, threatens human health, endangers the stability of ecosystems [5, 6].

The goal of the eco-economic planning of agricultural production is not only the determination of the conditions necessary to increase production, but also ensuring a balance between the production and environment, where land plays a major role. In the agricultural natural resource management an increased focus should be put on the environmental aspects of land use such as the protection of land resources from contamination and the restoration of degraded lands, with the aim of their further effective use by current and future generations, that requires attracting investments.

3.2 Investment attractiveness of agro-ecological entrepreneurship

The strengthening of government control over land use and environmental conditions leads to the development of environmental entrepreneurship – specific form of socially useful activities, the essence of which lies in the reproduction and rational use of natural resources involved in the production. The economic expediency of agro-ecological entrepreneurship is obvious: first and foremost, it is the return of free goods, 'lost' for the society, to agricultural resource circulation, attraction of previously inaccessible resources.

Considering the world experience in economic solutions to environmental problems of agriculture, the impetus for the development of agro-ecological entrepreneurship in Ukraine should be tax incentives for environmental investment – investment in ecologically sound technologies and equipment through tax methods [5, 7].

Evaluation of economic efficiency of investment plays an important role in determining the investment attractiveness of agro-ecological farms. An integrated assessment of social, ecological and economic results of investment activities in agricultural production can be developed, provided that social and environmental outcomes expressed in physical indicators, can be denominated in value terms. However, the specific features of agricultural production complicate social, environmental and economic evaluation of the effectiveness of investment in agriculture. The most important ones include the uneven machinery use over time, structural changes in soil (water-air regime, erosion resistance, microflora and zoofauny development), the impact of a number of biological factors, especially the chemical composition of the air in livestock houses, accompanied by a high content of pollutants (ammonia, hydrogen sulphide, carbon dioxide and others). Social outcomes in agricultural production are sometimes manifested not immediately, but after a lag. In addition, they are formed indirectly through flora and fauna. Environment is the first to be affected: air, soil, water, which in turn in modified form influence the flora and fauna, and the last have impact on a person through food consumption.

A crucial prerequisite for successful investment activity of enterprises in market conditions is the efficient investment of additional capital, as investments are subject to control. Therefore, the intensification of investment activity in the agribusiness on the micro and macro levels should be controlled. Enhancement of investment is possible on the basis of planning, which is an important element in the structure of the organizational and economic mechanism. In order to attract investment it is necessary to develop an investment project or program. Potential investor assesses production process in terms of environmental protection and conservation measures which should ensure compliance with the project:

- reduction in emissions that meet environmental standards in Ukraine;
- lack of toxic and hazardous substances;
- maintenance and improvement of land quality;
- rational use and decline in the amount of natural resources involved in the exploitation;
- decrease in emissions of pollutants into the environment.

The high development of the state soil reserves in Ukraine (70.9%, the level of tilled soil is 53.8%, including 77.9% of farmland) and extensive use of productive land result in the increased land degradation, soil loses its fertility and becomes depleted. According to the Ministry of Agrarian Policy and Food of Ukraine 13.2 million hectares of agricultural land suffer from water erosion, 19.3 million hectares from wind erosion, another 10.7 million hectares degrade due to high acidity, 3.9 million hectares are saline lands, 3.6 million hectares are wetlands. The area of arable land that has been degrading due to erosion for the last 25 years has increased by one-third. And every year this degradation is expanding at 90-100 ha, and humus content in soil reduces by 20% [4]. Cuts in financing of erosion preventive measures result in reduction in land protection measures. Therefore, trends of increasing efficiency of agricultural land use should be as follows:

- improvement in quality of soil surface (maintenance of physical and mechanical soil conditions, water regulation), which undergoes changes as a result of improper agricultural use;
- prevention of erosion processes with simultaneous improvement in land fertility (reducing the intensity of mechanical treatment of soil with heavy machinery, reducing the number of passes across the field);
- implementation of reclamation measures on acid soils and salt water marshes, increasing humus in soils by proper usage of organic and mineral fertilizers;
- prevention from soil contamination by technogenic factors and agricultural use;
- introduction of science-based systems of crop rotation by various entities in the agricultural sector;
- improvement in the structure of crop acreage.

Protection of agricultural land is based on a combination of organizational, economic and environmental elements of influence on businesses in order to comply with the requirements of the rational use and protection of land as the most essential element of the economic system, universal means of production and unique natural resource (Figure 1).

Under conditions of transformation of agricultural relations the role of organizational and economic mechanism of the efficient agricultural land use such as: financing, pricing, lending, taxation and insurance coverage increases dramatically in the presence of sound organizational and administrative and legal support [3]. Sources of financing the systems of measures aimed at land use and protection should be formed at the expense of state and regional budgets, administrative areas and farms, special off-budget state and other funds [8].

It is expedient to create a flexible system of incentives for improving the ecological condition of land by establishing tax and other benefits granted to the enterprises, companies, institutions and citizens while ensuring the quality of land that meets the environmentally sound standards, to release the enterprises from taxation of profit obtained from the investments allocated for the construction of soil and environmental protection facilities, introduce a system of benefits for producers of environmentally friendly products [9].

Land use in Ukraine is paid and land rental is in the form of land tax or rent for landowners and land users. These funds should be used exclusively to finance the management of soil fertility. Moreover, additional sources of stimulation for improving the efficiency of land use may be payments to the budget when removing unproductive lands from agricultural turnover and funds from the payment of penalties.

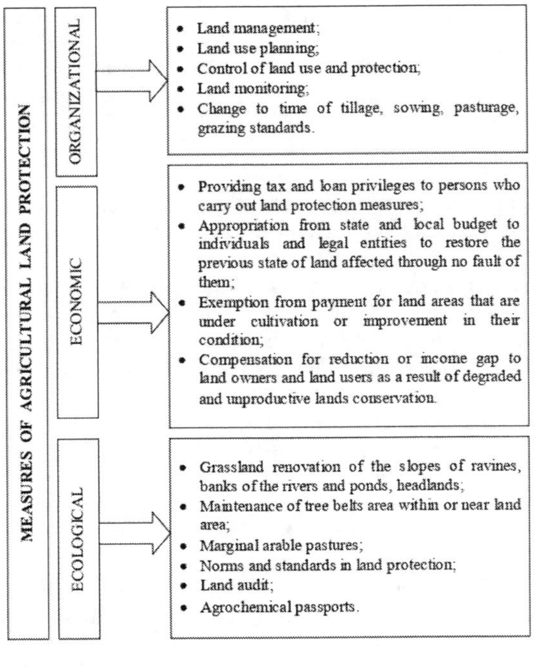

Fig. 1. System of Organizational, Economic and Ecological Measures of Agricultural Land Protection,

Source: Authors' presentation

Economic incentive for improving agricultural and ecological status of agricultural land is based on the methods of monetary valuation of land. Motive of economic incentives for landowners and land users is to increase agricultural and ecological score of land, which is determined at regular agrochemical examination. A necessary condition for stimulating is a growth of agrochemical indicators of soil fertility [10, 11].

Economic lever of influence on the land use efficiency is ecological and agrochemical score of land. Penalties should be applied for deterioration of ecological conditions of land. In the event that agricultural and ecological score of soils did not change during the latest agrochemical examination, it would be inexpedient to conduct stimulation. Economic incentives should not exceed the penalties when agricultural and ecological score tends upwards or downwards. It will promote the economic interests of landowners and land users in increasing land fertility, increase the value of agricultural and ecological score.

Therefore the main ecological and economic principles of rational use, restoration and protection of productive land, including eroded soils are: the advantage of long-term social and economic programs and approaches over the short-term needs; the complexity of solving environmental problems and issues of fertility restoration and preservation of eroded land, rational use of land resource potential, integration, interrelation of natural resources and logistical components of the total resource potential of agricultural production.

4 Investments in soil fertility recovery

Investments in the soil improvement expand the range of the capabilities of agricultural enterprises and farms, taking into consideration the location of products and resources. In addition to choosing technologies and crops, the optimization problem involves the allocation of resources between different types of products and future land quality [12]. Earlier on, the models used to represent the special cases, when agricultural enterprises were believed to be able to deplete the soil, though to different extents, depending on the productive and technological choices. The empirical significance of this "approach" to the soil degradation becomes apparent, and the need for their ecological and economic development is considered as well.

The problem of the soil fertility management is quite complicated and poorly studied, but it is the most rational way to produce cheap and ecologically pure products without outward investments under modern conditions in Ukraine. The study has shown the possibility of replacing missing or insufficient quantities of organic fertilizers of animal origin in a vegetable way [13, 14]: the replenishment of humus in the soil through the use of grain crop straw and other by-products; changes in the structure of sowing areas to increase the acreage of perennial grasses; the cultivation of green manure.

Our study has shown that about 1–1.5 million tons of straw stored in stacks and remaining unused are annually accumulated under the conditions of Sumy region (Ukraine). The wasted labor and means of production for growing and stacking straw turn to be useless, but, by the way, they are significant. Thus, if the estimated cost of 1 t of straw is 1.16 USD, such costs will be from 1.16 to 1.74 million USD as a whole in Sumy region. On the other hand, all the prepared straw in the accounting registers of farms is recorded as the straw used for the production needs, i.e. the actual cost of agricultural production is overestimated.

We believe that it is essential to store the optimal, economically justified amount of straw for the production purposes, and the rest of it should not be burnt, but used to replenish soil with organic matter. The evaluation of costs per unit of humus is made depending on the used vegetable or animal organic fertilizers (Table 2).

Table 2. Calculated data for the formation of humus and its cost under the conditions of Sumy region (Ukraine),

Types of organic fertilizers	Cost of 1t of organic fertilizers, USD	Coefficient of humification	Amount of humus formed, c	Cost of 1c of humus, USD
Green manure	0.52	0.15	1.5	0.35
Litter manure*	1.55	0.10	1.0	1.55
Peat festering compost	1.55	0.15	1.5	1.03
Straw	1.16	0.11	1.1	1.07
On average	-	-	-	1.00

* Adjusted for the cost of work on transportation and fertilizer application

Source: Authors' presentation

The above calculations have shown that the humus formed from vegetable organic fertilizers is cheaper than the humus formed from manure. The determination of the evaluation of 1 c of humus enables to make a comparative assessment of the economic feasibility of certain options for enhancing fertility, depending on the farm specialization. The results of these calculations are given in Table 3.

Table 3. Cost of humus depending on cost of raw materials for its formation.

Features of options	Humus formed, kg/ha		*Cost of humus, USD/ha		Cost of the received humus, total USD/ha
	from manure + SRR**	from manure	from manure + SRR**	from manure	
Forest-Steppe Zone					
Grain specialization, 0.3 head of cattle per 1 ha	1071	250	11.41	3.88	15.29
Grain-dairy-vegetable, 1 head of cattle per 1 ha	454	850	4.84	13.18	18.02
Grain-beet-dairy, 0.3 head of cattle per 1 ha	1361	250	14.51	3.88	18.39
Farm. Grain specialization	1420	-	15.14	-	15.14
Transition Zone					
Grain-beet-dairy. 1 head of cattle per 1 ha, perennial herb up to 30% of arable land	678	1000	7.23	15.51	22.74
Polissia Zone					
Grain-meat- dairy. 1.5-2 heads of cattle per 1 ha, perennial herb up to 30% of arable land	737	1200	7.85	18.61	26.46

* Average calculated price of one centner of humus from manure is 1.55 USD; 1c humus from straw and stubble remains – 3.0 USD

** SRR – stubble-root remains

Source: Authors' presentation

The data of Table 3 enable to determine the proportion of newly formed humus, which can be obtained as a result of the use of plant organic fertilizers, and to define the amount of humus to be supplemented by other sources of organic matter, e.g. manure. This allows for the estimated valuation of the humus balance per 1 ha of arable land, depending on the specialization and agro-climatic zones under the conditions of Sumy region (Ukraine). The figures are as follows: in the Forest-Steppe Zone the mineralized humus may be replenished by 68.7% at the expense of any alternative means of vegetable origin; in the Transitional Zone – by 40.4%; in Polissia Zone – by 38%.

Manure can be used to restore the shortage of organic substances, but an increase of its share in the total mass of organic fertilizers leads to the increase in expenditure on the humus formation on 1 ha from 15.14USD/ha in the Forest-Steppe Zone to 26.46 USD/ha in Polissia Zone.

Natural fertility is the fertility, by means of which it is possible to get a certain harvest, without any additional measures of the fertility increase, subject to the requirements of the technologies. Under favorable agro-climatic conditions, the natural capacity of grain crops grown in Sumy region (Ukraine) in the Forest-Steppe Zone is 26.4 c/ha, 19.5 c/ha in the Transition Zone, in Polissia Zone – 13.6 c/ha [15]. The calculation of the economic efficiency of applying the agricultural technologies of deficit-free humus balance shows that the greatest efficiency is observed in the Forest-Steppe Zone compared to the Transitional Zone and Polissia Zone (Table 4). At the same time, the cost of additionally produced products is greater on the less fertile soils of Polissia and Transitional Zones than in the Forest-Steppe Zone.

Table 4. The estimated economic evaluation of the agricultural technologies of deficit-free humus balance per 1 ha when growing winter wheat in Sumy region (Ukraine)*.

Agro-climatic zones	Evaluation of the products due to the natural fertility, USD	Economic indicators of agricultural technologies of deficit-free humus balance			
		Valuation, USD			Coefficient of agricultural technologies (hr. 5: hr. 3)
		deficit-free humus balance	additionally produced products	profit by means of agricultural technologies	
1	2	3	4	5	6
Polissia	73.39	18.93	81.11	62.18	3.28
Transitional	79.95	16.61	75.71	59.10	3.56
Forest-Steppe	142.91	13.91	70.30	56.39	4.05

* Calculated by the author according to the data of the Office for National Statistics of Ukraine [2]

It should be noted that keeping a deficit-free humus balance creates favorable conditions to obtain stable yields, enables to increase yields not only in the year of organic matter application to the soil, but in subsequent years, that is the goal of the effective and qualitative organizational-economic mechanism of fertility management for the proper provision of society with foodstuff.

The effectiveness of the soil fertility management largely depends on the applied methods and capabilities of agricultural enterprises and farms on providing them with the means of the fertility increase, the level of the adequate management of these processes, the attraction of investments to grow ecologically clean farm products.

5 Conclusions

The challenging ecological situation in the Ukrainian agriculture is characterized by the lack of targeted financial resources required for the reproduction of natural resources. The environmental and economic problems in the agrarian sector are the result of the overloading of the extensive development of agriculture in the past, the lack of effective policy of its ecologization. The most effective solution to the problems of ecological development of agricultural production should become the intensive investment activities aimed at financing the acquisition of environmental and resource-saving technologies, the implementation of other measures for the conservation of natural resources in the agrarian sector.

The environmental and economic investing is aimed at increasing economic efficiency of agricultural production with the consideration of environmental interests that enables agricultural production to get to a whole new level, to develop agro-ecological farming. The essence of this transition is the quantitative accumulation of qualitative changes in the elements of the agricultural production ecologization system, the new structural combination of which in their interrelation and interaction will lead to the more effective way of the natural resource management in this industry. The development of the investment management in agro-ecological farming will enable to expand the production of environmentally friendly farm products, to improve the fertility of agricultural land, to find internal investment resources for farms in order to reduce the cost of agricultural products.

References

1. O. Sobkevych, V. Rusan, A. Yurchenko et al., *Development of agricultural production as a prerequisite for food security in Ukraine*, NISS, 39 (2017)
2. [online] [access: date] State Statistics Service of Ukraine: http://www.ukrstat.gov.ua/
3. O. Dudchyk, Global and national problems of economics, **4**, 826–829 (2015)
4. S. Pyrozhkov, M. Khvesyk, *Economic evaluation of the Ukrainian natural wealth*, Public Institution "Institute of Environmental Economics and Sustainable Development of the National Academy of Sciences of Ukraine", 396 (2015)
5. Yu. Lupenko, M. Kisil, *Strategic directions of the investment provision of the Ukrainian agricultural development till 2020*, NSC IAE, 66 (2012)
6. A. Sahrbacher, J. Hristov, M. Brady, *A combined approach to assess the impacts of Ecological Focus Areas on regional structural development and agricultural land use*, Review of Agricultural, Food and Environmental Studies, **98** (3), 111-144 (2017)
7. A. Standar, J. Średzińska, *Wykorzystanie analizy czynnikowej do badania determinant dochodów gospodarstw rolnych (na przykładzie krajów Europy Środkowo-Wschodniej)*, Zeszyty Naukowe SGGW – Ekonomika i Organizacja Gospodarki Żywnościowej, **118** (4), 5-17 (2017)
8. J. Tomaszewski, *Fundusze inwestycji odpowiedzialnych społecznie w Europie – struktura i trendy rozwojowe*, Zarządzanie finansami i rachunkowość, **4** (2), 17-26 (2016)
9. Yu. Lupenko, *Development of agrarian sector of the Ukrainian economy: forecasts and prospects*, Scientific Bulletin of Mukachevo State University, Series "Economics", **2** (4), Part 2, 30-34 (2015)
10. V. Kutsenko, Ya. Ostafiichuk, M. Ilyina et al., *Socio-ecological priorities for sustainable rural development*, Public Institution "Institute of Environmental Economics and Sustainable Development of the National Academy of Sciences of Ukraine", 351 (2016)
11. Yu. Aviati, R. Karsidi, D. Tri Kartono, S. Anantanyu, *The Development of Corn Farmers'*

 https://doi.org/10.1051/bioconf/20181001003

Entrepreneurial Competencies in Indonesia, International Journal of Ecology & Development, **31** (4), 78-83 (2016)

12. St. Brosig, T. Glauben, I. Levkovych, S. Prehn, R. Teuber, *Are We Moving Towards Functioning Agricultural Markets and Trade Relations?* Journal of Agricultural Economics, **67** (3), 677-684 (2016)
13. P. Sabluk, M. Kodenska, *Conceptual framework for the development and implementation of investment programs in agri-industrial production*, NSC IAE, 46 (2012)
14. S. Sharybar, Actual Problems of Economics, **10** (160), 270-275 (2014)
15. Yu. Lupenko, V. Mesel-Veseliak, *Strategic directions of the Ukrainian agricultural development till 2020*, NSC IAE, 218 (2012)

Application of extrusion-cooking technique for foamed starch-based materials

Maciej Combrzyński[1,*], *Leszek* Mościcki[1],*Marcin* Mitrus[1], *Karol* Kupryaniuk[1] and *Anna* Oniszczuk[2]

[1]University of Life Sciences in Lublin, Department of Food Process Engineering, ul. Doświadczalna 44, 20-950 Lublin, Poland
[2]Medical University of Lublin, Department of Inorganic Chemistry, ul. Chodźki 4a, 20-093 Lublin, Poland

Abstract. Foamed materials are widely used, mainly as a protection objects during transport of various products. Traditionally foams are produced from plastics so they are very difficult for waste management. It is the challenge for many scientific centres to develop a technology for the production of bio-based materials which can be rapidly decomposed. The task for the researcher is to obtain a relatively cheap, easy to use and completely biodegradable materials. The aim of this work was the selection of the main raw materials, functional additives and process parameters to obtain the most effective parameters of extrusion-cooking process for foamed starch-based materials. Properties of the products and processing costs were taken into account. During the study, the extrusion-cooking process was performed under various conditions: temperature, humidity, type of the die, screw rotational speed, various raw materials and additives blends. The best results were obtained for mixtures based on potato starch and with addition the foaming agent Plastron foam PDE and poly(vinyl) alcohol PVA.

1 Introduction

Producers of foamed packaging materials are under increasing pressure of law regulations concerning environmental protection and waste disposal [1-4]. Packaging materials are a serious problem for businesses and local community, because their waste are difficult to manage and decompose in natural environment. Also recycling process is often unprofitable (too expensive transport and handling costs). More and more attention is being paid to the possibility of reprocessing of packaging materials, especially one-time use. The market is starting to require from manufacturers environmentally safe materials which have obviously the utility requirements. The interest in environmentally friendly materials stimulates the development of the production of raw materials from natural raw materials for packaging purposes[5,6].

Currently it is not possible to reuse all produced materials based on plastics. Hence, attempts were made to use agricultural crops in the production of biopolymers [7-10]. It is possible to use starch, protein or cellulose in the processing of biodegradable packaging materials. Biopolymers based on starch and cellulose have the great advantage – they are suitable for composting or reuse. A successful substitution of polymer-based porous materials with starch-based materials is being sought.

Starch-based foamed materials are lightweight and can be used successfully in packaging [11-14]. However biodegradable foamed materials, based on natural raw materials, are characterized by brittleness and poor water resistance. These materials dissolve completely, as opposed to synthetic foams which are neutral to moisture [15-19]. The additional disadvantage of this type of materials is the difficulty and limitations in creating the optimal cell-like structures which are required by potential customers. The lack of industry-ready solutions is a challenge for research institutions to obtain the possibility of producing fully functional, biodegradable foamed materials that can be widely used in packaging production.

Extrusion-cooking is a process of extrusion of loose material (mostly raw origin materials) under the influence of high pressure and high temperature [20-25]. It leads to the development of significant physical and chemical changes in the processed material. The process is carried out in devices called extruder-cookers [26-28]. The main working part of each machine is a screw or a pair of screws (hence the division into single and twin screw extruders), which are located in a cylinder (special casing). During the extrusion-cooking process working part squeezes the raw material and extruded it through a die installed at the end of the cylinder. Extrusion-cooking technology is widely used in agri-food processing because of its versatility and very high efficiency.

The extrusion-cooking process can be used to mold plastic with the addition of natural starch. As it was said in many research, starch can be a component of biodegradable materials and be exist in native or extruded form [29-34]. Starch which is processed by

*Corresponding author: maciej.combrzynski@up.lublin.pl

extrusion-cooking goes into destructuring and has gelatinise form. It has thermoplastic properties which allows starch to be molded and blended with other components [35-38], including traditional polymers. In relation to native starch, the extruded version has other rheological properties and is more sensitive to water. The properties of starch can be created by selecting the appropriate extrusion-cooking process parameters, as well as the selection of natural raw materials with the required properties and the use of various functional substances and functional ingredients. Synthetic materials with added starch are primarily used for the production of films, containers and foams. The thermoplastic starch obtained by extrusion-cooking is referred to as TPS. The advantage of using extruded-cooking starch in packaging is its cost and price of ready to use material.

The aim of this work was the selection of the main raw materials, functional additives and process parameters to obtain the most effective parameters of extrusion-cooking process for foamed starch-based materials.

Fig. 1. Prepared mixtures.

2 Materials and methods

2.1 Materials and mixtures preparation

During tests four types of starches (different botanical origin) were used: potato, wheat, corn and chemically modified corn starch. Various functional additives were used: 99.5% pure pharmaceutical glycerol, Plastron foam PDE, poly(vinyl) alcohol, guar gum, xanthan gum, food gelatine, baking soda, albumin, carrageenan, E471 monoglyceride, technical talc. These additives are used as foaming, emulsifying, stabilizing products [39-41]. They are used in plastic production and also used in bakery, confectionery and other fields of the food industry. Glycerin is a popular plasticizer used to produce TPS starch granules, which are a half-products to obtain biodegradable packaging materials.

Pharmaceutical glycerol was used in an amount from 5 to 25%. The remaining functional additives were used in an amount of 0.5 to 10% by weight of the total raw material mixture. A humidity range 15-20% of processed blends was used.

In the first stage, the starch with functional additives was mixed using a premix mixer produced by Rowag (Rogoźno, Poland). The each mixture has weighed of 5 kg (Figure 1). In order to homogenize all blends, the samples were kept in closed plastic bags for 24 hours in laboratory rooms. Before the extrusion-cooking process, all samples were mixed again.

2.2 Extrusion-cooking process parameters

The extrusion-cooking process was carried out using a single screw extrusion-cooker TS-45 (produced by Z.M.Ch. Metalchem, Gliwice, Poland) with L/D = 12 (Figure 2). The extruder was equipped with a water cooling system and temperature sensors for each section, engine load and screw rotation speed.

Fig. 2. The single screw extruder-cooker TS-45 (Z.M.Ch. Metalchem, Gliwice, Poland).

The screw rotational speed was set at 100 and 130 rpm, the temperature profile along the barrel sections (from the feeding section (I) to the head) was relative to type of the starches and varied from 80 up to 140°C (Table 1).

The product was cut using a high-speed knife mounted at the die. The speed was adjusted to the dimensions of the extrudates.

The extrusion-cooking process and the various parameters were continuously monitored and recorded.

Table 1. The temperature profile along the barrel sections.

Barrel section	Blends based on potato starch	Blends based on wheat starch	Blends based on corn starch and corn starch chemically modified
I	80	80	80
II	120	160	140
Head	80	120	100

Due to the specificity and intensity of expansion process, the size of the die was individually selected based on the operator's experience. The aim was to achieve optimum and stable process parameters. In the first stage three different diameters of the die were used: 3, 4 and 5 mm.

Moreover, circular die with the internal diameter of 5 mm was selected to obtain annular cross-section samples.After the extrusion-cooking, the foams were cooled down at room temperature and dried in an air oven at 40°C for 24h (Fig. 3).

Fig. 3. Air oven.

Various mixtures, humidity, temperature ranges, die size were used to see how the extrusion-cooking process and the properties of starch-based foamed packaging materials are changed.

3 Results and Discussion

Preliminary tests have shown that potato starch is susceptible to dextrinisation in higher temperature (over 120°C). It caused sticking and burning of the extrudates (Figure 4). So a decision was made to process potato starch at temperatures below 120°C.

Fig. 4. Dextrinisation of starch.

The results of the first stage also showed that the use only a two additives allowed to obtain satisfactory results in starch-based foamed products processing. Ingredients such as glycerin, guar gum, xanthan gum, food gelatin, baking soda, albumin, carrageenan monoglyceride E 471 and technical talc did not provide the expected results to obtain right form of foamed materials at established process configuration.

Materials produced with these substances were characterized by a compact structure, a small, uneven shape and lack of pores. In addition the extrusion-cooking process was disrupted - sticking of the screw and cylinder, uneven flow of raw material, disturbance of the process expansion. Based on the obtained results, a decision was made to reject most of these additives and use only two components in the further study: Plastronfoam PDE (Figure 5) and poly(vinyl) alcohol (Figure 6).

Fig. 5. Starch-based foam materials with Plastronfoam PDE addition.

Fig. 6. Starch-based foam materials with poly(vinyl) alcohol addition.

Plastronfoam PDE foaming agent is used for foaming thermoplastic elastomers and PVC. The supplier of this ingredient was VGT Polska Sp.zo.o. from Cracow, Poland. Based on the analysis of the first stage and product card, it was decided to use a PDE foaming agent in an amount from 1 to 3% in the mixture. The addition of PDE more than 3% affected negatively on the quality of the extrudates and the extrusion-cooking process. The basic properties of the Plastronfoam PDE are presented in Table 2.

Table 2. Basic properties of the Plastronfoam PDE foaming agent used in the extruded foam production process (data based on producer information card)

Characteristic	**The result**
Form	White powder
Active substances	Multi-component system
Content of active substances [%]	100
Beginning of decomposition [°C]	>140
Net price for 1 kg [€]	4.5

Poly(vinyl) alcohol is widely used in the production of biopolymers. The supplier of this ingredient was POCH S.A. from Lublin. As in the case of PDE, the proportion of this component in selected blends was from 1 to 3%. The basic features of poly(vinyl) alcohol are presented in Table 3.

Table 3. Basic properties of the poly(vinyl) alcohol used in the extruded foam production process (data based on producer information card).

Characteristic	**The result**
Form	White color granules
Smell	Without smell
Density [g·cm^{-3}]	1.2 – 1.3
Melting temperature [°C]	160 – 200
Net price for 1 kg [€]	45

During the first stage the influence of the humidity content in the mixture on the extrusion process and the quality of the extrudates were observed. It was noted that the humidity less than 17% and over 19% significantly have a significant impact on the extrusion-cooking process. This made the material more difficult to process by the extruder-cooker, the screw was wrapped at feeding section which finally led to stop the process. Such humidity values not allow to obtain good foamed extrudates. So there was decision to use three different humidity levels: 17, 18 and 19%.

Reducing of the die diameter allowed to check how compression ratio of the process and expansion ratio of the product will changed. Using the smallest die diameter allowed to obtain the best expansion ratio of the foamed starch-based materials. So there was decision to use a die with a diameter φ = 3 mm and a ring die. These kind of products shape is one of the most popular on the market.

The higher screw speed (130 rpm) allowed to obtain products with higher expansion ratio. The extrusion-cooking process was also milder and more effective.

Practically most of the starch types used in the tests allowed to produce foamed starch-based materials using the Plastronfoam PDE and the poly(vinyl) alcohol (Figure 7).

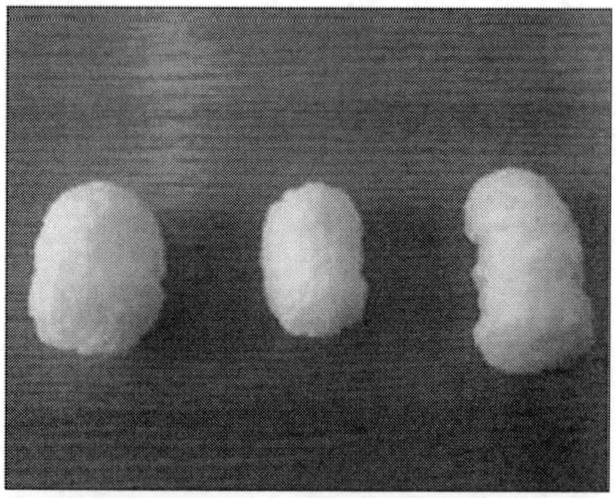

Fig. 7. Foamed starch-based materials with poly(vinyl) alcohol addition, from the left: wheat, corn and potato starch.

The exception was chemically modified corn starch, which was not processed during the trials. Finally, based on the price of basic starch raw material (Table 4), potato starch was recommended for further research.

Table 4. Price for 1 kg each starches which were used during tests.

Starch	Price (€) for 1 kg
potato	0.6
wheat	0.9
corn	0.8
corn chemically modified	2.2

4 Conclusions

Preliminary studies of the extrusion-cooking process of different mixtures allowed to select the best main raw materials and functional components. The most effective process parameters was obtain too. The attention was also paid to the properties of the final product and possible lower cost of raw materials.

Based on the results and observations, the following conclusions were made:

1. A single-screw extruder-cooker type TS-45 can be used for the production of starch-based foamed materials.
2. Different types of starch were used during the study and most of them allow to obtain foamed materials.
3. Extrusion-cooking of potato starch at temperatures higher than 120°C resulted dextrinisationof the starch.
4. Based on the results the most functional additives were rejected and only two components were used in the further study: Plastronfoam PDE foaming agent and poly(vinyl) alcohol.
5. Foaming process of the extrudates was achieved with a small addition of the synthetic additives. Both of them: the Plastronfoam PDE foaming agent and the poly(vinyl) alcohol are the substances which can have a contact with food.
6. Humidity content of the mixtures has a significant impact on the extrusion-cooking process and on the quality of the foamed extrudates.
7. Using the smallest die diameter allowed to obtain the extrudates with the best expansion ratio. In addition, the foaming material was obtained using a ring die.
8. The results of the research have made it possible to obtain starch-based foamed materials which can be tested to check basic properties and compare them with traditional polymeric materials.

5 References

1. Y. U. Nabar, D. Draybuck, R. Narayan, J. Appl. Polym. Sci.**102**,1(2006)
2. Y. Nabar, R. Narayan,J. Appl. Polym. Sci. **101**, 6 (2006)
3. Z. Yang, D. Graiver, R. Narayan.,Polym. Eng. Sci.**53**, 4 (2013)
4. J. Zhou, J. Song, R. Parker, Carbohyd. Polym.**63**, 4 (2006)
5. P. D. Tatarka, R. L. Cunningham,J. Appl. Polym. Sci.**67**, 7(1998)
6. J. L.Willett, R. L. Shogren, Polymer **43**, 22(2002)
7. C. Bastioli,*Handbook of Biodegradable Polymers*(Rapra Technology Limited, UK, 2005)
8. P. Cinelli, E. Chiellini, J. W. Lawton, S. H. Imam, Polym. Degrad. Stabil.**91**, 5 (2006)
9. H. A. Pushpadass, S. G. Suresh Babu, R. W. Weber, M. A. Hanna,Packag. Technol. Sci.**21**, 3(2008)
10. Z. Yang, D. Graiver, R. Narayan, Polym. Eng. Sci.**53**, 4 (2013)
11. L. Wang, G. M. Ganjyal, D. D. Jones, C. L. Weller, M. A. Hanna,Adv. Polym. Tech.**24**, 1 (2005)
12. J. W. Lawton, R. L. Shogren, K. F. Tiefenbacher, Ind. Crop. Prod. 19, 1 (2004)
13. J. F. Zhang, X. Sun, J. Appl. Polym. Sci. 106 (2007)
14. M. Mitrus, L. Mościcki, Chemical Engineering Research and Design 92 (2014)
15. A. L. Chaudhary,M. Miler, P. J. Torley, P. A. Sopade, P. J. Halley,Carbohyd. Polym.**74**, 4(2008)
16. El-Sonbati, Thermoplastic Elastomers (InTech, Croatia, 2012)
17. P. R. Salgado, V. C. Schmidt, S. E.Molina Ortiz, A. N.Mauri, J. B.Laurindo,J Food Eng.**85**, 3 (2008)
18. R. L. Shogren, J. W. Lawton, K. F. Tiefenbacher, Ind. Crop. Prod.**16**, 1(2002)
19. Y. X.Xu, Y.Dzenis, M. A. Hanna, Ind. Crop. Prod.**21**, 3 (2005)
20. R.Agbisit, S.Alavi, E.Cheng, T.Herald, A. Trater, J. Texture Stud.**38**, 2 (2007)
21. H. Chanvrier, F. Desbois, F. Perotti, C. Salzmann, S. Chassagne, J. C. Gumy, I. Blank,Carbohyd. Polym.**98**, 1 (2013)
22. A. Wójtowicz, A. Kolasa, L. Mościcki, Pol. J. Food Nutr. Sci.**63**, 4(2013)
23. M. Wu, D. Li, L. J. Wang, N. Özkan, Z. H. Mao.,J. Food Eng.**98**, 4(2010)
24. A. C. Conti e Silva, R. J. Da Cruz, J. A. Gomes Arêas, Meat Science 84 (2010)
25. Ş. İbanoğlu, P. Ainsworth, E. A. Özer, A. Plunkett, J. Food Eng. 75 (2006)
26. A. Bouasla, A. Wójtowicz, M. Nasereddine Zidoune, LWT - Food Science and Technology**75**(2017)
27. N. D. Frame,*The Technology of Extrusion Cooking*. (Blackie Academic & Professional, New York, USA 1994)
28. R. Guy, *Extrusion Cooking*. Technologies and Applications (Woodhead Publishing Limited, Cambridge, U.K. 2001)
29. A. Altskär, R. Andersson, A. Boldizar, K. Koch,M. Stading, M. Rigdahl, M. Thunwall, Carbohyd. Polym.**71**, 4(2008)
30. J. M. Ramos Diaz, S. Kirjoranta, S. Tenitz, P. A. Penttilä, R. Serimaa, A. M. Lampi, K. Jouppila, J. Cereal Sci.**58**, 1(2013)
31. L.Averous, N. Boquillon,Carbohyd. Polym.**56**, 2(2004)
32. L. P. B. M. Janssen, L. Mościcki, *Thermoplastic Starch* (Wiley-VCH Verlag GmbH & Co. KGaA, Weinheim, Germany 2009)
33. H. F. Mark, *Encyclopedia of Polymer Science and Technology* (Concise, Third Edition, Wiley–Interscience, A John Wiley & Sons, Inc., Hoboken, New Jersey, USA 2007)

34. J. Ruiz-Ruiz, A. Martínez-Ayala, S. Drago, R. González, D. Betancur-Ancona, L. Chel- Guerrero, Food Science of Technology 41 (2008)
35. T. Oniszczuk, R. Pilawka, Przem. Chem. **92**, 2(2013)
36. T. Oniszczuk, A. Wójtowicz, L. Mościcki, M. Mitrus, K. Kupryaniuk, A. Kusz, G. Bartnik, Int. Agrophys. **30**, 2 (2016)
37. T. Oniszczuk, M. Mitrus, A. Wójtowicz, L. Mościcki, Przem. Chem. **94**, 10 (2015)
38. T. Oniszczuk, R. Pilawka, A. Oniszczuk, Przem. Chem**92**,8 (2013)
39. A. Wójtowicz, Polish Journal of Food and Nutrition Sciences **57**, 3A (2007)
40. A. Rutkowski, S. Gwiazda, K. Dąbrowska: Kompendium Dodatków do Żywności (Wydawnictwo Hortimex, Polska, 2003)
41. L. Mościcki, Extrusion-Cooking Techniques (Wiley-VCH Verlag GmbH & Co. KGaA, Weinheim, Germany2011)

Ozone aeration impact on the maturation phase in the process of green waste composting

Maciej Gliniak[1,]*, *Łukasz* Grabowski[2], *Daria* Polek[2]

[1]University of Agriculture in Krakow, Institute of Agriculture Engineering and Informatics, Faculty of Production and Power Engineering, Mickiewicza 21 Av., 30-120 Krakow, Poland
[2]AGH University of Science and Technology, Faculty of Mining and Geoengineering, Department of Environmental Engineering and Mineral Processing, Mickiewicza 30 Av., 30-059 Krakow, Poland

Abstract. The paper presents work results on optimization of stabilization phase in the biomass composting process. In these studies, it was examined the influence of two doses of ozone (10 and 20 $mgO_3 \cdot dm^{-3}$) in the air used for aeration of stabilization. The results showed the ability to reduce compost maturation time by more than 50%. Application of these ozone doses resulted in a reduction of organic matter content in the stabilizer by 30 to 60%, while reduction of moisture in the material by 20%.

1 Introduction

Intensive socio-economic development observed in Poland in recent years, resulted in a significant increase of biodegradable waste volume. In accordance with applicable law and the EU concept called circular economy, one should strive to make the best use of raw materials obtained in the process of waste recovery and recycling. In Poland, the material recovery has been regulated at Regulation of the Minister of the Environment from 2015 on the recovery process R10, in the provisions of the Waste Act of December 2012 and in guidelines of the Department of Waste Management in the Ministry of Environment from December 2008. After analysis of these documents, it was found that the main process of recovering raw materials from biomass is the composting process [1-6].

In the moderate climate conditions, the composting process is divided into two phases, which both should not take less than 8 weeks [2-4]. The duration of the process depends on the choice of composting methods (active system with aeration of the material or passive without aeration) and the material properties which is going to be processed. Beside these factors the impact on composting process have also pH, ambient air temperature, amount of microorganisms, components containing auxiliary source of carbon [7]. Based on the studies carried out in different research centres, it was found that the first phase of the composting process should last about 2 weeks. Then the material is directed for the stabilization, which takes 6-8 weeks [2-4,8], the obtained compost must meet the following criteria [2]:

a) loss of ignition (LOI) - below 35% of the dry mass,
b) total organic carbon (TOC) - less than 20%,
c) AT4 respiration activity - not exceeding 10 $mgO_3 \cdot g^{-1}$ of dry mass.

In establishments, which carry out composting waste, there is a need to comply with quality standards, which often forces to extend the process length. It also affects the decline in their production capacity. Currently in the world there are many researches over the optimization of the composting process kinetics in its first phase and elimination of associated odorous emissions [8-14]. Optimization of stabilization phase in the composting process is mainly focused on the use of various substances for neutralizing additional contaminants (eg. toxic metals, petroleum products, etc.). In addition, in the literature there are several studies that suggest an additional hygienization of stabilized material and the possibilities of using gases generated in the process [15-20].

Literature review showed that the main problem of the composting process is a lenght of its stabilization phase. Previous studies conducted in the world related to the use of ozone in the composting process, were focused on the disposal of leachate [21,22]. Due to the lacks in research and data in the literature, which regards to possibilities of shortening phase of aerobic stabilization, it was decided to investigate whether it is possible to use ozone as a substance, which affects both the additional hygienization of the material and acceleration of stabilization in composting process.

2 Material and Methods

The composting process was carried out in the isothermal bioreactor BKB 100 used for testing of digestibility of organic material. Based on previous laboratory tests it was found that the course of the decomposition of organic matter in the bioreactor is similar to the actual conditions occurring in the

* Corresponding author: maciej.gliniak@urk.edu.pl

composting containers [1,23,24]. The study used the mixture of biomass consisting: grass (50%), sawdust (25%) and food waste (25%) - in the total fresh weight of 30 kg. The composition of the mixture was selected to obtain the appropriate structural properties of compost, while ensuring: a proper flow of air in the bioreactor, the humidity of mixed biomass used for the process was 63% and the C:N ratio was 23.8. In order to obtain the proper process of composting and trigger clear thermophilic phase, one set air flow in the bioreactor at a level of 0.035 $m^3 \cdot h^{-1} \cdot kg_{dw}^{-1}$ (which based on the weight of the composted waste was about 0.4 $m^3 \cdot h^{-1}$). In a study conducted by Dziedzic et al. [25] and Malinowski and Wolny-Koladka [26], it was demonstrated that such flow provides adequate aeration of biomass (>10% O_2 content) and does not lead to supercooling of material.

As soon as thermophilic phase of composting process has been completed (material reached the temperature of 45°C after the passage on the inflection point which defines the maximum temperature of the process) one collected three representative samples of the material from the bioreactor with a mass of 1500 g for the air stabilization process with ozone content of 0, 10 and 20 $mgO_3 \cdot dm^{-3}$. Each representative sample was then divided into three equal parts weighing 500 g each, for a total of three sets of samples for the ozone aeration. To produce air containing ozone used ozone generator O3PRO DRP-30,7VW by Prozonex (Poland). The stabilization was carried out in round-bottom flasks having a volume of 500 ml, equipped with the air diffuser at the bottom of the flask, fitted with a sensor PT-1000 for measuring the temperature connected to a recording computer. During aeration, flasks were assembled in a laboratory shaker type LP-Shaker360 (simulation of compost flipping on a pile). Aeration process of the samples took 28 days (achieved ambient stabilization temperature), and was divided into seven four-day cycles:

– day 1: measuring the initial temperature and shaking the sample with ozone aeration for 5 minutes,

– day 2: break,

– day 3: measuring the initial temperature and shaking the sample for 5 minutes,

– day 4: break.

After the end of the stabilization process, obtained compost was sieved through a sieve with an aperture of 2 mm and analyzed in terms of moisture content using the gravimetric method according to ISO 11465:1999 and the loss on ignition according to EN 15169:2011. After stabilization process, from each sample was collected 40 g of material in 2 replicates to determine the aerobic activity of the material according to the Richtlinie für die mechanisch-biologische Behandlung von Abfällen.

3 Results

Due to the fact that the thermophilic microorganisms are the most active at temperatures above 45°C, such value was adopted as the lower boundary for thermophilic phase of the composting process in the experiment and the beginning of the compost stabilization process [3,9]. As a point of achieving thermal stability of the composted material one assigned the ambient temperature which was equal to 23°C maintained for a minimum 2 consecutive nights. Thermal stabilization of blank test was reached after 26 days from the start of measurement procedure (Fig. 1a). Samples aerated with air containing 10 $mgO_3 \cdot dm^{-3}$ resulted in shorter stabilization time for 10 days (40% of the standard stabilization time) compared to blank sample (Fig. 1b). Air applied for stabilization with the content of 20 $mgO_3 \cdot dm^{-3}$ (Fig. 1c) effected in compost thermal stability after 12 days of stabilization (54% to 40% of the standard stabilization time). While analyzing the curves shown in Fig. 1 there is an increase in slope of the temperature with increasing concentration of ozone. The presence of this relationship may indicate faster hygienization of obtained stabilizer, so the decline of thermophilic microorganisms in the sample.

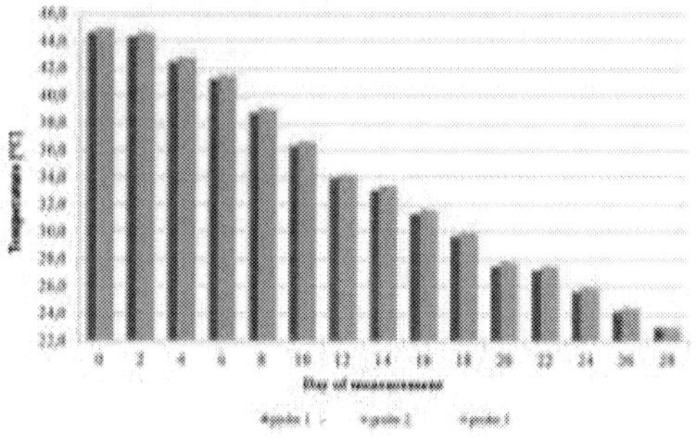

a)

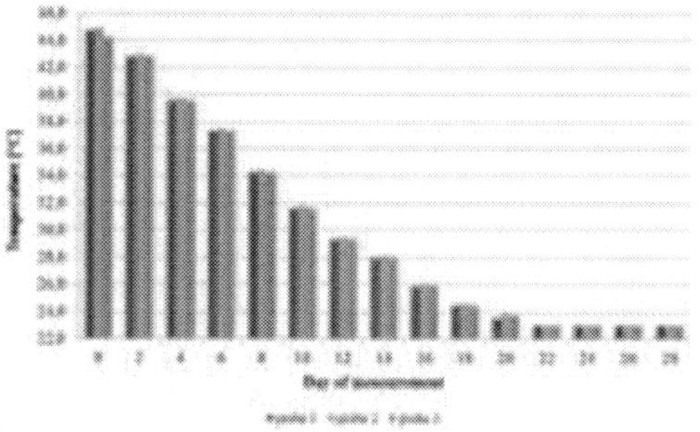

b)

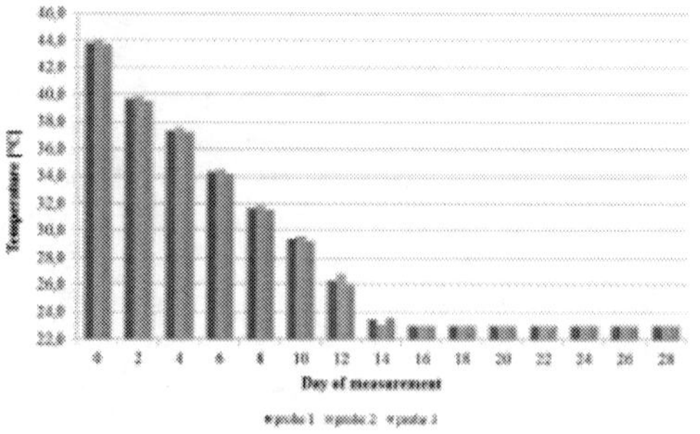

c)

Fig. 1. Changes of temperature in stabilization phase of the composting process depending on the concentration of ozone in the air: (a) the blank, (b) air containing 10 $mgO_3 \cdot dm^{-3}$, (c) air containing 20 $mgO_3 \cdot dm^{-3}$ [author's work]

Relevance analysis of the results obtained for each sample was performed based on Pearson's linear correlation (n=9), using a significance level of p=0.05.

Pearson coefficient used for concentrations of ozone in the air used for the stabilization of the samples reached the highest value for the blank test (0.99). Samples aerated with ozone also reached higher degree of correlation of 0.92 (10 $mgO_3 \cdot dm^{-3}$) and 0.81 (20 $mgO_3 \cdot dm^{-3}$). In all analyzed cases, the correlation had strongly positive nature. Analysis of standard deviations from medium temperature for ozonized samples (Fig. 2) showed variability in the range of 0.12÷0.14°C.

By monitoring the average changes in temperature of the samples aerated without the ozone addition, one can observe the presence of 3 stabilizing points after the material temperature reached 45, 34 and 28°C. Samples aerated with 10 $mgO_3 \cdot dm^{-3}$ also showed the presence of stabilization points, but the temperature was reduced by about 2.5°C in relation to the blank. The analysis of the average temperature for samples aerated with 20 $mgO_3 \cdot dm^{-3}$ (Fig. 2) showed the disappearance of the stabilization points, which can be explained by reaching toxic concentrations of ozone in the air used for the analysis for micro-organisms present in the stabilization process.

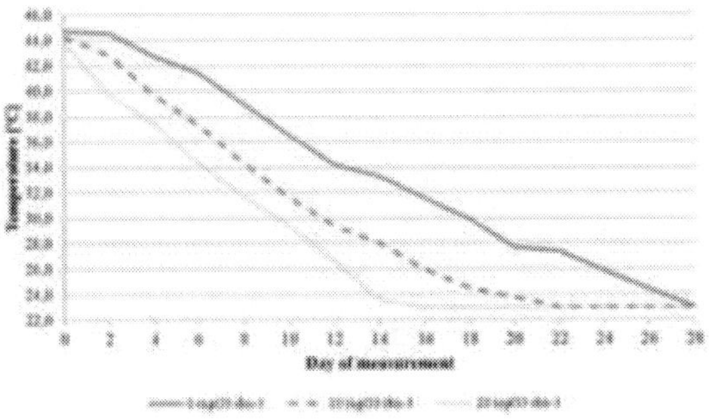

Fig. 2. Variability of average temperature in the stabilization phase of the composting process depending on the concentration of ozone in the air [author's work]

The use of ozone in the stabilization process of the samples did not affect the content of unprocessed biomass in the resulting compost, sifted through a sieve with an aperture of 2 mm. Analysis of the quality parameters of obtained compost indicates the existence of relationship between the amount of ozone applied to the stabilization and its moisture and organic matter content (expressed as loss on ignition). The use of air containing ozone at 20 $mgO_3 \cdot dm^{-3}$ resulted in the loss of moisture content of 8% and a decrease in organic matter content of 58% compared to the blank.

The experiment proved a positive ozone influence on accelerating the stabilization phase in the composting process. The proposed solution may be a contribution in the ongoing work on the intensification of thermophilic phase during processing of biomass. Analysis of the results showed that the increase in the concentration of ozone in the air deployed to stabilize the compost results in proportional decrease of its duration. In both analyzed cases (addition of 10 $mgO_3 \cdot dm^{-3}$ and 20 $mgO_3 \cdot dm^{-3}$) one showed a decrease in moisture content from 32% to about 26%, and organic matter content of the product from 26% to respectively 19% and 11% in comparison to the sample aerated without the use of ozone. pH of obtained compost did not depend on ozone concentration in the air and was at the level of 6.1. Oxygen activity of obtained compost was 10.01, 8.56 and 8.17 $mgO_2 \cdot g_{dw}^{-1}$ (for respectively aeration 0, 10, 20 $mgO_3 \cdot dm^{-3}$).

References

1. K. Grzesik, M. Malinowski, Energy Sources, Part A: Recovery, Utilization, and Environmental Effects, **38**, 21, 3150-3157 (2016)
2. Guidelines of the Department of Waste Management in the Ministry of Environment in Poland from December 2008, pp 35.
3. Waste Act in Poland from December 2012 (Dz. U. 2013 poz. 21).
4. P. Ghisellini, C. Cialani, S. Ulgiati, Journal of Cleaner Production, **114**, 11-32 (2016)
5. A.B. Morales, M.A Bustamante, F.C. Marhuenda-Egea, R. Moral, M. Ros, J.A. Pascual, Journal of Cleaner Production, **121**, 186-197 (2016)
6. M. Mourad, Journal of Cleaner Production, **126**, 461-477 (2016)
7. A.D. Neklyudov, G.N. Fedotov, A.N. Ivankin, Applied Biochemistry and Microbiology, **44**, 1, 9-23 (2008)
8. R. Lebrero, E. Rodríguez, P.A. García-Encina, R. Muñoz, Journal of Hazardous Materials, **190**, 1-3, 622-630 (2011)
9. M.C. Gutierez, M.A. Martin, A. Serrano, A.F. Chica, Journal of Environmental Management, **151**, 531–539 (2015)
10. J. Yuan, Q. Yang, Z. Zhang, G. Li, W. Luo, D. Zhang, Journal of Environmental Sciences, **37**, 83-90 (2015)
11. C. Fernández, C. Mateu, R. Moral, F. Sole-Mauri, Environmental Modelling & Software, **79**, 156-166 (2016)
12. G. Jinyi, H. Guangqun, H. Jing, Z. Jianfei, H. Lujia, International Journal of Heat and Mass Transfer, **97**, 735-741 (2016)
13. J.A. Siles, F. Vargas, M.C. Gutiérrez, A.F. Chica, M.A. Martín, Bioresource Technology, **211**, 173-182 (2016)
14. W. Yongjiang, P. Li, L. Xinyu, W. Yuansheng, Z. Kexun, L. Fei, Bioresource Technology, **206**,164-172 (2016)
15. F. Benlboukht, L. Lemee, S. Amir, A. Ambles, M. Hafidi, Ecological Engineering, **90**, 87-95 (2016)
16. Z. Junya, Ch. Meixue, S. Qianwen, T. Juan, J. Chao, L. Xueting, Z. Yuxiu, W. Yuansong, Water Research, **91**, 339-349 (2016)
17. Y. Lian, Z. Shihua, Ch. Zhigiang, W. Qinxue, W. Yao, Bioresource Technology, **204**, 185-191 (2016)

18. K.A. Mukesh, K.P. Akhilesh, S.B. Pushpendra, W.C. Wong, R. Li, Z. Zengqiang, Bioresource Technology, **213**, 181-189 (2016)
19. B. Vandecasteele, T. Sinicco, T. D'Hose, N.T. Vanden, C. Mondini, Journal of Environmental Management, **168**, 200-209 (2016)
20. W. Yuquan, W. Zimin, C. Zhenyu, Z. Yue, Z. Xinyu, L. Qian, W. Xueqin, Z. Xu, Bioresource Technology, **211**, 610-617 (2016)
21. N. Mokhtarani, A. Nasiri, H. Ganjidoust, S.Y. Yasrobi, The Journal of the International Ozone Association, **36**, 6, 540-548 (2014)
22. M.M. Amin, M.A. Moazzam, International Journal of Environmental Health Engineering, **3**, 1, 24-29 (2014)
23. S. Kasiński, I. Wojnowska-Baryla, Compost Science & Utilization, **21**, 3-4, 147-155 (2013)
24. M. Hurka, M. Malinowski, Infrastructure and ecology of rural areas, **4**, 1, 1127-1136 (2014)
25. K. Dziedzic, B. Łapczyńska-Kordon, M. Malinowski, M. Niemiec, J. Sikora, Chemical and Process Engineering, **36**, 4, 381-394 (2015)
26. M. Malinowski, K. Wolny-Koładka, Proceedings of ECOpole, **9**, 1, 256-261 (2015)

https://doi.org/10.1051/bioconf/20181001006

The effect of seed moisture and temperature on grinding characteristics of quinoa (*Chenopodium quinoa* Willd.)

Waleed Hameed Hassoon[1,2], and *Dariusz* Dziki[2,*]

[1]University of Al-Qasim Green, College of Food Sciences, Department of Food Technology, Babylon, Iraq.
[2]University of Life Sciences, Department of Thermal Technology and Food Process Engineering, Doświadczalna Str. 44, 20-280, Lublin, Poland

Abstract. Quinoa (*Chenopodium quinoa* Willd) is a native food plant of the Andean region of South America. Quinoa seeds have remarkable nutritional properties, not only from its high protein content, but also from its good amino acid balance. The aim of the study was evaluate the influence of quinoa seed moisture content (10, 12, 14, 16 and 18%) and temperature (-20, 3, 20 and 40°C, with the initial grain moisture content of 10.5%) on grinding process. Especially the particle size distributions and grinding energy indices were determined. The results showed that the increase of seed moisture content from 10 to 16% caused an increase the specific grinding energy from 6.9 to 8.3 kJ·kg-1, beside of this as the moisture increased the average particle also increased. The highest changes were observed in the fraction of coarse particles (above 1.0 mm). Interestingly, the mass fraction of fine particles (below 0.2 mm also increased). Other grinding indices also confirmed an increase the grinding energy requirements with the increase seed moisture content. The temperature of seed has little influence on quinoa grinding process. The highest grinding energy requirements were observed when the frozen seeds were ground.

1 Introduction

Quinoa (*Chenopodium quinoa* Willd) have been cultivated and consumed by humans for the last 5000-7000 years. Quinoa comprise a higher nutrition value than traditional grains and it is a promising worldwide cultivar for human consumption. Quinoa seeds have been traditionally consumed as food by the native population of the Andes region. Seeds of this plant contains of many health advantages and exceptional nutritional value such as: minerals, vitamins, fatty acids, antioxidants and a high concentration of protein in addition it is also considered as an important source for dietary fiber and other bioactive compounds [1]. Due to this fact the functional properties of quinoa grains can make contribution to avoid risk of many diseases by using it as food product for humankind. There are several developments with quinoa flour at a smaller scale, like bread, cookies, muffins, pasta, snacks, drinks, flakes, breakfast cereals, baby foods, beer, diet supplements, and extrudates [2]. Besides, quinoa is gluten free and it is an important feature that explains the baker's and brewer's increasing interest in using quinoa for production products for celiac people [3].

Nowadays, it became necessary to look at performance of the manufacturing processes due to increasing cost of energy and raw materials. The grinding process is one of the most important process in the seed industry. Hemery et al. [4] found that finer grinding gives with more available surface area may make these compounds more bioavailable and digestible for improved health benefits. From the other hand the grinding process is very energy consuming, for example during wheat flour milling it takes from 70% to 90% energy for flour production [5]. The grinding energy requirements depends on the size reduction method and on the properties of raw materials. The moisture content of seed strongly influences on mechanical properties of seed and thus affected on the grinding process [6]. Moistening is one of the most important pretreatments before the proper technological processes. The moisture of seeds has an effected on the grinding characteristics, and it causes an increase in seed elasticity and plasticity, which leads to an increase in the energy requirements in the course of grinding [7]. Dziki and Laskowski [8] found that increasing moisture content of wheat (range12-16%) leads to increase in the specific grinding energy (from 15 to 21%), while in contrary to these results Probst et al. [9] observed that there were no significant influence the moisture contents, range of 10% to 20%, for grinding energy of corn. Also the temperature of seeds changes the mechanical properties of kernel [10] and thus influences on the grinding process [11].

To the our best knowledge, there is no study concerning the grinding characteristics of quinoa seed. The aim of our work was to evaluate the influence of

* Corresponding author: dariusz.dziki@up.lublin.pl

https://doi.org/10.1051/bioconf/20181001006

quinoa moisture content and temperature on grinding energy requirements and particle size distribution.

2 Materials and methods

2.1 Sample preparation

The quinoa seeds were bought in the local supermarket. The seeds come from Argentina a and were characterized by: initial moisture content 10.5% (w.b), bulk density 735 kg·m^{-3}, mass of 1000 seeds 4.1g, fibre content 3.62%, fat content 6.68%, protein content 16.64% and ash content 1.99%.

The seeds were tempered by adding water or by air drying at 40°C to adjust moisture content (w.b.) to 10%, 12%, 14%, 16%, and or 18%. Besides, the seed with the initial moisture content (10.5%) were stored with different temperatures (°C): -20, 3, 2 and 40°C, using climatic chamber.

2.2 Grinding process

The samples of quinoa seed were ground by using the laboratory roller mill (Sadkiewicz Instruments, Bydgoszcz Poland). The 100-g samples were weighted just prior to grinding. The mill was equipped with a computer system that allowed the recording and analyzing the grinding energy consumption. The amount of energy consumed during grinding was obtained by means of a power transducer, a data acquisition and a computer system that recorded the data measured by the transducer. The grinding energy was calculated by using special computer software. The energy required to run the mill with no load was determined and subtracted from the total energy to obtain the grinding energy. The unloaded grinder current was monitored prior to grinding and remained constant over all testing. The specific grinding energy was determined as the ratio of the grinding energy to the mass of the material taken for grinding. The sieving test was used to determine the particle size distribution of the pulverized material. Sieving was carried out for 5 min, by using a laboratory screen (Thyr 2, SASKIA, Germany), and separated into fractions using sieves of sizes, 1.60, 1.00, 0.80, 0.63, 0.400, 0.315, and 0.200 mm.. On the basis of the particle size distribution, the average particle size (d_p) was calculated [12]:

$$d_p = \sum_{i=1}^{n} \phi_i d_i, \qquad (1)$$

where ϕ represents the differential weight fraction (kgkg-1) of particles passing through aperture size di (mm).

The grinding ability index (E_f) was calculated as a ratio of the grinding energy to the surface area of the pulverized material. The surface area of the pulverized material was evaluated according to the procedure described by Jha and Sharma [13]. The Sokołowski's grinding index (Ks) was calculated on the basis of the size reduction theory described by Sokołowski [14]. Details of the procedure used in determining these indices can be found in Dziki [15]. The distribution of the particle size was evaluated thrice and the values of grinding indices were calculated from the average particle size. Based on the literature data the values of specific grinding energy obtained for quinoa seed are several times lower in comparison to the values of this parameter obtained for wheat grain with different moisture. Werechowska et al. [16] found E_r values form 52 to 76 kJkg^{-1} depending on wheat variety and moisture content.

2.3 Statistical analysis

The analysis of basic properties of seed was performed in triplicate. The measurements of grinding energy were replicated ten times. The obtained data was further subjected to a statistical analysis and the consequent evaluations were analyzed for a variance analysis (one-way ANOVA). The statistical differences between the treatment groups were estimated through Tukey's test.. Statistical tests were evaluated by using the Statistica 6.0 software (StatSoft, Inc., Tulsa, USA). All the statistical tests were carried out at a significance level of $\alpha = 0.05$.

3 Results and discussion

The particle size distributions of the ground quinoa are given in Table 1 and 2. The results reveal that as the seed moisture content increased the fractions of the coarse (above 1.8 mm) and fine particles (below 0.2 mm) increased ground material was characterized mostly by the significantly higher fraction ($P < 0.05$) of the fine particles (below 0.2 mm). However the highest changes were obtained in the class of particles between 0.4-0.63 mm (decrease with the increase of seed moisture content from 53.3% to 29.3%). The distribution of the particle size depends on the method employed for grinding and the properties of the seed [17]. When the hammer mill is used for grinding, the degree of fineness of the grain depends mainly on the size of the opening of the screen. In the case of roller mill the particle size distributions depends mainly from seed moisture and hardness [18].

Table 1. The influence of seed moisture content on particle size distribution of ground quinoa (%).

Range of class (mm)	Seed moisture (%)				
	10	12	14	16	18
>1.6	0.02a	0.1b	0.1b	0.2c	1.9d
1.0-1.6	1.74a	2.4b	6.1c	11.0d	20.5e
0.8-1.0	5.25a	6.8b	8.2c	6.4a	11.4d
0.63-0.8	8.3a	8.7b	10.4c	16.9d	8.2a
0.4-0.63	53.3d	49.5c	43.9	33.5b	23.9a
0.315-0.4	11.4c	12.0d	10.3b	9.6a	10.5b
0.2-0.315	9.0a	9.3a	9.2a	8.9a	9.8b
<0.2	11.1a	11.2ba	11.9b	13.4c	13.6c

The average particle size (d_p) of ground quinoa was shown in Fig. 1. Generally, as the seed moisture content increased the average particle size of ground quinoa increased too. However there was no statistical differences between d_p for seed with 10% and 12% of moisture content and for seed with 12% and 14% of moisture. The d_p among other things, depends on the grain hardness. Soft seeds are more easy to size reduction and give during comminution different grinding pattern and lower grinding energy requirements [19]. Based on the wheat seed structure the soft grains are characterized by a lower degree of adhesion between starch granules and protein matrix and thus during grinding the higher mass fraction of fine particles is produced [20].

Table 2. The influence of seed temperature on particle size distribution of ground quinoa (%).

Range of class (mm)	Seed temperature (°C)			
	-20	3	20	40
>1.6	0.01a	0.05a	0.07a	0.21b
1.0-1.6	2.9a	4.3a	5.6b	5.2b
0.8-1.0	6,2a	7.1a	7.9b	8,0b
0.63-0.8	15.0c	9.10b	9.9b	9.4a
0.4-0.63	50.4c	47.9b	44.4a	43.1a
0.315-0.4	12.1c	11.6a	11.8ab	12.03bc
0.2-0.315	7.8a	8.6b	9.1c	10.0d
<0.2	11.3a	11.4a	12.0b	12.1b

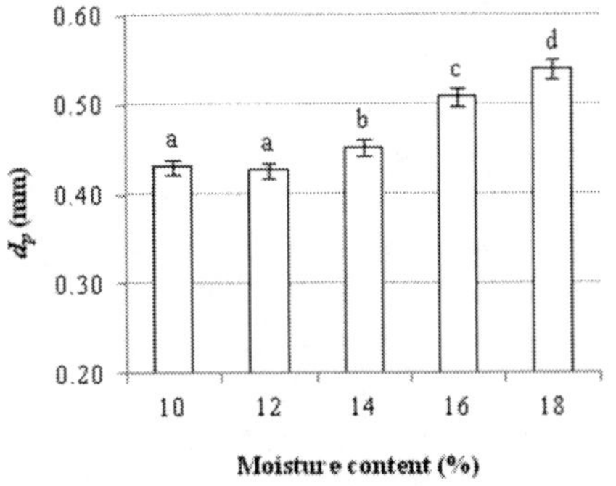

Fig. 1. Average particle size of ground of quinoa seed related to moisture content.

The specific grinding energy (E_r) ranged from 6.89 kJkg^{-1} to 8.07 kJkg^{-1} (Fig. 1). The incease of seed moisture form 10% to 16% caused an increase of E_r. The results of E_r obtained for 18% seed moisture content were not significanlty higher from seed with 14% and 16%. Similar tendency was obserwed by Dziki [15] for wheat seedss - as the seeds moisture increased from 10 to 18% the crushing energy also inreased. When the moisture was 20% a statistically significant decrease of crushing energy was observed. Mabille et al. [21] found that water affected both mechnical properties bran and endosperm properties. They found a reduction in stress to fracture in Young`s modulus, with an increase in moisture content.

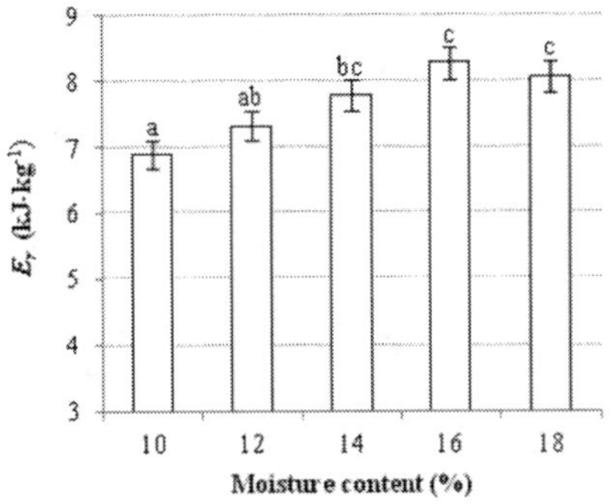

Fig. 2. Specific grinding energy of quinoa seed related to moisture content.

The Fig. 3. presents the results of grinding ability index (E_f) and Fig. 4. shows the Sokołowski's grinding index (E_s). These indices besides grinding energy take into account the effect of grinding expressed by increase of the seeds surface after grinding and the degree of fineness, respectively. These indices clearly show that moisture of seed caused an increase the grinding energy requirements. E_f changed with the increase quinoa moisture content from 0.644 kJ·m^2 to 0.946 kJ·m^2, whereas E_s ranged from 16.9 kJ·kg^{-1}·mm$^{0.5}$ to 33.1 16.9 kJ·kg^{-1}·mm$^{0.5}$. Seed temperature had little influence on the values of these indices. The highest values both E_f and E_s were found for the seed stored at the lowest temperature (-20°C) (Tab. 3). The values of these indices obtained for different kind of wheat are several times higher [15]. This indicated that quinoa seed required low grinding energy requirements in comparison to whet grain.

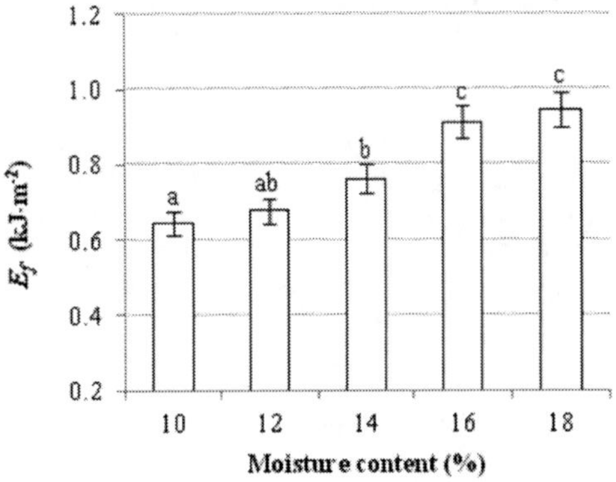

Fig. 3. Grinding ability index of quinoa seed related to moisture content.

Romański et al. [22] found relationship between grain temperature and wheat compression energy requirements. As the grain temperature decreased from 40 to -20°C energy requirements increased about 12%. However, Dziki [23] showed that grain temperature (range 0-40°C) has no significant influence on the specific grinding energy of wheat grain, but has an influenced on particle size distribution of the ground material.

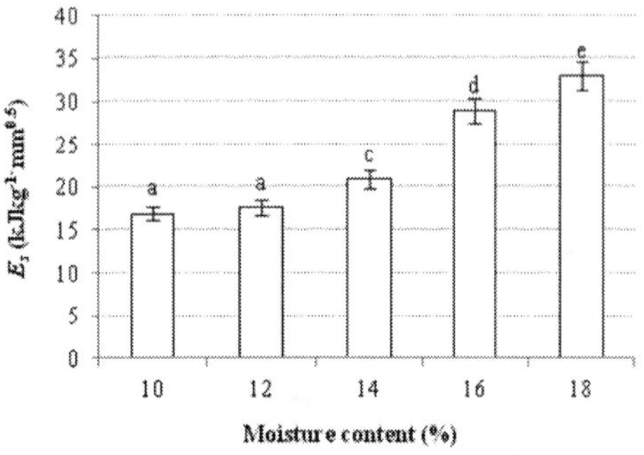

Fig. 4. Sokołowski's grinding index of quinoa seed related to moisture content,

Table 3. Influence of different seed temperature on average particle size and grinding energy indices

T (°C)	d_p (mm)	E_r (kJ·kg^{-1})	E_f (kJ·m^{-2})	E_s (kJ·kg·mm$^{0.5}$)
-20	0.48c	6.9a	0.71b	20.9c
3	0.44a	6.6a	0.63a	17.0a
20	0.45a	6.7a	0.65a	17.5a
40	0.46b	6.5a	0.68ab	18.2a

T – seed temperature, d_p – average particle size, E_r –specific grinding energy, E_f – grinding ability index, E_s – Sokolowski's grinding index

3 Conclusions

The effect of seed moisture content and temperature on the grinding behaviour of quinoa has been investigated. The increase of the seed moisture prior to roller mill grinding has a significant influence on the grinding process, both on specific grinding energy and particle size distribution of the ground material. The increase of seed moisture from 10 to 16% increases the specific grinding energy and particle size of ground material. Most importantly, the increase of seed moisture content up to 18% caused a decrease of specific grinding energy. The temperature of seed, from -20°C to 40°C (grain moisture content 10.5%) had no significant influence on the specific grinding energy and had little influence on the particle size distribution.

References

1. A. Vega-Gálvez, M. Miranda, J. Vergara, E. Uribe, L. Puente, E.A. Martínez. J. Sci. Food. Agric. **90**, 15 (2010)
2. R.A. Carciochi, L. Galván-D'Alessandro, P. Vandendriessche, S. Chollet. Plant Foods Hum. Nutr.**71,** 4 (2016)
3. A. Fardet. J. Cereal Sci. **60**, 3 (2014)
4. Y. Hemery, M. Chaurand, U. Holopainen, A.M. Lampi, P. Lehtinen, V. Piironen, A. Sadoudi, X. Rouau. J. Cereal Sci. **53**, 1 (2011)
5. D. Dziki, & J. Laskowski, TEKA Kom. Mot. Energ. Roln, **6A**, 45-52 (2006).
6. D. Dziki. Powder Technol. **185**, 2 (2008).
7. F. Mabille, J. Grilet, J. Abecassis. Cereal Chem. **78**, 3 (2001)
8. D. Dziki, & J. Laskowski, TEKA Kom. Mot. Energ. Roln, **5**, 56-64. (2005).
9. K.V. Probst, R.K. Ambrose, R.L. Pinto, R. Bali, P. Krishnakumar, K E. Ileleji. Trans ASABE, **56**, 1025-1033, (2013)
10. B.M. Ghodki, T.K. Goswami. J. Food Eng. **40**, 1 (2017)
11. K.K. Singh, T.K. Goswami. J. Food Eng. **22**, 3 (1999)
12. V. Velu, A. Nagender, P.G. Prabhakara Rao, D.G., Rao. J. Food Eng. **74**, 1 (2006)
13. S.N. Jha, R.S. Sharma. J. Food Sci. Technol. **47**, 5, (2010)
14. M. Sokołowski, Récents Progress en Génie Procédés **10,** 221–226 (1996).
15. D.Dziki. J. Food Eng. **104**, 4 (2011)
16. M. Warechowska, A. Markowska, A., J. Warechowski, A. Miś, A. Nawrocka, A. J. Cereal Sci. **69**, 306-312, (2016)
17. D. Dziki., J. Laskowski. J. Food Eng. **96**, 4 (2010)
18. V., Greffeuille, J. Abecassis, N. Barouh, V. Villeneuve, F. Mabille, C. Bar L'Helgouac'h, V. Lullien-Pellerina, V. J. Cereal Sci. **45**, 1 (2006)
19. D. Dziki, G.C. Pietrzak , A. Miś, K.Jończyk, U.G. Dziki. J Food Sci Technol. **51**, 10 (2014).
20. Y. Haddad, F. Mabille, A. Mermet, J. Abecassis, J.C. Benet. Powder Technol. **105,** 1-3 (1999).
21. F. Mabille, J. Grilet, J. Abecassis. Cereal Chem. **78**, 3 (2001)
22. L. Romański, R. Stopa, A. Niemiec, A. Wiercioch. Inż. Roln. **7**, 7 (2003)
23. D. Dziki. Inż. Roln. **9**, 51 (2003)

The opportunities for obtaining of the biogas on methane fermentation from marine algae biomass and water plant biomass

Ewa Jachniak[1,*], *Joanna* Chmura[2], *Mariusz* Kuglarz[1], *Józef* Wiktor[3]

[1] Institute of Environmental Protection and Engineering, University of Bielsko-Biala, Poland
[2] Student ofthe Environmental Engineering, University of Bielsko-Biala, Poland
[3] Institute of Oceanology of the Polish Academy of Sciences in Sopot

Abstract. The aim of the research was to try to obtain of the biogas on a laboratory scale from marine algae biomass and water plant biomass. The research was conducted in 2016 year and samples were taken from the Polish coast of the Baltic Sea. In laboratory work, algae and plant species were first identified. The next, in order to subject them to methane fermentation processes and to obtain biogas,partial mechanical treatment of the biomass was conducted. Dry matter content and dry organic matter content were also determined. The research has shown different production of the biogas depending on the various species of the algae and plants. The percentage composition of the biogas was also determined (% CO_2 and % CH_4). In this research some kinds and species of algae and aquatic plants were distinguished: *Scytosiphon cf. S. tortilis, Fucus vesiculosus, Cladophora, Audouinella, Potamogeton perfoliatus.* Production of biogas from selected algae and water plants oscillated between 0.023 $dm^3 \cdot g^{-1}$ and 0.303 $dm^3 \cdot g^{-1}$. The highest content of the methane in biogas was obtained from the mixture of *Ectocarpus* from spring and autumn harvest (values oscillated from 80.7 % to 81.2 %), while the highest percentage share of carbon dioxide in the biogas was characterized by the mixture *Fucus vesiculosus* and *Audouinella* (22 %). Due to a small amount of the research in this field, more research is needed.

1 Introduction

The marine and freshwater algae are used extensively for economic, consumer and industrial purposes *i.e.*Their role in the human economy is extremely large (in food industry and biofuels) [1,2,3]. In coastal areas algae are thrown ashore by the waves and are used as animal feed or fertilizer, because they contain lots of microelements and vitamins. Recently, much attention has been paid to the use of algae biomass as a potential source of renewable energy. There are many studies on the use of algae for the production of biogas, bioethanol and biooil all over the world [1, 2, 3, 4, 5, 6]. Algae have this advantage that they grow fast and they can double their biomass within 24 hours,with enough amount of the light, CO_2 and mineral salts (mainly inorganic nitrogen and phosphorus compounds) [7].

Currently, much attention has been paid to the use of the water vascular plants biomass for energy purposes, *i.e. Lemna minor, Lemna trisulica, Potamogeton perfoliatus, Myriophyllum aquaticum* and *Egeria densa* [3, 8]. The most popular is *Lemna*, because it grows fast (like algae) and its mass doubles every 1-3 days. It can also be easily collected from the surface of the freshwater reservoirs [3].

Excessive growth of algae is very unfavorable to the aquatic environment, because it disturbs its ecological balance. Nowadays the algae blooms are often happens, they arise as a result of the excessive discharge of the municipal and industrial wastewater and inappropriate farming economy. Algae blooms are observed in many water reservoirs, as well as in the Baltic Sea. They cause the water to become unfit for utility purposes (consumer and recreational), the aesthetic properties are also deteriorated. Algae blooms are a symptom of eutrophication of the water reservoirs and they carry many negative consequences, i. a. change in organoleptic properties of the water (taste, smell and colour), oxygen deficiency, death of the aquatic animals and plants, secretion of the toxins into the water by some cyanobacteria.These toxins are very dangerous for animals and human health [5, 9, 10, 11, 12, 13].

In the Baltic Sea,increased algae blooms are seen mainly in the Gulf of Gdańsk and in the east and south-east coast of Sweden. The Baltic Sea is an enclosed sea, characterized by a very large catchment area (1 729 000 km^2), of very intensively developed agriculture, inhabited by about 80 million people [14].

* Corresponding author: ejachniak@ath.bielsko.pl

All of these factors significantly leads to its eutrophication. The topic of eutrophication of the Baltic Sea was discussed by many authors, *i.e.* Wasmund 2002, Bień et al. 2010, Håkanson, Bryhn 2011, Leśniewska, Witak 2011, Michalak et al. 2016.

Many different methods are used to moderate the eutrophication of the Baltic Sea (construction of various sewage treatment plants, also home sewage treatment plants, construction and reconstruction of the small water retention).The surplus of the algae biomass can be used to reduce of eutrophication. A good way can be to collect algae from the beach and use for different purposes, *i.e.* for the biogas or bioethanol production. There are a variety of techniques to harvest of the excess algae from the sea, which are constantly being improved. Magnetite particles are often used and special nets are applied [3].

Water plants often overgrow large fragments of the water reservoirs and therefore their biomass can be used for energy purposes (f. e. biogas and biooil production) [3, 8, 15].

The aim of this research was to try to obtain biogas on a laboratory scale from selected marine algae biomass and aquatic plants biomass as a result of methane fermentation.

2 Methodology of research

The samples of selected algae and aquatic plants were collected from several points located at the Polish coast of the Baltic Sea (Fig. 1) in the period from May to September 2016. The Baltic Sea is an inland sea, which is located in the northern part of Europe. This sea connects with the North Sea through the Danish strait (f. e. Skagerrak and Kattegat).This sea is exposed to large amounts of the biogenic compounds, because it has a large catchment area. The biogenic substances are the main nutrients for algae. It causes that algae develop in excess and often create algae blooms [14, 16]. Algae blooms are most often observed in the coastal part of the Gulf of Gdańsk, so it was decided to collect samples from this part of the Baltic Sea in most terms (in May, July and September). The samples of the algae and water plants were collected in 12 May and 13 September from the beach in Brzeźno (northern district of Gdańsk), while 16 July from the beach in Mikoszewo (on the east of Gdańsk). In turn, in August, the samples were taken from Slowińskie Coast (26 August from the beach in Ustka and 29 August from the beach in Poddąbie) (Fig. 1).

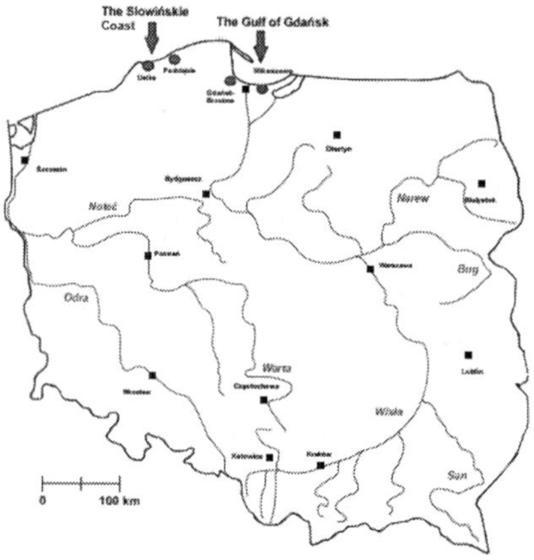

Fig. 1. The map of Poland and the places for taking samples [17, it was modified]

The collected samples were transported to ATH in Bielsko-Biala, where they were rinsed with tap water from sand and other contaminants and put on the tissue-paper to remove excess of the water and partial dry (for about 1 hour). After that, they were weighed, determined the species level, described and frozen. The analysis of the species composition of the collected algae and plants were conducted NIKON ECLIPSE E-200 microscope equipped with Delta Pix Invenio digital camera and digital image analysis software.

The key to identification of the macrophytes was done with help of identification key [18]. Additionally, the web pages of the Institute of Oceanography of the University of Gdansk were also used: http://www.hel.univ.gda.pl/atlas/organizmy.htm, http://www.hel.univ.gda.pl/. There was also used from the web page of theCultureCollection of Baltic Algae of the Institute of Oceanography of the University of Gdansk: http://ccba.ug.edu.pl/pages/pl/strona-glowna.php. In addition, some species designations were consulted with Professor Józef Wiktor from the Institute of Oceanology of the Polish Academy of Sciences in Sopot.

2.1 Determination of the dry matter and dry organic matter

Dry matter and dry organic matter measurements were carried out in accordance with Polish norms PN-EN 12880 and PN-EN 12879.The algae and water plants samples intended for the biogas production were fragmented into smaller parts (about 1 cm long) and then were weighed on an analytical balance with accuracy to 0.001 g. These samples were dried in ceramic evaporating dish in an electric oven for 4 hours at temperature of about 105 °C. After this time, evaporating dish with dried material was again weighed.Subsequently the samples were burn in a muffle furnace for 2 hoursat the temperature of about 550 °C and then evaporating dish with mineral residue was again weighed also with accuracy to 0.001 g.

2.2 Initial mechanical pretreatment of the substrates for the fermentation process

Initial mechanical pretreatment is carried out to provide better conditions for the work of anaerobic bacteria in the fermentation process. This is necessary, because cell walls of the algae and plants are poorly degradable (their main ingredient is cellulose) and it causes many obstacles in the work of the bacteria. The initial preparation of the collected samples for biogas production was done according method published in [5, 6,19,20]. In order to prepare algae and aquatic plants samples for fermentation, the frozen samples were thawed and then they were mechanically processed. Mechanical treatment consisted of cutting this material into lengths of 1 cm or smaller. Then the cut material was rubbed in a mortar to obtain pulp. The individual types and species of algae and plants were subjected to the mechanical pretreatment.

In each case 2 g/dm^3 of substrates used were added to the bioreactor. Assays for determination of practical methane potential of algae and aquatic biomass were carried out in 1.5 dm^3 bioreactors with working volume of 1 dm^3. After addition substrates to bioreactors 200 cm^3 of the distilled water, minus the volume of the substrates was added. In each case, methanogenic inoculum (800 cm^3), transported from Aqua S.A. Sewage Treatment Plant in Komorowice in Bielsko-Biala, were added to the bioreactors. Basic anaerobic media (BA) [21] were added to each bioreaktor. The bioreactors were sealed with butyl rubber stoppers and aluminum crimps and then purged with a mixture of CO_2 and N_2 (20/80%). The assays were undergoing digestion at 55 °C, until no significant amounts of CH_4 were produced (approx. 30 days). Every day, each bioreactor was mixed for about 60 seconds to ensure optimum conditions for the fermentation process. The amount of methane produced was expressed per unit of VS (d.o.m.) added. The amount of biogas produced by inocula (control vials without algae and plants) was subtracted from all tested assays. If the amounts of algae were insufficient for the fermentation process, they were used in the form of mixture (brown algae *Fucus vesiculosus* and red algae *Audouinella*).

2.3 Measurement of the biogas and methane content

The stand for biogas productionwas built of the equalizing reservoir and scaled reservoir. The scaled reservoir was filled with 5% sodium chloride. The both reservoirs were connected to each other and it caused the displacement of the surplus sodium chloride solution to the equalizing reservoir by forming fermentation gas. The volume of the biogas was converted to normal conditions, i.e. 0 °C and 1 atm.

The obtained biogas was collected in special bags and methane content of the obtained biogas was analyzed by using the ENVAG portable meter of the biogas composition.

The obtained methane wasex pressed as a unit production of this gas in relation to the organic dry matter unit used. The amount of the obtained methane from the control sample was subtracted from all tested samples, which were subjected to the fermentation processes.

The methane yield was calculated based on the volume of the methane produced per unit mass of volatile solids used, according to the following formula:

$$Y_{CH_4} = \frac{CH_4 obtained}{Ł\ d.o.m.} \quad (1)$$

where:

- Y_{CH_4}- unit production of the methane [dm^3CH_4·g^{-1} VS added]
- CH_4 obtained – the amount of the methane produced in the analyzed time [dm^3]
- Ł d.o.m. – the amount of the grams VS added to the bioreactor during the process [g].

3 Results

In all collected samples the following groups, kinds and species of algae and aquatic plants were distinguished:

a) brown algae [http://www.algaebase.org/search/species/detail/?species_id=291, 5]:

- *Scytosiphon cf. S. tortilis* – usually creates filamentous thallus with numerous lateral branches (phot. 1),
- *Fucus vesiculosus* - forms a ribboned and forked branched thallus, which has a numerous buoyant blisters,
- *Ectocarpus* – creates the brown filamentous and unbranched thallus,

b) green algae [22]:

- *Cladophora* - has filamentous thallus, which promotes the formation of the thick algae coat on the surface of the water (phot. 2),

c) red algae [23]:

- *Audouinella* –has a dark-reddish thallus, it creates dense and finely branched underwater turfs, which are sometimes visible on the water surface (phot. 3),

d)aquatic vascular plants [18]:

- *Potamogeton perfoliatus* - has a long stems, which can reach up to 6 meters in length. The rhizomes are durable and they have peak winter buds and the stem is flexible and strongly branched.

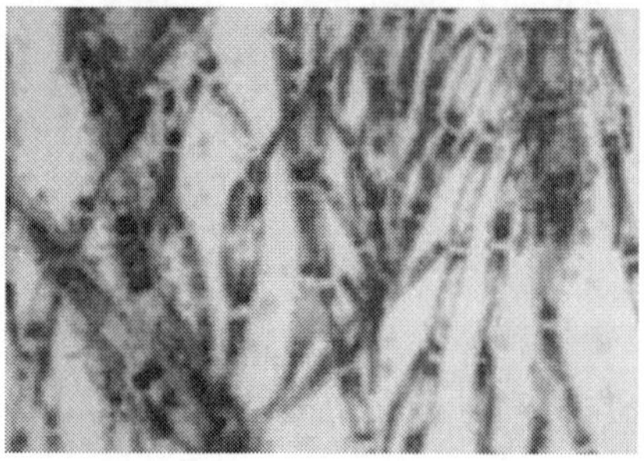

Phot. 1. *Scitosiphon cf. S. tortilis* visible under the microscope

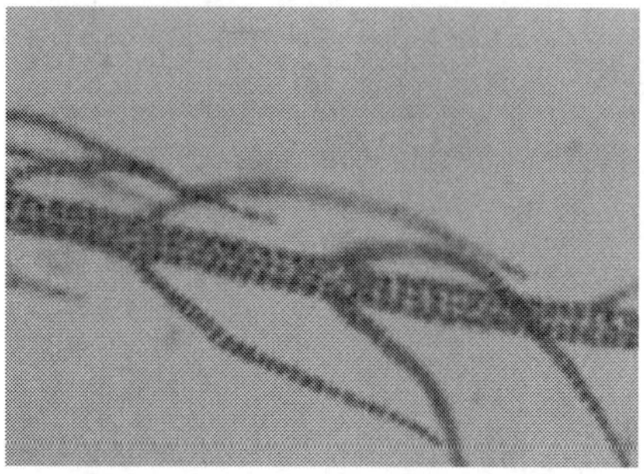

Phot. 2. *Cladophora* visible under the microscope

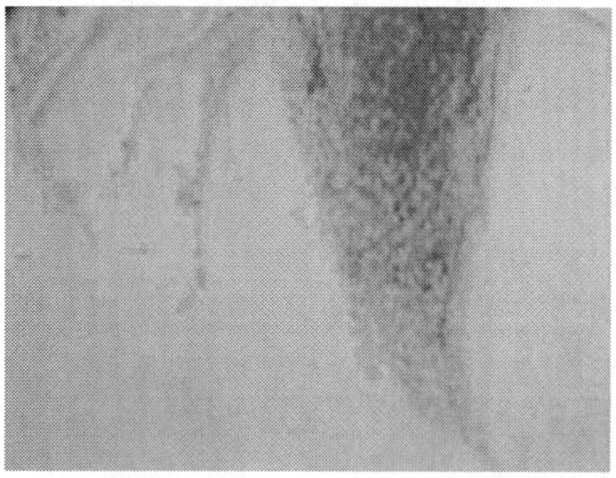

Phot. 3. *Audouinella* visible under the microscope

The content of the dry matter and dry organic matter of the identified algae and water plants is presented in below table (tab. 1).

Tab. 1. The content of the dry matter and dry organic matter in identified algae and water plants.

The identified algae and plants	d.m. ($g\cdot kg^{-1}$)	d.o.m. ($g\cdot kg^{-1}$)
Ectocarpus – collected in spring	220.84	218.16
Scitosiphon cf. S. tortilis	111.33	109.88
Potamogeton perfoliatus	120.21	113.53
Fucus vesiculosus	219.02	217.77
Cladophora	255.08	253.95
Ectocarpus – collected in autumn	186.65	185.82
Audouinella	570.09	569.63

The red algae *Audouinella* was characterized the highest content of the d. m. and d. o. m.(respectively 570.09 $g\cdot kg^{-1}$ and 569.63 $g\cdot kg^{-1}$). In turn, the brown algae *Scitosiphon cf. S. tortilis* contained the lowest content of the d. m. and d. o. m. (respectively 111.33 $g\cdot kg^{-1}$ and 109.88 $g\cdot kg^{-1}$).

The amount of the biomethane, obtained from selected algae and plants is presented in below table (tab. 2).

Tab. 2. The obtained amounts of methane from particular algae and aquatic plants.

The identified algae and plants	The obtained amounts of methane [$dm^3\cdot g^{-1}$]
Ectocarpus – collected in spring	0.303
Scitosiphon cf. S. tortilis	0.023
Potamogeton perfoliatus	0.076
The mixture of *Fucus vesiculosus* and *Audouinella*	0.271
Cladophora	0.198
Ectocarpus – collected in autumn	0.172
Audouinella	-
The mixture of *Ectocarpus* – from spring and autumn harvest	0.240

Production of biogas from selected algae and water plants oscillated between 0.023 $dm^3\cdot g^{-1}$ and 0.303 $dm^3\cdot g^{-1}$. The lowest production of biomethane was found for brown algae *Scitosiphon cf. S. tortilis*,the value of this production achieved only

$0.023\ dm^3 \cdot g^{-1}$.

In turn, the largest production was characterized by another brown algae - *Ectocarpus* from spring harvest.The value of the obtained methane was at the level of $0.303\ dm^3 \cdot g^{-1}$. Similarly, the high methane value was found when the mixture of these two brown algae (from spring and autumn harvest) was applied ($0.240\ dm^3 \cdot g^{-1}$).

The red algae *Audouinella* was not able to produce biogas, but mixed with *Fucus vesiculosus* has allowed to obtain the biogas on a fairly high level of $0.271\ dm^3 \cdot g^{-1}$.

The percentage of the biogas obtained is shown in the table 3.

Tab. 3. Percentage content of the methane and carbon dioxide in the obtained biogas.

Name of the substrate	The content of the methan (% vol.)		The content of the carbon dioxide (% vol.)
	min	max	
The mixture of the *Ectocarpus collected in spring and autumn*	80.7	81.2	18.8
The mixture of the *Fucus vesiculosus and Audouinella*	73.0	75.1	22.0
Potamogeton perfoliatus	69.9	70.9	17.2
Ectocarpus – collected in spring	79.4	79.8	20.4
Cladophora	73.3	73.3	12.8

The highest content of the methane in biogas was obtained from the mixture of *Ectocarpus* from spring and autumn harvest (values oscillated from 80.7 % to 81.2 %), while the minimum methane content was found in the biogas obtained from the *Potamogeton perfoliatus (*values oscillated from 69.9% to 70.9%).

In turn, the highest percentage share of carbon dioxide in the biogas was characterized by the mixture *Fucus vesiculosus* and *Audouinella* (22 %), while the lowest carbon dioxide content was found in the biogas obtained from the *Cladophora (*12.8 %).

From the *Scitosiphon cf. S. tortilis* wasn't read the composition of the obtained biogas, because there was too little biogas production. In turn, from the *Ectocarpus* collected in autumn wasn't also read the composition of the obtained biogas, because there were technical reasons.

4 Discussion

The influent of large amounts of the biogenic substances to the Baltic Sea causes increase its eutrophication and disturbs its ecological balance. Excessively growing algae often create of thick algae coat on the surface of the water. Algae blooms are especially visible at the edge of the sea, sometimes algae are thrown to the beach by the waves. Unfortunately, this greatly reduces the aesthetic and sanitary values of these places. Algae blooms also generate a threat to human health, therefore many different methods are used to eliminate algae from water reservoirs and Baltic Sea. Harvest of the algae from beaches and water for energy purposescould reduce their quantity in water and could improve the quality of Baltic Sea water [3, 5, 6, 16, 24].

The conducted research has shown, that the largest amount of the biogas was obtained from the biomass of the brown algae *Ectocarpus*, which was collected in spring ($0.303\ dm^3 \cdot g^{-1}$). The mixture of two algae: *Audouinella* and *Fucus vesiculosus* has allowed to obtain less quantity of the biogas ($0.271\ dm^3 \cdot g^{-1}$). Barbot et al. [5] obtained the less amount of the biogas from *Fucus vesiculosus*. Production of biogas from this algae oscillated between $0.04\ dm^3 \cdot g^{-1}$ do $0.11\ dm^3 \cdot g^{-1}$. In turn, the other authors Vivekanand et al. [19] found a little more of the biogas, which was produced from *Saccharina latissima*. The values oscillated between $0.22\ dm^3 \cdot g^{-1}$ and $0.26\ dm^3 \cdot g^{-1}$.

The presented research has revealed low value amount of the biogas ($0.076\ dm^3 \cdot g^{-1}$) with the lowest methane content (from 69.9 % to 70.9%) in *Potamogeton perfoliatus* biomass.This amount of the obtained biogas was comparable with other aquatic plants (i. a. *Elodea densa* and *Potamogeton malaianus*). Koyama et al. [15] found the methane production on a slightly higher level of $0{,}16\ dm^3 \cdot g^{-1}$and $0{,}36\ dm^3 \cdot g^{-1}$during the 14 day fermentation process. In turn, Kesaano [25] was found that the percentage content of the methane in the obtained biogas from *Lemna* biomass reached the level between 67,1% and 68,63%.The all authors used different pretreatment methods and depending on them they received different results.

Due to a small amount of the research in this field, more research is needed. The improvement of the technique of harvesting and initial pretreatment of the algae and water plants biomass is necessary. It would also be good to increase the number of studies on specific species and types of the algae and aquatic plants.

5 Conclusions

1. In all collected samples the following kinds and species of algae and aquatic plants were found: *Scytosiphon cf. S. tortilis, Fucus vesiculosus, Cladophora, Audouinella, Potamogeton perfoliatus.*
2. Production of biogas from selected algae and water plants oscillated between $0.023\ dm^3 \cdot g^{-1}$ and $0.303\ dm^3 \cdot g^{-1}$. The lowest production of biomethane was found for brown algae *Scitosiphon cf. S. tortilis* ($0.023\ dm^3 \cdot g^{-1}$). The largest production of biomethane was

characterized by another brown algae - *Ectocarpus* from spring harvest (0.303 $dm^3 \cdot g^{-1}$) and by the mixture of these two brown algae (from spring and autumn harvest) (0.240 $dm^3 \cdot g^{-1}$).

3. The highest content of the methane in biogas was obtained from the mixture of *Ectocarpus* from spring and autumn harvest (values oscillated from 80.7 % to 81.2 %). The minimum methane content was found in the biogas obtained from the *Potamogeton perfoliatus (*values oscillated from 69.9% to 70.9%).

Acknowledgments

I would like to thank the Professor Józef Wiktor from the Institute of Oceanology of the Polish Academy of Sciences in Sopot for help in identifying some species of algae.

References

1. A. Hossain, A. Salleh, Am. J. Biochem. Biotech. **4,** (3) (2008)
2. M. Frąc, S. Jezierska–Tys, J. Tys, Ac. Agroph. **13,** (3) (2009)
3. J. Bień, M. Zabochnicka-Świątek, L. Sławik, Inż. i Ochr. Środ. **13,** (3) (2010)
4. G. Jard, C. Dumas, J.P. Delgenes, H. Marfaing, B. Sialve, J.P. Steyer, H. Carrčre, Biochem. Eng. Jour. **79** (2013)
5. Y. N. Barbot, H. M. Falk, R. Benz, J. Appl. Phycol. **27** (2015)
6. Y. N. Barbot, L. Thomsen, R. Benz, Mar. Drugs **13** (2015)
7. M. Dębowski, A. Grala, M. Zieliński, M. Dudek, Archiv. of Environmental Protection **38,** (4) (2012)
8. T. Kobayashi, Y.-P. Wu, Z.-J. Lu, K.-Q. Xu, Energies **8** (2015)
9. W. W. Carmichael, Świat Nauki **3** (1994)
10. B. Rakowska, M. Sitkowska, E. Szczepocka, B. Szulc, Ocean. and Hydrobiol. Stud. **34,** (1) (2005)
11. H. Bucka, E. Wilk-Woźniak,*Glony pro- i eu-kariotyczne zbiorowisk fitoplanktonu w zbiorni-kach wodnych Polski Południowej* (Instytut Och-rony Przyrody – PAN, Kraków, 2007)
12. I. Michalak, R. Wilk, K. Chojnacka, Waste Biom. Valor. 8 (2017)
13. I. Sanseverino, D. Conduto, L. Pozzoli, S. Dobricic, T. Lettieri, *Algal bloom and its economic impact* (JRC Technical Reports, 2016)
14. N. Wasmund, Harmful algal blooms in coastal waters of the south-eastern Baltic Sea. In: G. Schernewski et al., *Baltic Coastal Ecosystems* (Springer-Verlag Berlin Heidelberg, 2002)
15. M. Koyama, S. Yamamoto, K. Ishikawa, S. Ban, T. Toda, Ecol. Eng. **69** (2014)
16. L. Håkanson, A. C. Bryhn, Controlling Eutrophication in the Baltic Sea and the Kattegat. In. A. A. Ansari et al. (eds.) *Eutrophication: Causes, Consequences and Control* (Springer Science+Business Media B.V., 2011)
17. E. Jachniak, *Wpływ czynników fizykochemicznych oraz hydrologicznych na przebieg procesów eutrofizacyjnych w wybranych zbiornikach zaporowych południowej Polski,* Praca doktorska (Uniwersytet Rolniczy, Kraków, 2010)
18. K. Szoszkiewicz, S. S. Jusik, T. Zgoła, *Klucz do oznaczania makrofitów dla potrzeb oceny stanu ekologiczmego wód powierzchniowych* (Biblioteka Monitoringu Środowiska, Warszawa 2010)
19. V. Vivekanand, V. G. H. Eijsink, S. J. Horn, J. Appl. Phycol. **24** (2012)
20. S. Tedesco, K.Y. Benyounis, A.G. Olabi, Energy **61** (2013)
21. I. Angelidaki, M. Alves, D. Bolzonella, L. Borzacconi, L. Campos, A. Guwy, S.V. Kalyuzhnyi, P. Janicek, J.B. Van Lier, Water Sci. Technol. **59** (2009)
22. W. Braune, M. D. Guiry, *A colour guide to common benthic green, brown and red algae of the world's oceans* (Königstein: Koeltz Scientific Books, 2011)
23. D.M. John,*The Freshwater Algal Flora of the British Isles: An Identification Guide to Freshwater and Terrestrial Algae, Tom 1,* (Cambridge University Press, 2002)
24. M. Leśniewska, M. Witak, Ocean. and Hydrobiol. Stud. Intern. Jour. of Oceanography and Hydrobiology **40,** (1) (2011)
25. M. Kesaano, *Sustainable Management of Duckweed Biomass Grown for Nutrient Control in Municipal Wastewaters.* All Graduate Theses and Dissertations, (Utah State University, DigitalCommons@USU 2011)

Web pages:

http://www.algaebase.org/search/species/detail/?species_id=291

http://www.hel.univ.gda.pl/atlas/organizmy.htm
http://www.hel.univ.gda.pl/
http://ccba.ug.edu.pl/pages/pl/strona-glowna.php

Biostimulants and the antiradical activity of soybean seeds

Anna Kocira[1,*]

[1]State School of Higher Education in Chełm, Institute of Agricultural Sciences, 54 Pocztowa Street, 22-100 Chełm, Poland

Abstract. In recent years, the interest in the application of biostimulants has increased, which positively affect the growth and development of plants and contribute to an increase in the yield and quality of crops. In Poland, the area of soybean cultivation is constantly increasing and thus the improvement of the quality of its seeds seems to be purposeful. In the available literature, there are few reports regarding the effect of biostimulants on the antiradical activity of plants. Therefore, studies on the influence of biostimulants on soybean seed antiradical activity seem to be justified. The study was carried out in 2014 - 2016 in Perespa, Poland. Annushka soybean seeds were sown in the third decade of April. During the growing season, four biostimulants: Kelpak SL, Terra Sorb Complex, Atonik, and Tytanit, were used in four combinations, using lower or higher concentrations and single or double spraying. After harvesting the plants, the antiradical activity of the seeds was evaluated by ABTS•+ assay. It has been found that the foliar application of biostimulants positively influenced the studied property. The highest antiradical activity of plants was found upon double spraying with lower concentrations of Atonik and Terra Sorb Complex.

1 Introduction

An increasing interest has recently been observed in the application of biostimulants, which positively affect the growth and development of plants and contribute to an increase in the yield and quality of crops. Their application is justified in the case of abiotic stress [1], especially in plants susceptible to adverse weather. Biostimulants used in crop cultivation are natural or synthetic origin and contain both organic and inorganic compounds. The natural biostimulants are based on free amino acids, humus compounds, extracts from fruit or marine algae, chitin and its derivative – chitosan, or effective microorganisms. In turn, the group of synthetic biostimulants includes preparations that contain growth regulators, phenolic compounds, inorganic salts, and nutrients (Al, Co, Na, Se, Ti, and Si) [2-6].

The use of biostimulants in the cultivation of crops with high thermal demands, like e.g. soybean, seems to be justified. In addition, the area of soybean cultivation in Poland is constantly increasing and therefore the possibility of improving the quality of its seeds is crucial. Literature data show that legumes are rich in antioxidants [7, 8], however only sparse reports are available on the effect of biostimulants on the antiradical activity of legumes [3, 4, 6]. Hence, research aimed at determining the effect of biostimulants on the antiradical activity of soybean seeds seems to be eligible.

2 Materials and Methods

2.1 Plant materials and growth conditions

The study was carried out in 2014-2016 in Perespa (50°66′N; 23°63′E), Poland. The soil type was characterized as Cambic Rendzic Phaeozems. It is alkaline (pH in 1M KCl 7.4–7.5) and rich in phosphorus, potassium, and magnesium. The experiment was established in a randomized block design in four replications with an elementary experimental plot area of 5 m^2. Seeds of soybean cultivar Annushka were sown in the third 10-day period of April with the spacing of 30 x 3.5 cm. The sowing rate was 80 plants·m^{-2} Over the growing season, four biostimulants: Kelpak SL (*Ecklonia maxima* extract, 11.0 mg/l auxin and 0.031 mg/l cytokinin), Terra Sorb Complex (20% free amino acids), Atonik (phenolic compounds, sodium para-nitrophenolate PNP (0.3%), sodium ortho-nitrophenolate ONP (0.2%) and sodium 5-nitroguaiacolate 5NG (0.1%)), and Tytanit (8.5 g Ti/l), were applied through leaf spraying according with the scheme presented in Table 1.

The biostimulants were applied with a GARLAND FUM 12B battery field sprayer (Lecher LU 120–03) at a pressure of 0.30 MPa, using 300 L liquid per hectare. All the results were compared to control, where plants were treated with the same volume of water (no biostimulant applied). Tillage of plants was done using good agricultural practices. No pesticides were used (pest

* Corresponding author: akocira@pwsz.chelm.pl

https://doi.org/10.1051/bioconf/20181001008

number did not exceed the thresholds of harmfulness) and plants were hand weeding.

The average temperature and rainfall in the soybean growing season are shown in Table 2.

Table 1. Scheme of biostimulants application.

Biostimulants	Number of applications Plant's stage	Concentration
Atonik	single spraying	0.1%
	BBCH 13-15	0.2%
	double spraying	0.1%
	BBCH 13-15 BBCH 61	0.2%
Tytanit	single spraying	0.07%
	BBCH 13-15	0.13%
	double spraying	0.07%
	BBCH 13-15 BBCH 61	0.13%
Kelpak SL	single spraying	0.7%
	BBCH 13-15	1%
	double spraying	0.7%
	BBCH 13-15 BBCH 61	1%
Terra Sorb Complex	single spraying	0.3%
	BBCH 13-15	0.5%
	double spraying	0.3%
	BBCH 13-15 BBCH 61	0.5%

Table 2. Temperature (T) and precipitation during the soybean growing season 2014-2016.

Month	Year						Average from 2002 to 2013	
	2014		2015		2016			
	T (°C) min max	Rainfall (mm)	T (°C) min max	Rainfall (mm)	T (°C) min max	Rainfall (mm)	T (°C)	Rainfall (mm)
April	9.4 -6.0 22.7	36.5	8.2 -1.7 24.3	30.1	9.2 -1.2 22.6	68.4	8.5	41.2
May	13.7 0.5 27.7	208.3	12.7 1.5 24.9	108.6	13.8 2.6 26.7	61.3	12.7	63.4
June	16.1 6.7 28.9	67.1	17.4 6.6 30.5	14.1	18.1 4.2 31.5	97.1	17.7	68.6
July	20.3 10.0 31.0	104.2	19.6 8.4 33.4	59.2	19.5 8.8 31.2	107.6	18.9	79.1
August	18.2 6.3 34.0	115.4	21.6 5.6 35.5	23.4	18.2 7.1 30.7	95.3	19.4	71.8
September	13.7 3.7 25.8	89.4	15.1 4.2 34.5	137.6	15.2 1.6 28.7	41.2	14.1	69.2
Average/ Total	15.1	620.9	15.8	373.0	17.1	470.9	15.2	393.3

2.2 Determination of the antiradical activity

After the harvest, seeds were dried, ground in a laboratory grinder, and sieved (mesh size 0.310 mm). Flour was stored at a temperature of -20°C until analyzed. The antiradical activity of soybean seeds against ABTS•+ was determined using an acetone extract prepared acc. to a modified method by Kumar et al. [9]. Ground soybean seeds (100 g) were weighed into a conical flask with a ground glass stopper, poured with 2 mL of 70% acetone, and shaken at a room temperature of 2 h. Next, the samples were centrifuged at the speed of 10,000 x g and temperature of 20°C for 15 min. The supernatant was collected and stored in the dark at a temperature of -50°C until analyzed.

2.2.1 Determination of the antiradical activity against ABTS •+cation radical

The antiradical activity was determined according to the method developed by Re et al. [10]. ABTS (diammonium 2,2'-azino-bis(3-ethylbenzothiazoline-6-sulphonic acid) was dissolved in distilled water until it has reached the concentration of 7mM. The ABTS•+ cation radical was obtained as a result of ABTS reaction with 2.45 mM potassium persulfate (K2S2O8). The reaction mixture was incubated at a room temperature for 12 h. The ABTS•+ solution was diluted with distilled water until its absorbance of 0.700 (±0.02) has reached at the wavelength of λ = 734. Determination of the antiradical activity consisted in the measurement of absorbance decrease during ABTS•+ cation radical reduction under the influence of antioxidants contained in the analyzed extracts, resulting in the formation of a colorless ABTS. To this end, 250 µL of the ABTS•+ solution were added to 5 µL of the extract, and the mixture was thoroughly mixed. Absorbance of the resultant solution was measured after 2h at the wavelength of λ = 734, using an EPOCH 2 microplate reader (BioTek, USA). The antiradical activity was expressed as Trolox equivalent (TE) w mg/g d.m.

2.2 Statistical analysis

Data on the antiradical activity of soybean seeds from four replicates of each combination were subjected to the statistical analysis. The Shapiro-Wilk test was performed for the normal distribution of data. The results were analyzed using one-way analysis of variance, ANOVA. The significance of differences between evaluated mean values was estimated by means of Turkey's test intervals of confidence at a significance level of $p < 0.05$. The

https://doi.org/10.1051/bioconf/20181001008

statistical analysis was performed with Statistica 12 (StatSoft, Inc.).

3 Results and Discussion

The study showed that foliar application of biostimulants were beneficially affected the antiradical activity of soybean seeds (Table 3). It was found that foliar application of Terra Sorb Complex in 2014, Atonik and Tytanit in 2015 or Kelpak SL in 2016 increased this property.

Table 3. Change in the antiradical activity of soybean cv. Annushka affected by the application of biostimulants (mg/g d.m.).

Year	Control	Application (B)	Biostimulant (A)				
			Atonik	Tytanit	Kelpak SL	Terra Sorb Complex	Mean
2014	5.65 bc	B1	5.60 ab	5.57a	5.54a	5.56a	5.57A
		B2	5.69c	5.68bc	5.72c	5.74c	5.71C
		B3	5.74c	5.75c	5.73c	5.68bc	5.72C
		B4	5.57a	5.69c	5.52a	5.72c	5.63B
		Mean	5.65AB	5.67AB	5.63A	5.68B	5.66
2015	5.18 de	B1	5.19de	5.30ef	5.25e	5.02abc	5.19C
		B2	5.41f	5.13cd	4.93a	5.10bcd	5.14AB
		B3	5.12cd	5.21de	4.96ab	5.07bc	5.09A
		B4	5.25e	5.27e	5.00ab	5.06bc	5.14AB
		Mean	5.24B	5.23B	5.04A	5.06A	5.14
2016	5.38 ab	B1	5.50cd	5.48bcd	5.48bcd	5.53de	5.50AB
		B2	5.41abc	5.44abc	5.66f	5.67f	5.54B
		B3	5.42abc	5.40abc	5.63ef	5.49cd	5.48A
		B4	5.40abc	5.35a	5.63ef	5.48bcd	5.46A
		Mean	5.43A	5.42A	5.60C	5.54B	5.50

Abbreviations: B1- single spraying with low concentration of biostimulant; B2- double spraying with low concentration of biostimulant; B3 - single spraying with high concentration of biostimulant; B2- double spraying with high concentration of biostimulant. Different letters within the same year denote significant differences between the treatments at $p<0.05$.

Considering the modes of biostimulants application, the best effects were achieved after double plant spraying with the lower concentration of the preparations in 2014 and 2016, when the antiradical activity of the seeds increased by 1.1 - 3.0% compared to the control seeds (Table 3). However, single spraying with the lower concentration of biostimulants in 2015 increased this property and gave a similar results as control. Both the double spraying with Atonik, Kelpak SL and Terra Sorb Complex in the lower concentration and Tytanit, Terra Sorb Complex in the higher concentration as well as single spraying with Atonik, Tytanit and Kelpak SL in the lower concentration increased the antiradical activity of soybean seeds in 2014. Also the best effect on this property had the double spraying with Atonik in the lower concentration in 2015 and the same mode of Kelpak SL and Terra Sorb Complex application in 2016.

It was no significant effect of the foliar application of the biostimulants on the antiradical activity of soybean seeds (means for 2014-2016) (Figure 1). However the plants double sprayed with Atonik and Terra Sorb Complex in the lower concentrations achieved the same and, simultaneously, the highest antiradical activity of the seeds. Single spraying with Atonik, regardless of its concentration, caused a similar antiradical activity of the seeds, but double spraying with this biostimulant in the higher concentration reduced it. Of all tested combinations, the lowest antiradical activity was noted upon single spraying with Terra Sorb Complex in the lower concentration. In turn, the higher concentration of this biostimulant increased the activity, however differences in the antiradical activity between the single and double sprayed plants were negligible. Plants single sprayed with Tytanit responded with an enhanced antiradical activity compared to the double sprayed plants, but still the highest value of the activity was obtained upon double application of the biostimulant in the lower concentration. Both the single spraying with Kelpak SL in both concentrations as well as double spraying with this biostimulant in the lower concentration had a positive effect on the antiradical activity of soybean seeds compared to the double spraying with Kelpak SL in the higher concentration.

The control plants were characterized by a lower antiradical activity which was, however, higher than that achieved upon single spraying with Terra Sorb Complex in the lower concentration and upon double spraying with Kelpak SL in the higher concentration. In addition, the single spraying of the plants with biostimulants in the higher concentrations resulted in a smaller scatter of results.

The positive effect of biostimulants on the quality of legume seeds was confirmed in earlier studies. Kocira et al. [3] demonstrated double application of Kelpak SL, in both lower and higher concentration, to significantly increase the antioxidative activity of seeds of *Phaseolus vulgaris* Toska cv. (with red seed coat) against ABTS•+. In the case of bean of Aura cv. (white seed coat), this activity was more beneficially affected by double spraying with Kelpak SL in the higher concentration. Single application of a biostimulant based on nitrophenolic compounds (Atonik) in the lower concentration had a positive effect upon antiradical activity (ABTS) of Aura cv. common bean seeds, causing a significant increase in the activity (by 64%). In turn, common bean of Toska cv. was more positively responding to the single spraying with Atonik in the higher concentration; the antiradical activity of its seeds increased by 9%, though differences were statistically insignificant [4]. A 5-6% increase in the antioxidative activity of seeds of common bean of Toska cv. was also obtained upon the application of an organic growth stimulant Nano-Gro, regardless of the form of its application [6].

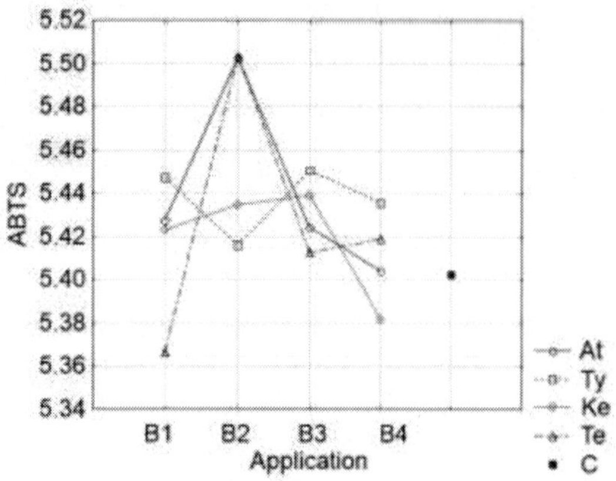

Fig. 1. Antiradical activity of soybean cv. Annushka affected by the application of biostimulants (means for 2014-2016) (mg/g d.m.). Abbreviations: see table 2; At-Atonik; Ty-Tytanit; Ke-Kelpak SL; Te-Terra Sorb Complex; C-Control.

4 Conclusions

The foliar application of both natural and synthetic biostimulants improved the quality of soybean seeds. In addition, it positively influenced the antiradical activity of soybean seeds, especially upon the double spraying with lower concentrations of the preparations.

References

1. M. Kozak, A. Wondołowska-Grabowska, M. Serafin-Andrzejewska, M. Gniadzik, M.K. Kozak, *Rolnictwo XXI wieku – problemy i wyzwania* (Idea Knowledge Future), Wrocław (2016)
2. E. Matyjaszczyk, Przem. Chem. **94,** 1841-1844 (2015)
3. A. Kocira, M. Świeca, S. Kocira, U. Złotek, A. Jakubczyk, Saudi J. Biol. Sci. (2016), https://doi.org/10.1016/j.sjbs.2016.01.039.
4. A. Kocira, S. Kocira, M. Świeca, U. Złotek, A. Jakubczyk, K. Kapela, Sci. Hort. **214**, 76–82 (2017)
5. S. Kocira, A. Kocira, M. Szmigielski, A. Piecak, A. Sagan, U. Malaga-Toboła, Przem. Chem. **94**, 1732-1736 (2015)
6. A. Kocira, S. Kocira, U. Złotek, R. Kornas, M. Świeca, Fres. Environ. Bull. **24**, 4034-4041 (2015)
7. I. Kowalska , Ł. Cieśla, T. Oniszczuk, M. Waksmundzka-Hajnos, W. Oleszek, A. Stochmal, J. Liq. Chromatogr. Relat. Technol. **36**, 2387–2394 (2013)
8. S.A. Marathe, V. Rajalakshmi, S.N. Jamdar, A. Sharma, Food Chem. Toxicol. **49**, 2005-2012 (2011)
9. V. Kumar, A. Rani, A.K. Dixit, D. Pratap, D. Bhatnagar, Food Res. Int. **43**, 323-328 (2010)
10. R. Re, N. Pellegrini, A. Proteggente, A. Pannala, M. Yang, C. Rice-Evans, Free Radic. Biol. Med. **26**, 1231-1237 (1999)
11. Biesaga-KościelniakJ., Kościelniak J., Filek M., Marcińska I., KrekuleJ. Machackova I., Kuboń M. Acta Physiol Plant.32**(5),** 987-995, (2010)

Improvement of the photosynthetic activity of Moldavian dragonhead (*Dracocephalum moldavica* L.) through foliar application of a nitrophenolate–based biostimulant

Sławomir Kocira[1,*], *Agnieszka* Sujak[2], *Tomasz* Oniszczuk[3], *Agnieszka* Szparaga[4], *Mariusz* Szymanek[5], *Hanna* Karakuła-Juchnowicz[6], *Anna* Krawczuk[1], and *Karol* Kupryaniuk[3]

[1]University of Life Sciences in Lublin, Department of Machinery Exploitation and Production Process Management, Głęboka 28, 20612 Lublin, Poland
[2]University of Life Sciences in Lublin, Department of Biophysics, Akademicka 13, 20950 Lublin, Poland
[3]University of Life Sciences in Lublin, Department of Food Process Engineering, Doświadczalna 44, 20280 Lublin, Poland
[4]Koszalin University of Technology, Department of Agrobiotechnology, Racławicka 15-17, 75-620 Koszalin, Poland
[5]University of Life Sciences in Lublin, Department of Agricultural Machinery, Głęboka 28, 20612 Lublin, Poland
[6]Medical University of Lublin, Department of Clinical Neuropsychiatry, Głuska 1, 20439 Lublin, Poland

Abstract. Application of biostimulants instigates many physiological processes that enhance nutrition efficiency, abiotic stress tolerance, and quality traits of crops, regardless of their nutrient content. One of such preparations is Atonik which contains nitrophenol compounds naturally occurring in plant cells. Several studies have confirmed its beneficial effect on the growth, development, and improved metabolic activity of plants. Therefore, it seems advisable to investigate the effect of Atonik preparation on the photosynthetic activity of Moldavian dragonhead (*Dracocephalum moldavica* L.). The reported study was carried out in 2014 in Perespa, Poland. Over the growing season, Atonik was foliar-applied at a dose of 0.3 L/ha (0.1%) and 0.6 L/ha (0.2%) by single and double spraying of plants. Chlorophyll content and nitrogen status (N) were estimated by a Chlorophyll Meter SPAD-502 Plus. The foliar application of Atonik was found to improve the efficiency of the photosynthetic apparatus and chlorophyll content in the leaves of *Dracocephalum moldavica* plants, but the results were dependent on biostimulant concentration and number of its applications, and on the date of measurement. To conclude, Atonik is an environmental-friendly preparation which has a positive impact on the metabolic processes of plants.

1 Introduction

It is commonly believed today that high yields of crops results from agriculture intensification. However, attempts have been undertaken in recent years to apply preparations that would ensure high yields of high-quality crops under minimized use of pesticides and fertilizers. These preparations include, e.g., biostimulants which aside the above benefits are also environment-friendly. The application of biostimulants accelerates many physiological processes that enhance nutrition efficiency, abiotic stress tolerance, and quality traits of crops, regardless of their nutrient content; and this stimulation effects improvement in crop yield [1, 2, 3]. The use of biostimulants has been reported to influence the metabolic processes in plants by stimulating the synthesis and enhancement of phytohormones, and also to improve their photosynthetic activity [4, 5]. One of the biostimulants is Atonik preparation which contains nitrophenolic compounds naturally occurring in plant tissues: sodium para-nitrophenolate PNP (0.3%), sodium ortho-nitrophenolate ONP (0.2%), and sodium 5-nitroguaiacolate 5NG (0.1%). Several studies have confirmed the beneficial effect of this biostimulant on the growth, development, and improved metabolic activity of plants. [6, 7, 8, 14]. Therefore, it seems advisable to investigate the effect of Atonik preparation on the photosynthetic activity of Moldavian dragonhead (*Dracocephalum moldavica* L.).

2 Materials and Methods

2.1. Plant materials and growth conditions

The study was carried out in 2014 in Perespa (50°66′N; 23°63′E), Poland. The soil type was characterized as Cambic Rendzic Phaeozems. It is alkaline (pH in 1M KCl 7.4–7.5) and rich in phosphorus, potassium, and magnesium. The experiment was established in a randomized block design in four replications with an elementary experimental plot area of 5 m^2. The sowing

* Corresponding author: slawomir.kocira@up.lublin.pl

rate was 15 plants·m^2. Moldavian dragonhead seeds were sown in the third 10-day period of April at a depth of 1-1.5 cm, with the spacing of 45 x 20 cm. over the growing season, Atonik was foliar-applied at a dose of 0.3 L/ha (0.1%) and 0.6 L/ha (0.2%) by single spraying (at the 10-12 leaf stage – 20th June) and double spraying of plants (at the 10-12 leaf stage - 20th June and at the beginning of plant blossoming – 4th July). The preparation was applied using a GARLAND FUM 12B battery field sprayer (Lecher LU 120–03) at a pressure of 0.30 MPa, using 300 L of liquid per hectare. The following experimental combinations were investigated: control: where plants were treated with the same volume of water (no biostimulant), 1- or 2-fold spraying with 0.3% Atonik, and 1- or 2-fold spraying with 0.6% Atonik. All results from experimental plots were compared to these from the control plots. Tillage of the plants was done following good agricultural practices. No pesticides were used (number of pests did not exceed the thresholds of harmfulness).

The average temperature and rainfall in the soybean growing season are shown in Table 1.

Table 1. Temperature (T) and precipitation during the soybean growing season 2014-2016.

Month	Year 2014 T (°C) min max	2014 Rainfall (mm)	2015 T (°C) min max	2015 Rainfall (mm)	2016 T (°C) min max	2016 Rainfall (mm)	Average from 2002 to 2013 T (°C)	Average from 2002 to 2013 Rainfall (mm)
April	9.4 -6.0 22.7	36.5	8.2 -1.7 24.3	30.1	9.2 -1.2 22.6	68.4	8.5	41.2
May	13.7 0.5 27.7	208.3	12.7 1.5 24.9	108.6	13.8 2.6 26.7	61.3	12.7	63.4
June	16.1 6.7 28.9	67.1	17.4 6.6 30.5	14.1	18.1 4.2 31.5	97.1	17.7	68.6
July	20.3 10.0 31.0	104.2	19.6 8.4 33.4	59.2	19.5 8.8 31.2	107.6	18.9	79.1
August	18.2 6.3 34.0	115.4	21.6 5.6 35.5	23.4	18.2 7.1 30.7	95.3	19.4	71.8
September	13.7 3.7 25.8	89.4	15.1 4.2 34.5	137.6	15.2 1.6 28.7	41.2	14.1	69.2
Average/ Total	15.1	620.9	15.8	373.0	17.1	470.9	15.2	393.3

2.2 Photosynthetic activity of plants

The photosynthetic activity of plants was measured over the entire growing season starting from one week after biostimulant application. For the single spraying combination, the second measurement was carried out in the second term at the appropriate plant development stage. To estimate the actual photochemical activity of PSII in situ, chlorophyll a fluorescence was measured in plants using a pulse amplitude-modulated (PAM) fluorimeter (Mini PAM, Walz, GmbH, FRG, Germany) with a light emitting diode at 650 nm and a standard intensity 0.15 μmol m^{-2}s^{-1} PAR (photosynthetically-active radiation). A Leaf-Clip holder 2030-B was used in all measurements. Standard settings optimized for measurements with leaf samples at approx. 12 mm distance between fiberoptics and leaf surface were used for measurements. The parameters listed hereinafter were set. Measuring light: on; Measuring pulse frequency: 0.6 KHz; Actinic illumination time: 30s; Actinic light intensity: 5 (relative intensity 5.5); Actinic light factor: 1.00; LC-WIDTH: 10s; LC INT: 3; Int-Temp: on; Light gain: 1.00; Temp. gain: 1.00; Saturating light pulse width: 0.8; Saturation pulse intensity: 8; Electronic signal damping: 2 (0.2s); Measuring light intensity: 8. The Auto-Zero function was applied in order to suppress any unavoidable background signal. All the measurements were done under field condition with the assumption that a photon flux density is below the saturation level. The following fluorescence parameters were measured: F - the present fluorescence yield, M - fluorescence after saturation pulse was applied, Yield = (M – F)/M, PAR – photosynthetically-active radiation, and ETR-electron transfer rate calculated as: ETR = Yield · PAR · ETR – factor; i.e. ETR = Yield · PAR · 0.5 · 0.84. The standard factor 0.84 corresponds to the fraction of incident light absorbed by a leaf [5]. In turn, Fm' stands for the maximum yield of fluorescence of an illuminated sample. Based on the above information, various parameters of fluorescence were determined which are useful in photosynthesis yield determination. The maximum Photosystem II (PS II) quantum yield of the dark-adapted samples and the effective PS II quantum yield of the illuminated samples, Y(II) = (Fm'-F)/Fm' = ΔF/Fm' were determined [9]. The Y (II) parameter corresponds to the fraction of energy which is photochemically-conversed into PS II (consumed during photosynthesis).

Chlorophylls and nitrogen status (N) were estimated by Chlorophyll Meter SPAD-502 Plus (Konica Minolta) (ten plants of each experimental combination). SPAD is the chlorophyll content

2.3 Statistical analysis

Data on photosynthetic parameters of four replicates of each combination were subjected to statistical analysis. The Shapiro-Wilk test was performed for the normal distribution of data. The results were analyzed using one-way analysis of variance, ANOVA. The significance of differences between evaluated mean values was estimated by Tukey's test intervals of confidence at a significance level of $p < 0.05$. Statistical analysis was performed with Statistica ver. 12 software (StatSoft, Inc.).

3 Results and Discussions

Fluorescence measurements conducted in the first term (measurement I) demonstrated the single foliar application of a 0.1% solution of Atonik to cause over three-fold increase in photosynthetically-active radiation (PAR) of the plants, compared to double spraying of the biostimulant in the same concentration (table 2). Kocira et al. [5,12,13] reported that single foliar spraying of a biostimulant, based on algae and amino acids (Fylloton) used in a lower concentration, had the most beneficial effect on PAR enhancement. This parameter indicates the amount of energy accumulated by a plant in a specified time of measurement. Result of this measurement obtained in the current study suggests that considering no distinct differences in weather conditions, the differences occurred probably at the molecular level and were additionally affected by the physical status of Moldavian dragonhead plants (energy levels in the photosynthetic apparatus). In addition, an increase in the photosynthetically-active radiation informs indirectly about the consumption of solar energy by plants. The photosynthetically-active radiation (PAR) is linked with the maximum yield of sample fluorescence under natural/field (F) conditions; namely an increase in F is accompanied by a decrease in PAR. Usually this parameter indicates the fraction of energy that is not consumed in the process of photosynthesis in the course of the experiment.

Results obtained demonstrated a 16% increase in the effective photochemical yield of PS II (Y (II)) upon the single spraying of 0.1% Atonik solution, compared to the double spraying of the biostimulant in the higher concentration (measurement II) (table 3). The single spraying of Moldavian dragonhead plants with 0.1% Atonik solution increased the electron transport rate (ETR) by almost 80% compared to the control (measurement I). However, an increase in biostimulant concentration to 0.2% (at the same stage of plant growth) caused over threefold decrease of ETR. In turn, during measurement II the highest value of ETR was determined as a result spraying the plants with Atonik in a higher concentration, regardless of the number of applications. The highest value of ETR was determined in plants during measurement I, when plants of Moldavian dragonhead were at the stage of 10-12 developed leaves, which is indicative of their more intensive growth at this stage than at the stage of blooming.

The foliar application of Atonik preparation caused an increase in chlorophyll content in leaves of Moldavian dragonhead plants in both measurement I and II. This observation confirms findings reported in the study by Kocira et al. [5], where spraying plants of Moldavian dragonhead with a Fylloton biostimulant increased chlorophyll content in leaves. The stimulating effect of Atonik preparation on chlorophyll content increase in plant leaves was also confirmed in investigations conducted with *Arabidopsis thaliana*, oilseed rape, and common osier [6, 8].

Table 2. Effect of Atonik on the PS II chlorophyll fluorescence and photosynthetically-active radiation of Moldavian dragonhead (*Dracocephalum moldavica* L.) plants.

Parameter		Control	Single spraying with 0.1% Atonik	Double spraying with 0.1% Atonik	Single spraying with 0.2% Atonik	Double spraying with 0.2% Atonik
		Experimental combination				
F	I	399.63a	422.88a	350.75a	430.25a	432.25a
	II	364.12a	342.14a	324.29a	324.00a	367.57a
Fm'	I	1028.50a	1069.25a	1473.75b	843.13a	811.88a
	II	1434.43a	1744.71b	1330.71ab	1479.43a	1204.14a
PAR	I	449.25a	919.50b	242.63a	376.00a	424.13a
	II	78.57a	113.00ab	77.00a	178.00bc	209.85c

I – measurement at the 10-12 leaf stage; II - measurement at the beginning of plant blossoming.
Means in the same column with different letters are significantly different (α=0.05).

Percival et al. [10] demonstrated chlorophyll content in leaves to be associated with leaf photosynthetic rates. Like in the previous study [5], the correlations between values of SPAD and F-Fm'/Fm' as measures of photosystem II efficiency were weak. The F-Fm'/Fm' ratio is positively correlated to the PSII quantum yield and an indirect measurement of plant physiological status for which the values of 0.8 ± 0.05 correspond to highly efficient use of the excitation energy in photochemical processes. In our study, such a value was obtained only upon single spraying the plants with Atonik in the lower concentration (II measurement). In addition, the F-Fm'/Fm' ratios of 0.6-0.8 were associated with SPAD values between 50.5 and 63.8.

Many authors confirmed the beneficial effect of biostimulants on enhanced intensity of photosynthesis of cultivable plants like *Dracocephalum moldavica*, basil, rocket, *Arabidopsis thaliana*, and oilseed rape [5-f7, 11].

Table 3. Effect of Atonik on the photosynthetic and chlorophyll parameters of Moldavian dragonhead (*Dracocephalum moldavica* L.) plants.

Parameter		Control	Single spraying with 0.1% Atonik	Double sprayin g with 0.1% Atonik	Single sprayin g with 0.2% Atonik	Double spraying with 0.2% Atonik
		Experimental combination				
Y(II)	I	0.597a	0.583a	0.683a	0.596a	0.583a
	II	0.742ab	0.804b	0.752ab	0.757ab	0.693a
ETR	I	110.79a	198.71b	107.88a	54.08a	121.10ab
	II	24.63a	38.07ab	24.11a	58.90b	60.37b
SPAD	I	51.1a	63.8d	62.8cd	60,0c	60.0b
	II	50.5a	61.3d	59.7cd	56.7bc	55.0ab

I – measurement at the 10-12 leaf stage; II - measurement at the beginning of plant blossoming.
Means in the same column with different letters are significantly different (α=0.05).

https://doi.org/10.1051/bioconf/20181001009

3 Conclusions

The foliar application of Atonik preparation improved the efficiency of the photosynthetic apparatus and chlorophyll content in the leaves of *Dracocephalum moldavica* plants, but the results were dependent on biostimulant concentration and number of its applications, and on the date of measurement (growth stage of plants). The photosynthetic activity of Moldavian dragonhead leaves was most positively affected by the single spraying with Atonik preparation in the lower concentration. Foliar application of Atonik preparation increased chlorophyll content in leaves of *Dracocephalum moldavica*. Atonik was found to be an environmental-friendly preparation which has a positive impact on the metabolic processes of plants.

References

1. P. Du Jardin, Scientia Horticulturae, **196,** 3-14 (2015)
2. A. Kocira, M. Świeca, S. Kocira, U. Złotek, A. Jakubczyk, Saudi Journal of Biological Sciences (2016), https://doi.org/10.1016/j.sjbs.2016.01.039
3. A. Kocira, S. Kocira, M. Świeca, U. Złotek, A. Jakubczyk, K. Kapela, Scientia Horticulturae **214,** 76–82 (2017)
4. P. Calvo, L. Nelson, J.W. Kloepper, Plant Soil, **383**, 1-2, 3–41 (2014)
5. S. Kocira, A. Sujak, A. Kocira, A. Wójtowicz, A. Oniszczuk, Agriculture and Agricultural Science Procedia, **7,** 108-112 (2015)
6. A. Przybysz, H. Gawrońska, J. Gajc-Wolska, Frontiers in Plant Science, **5, 713,** 1-15 (2014)
7. E. Borowski, Z.K. Blamowski, Folia Horticulturae, **21,** 39–48 (2009)
8. J. Wróbel, A. Woźniak, Biostimulators in Modern Agriculture. Ornamental and Special Plants, in: Łukaszewska, A. (Ed.). Editorial House Wieś Jutra, Warsaw, 47-55 (2008)
9. B. Genty, J.M. Briantais, N.R. Baker, Biochimica et Biophysica Acta, **990,** 87-92 (1989)
10. G.C. Percival, I.P. Keary, K. Noviss, Arboriculture and Urban Forestry, **34, 2,** 89-100 (2008)
11. R. Janas, Biuletyn Instytutu Hodowli i Aklimatyzacji Roślin, **262,** 197-206 (2011)
12. Niemiec M, Sikora J., Szeląg-Sikora A., Kuboń M., Olech E., Marczuk A. Przemysł chemiczny 96(**3**),685-688 (2017)
13. Niemiec M, Sikora J., Szeląg-Sikora A., Kuboń M., Olech E., Marczuk A. Przemysł chemiczny 96(**11**),1240-1243 (2017)
14. Biesaga-KościelniakJ., Kościelniak J., Filek M., Marcińska I., KrekuleJ. Machackova I., Kuboń M. Acta Physiol Plant.32**(5),** 987-995, (2010)

Investigation of the process of separation of hard-to-divide weeds from seed mixtures of perennial grasses

Stepan Kovalyshyn[1] and *Viktor* Dadak [1]
[1]Lviv National Agrarian University, 80381, Dublyany, str. Volodymyra Velykoho 1, Ukraine

Abstract. The work is devoted to the study of seed separation process of perennial grasses on a pneumatic-electric separator. It highlights the main problems of their post-harvest treatment. The influence of the regulated parameters - the magnitude of the supply of seeds to the air-separator, the air flow rate and the electric field strength on the quality from the segregated seed material of the perennial rye-grass were determined by conducting a multifactorial experiment.
The optimal values of the regulated parameters of the pneumatic electric separation are substantiated, in which the content of heavy separating admixture of weeds in the seed mixture of perennial rye-grass is the least. It is proved that the use in the technological line of seed preparation of perennial grasses of the pneumatic electric separator allows to ensure the quality of the seed material of the perennial rye-grass in accordance with the requirements of the existing standards.

1 Analysis of recent researches and publications

In the structure of agricultural lands of Ukraine about 5 million hectares or 20% occupy areas under the cultures, which belong to the small-seeded. These include those whose masses of thousands of seeds are less than 5 grams [1]. These, in the main, are vegetable, oil and overwhelming majority of perennial legumes and herbs. The loss of high yields of these crops is not possible without the presence of sufficient quantities of high-quality seeds obtained in the process of post-harvest treatment [2].

However, it is difficult to obtain seed material of perennial grasses of high quality in many cases, since it contains a large number of hard-to-divide impurities and biologically defective dead seeds, which differ little from the seeds of the culture by their basic physical and mechanical properties [2]. As a result, seed clearing machines used in the process of post-harvest treatment of mixtures of perennial grasses, can not provide for their qualitative cleaning. As a result, in some cases, during separation, a large number of conditioned crop seeds fall into waste, and in others it is impossible to separate the impurities of weeds.

The main reason for this situation is that in the vast majority of cleaning machines, separation of seed mixtures on the air-lattice-trier or friction working bodies is carried out. However, seeds of perennial grasses, especially ryegrass and its severely dispersed impurities of couch grass or lungwort, do not have a dominant feature of divisibility, since they differ little in size, form, and coefficient of friction. The lack of sufficiently effective means for their separation leads to a loss in the process of post-harvest processing of conditioned (full-fledged) seeds, which reaches 20 ... 30% [3]. The increase in the yield of purified seeds, for example, perennial grasses only for 1% on a country scale is equivalent to an expansion of crops more than 20 thousand hectares [4].

In this regard, there is a need to develop machines that take into account not only the physical and mechanical properties of seeds, but also biological. To solve these problems during separation of the seed material of cereal grasses becomes possible by use of electric field in seed post-harvest separators, as an additional working body. According to Dadak [7], one of such machines can serve as a pneumatic electric separator, in which the separation of components of seed mixtures is carried out in aggregate of their electrical and aerodynamic properties [8].

2 Setting objectives

The aim of the work is to improve the quality of sowing material of perennial grasses due to the study of the process of pneumatic electric separation and the substantiation of the optimal parameters of the separator itself.

3 The main material

For experimental researches on separation of seed mixtures of perennial grasses of heavily dispersed impurities, an improved design of a pneumo-electric separator [8] was used, the general form of which is shown in Fig. 1a, and the technological scheme of his work - in Fig. 1b.

* Corresponding author: stkovalyshyn@gmail.com

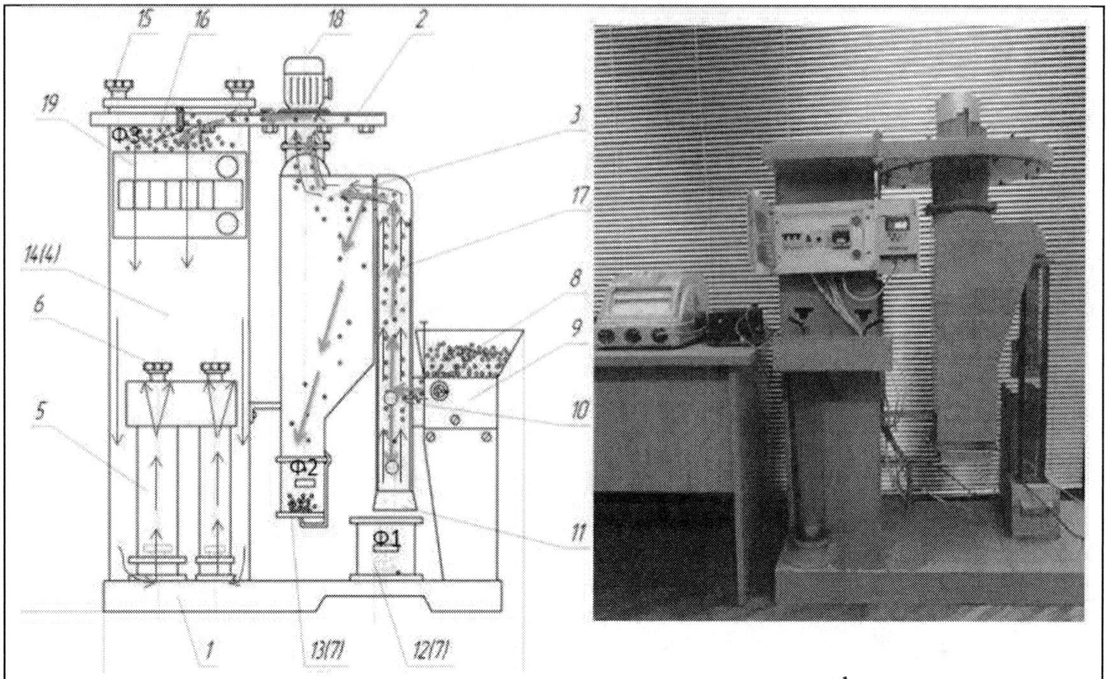

a b

Fig. 1a and 1b. General view (b) and technological scheme (a) of the pneumo-electric separator seed fractions: F1 – weed, F2 - main crop, F3 - light impurities; directions of movement : air ⟶ seeds ➡
1 - bed 2 - fan 3 - towering separator, 4 - filter section, 5 - air flow measuring devices , 6 - speed regulation device, 7 - seed containers, 8 - loadingbunker, 9 - vibration feeder, 10 - separation channel, 11 - retractor hole, 12,13 - containers for seeds, 14 - filter section, 15 - the bolts, 16 – cover of the filter section, 17 – electrodes, 18 – electric engine, 19 - control unit.

The peculiarity of this separator is the presence in theseparation channel 10 two electrodes 17 located opposite each other, which are fed with high voltage. Feedingof the separating mixture inloading bunker 8 to the separation channel 10 is carried out using a regulated vibration feeder 9.

Investigation of the process of pneumatic electric separation of the seed material of perennial grasses was carried out through a multifactorial experiment. It determined the influence of the regulated parameters - the values of the seeds feed Q (kg□h^{-1}), the air velocity in the vertical separation channel V_n (m/s), the electric field intensity E (kV/cm) in the separation zone on the quality indices of the sowing material that was expressed as a percentage of K (%) in it of qualitative seeds of the main culture.

Experiments were carried out using a seed mixture of perennial ryegrass, sort "Osip" II reproduction with a moisture content of 10% on a pneumatic electric separator with a vertical air channel [9].

The rate of air flow in the channel and its rotational speed is changed by the fan with frequency converter ASF 110.

The required electric field strength between the electrodes was achieved by changing the voltage applied to the electrodes 17.

During the experiments, the seed mixture was fed directly to the separating channel through a vibration dosing unit, which ensured a uniform initial rate of movement of individual seeds.

In the process of pneumatic electric separation, fractions of the separated seeds were obtained. Of these, samples were taken per 1000 seeds, counting the number of seeds of hard-to-divide weeds and determining their percentage content. The research was repeated - threefold. Each experiment corresponded to the specific values of the supply of seed material to the separator channel, the velocity of air in it and the intensity of the electric field.

Having analyzed the preliminary theoretical and experimental studies [15, 17], the zero levels of the investigated parameters of the pneumatic electric separation were established: the seed supply X_{10} = 3 kg□h^{-1}; speed of air flow X_{20} = 6 m/s; the intensity of the electric field X_{30} = 1 kV/cm. The results of their coding are shown in Table 1.

Table 1. Results of coding of investigated factors.

Factors and units of measurement	Natural designation	Coded designation	Variable interval	Levels of variation					
				natural			coded		
				upper (+1)	zero (0)	lower (-1)	upper (+1)	zero (0)	lower (-1)
1	2	3	4	5	6	7	8	9	10
The seed supply, Q, kg□h^{-1}	X_1	x_1	1	4	3	2	+1	0	-1
Speed of air flow V_n, m/s	X_2	x_2	0,5	6,5	6	5,5	+1	0	-1
Intensity of the electric field E, kV/cm	X_3	x_3	0,15	1,15	1	0,85	+1	0	-1

During the multifactorial experiment, the controlled parameter was the content of quality seeds of the main culture in the material under study. To find the coefficients of the polynomial, a three-level design of the second order of Box-Behnken was used. The matrix of the plan and the results of the multivariate experiment are shown in Table 2.

Table 2. Matrix plan and results of multi-factor experiment.

№	Value of coded factors			Results of determining the content of the main culture in the separated samples, %			Average value
	x_1	x_2	x_3	Sample 1	Sample 2	Sample 3	
1	-1	-1	0	90	89	88	89
2	1	-1	0	86	88	90	88
3	-1	1	0	96	95	94	95
4	1	1	0	80	82	84	82
5	-1	0	-1	86	87	88	87
6	1	0	-1	88	90	92	90
7	-1	0	1	96	96	96	96
8	1	0	1	89	88	90	89
9	0	-1	-1	90	91	89	91
10	0	1	-1	93	92	92	92
11	0	-1	1	94	88	88	90
12	0	1	1	95	92	92	93
13	0	0	0	88	89	90	89
14	0	0	0	91	93	89	91
15	0	0	0	90	89	91	90

As a result of processing of the obtained data of the multivariate experiment and the use of software Statistica 8.0, the regression equation for the content of seeds of the main crop in the sowing material of the perennial ryegrass was obtained:

- in coded notation:

$$K = 97{,}45 + 0{,}0412x_1 - 0{,}0425x_2 - 0{,}5838\,x_3 + 0{,}4175\,x_1x_2 + 0x_1x_3 + 0{,}1675x_2x_3 - 0{,}596x_1^2 + 0{,}0717x_2^2 - 0{,}0171x_3^2; \quad (1)$$

- in natural values:

$$K = 91{,}54 + 4{,}21Q + 3{,}72V - 18{,}85E + 0{,}417Q^2 - 0V^2 + 7{,}444E^2 - 1{,}192QV + 0{,}478QE - 0{,}228VE. \quad (2)$$

The randomized block designs of the experimental results were checked by the Cochran criterion. The hypothesis of the randomized block designs of experiments was taken if the calculated value of the G_p criterion was less than its table value. In our case, the calculated value G_p= 0,173, which is less than the table value (G_m= 0,3346). This indicates the randomized block designs of the experiment.

The equilibrium surface (2, 3, 4) is constructed for equation (1), which determines the optimal parameters of the process of pneumatic electric separation. The function $K = f\,(E, Q)$ of the content dependence in the decoupled material of the seed of the main crop from the intensity of the electrostatic field in the separator channel and the supply of the seed mixture (Figure 3) acquires the maximum value at 0,4> Q> 0,4 and E = -1, which in natural quantities is 3,4>Q>2,6 $kg\square h^{-1}$ and E = =0,85 *kV/cm*.

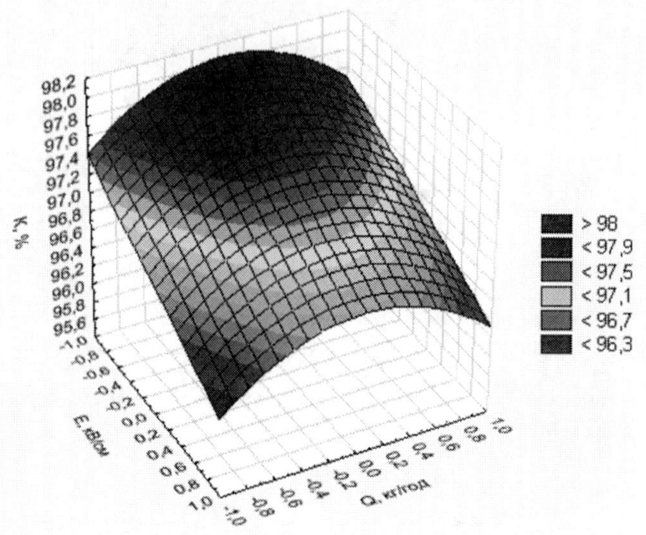

Fig.2. The equilibrium surface $K = f(E, Q)$.

Under such a separator operation, the seed content of the main culture reached $K = 97 \ldots 98{,}2$ %.

The function $K = f(Q, V)$ of the dependence of the content of seeds of the main crop on the velocity of the air flow in the separator channel and the supply of the seed mixture (Figure 3) acquires a maximum value at 0,2> Q > 0,2 and V = 0,8 ... 1, which in physical quantities is 3,2> Q> 2,8 $kg\square h^{-1}$ and V = 6,2 ... 6,5 *m/s*. For such parameters of the separator, the content of seeds of the main culture reaches K = 97,6%.

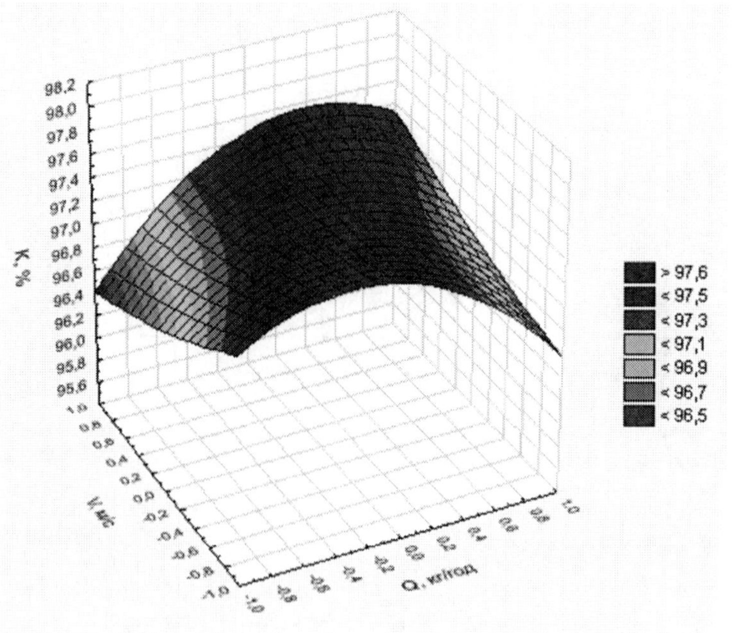

Fig.3. The equilibrium surface $K = f(Q, V)$.

The function K = f (E, V) of the dependence of the content of seeds of the main culture on the strength of the electrostatic field and the air flow velocity in the separator channel (Figure 4) acquires the maximum value at E = -1 and V = -1, which in natural quantities makes E = 0,85 *kV/cm* and V = 5,5 *m/s*.

Under such a separator operation, the seed content of the main culture reached K = 98%.

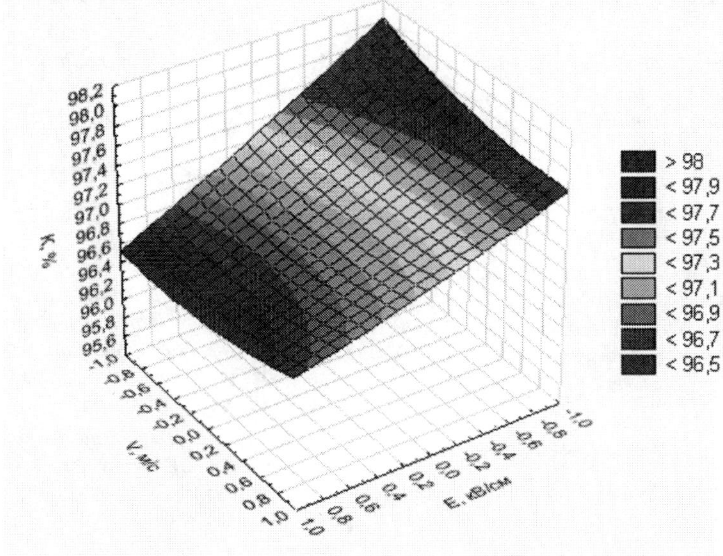

Fig.4. The equilibrium surface $K = f(E, V)$.

The adequacy of the results obtained is checked by Fischer's criterion. Estimated value of which is F = =0,877. Its tabular value for this case will be F_m = 2,12. Since the calculated value of Fisher's criterion is smaller than the tabular one, it can be argued that this model is adequate.

The analysis of the above results allowed to determine the optimal values of the regulated parameters of the pneumo-electric separator, which yields the best-quality seed material. In order to ensure the maximum efficiency of the separation from the perennial ryegrassof hard-to-divide impurities, it is necessary to ensure the velocity of the air flow V_n = 5,5...6,2 *m/s*, the electrostatic field strength in the separation channel E = = 0,85 ... 0,92 *kV/cm*, the supply of a seed mixture Q = 2 ... 3,2 $kg\square h^{-1}$.

For the indicated parameters of the separator, the content of seeds of the main culture was K = 97 ... 98,2%, and was in compliance with the requirements of DSTU 7160: 2010 "Seeds of vegetable, melon, coriander and spicy aromatic cultures. Varietal and sowing quality. Specifications ", according to which this indicator should be 90 ... 95%.

4 Conclusions

1. The quality of the process of pneumatic electric separation of seed mixtures of perennial ryegrass has a significant influence on the investigated adjustable parameters: air flow velocity, seed feed rate, and the strength of the electrostatic field.
2. By conducting a multifactorial experiment, optimal modes of operation of the separator were determined, in which the maximum separation of hard-to-divide separating impurities from the investigated seed mixture is achieved: the velocity of the air flow V_p = 5,5 ... 6,2 *m/s*; The intensity of the electrostatic field in the separation channel E = 0,85 ... 0,92 *kV/cm*; supply of seed mixture Q = 2 ... 3,2 $kg\square h^{-1}$.
3. The use of the proposed pneumatic separator in the production line for the preparation of seeds of perennial grasses, in particular, the perennial ryegrass, allows us to significantly improve the quality of the received sowing material and to bring the content of hard-to-divide weeds in accordance with DSTU 7160: 2010 "Seeds of vegetable, melon, coriander and spicy aromatic cultures. Varietal and sowing quality. Specifications ".

References

1. S.Y. Kovalyshyn. *The Bulletin of Agrarian Science*. - №10. - P. 71-74. (1998)
2. S.Y. Kovalyshyn. *Agricultural machines. - Collection of scientific articles*. - Issue 4. - Lutsk, P. 69-73. (1998)
3. S.Y. Kovalyshyn. Substantiation of technological parameters of seed treatment of perennial grasses by electrovibrator separator: *Dis. Cand. Tech Sciences*: 05.20.01. - Lviv. 178p. (1999)
4. V.O. Paranyuk. Separation of seed mixtures in an electric field on a moving with constant speed of an inclined plane: *Dis. Cand. Tech. Sciences*: 05.20.02. – Chelyabinsk. 152p.(1983)
5. P. M. Zaika. Moscow: *Kolos*. 280 p.1978.
6. S. Y. Kovalyshyn, V.O. Dadak. *Bulletin of the Kharkiv National Technical University of Agriculture named after Petro Vasilenko, "Technical systems and technologies of animal husbandry"*. Ed. 144. P. 225-232. (2014)

7. V.O. Dadak. *Mechanization and electrification of agriculture: Inter-departmental thematic scientific collection*. Ed. 97, Volume 2, Glevaha. P. 495-501.(2013)
8. S. Y. Kovalyshyn. V.O. Paranyuk, V.O. Dadak. *Scientists of Lviv. Nats Agr University of Production: catalog of innovative developments*. Ed. XIII. Lviv.P. 48.(2013)
9. V. V. Shmigel. *Mehanization and electrification of agriculture №2 / power supply, electric drive and electrotechnology*. P. 37-38.(1978)
10. O. P. Shvets. Justification of parameters and modes of operation of separator of winter rape seed. *Author's abstract Dis For obtaining sciences. Candidate Degree Tech Sciences*. Lviv. 20 p.(2012)
11. V. G. Khamuyev. *Engineering in agriculture*. No. 5. P. 23-26.(2008)
12. A.K. Turov. *Engineering in agriculture*. № 2, (2013)
13. V.O. Paranyuk, S.Y. Kovalyshyn, V.I. Melnychuk. *Techno-technological aspects of development and testing of new technology and technologies for agriculture Of Ukraine. Zb Sciences Etc. UkrNDIPVT them. L. Burned*: Vip. 10 (24), K. 1. - Research. - P. 77-86. (2007)
14. *Seeds of vegetable, melon, fodder and spicy-aromatic crops. Varietal and sowing quality. Technical specifications*: DSTU7160: 2010 [Effective from 2010-07-01]. - K .: National standard of Ukraine. 19 p. (2010)
15. S. Y. Kovalyshyn, V.O. Paranyuk, V.O. Dadak. *Motrol: Motorization and power industry in agriculture - Lublin: Comission of motorization and power industry in agriculture* - Vol. 14D. P. 95-103. (2012)
16. V. P. Yermak, Ye. V. Bogdanov, A. A. Ilchenko. *Collection of scientific works of Luhansk National Agrarian University.Series (Technical sciences.)* - Lugansk: LNAU - P. 127-132. (2012)

Estimation methods of the agricultural equipment value with regard toevaluation of Alina carrot harvester

Zbigniew Kowalczyk[1,]*, *Michał* Cupiał[1]

[1]Agricultural University in Krakow, Faculty of Production and Power Engineering, ul. Balicka 116B, 30-149 Kraków, Poland

Abstract. The objective of the paper is to present methods of evaluation of real estates used for evaluation of agricultural equipment and types of value related thereto. Alina carrot harvester by Waremczuk company served as an example, for which a detailed methodology and the results of valuation were presented.
Evaluation was carried out based on two approaches namely a comparative and cost approach. In the case of the comparative approach, the average price adjustment method was used and in the case of the cost approach - a replacement cost method.As a result of using various approaches and methods of evaluation two slightly different values of the evaluated harvester were obtained. It was respectively: EUR16,406 - at the use of the average price adjustment method and EUR 17,678 - the estimated net replacement value in the cost approach for the replacement method.

1 Introduction

In the developed market economy, one may observe even higher demand for precise estimation of values of technical production means used in agriculture [1]. The objectives of the fixed assets evaluation are varied: to begin with the need to determine the value on account of a possible sale transaction, where a seller and a buyer may be interested parties, and to end with the need to determine the value for the credit or loan guarantee. Besides, the evaluation is carried out in the case of lease contracts concerning fixed assets, insurance, tax, accounting purposes, etc. In relation to the objective and evaluation purpose, varied procedures of estimation are applied. As a result various types of values are obtained. The most popular categories are: the market value and the replacement value, but each of them may be of many types.

In the case of new machines, estimation of their value is not troublesome. However, estimation of used machines is different. With years of exploitation, this exploitation potential gets depleted and thus values of machines decrease as a result of:
- technical wear (physical),
- functional wear (moral),
- external wear (environmental).

The issue of the wear of the agricultural equipment (particularly in the aspect of the value loss) is relatively rarely presented in the scientific literature. This subject has been mentioned by, inter alia, [1-12].

The objective of this paper is to present the most frequently used methods of evaluation of real estatesfor estimation of agricultural devices and the types of value related thereto. Alina carrot harvester by Waremczuk company was used as an example. A detailed methodology of estimation acc. to several methods supported by examples in the form of calculations will be presented.

2 Material and methods

The evaluation procedure of a harvester in the comparative approach acc. to the average price adjustment method will be made based on the methodology acc. to the the following stages [4]:
1. Determination of the local market through defining its type, area and the period of investigation of transaction prices.
2. Description of harvesters, which are the object of the market turnover (collected as a consequence of the market analysis) and the evaluated harvester with emphasising their features (attributes).
3. Determination of the type and number of market features which influence the level of prices on the local market along with determination of the weight of market features (k_i), $k_i = r_i^2/\Sigma\ r_i^2$, (r - correlation coefficient between market features and prices),
4. Calculation of the average price ($C_{śr}$) based on the representative sample defined earlier (the price bases).
5. Calculation of the correction coefficients value (U) as a quotient of extreme prices to the mean price:
$U_{min} = C_{min} / C_{śr}$

*Corresponding author: Zbigniew.Kowalczyk@ur.krakow.pl

$U_{max} = C_{max} / C_{śr}$

6. Calculation of the variability scope of the correction coefficient (Δu_i) for particular features (attributes):

$k_i \times U_{min} \leq u_i \leq k_i \cdot U_{max} \Leftrightarrow \Delta u_i = k_i \cdot U_{max} - k_i \cdot U_{min}$

7. Determination of the maximum scope of scales of particular features (attributes), (Δa_i):

$\Delta a_i = a_{max} - a_{min}$

8. Calculation of the value of the correction coefficient per a unit of a feature (attribute) of a harvester ($u_{i/jed}$):

$u_{i/jed} = \Delta u_i / \Delta a_i = (k_i \cdot U_{max} - k_i \cdot U_{min}) / (a_{max} - a_{min})$

9. Calculation of the difference of the feature of the evaluated harvester and the minimum value of this feature ($\Delta a_{w\text{-}min}$):

$\Delta a_{w\text{-}min} = a_w - a_{min}$

10. Calculation of the increase of the correction coefficient value for the evaluated harvester in comparison to the minimum value (u_i):

$u_i = k_i \cdot U_{min} + \Delta a_{w\text{-}min} \cdot u_{i/jed}$

11. Calculation of the sum of correction coefficients from all prices (Σu_w):

$\Sigma u_w = u_1 + u_2 + u_3 + \ldots\ldots u_n$

12. Calculation of the harvester value, which constitutes a product of the average price of a harvester from a representative sample (base) and the sum of correction coefficients u_w

$W = C_{śr} \cdot \Sigma u_w$

The procedure for the evaluation of a harvester in the cost approach acc. to the method of replacement costs will be made based on the methodology acc. to the the following stages [7]:

1. Determination of the gross replacement value based on the replacement costs of a new technical mean or the cost of replacing a new technical mean
2. Specification of particular, more important components of the evaluated harvester, assessment of their technical condition and determination of their weights.
3. Determination of the loss of the value of a harvester due to physical reasons (the physical wear),

$$S_{fz} = [1 - \sum_{i=1}^{n} (\frac{o_i}{p_i} \cdot w_i)] \cdot 100\%$$

where:
S_{fz}– physical wear %,
o_i– obtained grade,
p_i– the number of points in the scale of grades,
w_i– weight of a parameter i.

4. Determination of the loss of the value of a harvester dueto internal reasons (the functional wear),

For determination of the degree of the functional wear proportions of changes of technical parameters of the evaluation object and modern, presently manufactured technical means may be helpful e.g. proportions which result from performance, efficiency etc. The above guidelines were used for assessment of particular functional and utility features of a harvester and as a result its functional wear, which finally was estimated acc. to the relation:

$$S_{fn} = [1 - \sum_{j=1}^{m} (\frac{o_j}{p_j} \cdot w_j)] \cdot 100\%$$

where:
S_{fn} – functional wear %,
o_j – obtained grade,
p_j – the number of points in the scale of grades,
w_j – weight of a parameter j.

5. Determination of the loss of the value of a harvester dueto external reasons (the environmental wear),
6. Calculation of the total loss of the value related to the wear of a harvester
7. Calculation of the net replacement value of a harvester

$W_N = W_B \cdot (1 - S_{fz}) \cdot (1 - S_{fn}) \cdot (1 - S_z)$ EUR

where:
W_N – net replacement value EUR
W_B – gross value of a machine EUR,
S_{fz} – physical wear,
S_{fn} – functional wear,
S_z – external wear (environmental).

3 Results and discussion

Table 1 presents information concerning the market transactions in 2014-2015 including attributes of harevsters and transaction prices. It was found out that the most significant attributes of harvesters, which influence the level of their transaction prices are: age, version, technical condition of sub-assemblies, standard of equipment, origin and a general appearance

Table 1. Information from the local market on the transaction of Alina harvesters by Waremczuk company.

Item	Age(a_1) (years)	Version(a_2) (1 – 2)	Technical condition(a_3) (0 – 3)	Standard of equipment (a_4) (0 – 3)	Origin(a_5) (0–2)	General appearance (a_6) (0 – 3)	Price EUR
1.	1	2	3	2	2	2	24,519
2.	3	1	2	1	2	2	16,346
3.	1	2	3	3	2	3	29,567
4.	4	2	1	1	1	2	11,298

5.	6	2	1	2	1	2	6,370
6.	2	1	2	0	2	2	13,053
7.	4	1	1	2	0	1	5,048
8.	2	2	3	2	1	3	21,635
9.	7	1	1	2	2	1	2,332
10.	3	1	1	1	2	2	14,904
11.	3	2	2	0	1	1	14,267
12.	5	1	2	0	2	3	5,264
13.	7	2	1	0	2	1	2,404
14.	2	2	3	1	2	2	24,279
15.	1	1	2	3	0	3	22,308
16.	4	1	0	1	0	1	9,387
17.	2	2	3	3	2	2	22,224
18.	1	2	2	3	1	3	22,548
19.	5	1	1	1	2	1	7,740
20.	3	2	2	3	1	2	14,663
21.	2	2	2	0	1	1	17,500
22.	6	1	1	0	0	1	4,615
23.	3	2	2	2	1	2	14,423
24.	1	2	3	1	2	2	23,425
25.	6	1	1	0	2	1	4,567
26.	3	2	3	3	2	3	17,380
27.	5	1	1	0	0	1	2,885
28.	4	1	1	3	1	2	12,452
29.	1	2	3	2	2	2	28,365
30.	3	1	2	0	1	2	13,550
Evalua-ted	**2**	**2**	**2**	**2**	**0**	**1**	**???**

Each of the attributes listed in table 1 was graded in an appropriate scale, which was as follows:

- Age (years)
- Version (1-2): 1- Eco version with lower performance, adjusted to tractors with power 70-90 KM, 2 - Supernova version with higher performance adjusted to tractors with power 90-110 KM
- Technical condition (0-3): 0 - needs major repair, no possibility of use, some sub-assemblies are incomplete or need a major repair 1 - average, a complete harvester, used but needs to be repaired, 2 - good, a harvester is suitable for regular operation, 3 - very good, a new harvester, almost new or after major repair,
- Standard of equipment (0-2): 0 - basic version, 1 - a machine which has additional equipment, 2 - a machine with full additional equipment. A harvester may be additionally equipped with inter alia: haulm catching automatic control module, a conveyor that removes remaining soil from carrots, second direction of conveyor for cleaning crops, rotary haulm gatherers, additional control of a conveyor form a working platform.
- Origin (0-1): 0 - a harvester operated in difficult conditions picking carrot from big acreage, 1 - a combine was used in quite good conditions and picked carrot from small areas.

The impact of attributes on the shaping of the market value was determined with the use of statistical methods, namely the analysis of correlation [6], which is presented in table 2.

Table 2. Coefficients of total correlation and their squares as well as percentage share (weights) of their features for explaining variability of prices on the market

Specification	Age(a_1)	Version(a_2)	Technical condition(a_3)	Standard of equipment(a_4)	Origin(a_5)	General appearance (a_6)
r_i	-0.931	0.552	0.825	0.494	0.275	0.613
r_i^2	0.866	0.305	0.680	0.244	0.076	0.375
Weight of a feature % $k_i = r_i^2/\Sigma r_i^2$	**-34.0**	**12.0**	**26.7**	**9.6**	**3.0**	**14.7**

One of the evaluation methods as a part of the comparative approach is the average prices adjustment method. In this method the market analysis and data presented in tables 1 and 2 were used.
According to the formula presented in the methodology, the value of the correction coefficients (U) constitutes a quotient of extreme prices and the average price:
$U_{min} = C_{min} / C_{śr}$ = EUR2,332/ EUR14,311 = 0.1629
$U_{max} = C_{max} / C_{śr}$ = EUR29,567 / EUR14,311 = 2.0660
Evaluation procedure was presented in table 3.

Table 3. Evaluation of Alina harvester acc. to the average price adjustment method

Specification	**Age(a_1)**	**Version(a_2)**	**Technical condition(a_3)**	**Standard of equipment(a_4)**	**Origin(a_5)**	**General appearance (a_6)**
Weight of a feature:k_i (%)	34.0*	12.0	26.7	9.6	3.0	14.7
$k_i \cdot U_{min}$	0.0544	0.0195	0.0435	0.0156	0.0048	0.0240
$k_i \cdot U_{max}$	0.7026	0.2475	0.5519	0.1983	0.0614	0.3045
Spread of attributes:Δa_i	6	1	3	3	2	2
Value of the correction coefficient per a unit of a feature: $u_{i/jed}$	0.108	0.228	0.169	0.061	0.028	0.140
Attributes of the evaluated harvester:a_w	2	2	2	0	1	3
Difference in attributes of the evaluated harvester and the minimum attribute:Δa_{w-min}	1	1	2	0	1	2
Increase of the correction coefficient towards the minimum value:u_i	0.163	0.247	0.382	0.016	0.033	0.304
W	16,406 EUR					

where * - negative correlation

The second approach acc. to which evaluation of the particular harvester was made is the cost approach and the applied method is called a replacement method. The gross replacement value was determined based on prices of brand new harvesters on the original market. The value of a harvester with an identical additional equipment is W_B = EUR31,490.

In this paper, a technical condition of agricultural equipment was assessed based on the detailed inspection and interview with an owner and sometimes with other users.

During determination of the physical wear of a harvester, a condition of each of more important sub-assemblies was assessed. Weights for particular sub-assemblies were also determined taking into consideration the participation of the replacement costs of sub-assemblies in the total cost of the object and observed preferences of possible purchasers on the market (table no. 4)

Each object (sub-assembly) of the evaluated harvester was assessed from 0 to 3, where:
0 - a technical condition of a particular sub-assembly rendered the regular machine operation impossible and a sub-assembly required a general renovation (or possibly a replacement),
1 - average condition, degree of wear of a sub-assembly makes the machine operation possible but a sub-assembly should be repaired,

2 - good condition - without any bigger traces of wear, possibly a sub-assembly after repair
3 - very good condition - a new sub-assembly or almost new, without traces of use.

Table 4. Sub-assemblies of a harvester, their price and ordered weights

Item	Sub-assembly	Weight %	Grade (0-3)
1.	Frame and driving system	20	2
2.	Ploughing and topping unit	30	2
3.	Separating and loading unit	35	2
4.	Hydraulic system	15	3

Based on the obtained description of the technical condition of a harvester, thephysical wear was calculated (according to the formula presented in methodology) which is:S_{fz} = 28.3%

Table 5. Functional and utility features of a harvester, their price and ordered weights

Item	Sub-assembly	Weight %	Grade (0-3)
1.	Performance	30	2
2.	Power demand	15	2
3.	Failure frequency	20	2
4/	Availability of spare	20	3
5.	Simplicity of operation	15	3

The functional wear of a harvester was determined according to the relation presented in the methodology and it is: S_{fu} = 21.7%.

The degree of wear due to the external reasons may not appear and in case of the evaluated harvester - did not occur.

Net replacement value of a harvester was calculated acc. to the formula presented in the methodology and finally it was: W_N = EUR17,678

4 Closing remarks

The paper herein includes two approaches used in the evaluation of technical fixed assets, namely: the comparative approach and the cost approach. The comparative approach is used for determination of the market value of machines and devices only when the objects similar to the object, which is the subject of evaluation, were in the turnover on the determined market in the period of the last two years, preceding the evaluation and also when their prices and features ,which influence those prices and conditions for transactions are known.

The cost approach is based on the assumption that the purchaser, which is aware, so-called informed and oriented in the market conditions, will not pay more for a technical mean than the manufacture cost of the substituting mean of the same utility as the evaluated one. This approach assumes that the amount equal to the price of construction of a new object of the same utility is the maximum value of the technical mean.

A market value - constitutes the most probable price which may be obtained on the market when the following assumptions have been made:

a) the parties to the contract do not depend on each other and act rationally without any specific motives,

b) intend to conclude a contract,

c) are aware of the co-existing circumstances which have an impact on the value of the transaction object,

d) do not act in forced circumstances,

e) period of indispensable exposure of the object on the market with the use of appropriate advertisement and the time needed for negotiation of the contract conditions have lapsed, taking into account the nature of the real estate and the market condition.

The replacement value reflects the cost of manufacturing of a substitutive mean of the same utility as the evaluated mean.

The marked value of the evaluated harvester in the average price adjustment method is: EUR 16,406. The evaluated net replacement value in the cost approach, the replacement method is EUR17,678 and is close to the market value obtained in calculations.

The procedures used in the paper may be used for estimation of the value of machines, tools and agricultural devices.Due to the used methodology, a result obtained in case of the average price adjustment method, which is generally used in case of a higher number of market transactions, is more precise. In case, there is no sufficient number of market transactions a method different than a statistical one shall be used for determination of weights of attributes of the evaluated objects.

The procedures used in the paper may be used for estimation of the value of machines, tools and agricultural devices.

References

1. Z. Kowalczyk, Inżynieria Rolnicza, **6(94)** 89-95 (2007)
2. Z. Grześ, Prace Przemysłowego Instytutu Maszyn Rolniczych, **1**, 84-85 (2001)
3. J. Borcz, Z. Kowalczyk, Inżynieria Rolnicza, **6**, 49-56 (2005)
4. J. Borcz, Z. Kowalczyk, Inżynieria Rolnicza, **6**, 57-63 (2005)
5. J. Borcz, Z. Kowalczyk, Inżynieria Rolnicza, **11**, 77-83 (2006)
6. Z.Kowalczyk, Inżynieria Rolnicza, **9(107)**, 151-156 (2008)
7. Z. Kowalczyk,The level and the structureof the wear of technical production means in different types of agricultural farms, Rozprawa habilitacyjna, Inżynieria Rolnicza, **2**,(2011)

8. R. Michalski, J. Napiórkowski, Technical Sciences. University of Warmia and Mazury in Olsztyn, **1**: 139-147(2004)
9. J. Borcz, J. Kosek, Estimation methods of machines' value, Vademecum Rzeczoznawcy majątkowego, Kraków,(1994)
10. H. Macniak, Z. Makowicz, Vademecum of the machines' valuation, PFSRM, Gdańsk, (1996)
11. J. Napiórkowski, R. Źróbek, Estimation methods of machines' value, Zachodnie Centrum Organizacji, Zielona Góra,(2001)
12. T. Klimek, Valuation of the machines. Stowarzyszenie Rzeczoznawców Majątkowych w Katowicach,Katowice, (1995)

The influence of mixing inside the fermentation reactors during the study of biogas efficiency of the substrates in mesophilic technology

Kamil Kozlowski[1,*], *Marta* Cieslik[2], *Jakub*Mazurkiewicz[1], *Anna* Smurzynska[1], *Michal* Brzoski[1], *Dawid* Chelkowski[1], *Aleksandra* Jezowska[1], *Maciej* Neugebauer[3], *Tomasz* Oniszczuk[4], *Karol* Kupryaniuk[4]

[1]Poznan University of Life Sciences, Institute of Biosystems Enginering, Wojska Polskiego 50, 60-637 Poznan, Poland
[2]Poznan University of Life Sciences, Department of Biotechnology and Food Microbiology, Wojska Polskiego 48, 60-637 Poznan, Poland
[3]Faculty of Technical Sciences University of Warmia and Mazury in Olsztyn, Department of the Electrotechnic Energetics Electronics and Automatics, Oczapowskiego 11, 10-719 Olsztyn, Poland
[4]Lublin University of Life Sciences, Department of Process Engineering, Gleboka 28, 20-612 Lublin, Poland

Abstract. Along with the development of environmental friendly technologies, an increasing interest in generating the electricity and heat from renewable sources has been observed in Poland. Perfect example of such installations are agricultural biogas plants, where methane fermentation produces high-energy gas fuel i.e. biogas. Before investing, it is necessary to perform a detailed studies of the biogas efficiency of the substrates used. The proper run of fermentation process, as well as the high biogas production related to the operation, depend on many factors, including the mixing inside the reactor. The DIN 38 414-S8 standard, commonly used in the tests, assumes the mixing of the whole eudiometric sets, used in accredited laboratory units, which due to their uniform construction may cause problems. The aim of the study was to determine the effect of laboratory mixing in the fermentation reactor on the biogas efficiency of the maize straw. The experiment tests were performed in the Laboratory of Ecotechnologies, in the Poznan, in accordance with the current standard DIN 38 414-S8.The obtained research results confirmed the effect of mixing inside the fermentation reactors on the biogas efficiency of the maize straw.

1 Introduction

Constant increase in demand for electricity, heat and fossil fuels results primarily from intensive development of civilization in the last century and the observed increase in the population of the world. These factors also cause a noticeable increase in environment pollution [1]. Bearing in mind these unfavourable changes, the use of new technologies for the production of environmentally friendly energy becomes more important [2, 3]. During the ongoing energy crisis, the use of alternative energy sources such as hydrogen or methane seems to be the most advantageous solution [4, 5].

Currently, it is believed that the most significant contribution to the production of unconventional fuels is likely to be substrates with relatively low purchase costs, which are waste from agriculture and the agro-food industry [6-8]. These materials can be easily used in composting and methane fermentation [2, 9, 10] and methane fermentation [11]. It should be noted that Poland is a country with a high potential for energy from agricultural biogas [7, 12]. Unfortunately, this potential is not fully exploited. According to the register of producers of agricultural biogas, as of 25.07.2017, there are 95 agricultural biogas plants in Poland [13]. However, anticipate that with the development of renewable energy sources new installations will be developed and produced using agricultural biogas [14].

Biogas is included in the group of renewable energy sources [15]. For many years, intensive methane fermentation and the use of biomass and its decomposition products in the industry have been carried out worldwide [1, 8, 16, 17, 19, 20]. The efficiency of the fermentation process depends on many process factors such as temperature [16], pH [20], C/N ratio [21, 22], hydraulic retention time (HRT) [21] and reactor organic load rate (OLR) [23].

However, one of the equally important factors affecting the final yield of biogas is the mixing of the contents of the fermentation reactors [24, 25]. DIN 38 414-S8 standard is commonly used for the biogas substrate tests [26-28]. This standard takes into account the mixing of the eugenic sets by careful movement of the circular motions [29, 37]. These sets are used by many accredited laboratory units. Unfortunately, due to the uniform construction, mixing of their contents can be

* Corresponding author: kamil.kozlowski@up.poznan.pl

difficult, which can significantly affect the final result of the biogas and methane yields.

A possible solution is to use sets equipped with a mechanical mixing system or reactors to allow accurate manual mixing, with an independent gas storage system. The first solution is mainly used in fermentation stations used in continuous process studies [24, 30, 36]. An additional problem may be the gas seal introduced into the inside of the mixer, which prevents oxygen from entering the reactor. In this case, it is important to use fermentation kits equipped with an independent gas storage system and the possibility of free manual mixing of the contents take place. Such stations were constructed in the Department of Environmental Engineering at the Institute of Biosystems Engineering at Poznan University of Life Sciences (Figure 1).

Fig. 1. The biofermenter for biogas production research in Laboratory of Ecotechnology at Poznan University of Life Sciences.

The aim of the study was to determine the effect of laboratory mixing in the fermentation reactor on the biogas efficiency of the maize straw.

2 Materials and methodology

2.1 Analysis of basic parameters

The analyses of basic physicochemical parameters of the substrates and inoculum were conducted according to Polish standards: dry matter (PN-75 C-04616/01), dry organic matter (PN-Z-15011-3) and pH (PN 90 C-04540/01). These parameters enabled the selection of appropriate proportions of fermentation mixtures and subsequent calculation of the biogas efficiency calculated on Mg of fresh matter, dry matter and dry organic matter of the substrate.

2.2 Biogas efficiency

The research tests on methane efficiency of the substrates in batch culture technology were carried out in the Laboratory of Ecotechnologies at the Institute of Biosystems Engineering at the Poznan University of Life Sciences on basis of internal procedures, based on adapted standards: DIN 38 414-S8 and VDI 4630, commonly used in Europe. Detailed methodology of performed research was presented by Cieslik et al. [26]. The tested material was maize straw, which obtained from Przybroda Agricultural-Orchard Experimental Farm belonging to the Poznan University of Life Sciences (PULS). The fermentative inoculum was separated liquid fraction (after dry mass separator) taken from operating agricultural biogas plant in Dzialyn (Poland). After sampling from dig estate tank, Inoculum was storing in anaerobic conditions in room temperature.

To determine the effect of stirring of fermentation reactors on the biogas yield of corn straw, the individual reactors were subjected to circular stirring at different intervals: daily, twice a week (Monday and Thursday) and without mixing throughout the experiment.

3 Result and discussion

3.1 Physical and chemical parameters

In the first stage of the experiment, basic physicochemical parameters of the substrate and two grains were analysed, the results of which are presented in Table 1. These parameters were necessary to determine the appropriate fermentation mixtures and subsequent calculations of the biogas yield

Table 1. Physical and chemical parameters of maize straw and inoculum.

Substrate	pH	Conductivity [mS]	Total Solids [%]	Volatile Solids [% TS]
Maize straw	9.66	1.41	77.80	89.78
Incoulum	7.87	13.10	2.91	70.04

The incubation used in the experiment was characterized by organic matter in fresh weight of about 2.1%. The total solids and volatile total solids content of maize straw used in the experiment were 77.80% FM and 89.78% TS, respectively.

3.2 Daily dynamic of methane production

Figure 2 shows the dynamics of methane production in fermentation reactors. During the process, differences in the dynamics of gas production were observed, which were influenced by the mixing frequency of the reactors.

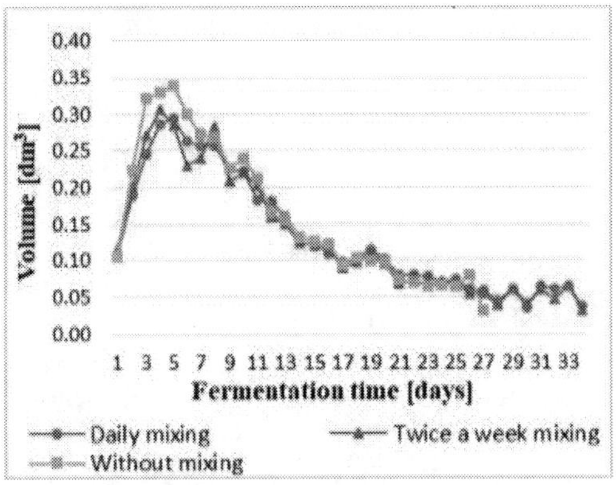

Fig. 2. Daily dynamic of methane production from reactors.

During the mesophilic fermentation, two distinct gas production peaks were observed. They were recorded on the 3rd and 8th day of the trial. The local peaks of biogas production are primarily due to the complex construction of the test material, since the simple sugars are converted to biogas faster than proteins or fat [26]. What is worth noting is the intensity of production peaks of gas in reactors that was mixed twice a week. This means that mixing of the fermentation reactor affects the change in the daily dynamics of gas production. In addition, shortening the total digestion time of the substrate in the case of daily mixing of the contents of the reactor was observed. This phenomenon may be due to the accurate distribution of fermentation microbes throughout the volume of the liquid, and consequently the potential for the formation of bacterial consortia.

3.3 Biogas and methane efficiency of the substrate

The following table 2. shows the biogas yield and methane concentration in biogas from the tested methane fermented substrates under mesophilic conditions.

Table 2. The biogas and methane efficiency.

		Fresh matter		Volatile Solids	
Sample	Methane percent [%]	Cumulated methane [$m^3 \cdot Mg^{-1}$ FM]	Cumulated biogas [$m^3 \cdot Mg^{-1}$ FM]	Cumulated methane [$m^3 \cdot Mg^{-1}$ VS]	Cumulated biogas [$m^3 \cdot Mg^{-1}$ VS]
Daily mixing	50.94	207.75	407.81	297.43	583.85
Twice a week mixing	50.24	196.50	391.10	281.33	559.93
Without mixing	50.05	186.26	372.17	266.66	532.83

In process of mesophilic fermentation of maize straw mixed every day, (as converted to Mg of fresh mass of substrate) 407.81 m^3 of biogas with average methane content of 50.94% was produced (table 2). However, in the case of corn straw mixed twice a week and without mixing, the yields of biogas and methane production were significantly lower by 16.71 $m^3 \cdot Mg^{-1}$ FM and 11.24 $m^3 \cdot Mg^{-1}$ FM and 35.64 $m^3 \cdot Mg^{-1}$ FM and 21.49 $m^3 \cdot Mg^{-1}$ FM respectively.

The results of the studies made so far confirmed that the mixing of the fermented tank influences the efficient production of biogas [31-34]. In the absence of mixing the fermentation substance, the formation of layers on the surface of the fermentation mixture can be observed. Mass containing methanogenic bacteria is deposited in the lower part of the tank due to higher density. On the other hand, the decayed substrate collects in its upper layer. [35]. However, it should be borne in mind that excessive mixing of the contents of the fermentation reactor can also adversely affect microorganisms [24, 30].

4 Conclusions

1. The studies have confirmed the effect of mixing in the fermentation reactor to increase the production of biogas and biomethane.
2. Mixing in the fermentation reactor contributes to shortening the digestion time of the substrate.
3. Lower biogas yields due to lack of mixing of reactors during testing may have a significant impact on the estimation of the economic viability of the planned industrial scale investment.

References

1. R. Chandra, H. Takeuchi, T. Hasegaw, Renewable and Sustainable Energy Reviews, **16**, 3, 1462–1476 (2012), DOI: 10.1016/j.rser.2011.11.035
2. A. Lewicki, J. Dach, P. Boniecki, W. Czekala, K. Witaszek, Advanced Materials Research, **909**, 455-462 (2014), DOI: 10.4028/www.scientific.net/AMR.909.455
3. D. Janczak, R. Mazur, P. Boniecki, J. Dach, J. Przybyl, P. Pawlak, K. Pilarski, W. Czekala, *Proc. SPIE 8878: Fifth International Conference on Digital Image Processing* (2013), DOI:10.1117/12.2031069
4. Budzianowski W.M., Renewable and Sustainable Energy Reviews, **16**, 342–349 (2012), DOI: 10.1016/j.rser.2011.07.161
5. E. Wicher, K. Seifert, R. Zagrodnik, B. Pietrzyk, M. Laniecki, Journal of Hydrogen Energy, **38**, 7767-7773 (2013), DOI: 10.1016/j.ijhydene.2013.04.008
6. M. Adamski, K. Pilarski, J. Dach, Journal of Research and Applications in Agricultural Engineering, **54**, 3 (2009)

7. A. Piwowar, M. Dzikuć, J. Adamczyk, Renewable and Sustainable Energy Reviews, **58**, 69-74 (2016), DOI: 10.1016/j.rser.2015.12.153
8. A. Lewicki, P.C. Rodriguez Carmona, J. Dach, P. Boniecki, D. Janczak, W. Czekala, M. Cieslik, J. Przybyl, *Proceedings of the 2nd International Conference On Energy & Environment: Bringing Together Engineering And Economics*, 714-720 (2015), ISBN: 978-989-97050-4-3
9. R. Caceres, N. Coromina, K. Malińska, O. Marfa, Bioresource Technology, **179**, 398-406 (2015), DOI: 10.1016/j.biortech.2014.12.051
10. J. Starzyk, W. Czekała, Archives of environmental protection, **40**, 4, 125-135 (2014), DOI: 10.2478/aep-2014-0044
11. J. Dach, P. Boniecki, J. Przybył, D. Janczak, A. Lewicki, W. Czekała, K. Witaszek, P.C. Rodriguez Carmona, M. Cieślik, Energy, **69**, 34-38 (2014) DOI: 10.1016/j.energy.2014.02.013
12. B. Igliński, A. Iglińska, W. Kujawski, R. Buczkowski, M. Cichosz, Renewable and Sustainable Energy Reviews, **15**, 2999-3007 (2011), DOI: 10.1016/j.rser.2011.02.037
13. Agricultural Agency Market - http://www.arr.gov.pl/data/02004/rejestr_wytworcow_biogazu_rolniczego_17052017.pdf(accesed: 2017.05.17)
14. M. Zaborowicz, J. Przybył, K. Koszela, P. Boniecki, W. Mueller, B. Raba, A. Lewicki, K. Przybył, *Proc SPIE. 9159: Sixth International Conference on Digital Image Processing* (2014), DOI:10.1117/12.2064066
15. Q. Zhang, J. Hu, D.J. Lee, Renewable Energy, **98** (2016), DOI:10.1016/j.renene.2016.02.029
16. L. Appels, J. Baeyens, J. Degrève, R. Dewil, Prog Energy Combust Sci, **34**, 755–781, (2008), DOI: 10.1016/j.pecs.2008.06.002
17. V.N. Gunaseelan, Biomass Bioenergy, **13**, 83–114 (1997), DOI: 10.1016/S0961-9534(97)00020-2
18. J. Mata-Alvarez, S. Macé, P. Llabrés, Bioresource Technology, **74**, 1, 3–16 (2000), DOI: 10.1016/S0960-8524(00)00023-7
19. J. Mata-Alvarez, J. Dosta, S. Macé, S. Astals, Crit Rev Biotechnol, **31**, 99–111 (2011), DOI: 10.3109/07388551.2010.525496
20. A.J. Ward, P.J. Hobbs, P.J. Holliman, D.L. Jones, Bioresour Technol, **99**, 7928–7940 (2008), DOI: 10.1016/j.biortech.2008.02.044
21. Y.Y. Choong, I. Norli, A.Z. Abdullah, M.F. Yhaya, Bioresource Technology, **209**, 369-379 (2016), DOI: 10.1016/j.biortech.2016.03.028
22. C. Mao, Y. Feng, X. Wang, G. Ren, Renewable and Sustainable Energy Reviews, **45**, 540-555 (2015), DOI: 10.1016/j.rser.2015.02.032
23. N. Nagao, N. Tajima, M. Kawai, C. Niwa, N. Kurosawa, T. Matsuyama, M.F. Yusoff, T. Toda, Bioresour. Technol., **118**, 210-218 (2012), DOI: 10.1016/j.biortech.2012.05.045
24. J. Lindmark, E. Thorin, R. BelFdhila, E. Dahlquist, Renewable and Sustainable Energy Reviews, **40**, 1030–1047, (2014), DOI: 10.1016/j.rser.2014.07.182
25. P. Satjaritanun, Y. Khunatorn, N. Vorayos, S. Shimpalee, E. Bringley, Biomass and Bioenergy, **86**, 53–64, DOI: 10.1016/j.biombioe.2016.01.018
26. M. Cieslik, J. Dach, A. Lewicki, A. Smurzynska, K. D. Janczak, J. Pawlicka-Kaczorowska, P. Boniecki, P. Cyplik, W. Czekała, K. Jóźwiakowski, Energy, **115**, 1495-1502 (2016), DOI: 10.1016/j.energy.2016.06.070
27. M. Oslaj, B. Mursec, P.Vindis, Biomass and Bioenergy, **34**, 11, 1538-1545 (2010), DOI: 10.1016/j.biombioe.2010.04.016
28. A. Wellinger, J. Murphu, D. Baxter, *The biogas handbook. Sciance, production and applications* (Woodhead Publishing Limited 2013), ISBN: 9780857097415
29. DIN 38 414, *Bestimmung des Faulverhaltens "Schlamm und Sedimente"* (BeuthVerlag GmbH, Berlin, 1985)
30. D. Deublein, A. Steinhauser, *Biogas from waste and renewable resources* (Wiley-VCH Verlag GmbH& Co. KGaA, Weinheim, Germany 2008), ISBN: 978-3-527-32798-0
31. T.J. Casey, *Elsevier App. Sci. Pub.*(90-103, London, 1986)
32. M. Kaltschmitt, H. Hartmann, *EnergieausBiomasse – Grundlagen, Techniken und Verfahren* (Springer Verlag Berlin, 2001), ISBN 978-3-540-85095-3
33. S.R. Lee, N.K. Cho, W.J. Maeng, J. Ferment. Bioengng., **80**, 4, 415-417 (1995), DOI: 10.1016/0922-338X(95)94215-D
34. Niemiec M, Sikora J., Szeląg-Sikora A., Kuboń M., Olech E., Marczuk A. Przemysł chemiczny 96(**3**),685-688 (2017)
35. Niemiec M, Sikora J., Szeląg-Sikora A., Kuboń M., Olech E., Marczuk A. Przemysł chemiczny 96(**11**),1240-1243 (2017)
36. L.C. Smith, D.J. Elliot, A. James, Wat. Res., **30**, 12 (1996), DOI: DOI: 10.1016/S0043-1354(96)00169-8
37. M. Maurer, J.P. Winkler, *Biogas – TheoretischeGrundlagen, Bau und Betrieb von Anlagen* (VerlagC.F.Müller, Karlsruhe, 1980), DOI: 10.1002/star.19810331013

BIO Web of Conferences **10**, 01013 (2018) https://doi.org/10.1051/bioconf/20181001013
Contemporary Research Trends in Agricultural Engineering

Selected parameters of maize straw briquettes combustion

Artur Kraszkiewicz[1,*], and *Ignacy* Niedziółka[2]

[1]University of Life Sciences in Lublin, Section of Management of Agriculture Machines and Eco-Energetic Systems, Głęboka 28, 20-612 Lublin, Poland
[2] University of Life Sciences in Lublin, Department of Agricultural, Forest and Transport Machinery, Głęboka 28, 20-612 Lublin, Poland

Abstract. An analysis of the process of burning briquettes made of maize straw was performed. A number of traits have been evaluated, including physical characteristics of the fuel through parameters describing combustion kinetics as well as products and combustion efficiency. The study was conducted in a grate boiler, during which the differentiating factor was the air velocity flowing to the boiler. It was observed that the obtained values of the considered parameters were different, particularly temperature of the flue gas and the amount of CO and SO_2 in the flue gas.

1 Introduction

Biomass fuels are gaining popularity due to their renewable nature and potential reduction of pollution generation. The biomass usually contains minor amounts of sulfur, heavy metals and is considered neutral for greenhouse gas emissions. The problem with its energetic use is its variability in the physical and chemical characteristics of the plant species and the forms of biofuels produced. At the same time, the conversion of biomass into heat energy, which is available in a wide range of heating devices, makes each of them requires separate operational settings dedicated to the fuel used so as to achieve high combustion efficiency with the highest possible heat transfer and the lowest emission of pollutants into the atmosphere, both gaseous and solid. Gases such as CO, NO, SO_2, that appear in the atmosphere due to badly run combustion process, adversely affect the human health. Incomplete combustion of carbonaceous material can lead to carbon monoxide (CO) emissions. Carbon monoxide is a colorless and odorless gas, and it is poisonous for living beings, because it binds to hemoglobin during contact with blood, to which it has 200-240 times higher affinity than oxygen. Concentration of 125 $mg \cdot m^{-3}$ CO in the air and the impact on humans in 1 hour (according to the World Health Organization, WHO) causes the binding of 4% blood hemoglobin and triggers cardiac disorders. The value of 250 $mg \cdot m^{-3}$ carbon monoxide (II) in the air can cause death [1, 2]. Another characteristic element in solid biomass is nitrogen (N). During fuel combustion, nitrogen is almost completely converted to gaseous N_2 and nitrogen oxides (NOx). The most dangerous is NO_2, which is characterized by acute odor with acid taste and reddish brown color. Passing a person in a NO_2 environment at the concentration of 0.5-1.5 ppm within 1 hour will cause breathing difficulties. On the other hand, NO is a poorly reactive, odorless and colorless gas, which, unfortunately, in the Earth's atmosphere, under the influence of UV rays, is converted into NO_2. A similar reactivity is revealed by nitrous oxide, called the laughing gas, generated during low-temperature fuel combustion. It is counted among the most active gases that prevent sudden heat radiation of the Earth surface into space, thus contributing to the creation of the greenhouse effect [1, 2]. Sulfur (S) contained in the fuel during combustion is bound in ash or enters the atmosphere with exhaust gases mainly in the form of sulfur dioxide (SO_2) and less often as sulfur trioxide (SO_3). SO_2 is a colorless gas with a sharp and pungent smell, strongly irritating the respiratory tract. Sulfur dioxide is poisonous to animals and harmful to plants and contributes to atmospheric pollution when generated during the fuels combustion (smog). Rybak [3] reported that monitoring of combustion temperature in the furnace plays a fundamental role in modifying the combustion mechanism, controlling the emission of harmful substances, and optimizing the efficiency and operation of the whole system. The furnace temperature can be controlled by manipulating a number of parameters, with the most important and not relevant to the combustion chamber and related to the physical and chemical characteristics of the fuel being the value of the air ratio excess, the oxidation temperature, and the moisture content.

Among the wide range of biomass fuels, maize straw is an interesting raw material. Maize is one of the main cereal crops grown in Poland, and its straw is considered an interesting source of a biomass for energy purposes [4, 5]. The analysis of the combustion process and explanation of the mechanism of maize straw combustion on a fixed bed will contribute to the clean and efficient use of maize straw as a source of energy.

*Corresponding author: artur.kraszkiewicz@up.lublin.pl

The purpose of the study was to analyze the process of maize straw briquettes combustion in terms of the dependence of the influence of air velocity on the achieved combustion temperatures as well as on the emission of carbon monoxide, nitrogen and sulfur.

2 Material and methods

2.1 Biomass types and sample preparation

- Maize straw briquettes were used for the study. The selected physical and chemical parameters of this fuel were determined applying the following research methods:
- external dimensions of the fuel - by direct measurement of 10 randomly selected briquettes;
- moisture content - weight method according to the standard PN-EN 14774-1:2010 [6];
- density - indirect method using the mass and volume of 10 randomly selected briquettes and calculated according to Annex A of the standard PN-EN 15150:2012 [7];
- carbon, hydrogen, sulfur - by IR absorption method;
- nitrogen – katharometric method;
- calorific value - calculated after prior determination of combustion heat according to the standard PN-EN 14918:2010 [8];
- ash content – according to the standard PN-EN 14775:2010 [9].

Mean values of the obtained results are shown in Table 1.

Table 1. Physical and chemical properties of maize straw briquettes.

Parameter		Unit	Maize straw briquettes (average values)
Length		mm	23
Diameter		mm	50
Total moisture		%	7.92
Density		$kg \cdot m^{-3}$	900
Elementary composition	C	%	45.0
	H	%	5.50
	N	%	0.70
	S	%	0.18
Calorific value		$MJ \cdot kg^{-1}$	16.43
Net calorific value		$MJ \cdot kg^{-1}$	15.10
Ash content		%	5.77

2.2 Combustion process conditions

During the study, a test bench with a water-jacketed, rectangular combustion chamber closed from the bottom with the water grid (Figure 1), was used. The capacity of the combustion chamber to the exchanger part was 3.51 dm^3 (0.26×0.3×0.45 m). Above the combustion chamber, there was a heat exchanger with horizontal divisions with water channels. The ash chamber was below the water grid, in which the ash container was placed. The whole device was extended by an air supply fan directed beneath the grate and the circulation fluid pump placed between the boiler and the storage tank working in the open system. The water jacket capacity was 30 dm^{-3}, and that of accumulation tank 400 dm^{-3}. This system was supposed to provide a similarity to the actual chambers and allow the combustion process to be carried out as in low-power heating devices.

The exhaust gases were collected from the chimney at a distance of 1 m from the flue boiler. The measuring probe was connected to the PGD-100 exhaust gas drier (Madur Eljack Electronics), from which exhaust gases were sent to the exhaust gas analyzer. During the tests, Photon's portable flue gas analyzer was used. This is a device based on infrared sensors (NDIR) for the following gases: CO, CO_2, NO, SO_2. The temperature was measured using a K type thermocouple. The first one was located in the middle of the height of the combustion chamber at its outer wall, while the second one performed measurements of the flue gas temperature in the chimney and was located at the exhaust gas collection point (Figure 1).

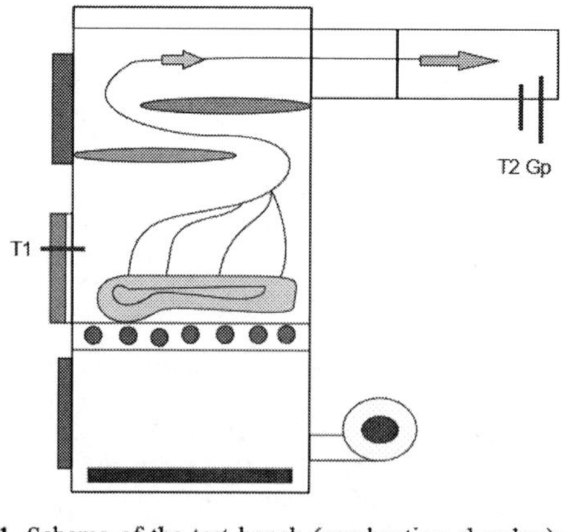

Fig. 1. Scheme of the test bench (combustion chamber): T1 - temperature measurement in the combustion chamber, T2 - temperature measurement in the chimney, Gp - exhaust gas consumption.

The study consisted of combusting the 1 kilogram samples of maize straw briquettes in triplicates at four velocities of the supplied air stream – 0.5 $m \cdot s^{-1}$ (A), 1.0 $m \cdot s^{-1}$ (B), 1.5 $m \cdot s^{-1}$ (C), 2.5 $m \cdot s^{-1}$ (D). Fuel loading and ash removal were handled manually. Measurements of the exhaust gas composition were carried out continuously from the time the fuel was laid on the stabilized layer of heat until the reaction complete. The results of these parameters were recorded automatically to the analyzer database every 2 seconds with simultaneous recording of data recording duration. The database created was transferred from the analyzer to a PC after the tests. The results of the study were statistically analyzed in Statistica 10.0 software. The

Shapiro-Wilk test verified compatibility of the results with the normal distribution, and the Brown-Forsyth test estimated variance homogeneity. When homogeneity of variance was absent, the Kruskal-Wallis test was used. In contrast, the Spearman correlation test was used to describe relationships between individual variables. Moreover, the observed differences were found to be statistically significant at the significance level $p < 0.05$.

3 Results

Average values of temperatures in the combustion chamber and in the chimney are shown in Figures 2 and 3, respectively.

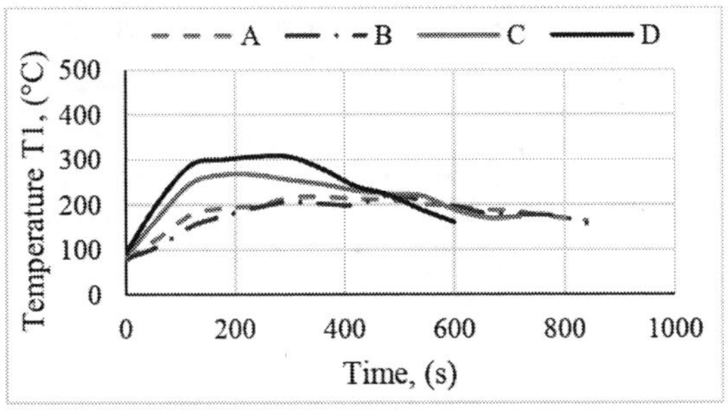

Fig. 2. Temperature changes in the combustion chamber during combustion of maize straw briquettes.

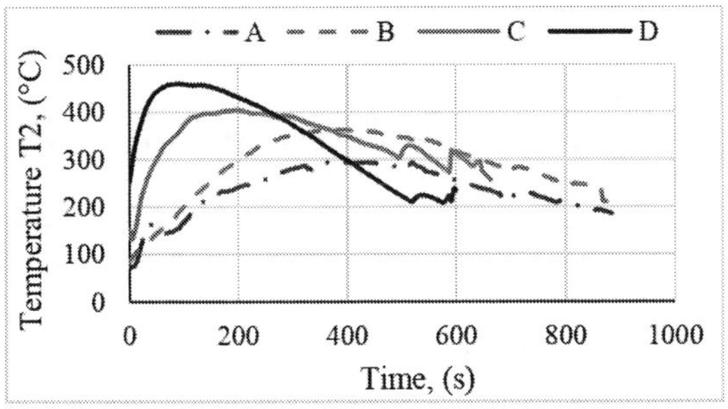

Fig. 3. Temperature changes in the chimney during combustion of maize straw briquettes.

As the air velocity increases from 0.5 $m \cdot s^{-1}$ to 2.5 $m \cdot s^{-1}$, the combustion temperature increases both in the flue channel and in the combustion chamber. At the same time, the burning time of the fuel sample is reduced from less than 15 minutes to 10 minutes, and for the fastest speed of this medium, much shorter time is required to reach the maximum temperature. Temperature T2 (in the flue channel) was about 100-150 °C higher than the temperature T1 (in the combustion chamber) - Figures 2 and 3.

Composition of the combustion gases generated during the combustion of maize straw briquettes under combustion conditions assumed in the tests is shown in Figures 4-6.

Only for the highest velocity of the air supplied to the boiler, the amount of NO particles in the exhaust gas reached the maximum value in the first minute of combustion. At this time, at the highest air velocity, a temporary increase in temperature T2 was also observed - Figures 3. There was a difference in SO_2 content in the exhaust gas. At the smallest airflow velocities in the first period, the SO_2 content of the exhaust gas amounted up to 120 ppm. After the first 3 minutes of combustion, these values were reduced and stabilized to a range of 40 to 60 ppm.

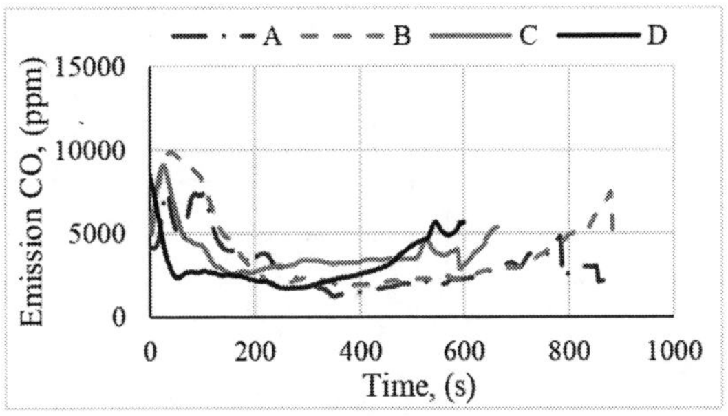

Fig. 4. Changes of the concentration of carbon monoxide during combustion of maize straw briquettes.

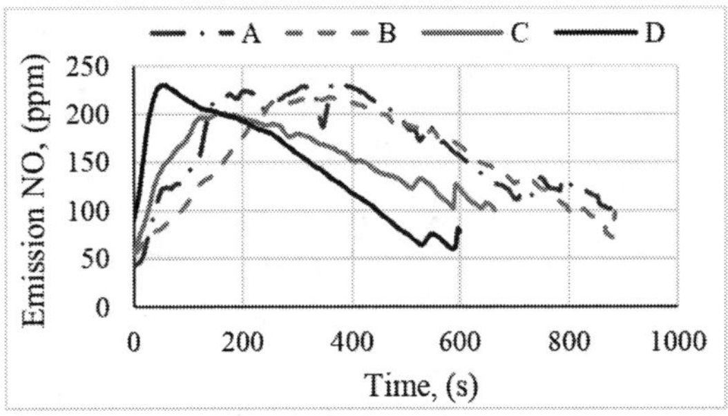

Fig. 5. Changes of the concentration of nitrogen oxide during combustion of maize straw briquettes.

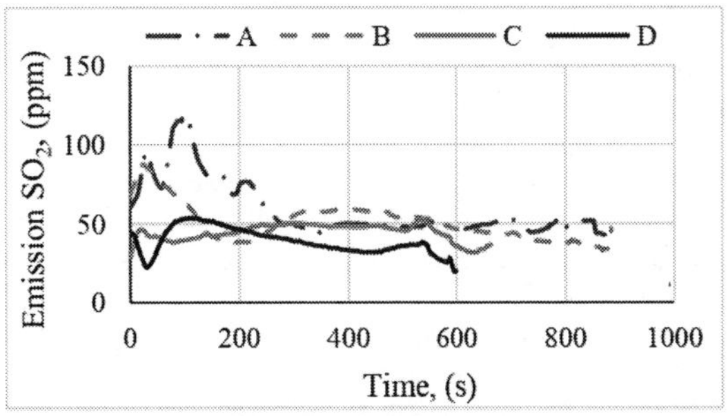

Fig. 6. Changes of the concentration of Sulphur dioxide during combustion of maize straw briquettes.

The statistical analysis of the obtained results indicated the use of the non-parametric Kruskall-Wallis test. Results of this test indicated significant differences between values obtained for the following components of exhaust gas: CO, SO_2 as well as between the temperature in the chimney. No significant differences were found between the number of NO particles in the flue gases that were recorded during tests at varying velocity of supply air to the boiler. Probability was 0.7193. The results of the T2 temperature correlation tests performed with the combustion products are shown in Table 2.

Table 2. Spearman's correlation coefficients between temperature and gaseous combustion products.

Variables	Air velocity ($m \cdot s^{-1}$)							
	0.5 (A)		1.0 (B)		1.5 (C)		2.5 (D)	
	R	p	R	p	R	p	R	p
T2 (°C) vs CO (ppm)	-0.74	0.00	-0.77	0.00	-0.30	0.00	-0.40	0.00
T2 (°C) vs NO (ppm)	0.75	0.00	0.92	0.00	0.93	0.00	0.94	0.00
T2 (°C) vs SO_2 (ppm)	-0.46	0.00	0.19	0.00	0.44	0.00	0.62	0.00

4 Discussion

Maize straw, or rather briquettes, made from this raw material used for testing, have typical properties corresponding to such form and type of biomass [10-14]. Burning the plant biomass such as wood and wheat or rye straw compacted into briquettes and pellets in heating devices has been well researched [15-19]. However, maize straw causes burning problems as recorded in the work by Zhao et al. [20].

Fuel characteristics and process conditions such as fuel type, particle size, airflow and fuel humidity affect the combustion characteristics by changing the heat generation, heat transfer rate, and reaction speed. The properly performed combustion process is dependent on the amount of oxygen supplied and actually on the amount of air [21-23]. Airflow is a key parameter in the process that determines the amount of available oxygen and convection heat transfer. Process indicators are classified in three successive groups depending on the air velocity: oxygen limitation, reaction limitation and extinguishing by convection systems [20, 24, 25].

However, variability of the recorded data (temperature, gaseous combustion products) identifies two distinct stages as the bed combustion progresses: (1) evaporation of the volatiles with transfer of the afterburning zone to the exchanger and flue gas zone as well as (2) oxidation of the coal residue.

Results obtained under the test conditions indicate that the NO concentration in the exhaust gas initially reaches a maximum and then decreases to a stable value after the bed is lit. Similar results were obtained by [22, 26]. The statistical analysis indicated that the NO content of the flue gas is insensitive to the change in velocity of air flowing into the combustion chamber. Other chemicals, CO and SO_2, have such a relationship. On the other hand, the relationship between the temperature values recorded in the chimney and the combustion products showed a strong, positive dependence on the NO content of the flue gases and poor CO and SO_2 dependence, especially when changing the velocity of supplied air. For CO, the force of temperature dependence decreased as the air velocity increased, and for SO_2 this was the inverse relation.

5 Conclusions

As a result of the research, the following observations and conclusions were drawn up:

1. Maize straw briquettes, with respect to air velocity as the primary factor of differentiating the combustion process, generally show different parameters of this process. Only the NO content is not dependent on air velocity.
2. Effective ecological combustion of maize straw briquettes requires the use of heating devices with the possibility of adjusting the air supply depending on the demand of combusted fuels. However, using higher air velocities, the combustion zone must be shifted towards the chimney and the temperature of the flue gas will probably increase, especially in the initial combustion phase.
3. Due to the remarkable variation in physicochemical characteristics of solid biofuels and variety of technical parameters of heating equipment, it would be appropriate to extend the research to other species, and above all, forms of plant biomass and their impact on the combustion process in order to achieve tangible ecological and energy effects.

References

1. M. Domański, L. Dzurenda, M. Jabłoński, M. Osipiuk, *Drewno jako materiał energetyczny* (SGGW, Warszawa, 2007) *in Polish*
2. Y.S.H. Najjar, Ashdin Publishing Innovative Energy Policies, Vol. **1** (2011)
3. W. Rybak, *Spalanie i współspalanie biopaliw stałych* (Wyd. Politechniki Wrocławskiej Wrocław 2006) *in Polish*
4. H. Karcz, M. Kantorek, M. Grabowicz, K. Wierzbicki, Piece przemysłowe i kotły, **XI-XII** (2013) *in Polish*
5. I. Niedziółka, A. Zuchniarz, Motrol, Nr **8A** (2006) *in Polish*
6. PN-EN 14774-1:2010 Solid biofuels - Determination of moisture content - Oven dry method - Part 1: Total moisture - Reference method
7. PN-EN 15150:2012 Solid biofuels. Determination of particle density
8. PN-EN 14918:2010 Solid biofuels - Determination of net calorific value
9. PN-EN 14775:2010 Solid biofuels - Determination of ash content
10. A. Demirbas, Prog. Energy Combust Sci., Vol. **30**, Issue 2 (2004)
11. M.H. Eisenbies, T.A. Volk, A. Biomass and Bioenergy, **86** (2016)
12. P. Mc Kendry, Bioresource Technology, **83** (2002)
13. I. Obernberger, T. Brunner, G. Bärnthaler, Biomass Bioenergy, **30** (2006)
14. M. Stolarski, M. Krzyżaniak, M. Śnieg, E. Słomińska, M. Piórkowski, R. Filipkowski, Int. Agrophys., **28** (2014)

15. J.J. Saastamoinen, R. Taipale, M. Horttanainen, P. Sarkomaa, Combust Flame. Volume **123**, Issues 1-2, (2000)
16. R.P. Van der Lans, L.T. Pedersen, A. Jensen, P. Glarborg, K. Dam-Johansen, Biomass and Bioenergy, Volume **19**, Issue 3 (2000)
17. Y.B. Yang, J. Swithenbank, V.N. Sharifi, Fuel, Volume **83**, Issues 11-12 (2004)
18. H.C. Smit, M. Meincken. Biomass and Bioenergy **39** (2012)
19. L. Carvalhoa, E. Wopienkaa, Ch. Pointnera, J. Lundgrenb, V. K. Vermaa, W. Haslingera, Ch. Schmidla, Applied Energy, Volume **104** (2013)
20. W. Zhao, Z. Li, D. Wang, Q. Zhu, R. Sun, B. Meng, G. Zhao, Bioresour. Technol.,**99**(8) (2007)
21. T. Nussbaumer, *Swiss Federal Office for the Environment (FOEN) as a contribution to the expert Group on Techno-Economic Issues (EGTEI) under the Convention on Long-Range Transboundary Air Pollution*, Verenum Press Zürich (Switzerland), (2010)
22. A. Gungor, Biomass and Bioenergy **34** (2010)
23. Ch. Ryu, Y.B. Yang, A. Khor, N.E. Yates, V.N. Sharifi, J. Swithenbank, Fuel, **85** (2006)
24. D. Shin, S. Choi, Combust Flame, Volume **121**, Issues 1-2 (2000)
25. H. Zhou, A.D. Jensen, P. Glarborg, P.A. Jensen, A. Kavaliauskas, Fuel, **84** (2005)
26. H. Zhou, A.D. Jensen, P. Glarborg, A. Kavaliauskas, Fuel, **85** (2006)

Marketing of organic products in southern Poland

Maciej Kuboń[1*], *Elżbieta Olech*[1]

[1]Department of Production Engineering and Power Energy at the University of Agriculture in Kraków, Poland

Abstract. The article presents an outline of the issue concerning formulation of a marketing strategy and the possibility of using the knowledge on consumers' preferences for organic development of farms and their products on the example of southern Poland. The paper analyses the distribution process of organic food in the aspect of developing innovative marketing strategies. The studies were performed in 50 organic farms and on the example of 100 respondents from the region of southern Poland. In the opinion of the surveyed representatives of the organic food producers, a competitive advantage of their offer depends the most on the health values, brand, reputation, and taste. Moreover, information on products and the form and place of their sale are significant. The analysis shows that the knowledge is the most eagerly obtained from the Internet. Thus, producers should publish their profiles and pages on social media and business portals.

1 Introduction

Over the years, a significant growth of some types of products in the food marketing has been observed. This long-term trend, with a varied degree of advancement with regard to the market, concerns, inter alia, functional, health, regional, traditonal and organic products [3]. Development of the organic products market has taken place as a result of the impact of many factors and events on the market, among others, food scandals, which caused that consumers choose natural products. They care more for the environment and health, they return to nature, which results in purchase-related decisions [4]. A raising demand for organic food is one of the components of deeper changes which take place in the modern consumption model [12, 15]. Numerous domestic and foreign research confirm development of purchase behaviour on the organic production market. Consumers' behaviour with regard to these products depend on various factors. They are mainly determined by the level of organic awareness, income level, knowledge level and marketing operations [8, 11,13, 16]. The research carried out in Great Britain proved that the care for food safety is the most insignificant determiner in relation to organic products as well as it confirmed the significance of the ethical stand of consumers in this context [2]. The marketing mix specificity for organic food may be analysed at the global level [9]. Obtaining the quality certificate triggers the success on the organic products market [1]. According to A. Graczyk and K. Mazurek-Łopacińska development of these markets in Europe is activated by accepting the sustained development strategy of the EU, realization of which depends on cultural factors [6]. In Poland, one can also observe a dynamic increase of the number of producers producing food with organic methods. In 2012, the number of organic food producers was 26 376 (including 312 food processing plants) and for comparison in 2011 - 23 847 (including 270 food processing plants) [9]. B. Pilarczyk and R. Nestorowicz show that the organic food market in Poland is at the initial stage of development and entities acting on this market only slightly utilize the experience of marketing-oriented enterprises [10, 17]. An agricultural producer, in order to successfully participate on the European market, must participate in pro-development activities, i.e. inter alia: active marketing, trainings, education, participation in fairs or implementation of innovative processes. Finally, agricultural farmers in Poland who want to obtain a high competitive position on the European market, must, first of all, have a wide range of operations which enable a continuous distribution of own products [5, 7, 18]. Organic food is not easily found in usual shops and distribution centres. Indeed, specialist shops with organic food and their networks are established but it takes place only in big cities. However, the scale of these activities is still too small to increase the sales in Poland. Certainly, high prices of organic products in comparison to traditional food impede obtaining wider consumers' interest [15, 17]. Thus, the need to implement a strategy that aims at improvement of conditions and facilitation of sale has occurred. Marketing strategies, which constitute a whole comprising price strategies related to each other, product strategies, distribution strategy and promotion strategy are examples of such actions[16]. The strategy elements form an integrated market operations system which ensures the enterprise with a competitive position. One of important elements of marketing, from which designing marketing strategies takes place, is a good, namely a product. The product should be attractive so that the consumer may maximize the sum of satisfaction

* Corresponding author: maciej.kubon@ur.krakow.pl

and advantages related to owning and consumption[17, 18, 19]. Having a product should thus provide a consumer with a relevant level of satisfaction, profit and credible image on the market for the enterprise. In order to undertake an initiative related to implementation of marketing strategies, the first step is to know the client's preferences concerning the product itself and elements related thereto such as the packaging form, preferred form of sale or the motive behind the selection of a given product.

2 Objective, methodology and scope of research

The objective of the research was to analyse the organic food distribution process through the prism of consumer's preferences towards the process in the south Poland region. Knowledge on them gives a chance to prepare suitable channels and develop a relevant marketing approach. In order to achieve this aim, a guided survey was made for the clients from the selected area and organic food farms were investigated. The investigated population included a mixed sample of 100 people and the number of investigated organic farms was 50. The prepared survey included questions which give the opportunity to collect information related to financial possibilities of families which they may designate for the purchase of organic food, frequency of consumption, properties which influence the selection of the product as well as a preferred form of sold products and packagings for organic food. Knowing this information will enable adjusting the production structure and the product itself to specific consumer's requirements and thus improvement of distribution channels and broadening the knowledge within this range in order to develop a relevant marketing strategy. The collected information was obtained on the basis of a guided survey carried out with a farm owner and randomly selected group of responders. The collected data allowed determination of the production size and structure of sale of organic products and clients' preferences at the purchase of organic products. The global production value was assumed as a sum of plant production, animal production produced in the production process and sold services and the commodity production value as a sum of agricultural production sold by a farm.

3 Characteristics of the investigated population

All respondents knew the concept of organic food. Majority wants these products to be more available on the market. From among the surveyed persons, 54% were women and 46% were men. The most numerous age group consisted of people within the age of 26-35 i.e., 33% of the respondents. People above 45 were the most numerous group - 11%. With regard to the financial situation of the households, the questioned respondents were divided into two groups. The first one (58% of respondents) admits that their financial situation is average namely the monthly gross income does not exceed PLN 3 thousand, while 42% declares that it is good, i.e., income per a family exceeds PLN 3 thousand. With regard to education, the biggest group of respondents included people with secondary school education - 52% and above 70% of respondents are people who are employed and have children.

4. Research results

A fast civilization progress and social requirements cause the necessity of extending the economic growth and satisfaction of people's needs. Furthermore, eating habits of citizens of particular countries as well as the food consumption structure depend mainly on the environmental conditions, which take place in the place of residence, tradition and the degree of economic development of the country and related financial resources of the society.

Analysis of the global markets shows that presently the organic food market is one of the most developing food markets in the European Union. This food more and more often becomes a substitute of the traditional food. For several years, the increase in the number of organic food shops can be observed. The Internet has become the most frequently chosen form of organic food purchasing. Following the example of other countries, it is necessary to develop detailed characteristics of the organic food buyer. It aims at providing entrepreneurs with this knowledge, so that they can precisely adjust their products to consumer's expectations. This characteristic should present specific consumers' preferences including organic production so that, recommendations, which will concern the possibilities of use and extending an offer on the market should be prepared. It should result in the increase of competitiveness and sale of organic products.

When mentioning competitiveness it should be emphasised that nowadays, it is strongly developed and depends on the consumers' changing needs and requirements. Proper adjustment of the market assortment to consumers' requirements may ensure demand on products as well as the increase of the enterprise competitiveness.

Organic farms in Małopolska region feature high distribution of the agrarian structure as well as production structure. It can be translated into the production scale and that on the size of commodity production which basically influences the farm income amount. Table 1 presents a unit value of organic production divided into production divisions.

Table 1. Unit value of organic production divided into production divisions.

Farm group	Parameter	Production					
		Plant		Animal		Total	
		Global	Commodity	Global	Commodity	Global	Commodity

		PLN·ha^{-1}	%	PLN·ha^{-1}	%	PLN·ha^{-1}	%
Total	average	5008.4	78.3	2626.9	79.5	7635.3	78.7
	standard deviation	6962.2	-	3027.2	-	6908.6	-
up to 5 ha	average	7526.6	85.4	2110.4	72.9	9636.8	82.7
	standard deviation	8486.5	-	2067.4	-	8811.2	-
5.01 – 10.00 ha	average	3856.7	64.6	3108.3	81.2	6965.0	72.0
	standard deviation	3498.5	-	4176.8	-	4959.5	-
10.01 – 20.00 ha	average	7497.3	91.7	1503.2	79.5	9000.5	89.6
	standard deviation	10043.1	-	2108.4	-	9231.5	-
area 20.00 ha	average	1425.0	20.8	3640.6	83.2	5065.6	64.1
	standard deviation	1060.1	-	1919.8	-	1932.4	-

Source: Kuboń, et al. 2014

The total commodity production in the investigated farms was 6009.73 PLN·ha^{-1}, which constituted at the average 78.7% of the global production. The commodity production in the group of objects from 5-10 ha was at a comparable level. In comparison to these objects, in the group of farms from 10-20 ha the commodity production was higher by 4 and 11% and in the biggest farms (area of 20 ha) - it was lower by almost 15%. Whereas, taking into consideration production departments, it was reported that the animal commodity production (79.5%) slightly prevailed over the plant commodity production (78.3%). The highest participation of animal commodity production occurred in the group of farms with the area of 20 ha - 83.2% and the lowest up to 5 ha - 72.9%

Distribution next to production is one of the most important elements in the logistic chain. Its main task is making the product available in place and time corresponding to the clients' needs and expectations. From the point of view of macroeconomics, distribution means the process and structure of moving goods from producers to target consumers. Thus, it constitutes a selected group of market channels and connections between them. Distribution systems of a particular organization structure and equipped with material-technical factors, which exist in economy, create conditions, which constitute the basis for selection of the manner of relocating products from the production zone to the consumption zone. Features of these systems cause that in a short period, they become an external factor for shaping ways of distribution. Decisions related to selection of the sale manner are included in each enterprise to strategic decisions, because they finally influence effectiveness of market activities and achieved economic effects [7].

Producers of organic food products should pay attention more often to the changes of consumers' likings and preferred manners of satisfying food needs, e.g. a demand for the so-called convenient food products, partially or totally ready for consumption and also to the need of manufacturing new products. This, on the other hand requires the use of new technologies, adjusted to the existing legal regulations with regard to the organic farming.

A requiring outlet of agricultural products forces organic food producers to numerous activities which aim at obtaining new and maintaining regular customers. Presently, it is very difficult, because, the present market provides a richer offer of products of foreign origin. Only thorough research of client's needs will create a demand for specific products and will allow loyalty to a known product. It should be also emphasised that consumers' awareness rising every year is reflected in the demand for goods which differ with original recipe, which results from the search of extraordinary composition.

A modern consumer, taking up a decision on the purchase of food, becomes more requiring, thus a relevant price and the standard quality does not suffice. A choice of products with suitable parameters, in particular the most important are: taste and health values, popular brand, production tradition, regional origin, more strict control of the production process, is a basic issue which has become more significant for the potential client.

Taking into consideration the criterion, namely, the form of organic products packaging, respondents preferred to buy loose products, in small packaging or in sealed containers. Women aged 26-35 prefer to buy loose products because of a wider choice of products. They are also interested in a small packaging within 1-2 kilo. A trend among men was similar. Respondents aged 18-25 were interested in the purchase of loose products or with a small weight - up to 2 kilo, and in products in sealed containers because of a longer shelf life of such products. Figure 1 presents preferred forms of organic products.

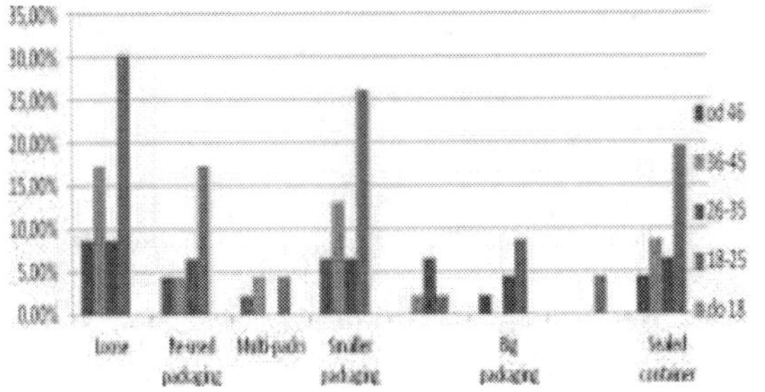

Fig. 1. Form of organic products packaging.

In case of the Poland's region it is difficult to talk about an organized sale of organic food products. Direct sale of organic food and sale through specialised channels, which is characteristic for the initial phase of the market development, prevails. Contrary to the Western Europe countries, in Poland, there is a small number of specialist shops (ca. 250) with a small variety of products. An insufficiently developed indirect sell is an additional weakness (number of specialistic warehouses is about 30) [14].

Basic information, significant for producers in order to develop marketing strategies consist of: sources of information on organic products, motive of selection and place of purchase of organic products. Knowing answers to those and other questions, they will be able to prepare a suitable marketing or promotional campaign.

Taking into consideration the age criterion in acquisition of knowledge on organic food, the Internet was prevailing both among women and men. Women within 26-35 years of age and 36-45 prefer searching for information on products in the Internet. Number of these answers exceeded 30%. However, among women information from press and radio is also valuable. The highest number of such answers was given by people aged 18-25 and 36-45. Men in the age of 18-45 ,similarly to women, obtain information on organic food most often in the Internet. Studies proved that this is the main source of obtaining information on organic products.

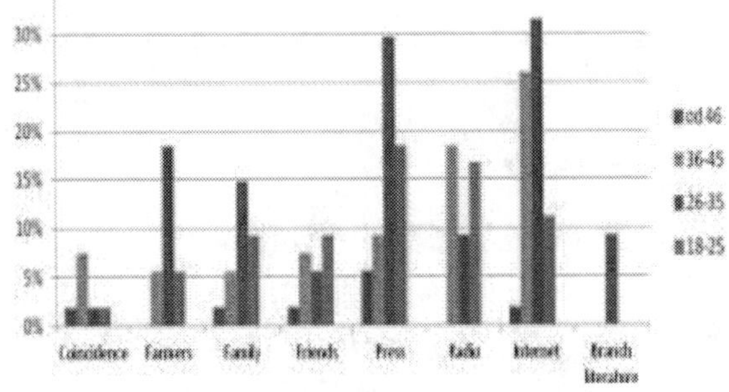

Fig. 2. Sources of knowledge on organic products in relation to age.

Clients' motives for choosing organic food are other significant factors. Women in the age 36-45 mentioned a popular source as the main motives for choosing organic food, no hazardous substances and low processing of products. More than 30% of respondents provided such answer. While, women aged 18-25 years mentioned known sources and no preservatives and dyes as the main motive of selection of an organic product. For men aged 26-45 it is the most important that organic food was without preservatives and dyes and it was rich in nutrients and processed to a small scale. It is also important that food has high taste values. Men, aged above 46 value such features as: no preservatives and dyes and the fact that it is rich in nutrients.

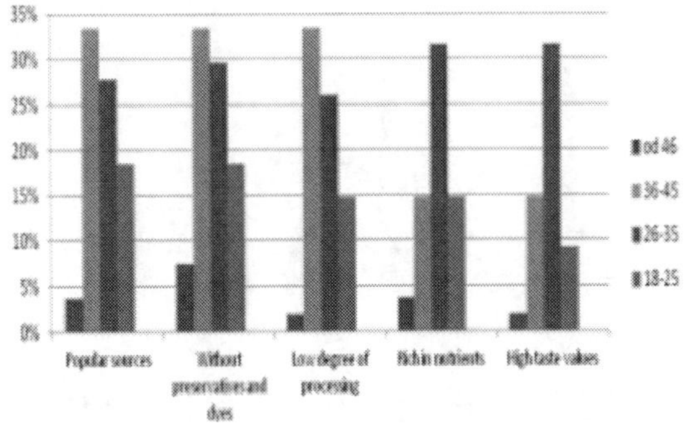

Fig. 3. Motives of choice of organic food.

With regard to the place of purchase of organic food, the most often women bought organic food at the healthy food stand of the shop - over 20% of answers in the age group 26-35. In that age group responses indicated that they also make shopping in healthy food shops and directly at a producer. Women do online shopping more rarely than men. A higher variety at a selection of the place of purchase was reported in case of men. The most often they buy at healthy food stands. Men aged 18-25 also often buy directly at a producer as well as in the Internet. Men aged 26-35 and above 46 years value also shopping in healthy food shops.

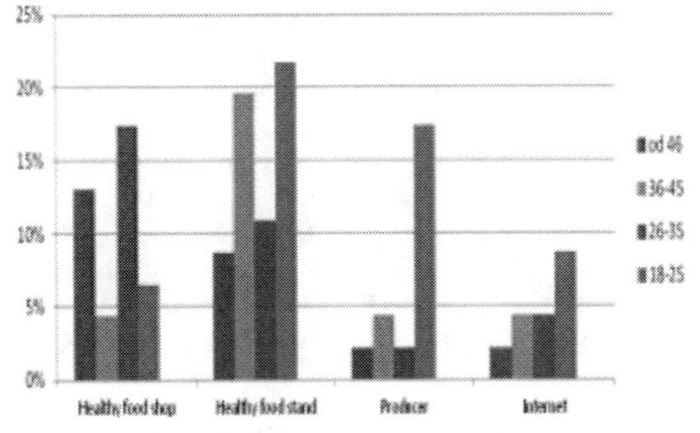

Fig. 4. Preferred places of organic food purchase.

With reference to the organic food purchase for all surveyed people, a price is important but according to the research, other factors also play a significant role. All mentioned properties are more or less important for respondents. For women aged 26-35 the product composition and its smell - 25 % and for women aged above 46 the shelf life, composition and labelling is important. Women pay less attention to the product appearance. Men indicated smell as one of the most important factors - over 35% of responses in age groups 18-25 and 26-35. The product composition and shelf life are important in the age group 18-25. Appearance is less important. Men decided that the least significant feature is product labelling.

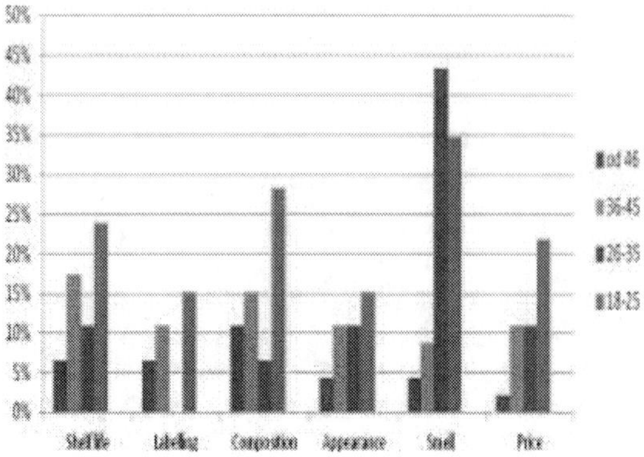

Fig. 5. Product features which determine its purchase.

Conclusions

In order to develop new distribution channels or marketing strategies of organic food sale, a periodical market analysis is necessary, especially if their main element is recognition of clients' needs and requirements. Results of such analysis are a great support for producers because they provide necessary knowledge which enables orientation of production to specific requirements, finding a production niche, and thus to be more competitive among other producers.

Authentic quality values of a product favour implementation of the marketing principles in the organic food sector. However, according to the theory, the product itself is not sufficient to become successful on the market. Except for relevant pricing, the product must be delivered optimally to the places of sale (distribution) and the customer must be informed on its values and encourage to buying (promotion).

In the opinion of the surveyed representatives of the organic food market, a competitive advantage of their offer depends the most on the health values, brand, reputation and taste. Current provision of information on products is also important. The analysis proves that such knowledge is the most eagerly obtained from the Internet because presently a majority of people in their free time broaden their knowledge by reading information on webpages. With reference to the entrepreneur's marketing they should publicise their profiles on social media pages and business portals.

Higher ecology awareness of producers and consumers may help to extend organic production and distribution of their products in Poland. Suitable application of the market research may support producers in extending their offer and enable fighting over outlets through adjustment of their behaviour and the product to the consumer's preferences.

References

1. K. Achilleas, S. Anastasios, British Food Journal, **8**, 110, 829 (2008)
2. N. Michaelidou, L. Hassan, International Journal of Consumer Studies, **32**, 163 (2008)
3. M. Strumioska–Kutra, *Ekologiczna świadomość Polaków. Analiza wyników badań ilościowych z lat 1992-2011,* Instytut na rzecz Ekorozwoju, Warszawa (2012)
4. S. Żakowska-Biemans, British Food Journal, 1, **113**, 122 (2011)
5. G. Wójcik, Wiadomości Zootechniczne, **2**, 29 (2012).
6. A. Graczyk, K. Mazurek-Łopacińska, *Badanie rozwoju rynków produktów rolnictwa ekologicznego i żywności ekologicznej w Polsce.* Wydawnictwo Uniwersytetu Ekonomicznego we Wrocławiu, Wrocław (2009)
7. M. Kuboń D. Kwaśniewski, U. Malaga-Toboła, S. Tabor, Agricultural Engineering **2** (150), 77 (2014)
8. B.J. Phillips, E.F. McQuarrie, W.G. Griffin, Psychol. Mark. **31**, 225 (2014)
9. L. Witek, Handel Wewnętrzny **1** (354) 281 (2014)
10. B. Pilarczyk, R. Nestorowicz, *Marketing ekologicznych produktów żywnościowych.* Wolters Kluwer Polska, Warszawa (2010)
11. C. Prentice, J. Chen, X. Wang, Journal of Retailing and Consumer Services (2017) - article in press
12. B. Chekima, A. I. Oswald, S. A. Wafa, S. Wafa, K. Chekima, Journal of Cleaner Production **166**, 1438 (2017)
13. Q. Zhu, J. Sarkis, Int. J.Production Economics **181**, 289 (2016)
14. W. Łuczka-Bakuła, J. Smoluk-Sikorska, Journal of Research and Applications in Agricultural Engineering, **55**(4), 12 (2010)
15. A. Szeląg-Sikora, M. Cupiał M. Niemiec, Agriculture and Agricultural Science Procedia, **7**, 94, (2015)
16. A. Szeląg-Sikora, M. Cupiał, M. Niemiec, Agriculture and Agricultural Science Procedia. **7**, 249 (2015)
17. A. Szeląg-Sikora. M. Niemiec, J. Sikora. Journal of Elementology, **21**, 3. 915 (2016)
18. Z. Gródek-Szostak, A. Szeląg-Sikora, J. Sikora, M. Korenko Business And Non-Profit Organizations Facing Increased Competition And Growing Customers' Demands, **16**, 427, (2017)
19. M. Kołtun, S. Kocira, Z. Krzysiak, M. Ćwiklińska A. Kocira, M. Koszel, Agriculture and Agricultural Science Procedia **7**, 113 – 118, (2015)

Content of Ni, Pb and Zn, in selected elements of ecosystem in three bays in the area of Sevastopol

Marcin Niemiec[1], *Anna* Szeląg-Sikora [2*], *Natalya* Kuzminova [3], *Monika* Komorowska [4]

[1]Department of Agricultural and Environmental Chemistry, University of Agriculture in Krakow, . Mickiewicza Ave. 21, 31-120 Krakow, Poland
[2] Institute of Agricultural Engineering and Informatics, University of Agriculture, ul. Balicka 116B, 30-149 Krakow, Poland
[3]Institute of Marine Biological Research A.O.Kovalevsky named (Russian Academy of Sciences), Crimea, 299011, Sevastopol, Nakhimov Ave. 2, Russia
[4]Department of Vegetable and Medicinal Plants, University of Agriculture in Krakow 29 Listopada Ave. 54, 31-425 Krakow, Poland

Abstract. The aim of the study was to evaluate the content of Ni, Pb and Zn in the water and larvae of fish of the genus *Atherinidae* L., caught in three bays in the area of Sevastopol. The study was conducted in 2012 in the three bays in the area of Sevastopol: Galubaya, Omega and Karantinna. The results indicate that the content of all elements studied was at a higher level than that observed in the uncontaminated basins. The average concentration of Ni, Pb and Zn in water was respectively: 1.199; 14.62 and 64.32 $\mu g \cdot dm^{-3}$. The content of the elements studied in the examined fish larvae Atherinidae ranged from 6.413 to 19.65 mg Ni kg^{-1}, from 1.749 to 5.393 mg Pb kg^{-1} and from 97.16 to 215.24 mg Zn kg^{-1}.

1 Introduction

One of the major effects of human impact on the environment is the degradation of water resources in quantitative and qualitative terms. The degree and rate of degradation of the aquatic environment depend on the human impact, the time of exposure, the degree of transformation of the environment and the degree of environmental buffers. The assessment of environmental hazard resulting from human impact is made based on the inventory of the degrading factors and of the monitoring of the environment. Monitoring of pollution of water bodies is an important part of environmental and natural resources management policies. The use of living organisms in the assessment of the level of pollution of the aquatic environment is one of the most effective methods of environmental quality assessment [1]. The level of bioaccumulation of trace elements in living organisms allows evaluating the risk associated with their excess to the functioning of ecosystems and the safety of food produced in such environments [2,3] Aquatic ecosystems play an important role in the proper functioning of life on Earth. They participate in the global cycle of water, as well as of many elements, and play an important role in the primary production on Earth and binding of carbon from the atmosphere. They are breeding place of many environmentally and economically valuable animals and plants. Bodies of water are strategic parts of the economy of many countries. They provide opportunities for the development of water transport, tourism, aquaculture food sourcing, and waste disposal. The benefits that come from the exploitation of bodies of water often cause their unreasonable exploitation, leading to their degradation in quantitative and qualitative terms. The degree and rate of degradation of the aquatic environment depend on the level and duration of human impact, as well as the effectiveness of the compensatory treatments [4]. The larvae and juvenile fish specimens are becoming more widely used in bioindicative research [5,6] Despite the difficulties associated with obtaining the testing material, determining the accuracy of growth and development of fish larvae, as well as their trace element accumulation levels, allows a more complete interpretation of the toxicological effects of environmental pollution. Determining the possibility of successful breeding of animal species is an important parameter, which reveals their position in the changing environmental conditions. The most important indicators of the ecological condition of a body of water are: the number of fish, their health, growth and reproductive success level [7].

The aim of the study was to evaluate the content of Ni, Pb, Zn in the water and larvae of fish of the genus L. *Atherinidae*, caught in three bays in the area of Sevastopol. The second aim was to determine the level of bioaccumulation of these elements in the organisms used in the study.

2 Material and methods

The study was conducted in 2012 in the three bays in the area of Sevastopol: Galubaya, Omega and Karantinna. Fish larvae of the genus *Atherinidae* were caught in July 2012. The larvae were caught in the coastal shallows, at a depth of 1 m, using fishing nets. The collected material was transported to the laboratory immediately after

* Corresponding author: anna.szelag-sikora@ur.krakow.pl

collection, where it was preserved by drying. Before drying, the fish samples were washed thoroughly with distilled water. Fish larvae samples were subjected to wet digestion in a closed system with the use of microwave energy. The digestion was carried out using a microwave system Anton Paar Multivawe 3000. The biological material was dissolved in a mixture of HNO_3 and H_2O_2 at a ratio of 5: 1, v/v. The concentration of the analysed elements in the samples was determined using an atomic emission spectrometry with inductively coupled plasma in an apparatus Optima 7600 DV Perkin Elmer. The obtained results of mean element content were compared using the student's t-test, at the significance level of p = 0.01. Based on the results, calculated were the bioaccumulation factors of individual elements. The bioconcentration factor was calculated by dividing the concentration of the element in the dry matter of the larvae used in the experiment by the content of these elements in the water.

3 Results and discussion

The results of the study indicate significant differences in the content of soluble forms of the analysed elements in the water taken from each of the bays in the area of Sevastopol. The concentration of nickel in the water sampled from the Karantinna Bay was almost double compared to the water from the other bays. Lead content was also the highest in the water from the Karantinna Bay, twice as high in comparison with the water sampled from the Galubaya Bay (Table 1). Concentration of zinc in the tested water samples ranged from 51.42 to 85.67 mg µg Zn · dm^{-3} (Table 1). There were no statistically significant differences between the content of this element in the water in the Karantinna and Galubaya bays. The amount of this element found in the water from the Omega Bay was higher by more than half.

Table 1. The content of elements in water from bays [µg· dm^{-3}].

Points of sampling	Ni	Pb	Zn
Karantinna	4.653b	19.47c	55.86a
Omega	2.561a	14.24b	85.67b
Galubaya	2.383a	10.15a	51.42a
Detection limit	0.502	2.100	0.295

Oberholster et al [8] report the content of zinc in water from the lake Loskop in southern Africa in the range of 30 to 90 µg Zn · dm^{-3}. The concentration of nickel in the water of the river Ripol in Spain, impacted by anthropogenic pollution, was between 1.47 and 14.2 µg Ni · dm^{-3} [9]. Kahle and Zauke [10] report the content of zinc in the water from the Arctic Ocean at 0.3 µg Zn dm^{-3}, which can be considered to be the natural content.

The use of fish larvae in this biomonitoring study was due to the fact that they are better indicators of changes in the quality of the environment in comparison to adult specimens, as pointed out by other authors [11,12].

Table 2. Statistical parameters of the results (mean, minimum value, maximum value mg·kg^{-1}).

	Ni	Pb	Zn	Ni	Pb	Zn	Ni	Pb	Zn
	Karantinnaya			Omega			Galubaya		
mean	15.49b	3.425b	122.0a	9.96a	2.301a	176.9c	8.71a	3.109b	148.0b
min	11.02	2.606	97.16	6.410	1.749	138.2	7.302	2.231	137.9
max	19.65	5.393	146.8	13.58	3.187	215.2	11.94	3.879	160.3
SD	3.299	1.145	22.56	2.754	0.561	37.20	1.843	0.720	9.237
%*	21.29	33.44	18.48	27.65	24.40	21.02	21.17	23.15	6.241

Different letters at mean values indicate the statistically significant differences between the samples from each of the bays at p = 0.01.
* coefficient of variation

The level of nickel in the surveyed fish larvae of the genus L. *Atherinidae* ranged from 6.410 to 19.65 mg Ni · kg^{-1}. The concentration of this element in the larvae caught in the bays Karantinna and Omega did not differ significantly and was approx. 9 mg Ni · kg^{-1}. The accumulation of nickel in larvae from the Karantinna Bay was almost twice as high (Table 2). The differences in the nickel content in the larvae from the individual bays were similar to those found in water, and therefore the bioaccumulation factors of this element were at a similar level. The bioaccumulation factor of nickel ranged from 3329 to 3890. Similar quantitative relationships between elements content in the tested larvae were found in the case of lead. Its average amount in organisms caught in the Karantinna Bay was 3.425 mg Pb · kg^{-1} and was higher by approx. half as compared to the amount of this element in the larvae of the Omega Bay, in which 2.301 mg Pb · kg^{-1} was determined (Table 2). The average content of this element in the studied organisms from the Galubaya Bay was 3.109 mg Pb · kg^{-1} and did not differ statistically from the values of this parameter found in the case of the Karantinna Bay. The zinc content in the surveyed fish larvae ranged widely from 97.16 to 215.25 mg Zn kg^{-1}. Its largest accumulation was found in the organisms caught in the Omega Bay, and the larvae caught in the Karantinna Bay contained the least amount of the metal. The value of the bioaccumulation factor of zinc ranged from 2065 to 2878. The greatest value of this parameter was found in the larvae from the Galubaya Bay, and the lowest bioaccumulation of zinc was found in the larvae from the Omega Bay.

The research has indicated considerable accumulation variability of the analysed elements in the investigated fish larvae. The value of the relative standard deviation of the content of Ni, Pb and Zn in all the studied samples was accordingly: 34.64; 31.36 and 22.29%. The observed differences in the content of the analysed elements are not large, and are much smaller than reported in the literature. Many authors in this field draw attention to much greater differences in the concentrations of the analysed elements in fish from the same area, as compared to the results of their own research [13, 14, 15, 16]. The greatest content variability in the samples taken from the respective bays was observed in the case of lead. The relative standard content deviation of this element in the larvae from the bays Karantinna, Omega and Galubaya was respectively: 33.44; 24.40 and 23.15% (Table 2). Of all analysed elements, the lowest value of the coefficient of content variation was found in the Galubaya Bay. Among the surveyed areas, the least impact of human pressure on the quality of the environment is observed in the Galubaya Bay. This is due to a lower amount of pollutants discharged directly into the bay, and a greater coefficient of water exchange with the open sea [17]. In the bodies characterized by higher levels of human pressure, usually observed are greater differences in the content of trace elements in fish, as compared to the uncontaminated areas [13]. In our study, the content of elements is high and typical of contaminated environments. The content of lead per dry weight of several species of fish caught in several places in coastal areas in the southern part of China ranged from 0.02 to 0.8 mg · kg^{-1}, of zinc: from 4 to 13 mg kg^{-1}, and of nickel: from approx. 0.5 to approx. 1.5 mg kg^{-1} [14]. On the other hand, the concentration of nickel in various species of fish from the Yangtze River ranged from 0.07 to 6.35 mg kg^{-1} DM [18]. Leung et al. [16] reported concentrations of zinc, nickel and lead in several species of fish caught in the area of the Pearl River Delta in China at a similar level to that obtained in the present study. The authors emphasize that the researched area is characterized by a high level of pollution. Canli and Atli, [19] report lead content at 2.98 to 6.12 mg · kg^{-1}, and zinc at 16.48 to 37.39, in the muscles of fish caught in the Mediterranean Sea off the coast of Turkey, in areas with high human pressure. The bioaccumulation factor of the elements is an important indicator of the environment quality from the point of view of the threat of including them into trophic chains. In the present study, the bioaccumulation coefficient expressed as the ratio of the content of the element in the dry weight of fish larvae to that in water. The values of the bioaccumulation coefficients in the organisms used ranged from 400 to 3890 (Figure 1). The lowest value of this parameter was found in the case of lead, and the largest bioaccumulation factor was found for nickel. The mean value of the bioaccumulation factor of nickel 3624, of lead 214, and of zinc 2376. It was found that in relation to the literature data the significant differences in the values of this parameter in the larvae of fish from individual bays were low. The variation of the bioaccumulation factor of the studied elements ranged from approx. 7 to 35%. The bioaccumulation factor of Pb and Zn in the muscle of *Cyphocharax voga*, a freshwater bentos-feeding fish living in the River Sinos in Brazil in areas with different levels of human pressure was respectively: from 2514 to 9090 for zinc and 141 to 247 for lead. The values of the bioaccumulation factor of these elements in pelagic fish were usually several times smaller [20]. The authors found a significant negative correlation between the level of environmental pollution and the bioaccumulation factor of trace elements.

The values of the bioaccumulation factor of zinc in the muscles of several species of fish from the Danube in Serbia were, respectively, approx. 500 and 6000 [21].

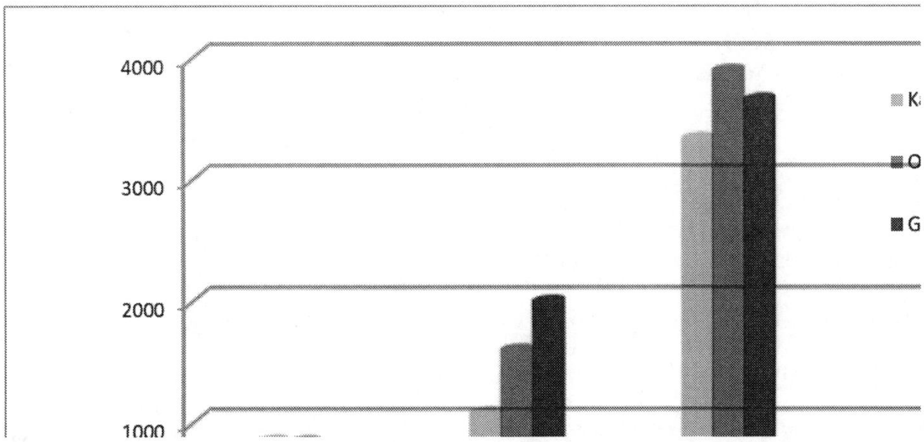

Figure 1. The bioaccumulation coefficient in the bodies of the larvae.

4 Conclusion

1. The largest nickel and lead content in the water were found in samples taken from the Karantinna Bay, and of lead, in the water sampled from the Omega Bay. The least amount of all the analysed elements was found in the Galubaya Bay.
2. The largest, nickel and lead content in the examined fish larvae were found in samples taken at the Karantinna Bay, and of lead, in organisms caught in the Omega Bay.

3. The concentration of the studied elements in water was, in sequence from the largest: Zn> Pb> Ni, while in the larvae of fish *Atherinidae L.*: Zn> Ni>Pb.
4. The value of the bioaccumulation coefficient of the analysed elements in *Atherinidae L.* fish larvae was, in sequence from the largest: Zn>Co>Pb.
5. The results of the study showed significant differences in the concentration of all elements in samples taken from the various bays, and large differences in the value of bioaccumulation factors.

References

1. Szeląg-Sikora, M. Niemiec, J. Sikora. J Elementol **21** (3), 915–926 (2016)
2. F. Galgani, F. Claro, M. Depledge, C. Fossi. Mar Environ Res **100**, 3–9 (2014)
3. N. Van Praet, M. De Jonge, R. Blust, R. Stoks, L. Bervoets. Sci Total Environ **484** (15), 92–101 (2014)
4. M.N. Filimon, R. Popescu, F.G. Horhat, O.S. Voia. Knowl Manag Aquat Ec **417** (30), 9 (2016)
5. M. Hundt, M. Schiffer, M.Weiss, B. Schreiber, C.M. Kreiss, R. Schulz, R. Gergs. Knowl Manag Aquat Ec **416** (27), 14 (2015)
6. M. Niemiec, N. Kuzminova, M. Chowaniak. J Elementol, **21**(3), 769–779 (2016)
7. M. Lenhardt, I. Jarić, Ž. Višnjić-Jeftić, S. Skorić, Z. Gačić, M. Pucar, A. Hegediš. Knowl Managt Aquatic Ecosyst **407**, 02 (2012)
8. P.J. Oberholster, J.G. Myburgh, P.J. Ashton, J.J. Coetzee, A.M. Botha. Ecotoxicol. Environ. Saf. **76** (1) 13 4 –141 (2012).
9. A. Maceda-Veiga, M. Monroy, E. Navarro, G. Viscor, A. de Sostoa. Sci Total Environ **449**, 9–19 (2013)
10. J. Kahle, G.P. Zauke.Mar Environ Res **55** (5), 359–384 (2003).
11. Kuźnia M., Jerzaka W., Łykob P., Sikora J. Journal of Power Technologies. **95(2)**,158-165, (2015)

K.W. Chu, K.L. Chow. Aquat Toxicol **61**, 53–64 (2002)

12. X. Kong, H. Jiang, S. Wang, X. Wu, W. Fei, L. Li, G. Nie, X. Li. Chemosphere **92** (11), 1458–1464 (2013)
13. F. Korkmaz, R. Keser, N. Akçay N, S. Dizman. Chemosphere **87**, 356–361 (2012)
14. S.L. Wang, X.R. Xu, Y.X. Sun, J.L. Liu, H.B. Li. Mar Pollut Bull 76, 7–15 (2013)
15. M. Niemiec, B. Wiśniowska-Kielian. J Elem **18** (3), 425-436 (2013)
16. H.M. Leung, Leung AOW, H.S. Wang, K.K. Ma, Y. Liang, K.C. Ho, K.C. Cheung, F. Tohidi, K.K.L. Yung. Mar Pollut Bull **78** (1–2), 235–245 (2014)
17. E.V. Pavlova, V.V. Murina, R.B. Kemp, J.G. Wilson, V.P. Parchevsky. Морський екологічний журнал, **2**, 63–77 (2007)
18. J. Fu, X. Hub, X. Tao, H. Yu, X. Zhang. Chemosphere **93**: 1887–1895 (2013)
19. M. Canli, G. Atli. Environ Pollut **121** (1), 129–136 (2003)
20. P. Weber, E.R. Behr, C. De Lellis Knorr, D.S. Vendruscolo, E.M.M Flores, V.L. Dressler, D.V.L. Baldisserotto. Microchem J **106**, 61–66 (2013)
21. S. Subotić, S. Spasić, Ž. Višnjić-Jeftić, A. Hegediš, J. Krpo-Ćetković, B. Mićković, S. Skorić, M. Lenhardt. Ecotox Environ Safe **98**, 196–202 (2013)

Methane fermentation as a possibility of utilization and energy – efficient management of sewage sludge

Anna Smurzyńska[1*], *Kamil* Kozłowski[1], *Marta* Cieślik[2], *Michał* Brzoski[1], *Dawid* Chełkowski[1], *Jakub* Mazurkiewicz[1] and *Ewa* Woźniak[1]

[1]Poznan University of Life Sciences, Institute of Biosystems Engineering, Wojska Polskiego 50, 60-637 Poznan, Poland
[2]Poznan University of Life Sciences, Department of Biotechnology and Food Microbiology, Wojska Polskiego 48, 60-637 Poznan, Poland

Abstract. Sewage sludge contains organic and biogenic substances that should be used. These materials may also contain toxic substances and heavy metals, and therefore require proper management in accordance with applicable law. Sewage sludge disposal is important for the protection of the environment, but also for the costs associated with the disposal of sludge. There are many technologies that are designed to use the properties of sewage sludge. It should be emphasized, however, that the proposed sewage sludge management technologies require investment outlays. In the present situation, research experiments have been carried out in the Environmental Technology Laboratory of the University of Environmental and Life Sciences in Poznan, indicating the possibility of efficient use of sewage sludge in the methane fermentation process. The use of sewage sludge in the fermentation mix allows the fertilizer to be obtained in the form of fermentation pulp, which also has the same properties as natural fertilizers. It should also be noted that the effect of physicochemical changes is the production of biogas, which is the source of electricity and/or heat. The economic sustainability of the use of sludge (taking into account the current costs of their disposal) is unequivocally positive when co-fermented in an appropriate proportion. The results show that sewage sludge can be successfully used in biogas plants as a co-substrate of fermentation mix. Therefore, the use of sewage sludge in biogas plants not only allows for the safe management and production of fertilizers, but also for the "green" energy.

1 Introduction

Biodegradable waste is a serious environmental hazard. Bearing in mind the protection of ecosystems, biodegradable waste require a rational economy. Examples may be studies showing the possibility of utilizing unsuitable plants from hybrid sewage plants for energy purposes [12]. Applicable law provides for a number of regulations on the proper management of waste from agriculture but also from industry. As indicated by data, there is a lack of interest and technological possibilities of using organic waste as a material for the production of other products. That is why a large part is stored or subjected to selected treatments for their further agricultural use [2].

In particular animal feces require proper management. They are produced in large quantities during intensive livestock production, which is present all over the world. Studies indicate that their improper management leads to contamination of soil but also to groundwater and surface waters [18]. This is due to accumulation in high concentrations of macro-contents especially nitrogen, but also from the presence of heavy metals, which poses a serious threat to the environment, animals and people [16,23,24]. In addition to livestock buildings, but also in the storage or use of animal manure, emissions of greenhouse gases and odors are generated [5]. Appropriate manure and manure management is governed by construction law specifying the way they are stored. On the other hand, their use is determined by the Fertilizer and Fertilizing Act or the Code of Good Agricultural Practices.

Another sludge that requires proper management is sewage sludge, the amount of which increases with the natural increase. This substrate has a high level of organic and biogenic substances, which makes its agricultural use as fertilizer or for the restoration of green areas is a good solution [13]. However, sludge from sewage treatment plants also contain toxic substances and heavy metals, pathogenic bacteria and viruses, and parasite eggs, which according to law requires their proper disposal of [9, 10,14, 22]. It should be stressed that since 2016 the legal regulations prohibiting the deposition of sewage sludge have been successfully implemented since 2016 [19]. In view of the situation, they are looking for possibilities of their development, which will allow them to use their properties and at the same time to neutralize harmful effects on the environment. The most commonly used solution is liming or burning, which effectively helps to

*Corresponding author: anna.smurzynska@up.poznan.pl

neutralize the hazards of sewage sludge. However, heat treatment involves other hazards, namely the emission of dangerous gases, including sulfur and nitrogen oxides [3].

However, it is stressed that the use of the mentioned processing of the onerous substrate requires high expenditure on the operation. In view of this, another possible technology is to subject the sludge to methane fermentation. The use of this technology not only allows us to dispose of the onerous substrate but also to produce electricity and/or heat [8]. In addition, as shown by methane fermentation studies, the resulting fermentation pulp exhibits very good fertilizer properties without compromising the environment as opposed to raw slurry or sludge [6,7].

The aim of this work was to carry out the process of fermentation of sewage sludge. This allowed us to indicate their biogas and methane yield, which is to confirm the potential for utilization of the substrate and to indicate the cost-effectiveness of using sewage sludge in an anaerobic process.

2 Materials and methodology

2.1 Sewage sludge used in the experiment

The research material was sewage sludge which came from domestic wastewater treatment plant located in Wielkopolska. In order to test the biogas and methane yield of the selected substrate, a fermentation broth containing the desired microbial groups was required. In the experiment, it was a warm separated fraction of fermentation pulp from one properly functioning agricultural biogas plant. Samples were prepared with accuracy 0.5 g. All samples needed to be analyzed in respect to the correct chemical and physical parameters. This was indispensable to calculate the contents in mixture (sewage sludge + innoculum). The pH was measured using laboratory multi-meter CP-411 (Elmetron). Moreover, dry matter and organic dry matter were determined. It was necessary to calculate biogas production efficiency in following units – m^3/t of dry matter in normal conditions.

In the experiment the following standard methodology established by Polish Norms (PN) has been used: pH - PN-90 C-04540/01, dry matter PN-75 C-04616/01 and organic dry matter PN-Z-15011-3.

2.2 Place of research and research station

The study was conducted in Laboratory of Eco technologies – the biggest biogas laboratory in Poland, working within the Institute of Biosystems Engineering. The experimental methods have based on modified German standard DIN 38 414, while physical and chemical analytical methods based on Polish Standard System [11]. The analytical procedures concerning biowaste were also developed within several scientific projects financed within EU 6th Framework Program and Polish Ministry of Science and Higher Education in the years 2006-12.

The experiment of biogas production was conducted through anaerobic digestion in the set of fermentation chamber (Figure 1). Samples were tested in three replications. This biofermentor is commonly used for testing biogas and biomethane production efficiency.

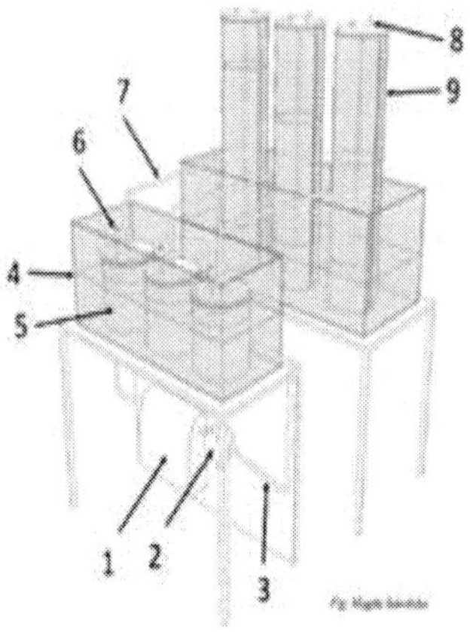

Fig. 1. Scheme of biofermentor for biogas production research:1 - water heater with temperature regulator, 2 - water pump, 3 - insulated conductors of calefaction liquid, 4 - water coat with temp. 39°C, 5 - biofermentor with charge capacity 2 dm^3, 6 - sampling tubes, 7 - biogas transporting tube, 8 - gas sampling valve, 9 - biogas volume-scale reservoir.

Methane fermentation experiments were carried out in stirred tank reactors constructed of glass in the Laboratory of Ecotechnologies. General rules for chamber work were based on the fermentation of organic substrate samples which were put into theglass chambers with 2 dm^3 capacity. Without oxygen presence and additive of fermentation inoculum the conditions present within the fermentation chamber allowed to create an ideal conditions for methane fermentation of the samples. Fermentors with samples were placed in water with regulated temperature (around 39°C) – similar to the real conditions of biogas plant. Biogas produced in each chamber was transferred to cylindrical store – equalizing reservoirs, filled in with liquid.

2.3 Biogas measurements

The volume of the produced gases has been measured every 24 hours. Gas composition has been checked out from at least each 1 dm^3 of the produced biogas. The concentration measurements of methane, carbon dioxide, oxygen, ammonia and hydrogen sulfide in the produced biogas have been carried out with the use of the absorption sensors working in an infrared and electrochemical sensor line. Consequently, each sample for biogas production was monitored for the gas compounds daily. The volume of biogas production and the methane content of biogas were calculated in the Excel sheet. Gas-monitoring system has been calibrated each week using calibration gases.

2.4 Methodology of calculations

Studies of sewage sludge biogas and methane were carried out in the Ecotechnology Laboratory of the Institute of Biosystems Engineering at the University of Life Sciences in Poznań. The experiments were performed on the basis of internal procedures based on the adopted DIN 38 414-S8 and VDI 4630 standards. A detailed description of the methodology of the performed tests was presented in Cieślik et al. [4].

A computational analysis was conducted to estimate the amount of electricity and heat production. The revenue from the energy sales and the balance of corn silage and sewage sludge were calculated. This paper uses the methodology of calculations presented by Szulc and Dach [20].

The following data were included in the calculation:

- purchase cost of corn silage - PLN 140, according to data provided by the owners of biogas plants using this substrate,
- yellow certificate price - 120 PLN [15],
- green certificate price - 33.55 PLN [1],
- electricity sales price – 169.57 PLN [17],
- heat sales price (calculated on the basis of the price of heat obtained from coal-fired coal furnace).

3 Result and discussion

3.1 Daily gas production

Figure 2 shows the dynamics of methane production in fermentation reactors with a fermentation mixture content of 1200 g.

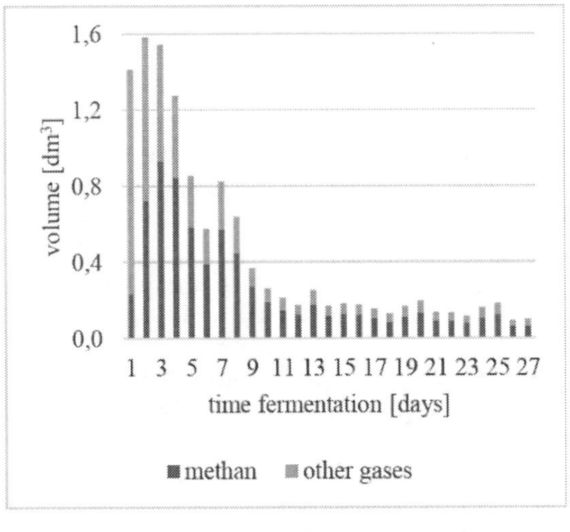

Fig. 2. Daily production of biogas.

The highest production of methane and other gases has already been observed in the second and third day of the experiment, which demonstrates the high content of easily biodegradable components in the recycled substrate.

By analyzing the course of biogas production, the influence of factors inhibiting methane fermentation of sewage sludge is not observed. Therefore, this substrate can be successfully used in biogas plants.

3.2 Cumulative emissions of methane and biogas

The following Table 1 summarizes the biogas, methane and methane concentration in biogas from sewage sludge undergoing mesophilic methane fermentation.

Table 1. Content methane, cumulated methane and biogas on dry matter basis.

Parameters	Unit	Value
Methane percent	[%]	61.27
Cumulated methane	[$m^3 \cdot t^{-1}$ TS]	160.00
Cumulated biogas	[$m^3 \cdot t^{-1}$TS]	261.14

In the mesophilic sludge process, 261.55 m^3 of biogas with average methane content of 61.27% was produced (Table 1), calculated as ton of dry substrate.
It is noted that the data collected are close to those of the biogas and methane yields of slurry, which is the basis for methane fermentation in many biogas plants (Table 2).

Table 2. Cumulated biogas and methane of slurry and maize silage.

Substrate	Methane percent [%]	Cumulated methane [$m^3 \cdot t^{-1}$TS]	Cumulated biogas [$m^3 \cdot t^{-1}$TS]
Slurry	77.66	208.53	268.31
Maize silage	53.4	261.50	489.74

On the other hand, the methane and biogas yields obtained are lower than the most commonly used substrate, enriching the fermentation mixture, in the forms of corn silage by 1.63 and 1.88 respectively [4]. Consequently, sludge can only be used as a co-substrate for the fermentation mixture. Another argument for using the substrate to be tested in methane fermentation is the cost of obtaining them.

Table 3. Cost, amount of electricity and heat Energy, revenue from the sale of Energy and heat for sewage sludge and maize silage.

The presented data demonstrate the high cost of obtaining maize silage for biogas plants. This is due to its many uses and thus becomes a substrate unattractive for biogas plants. In turn, sewage sludge is a substrate that can be obtained for free, due to the possibility of its safe disposal in biogas plants. The presented balance in Table 3 indicates that the total methane fermentation revenue in the case of using sludge is PLN 35.88, whereas in the case of using corn silage the amount of PLN 22.27 is to be paid.

Volume	Unit	Sewage sludge	Maize silage
Cost of substrate	PLN	0	140
Amount of electricity	$MWh \cdot t^{-1}$ FM	0.08	0.28
Amount of heat energy	$MWh \cdot t^{-1}$ FM	0.10	0.31
Revenue from the sale of electricity	$PLN \cdot t^{-1}$ FM	27.17	89.15
Revenue from the sale of heat	$PLN \cdot t^{-1}$ FM	8.71	28.58
Balance	$PLN \cdot t^{-1}$ FM	+35.88	-22.27

4 Conclusions

During the decomposition process under anaerobic sewage sludge, no process inhibitors are observed.
It is possible to dispose of sewage sludge in the methane fermentation process. Its biogas yield makes it a good feedstock for the fermentation mix consisting of other higher yields.

Methane fermentation turns out to be a good solution for the management of sewage sludge. It brings benefits to waste water treatment plant by efficiently disposing of unnecessary waste water treatment products. In addition, biogas plants extract sludge that enrich the fermentation mixture without incurring additional costs.

References

1. Act on renewable energy sources, art. 43 ust. 3
2. E. Adrjanowicz, M. Janczar, J. Pietkiewicz, Przemysł Fermentacyjny i Owocowo-Warzywny, **10**, 20-21(2000)
3. J. Bień, E. Neczaj, M. Worwąg, A. Grosser, D. Nowak, M. Milczarek, M. Janik, Inżynieria i Ochrona Środowiska, **14**, 4, 375-384 (2011)
4. M. Cieślik, J. Dach, A. Lewicki, A. Smurzyńska, D. Janczak, J. Pawlicka-Kaczorowska, P. Boniecki, P. Cyplik, W. Czekała, K. Jóźwiakowski, Energy **115**(2), 1495-1502 (2016),DOI: 10.1016/j.energy.2016.06.070
5. A.L. Collins, Y.S. Zhang, M. Winter, A. Inman, J.I. Jones, P.J. Johnes, W. Cleasby, E. Vrain, A. Lovett, L. Noble, Science of The Total Environment, **547**, 15, 269-281 (2016)
6. W. Czekała, J. Dach, J. Przybył, P. Boniecki, A. Lewicki, P.C.R. Carmona, D. Janczak, H. Waliszewska, *Proceedings of the 2nd International Conference on Energy & Environment: Bringing Together Engineering and Economics*, 547-553 (2015), ISBN: 978-989-97050-4-3
7. W. Czekała, S. Bartnikowska, A. Lewicka, A. Bugała, Z. Zbytek, A. Lewicki,*3rd International Conference on Structural Nonlinear Dynamics and Diagnosis*,83,04005 (2016), DOI: 10.1051/ matecconf/20166004005
8. W. Czekała, A. Smurzyńska, M. Cieślik, P. Boniecki, K. Kozłowski, *Energy And Clean Technologies Conference Proceedings*, **III**, 227-233 (2016)
9. A. Grobelak, A. Placek, A. Grosser, Bal Ram Singh, R. A. Åsgeir, A. Napora, M. Kacprzak, Journal of Cleaner Production, **155**, 1, 189-197 (2017), DOI: 10.1016/j.jclepro.2016.10.005
10. K. Gorazda, B. Tarko, Z. Wzorek, H. Kominko, A.K. Nowak, J. Kulczycka, A. Henclik, M. Smol, Environmental Research, **154**, 171-180(2017)
11. DIN 38 414, Bestimmung des Faulverhaltens "Schlamm und Sedimente" (BeuthVerlag GmbH, Berlin, 1985)
12. M. Gizińska-Górna, W. Czekała, K. Jóźwiakowski, A. Lewicki, J. Dach, M. Marzec, A. Pytka, D. Janczak, A. Kowalczyk-Juśko, A. Listosz, Ecological Engineering, **95**, 534-541 (2016),DOI:10.1016/j.ecoleng.2016.06.055.
13. E.Z. Harrison, S.R. Oakes, M. Hysell, A. Hay, Sci. Total Environ.,**367**, 481–497 (2006)
14. K. Ignatowicz, Environmental Research, **156**, 19-22 (2017)
15. Information of the President of the Energy Regulatory Office No. 26/2016 on unitary substitute fees for cogeneration in force in 2017
16. Leclerc, A. Laurent, Science of The Total Environment, **590–591**, 452-460 (2017)
17. Energy Law 2017. Information (No 6/2017) on the average quarterly price of electricity sold on terms other than those resulting from Art. 49a sec. 1 and 2 of the Act
18. G.S. Petkov, G.S. Kostadinova, S.A. Denev, G.S. Mihaylova, D.C. Pavlov, Applied Soil Ecology, **34**, 1, 10-18 (2006)
19. Regulation of the Minister of Economy of 16 July 2015 on the acceptance of waste for landfill. OJ 2015 pos. 1277.
20. R. Szulc, J.Dach, F. Adamczyk, D. Janczak, J. Lenarczyk, I. Rzeźnik, W. Rzeźnik, Z. Zbytek, Polskie Towarzystwo Inżynierii Rolniczej w Krakowie, 72-79 (2014)
21. K. Szwedziak, Inżynieria Rolnicza, **4**, 297-302 (2006)

22. T.P. Thomsen, Z. Sárossy, J. Ahrenfeldt, U.B. Henriksen, F.J. Frandsen, D.S. Müller-Stöver, Journal of Environmental Management, **198**, 1, 308-318 (2017)
23. X. Yang, Q. Li, Z. Tang, W. Zhang, G. Yu, Q. Shen, F.-J. Zhao, Waste Management, **64**, 333-339 (2017)
24. H. Wang, Y. Dong, Y. Yang, G. S. Toor, X. Zhang, Journal of Environmental Sciences, **25**, 12, 2435-2442 (2013)

Economic Assessment of the Technology Harvesting Maize Straw for Biogas Production

Dawid Wojcieszak[1,*], *Jacek* Przybył[1], *Jacek* Dach[1], *Maciej* Zaborowicz[1], *Żaneta* Staszak[2]

[1] Poznan University of Life Sciences, Institute of Biosystems Engineering, Wojska Polskiego 50, 60-625 Poznan, Poland
[2]Poznan University of Technology Faculty of Machines and Transport, Piotrowo 3, 60-965 Poznan, Poland

Abstract. The aim of the study was to determine and analyse the costs of harvesting and storage of maize straw for biogas production. The investigations showed that it was the most economical to harvest maize straw with a self-loading wagon equipped with a cutting system and to store it in a field prism. The cost of this technology amounts to €12.5 per Mg d.m. The cost of maize straw harvest with a field chopper and storage in a flexible silo was the highest, amounts to €126.9 per Mg d.m. The research findings can be used for estimating the profitability of harvesting maize straw for energy production and other industrial purposes.

1 Introduction

In the last decade there has been a 40% increase in the production of maize grains. At present the volume of production amounts to 1,030 million tonnes. In 2016 the total maize grain production in the European Union was 60.3 million tonnes. In comparison with 2015 it increased by 1.8 million tonnes [1, 2]. In 2015 maize grain production in Poland amounted to 3.16 million tonnes [3]. The production of maize grains leaves crop residue, such as stalks, leaves, cob cores and cob-covering leaves. The crop residue makes 47-50% of the dry weight yield of whole maize plants [4, 5]. There is 0.15 kg of cobs , 0.22 kg of leaves, 0.14 kg of husks and 0.5 kg of stalks per one kilogram of the dry weight of maize grains [4]. Our estimates show that the annual maize straw production in Poland may amount to as much as 4 million tonnes. For this reason recently there has been increasing interest in the production of biogas from maize straw [1, 6].

There is high potential of maize residue as a substrate forbiogas production in Poland [7-9]. In Poland maize is harvested for grain in late autumn (October-November), when the air temperature drops and the humidity rises. Under these conditions the moisture of maize grains usually ranges from 30% to 35%. The moisture of maize residue ranges from 47% to 66% [10]. This causes technological problems during harvesting. Apart from that, it is impossible to dry maize straw to the moisture which guarantees lossless storage of the biomass in the field. The maize straw moisture is most influenced by random factors related with the weather conditions. Ensilage is the only reasonable solution in the storage of maize straw.

The technology of harvesting and storage of maize straw to be used as a substrate for biogas production should guarantee that the biomass is free from contamination. Apart from that, it should be characterised by low costs per raw material weight unit.

The Institute of Biosystems Engineering, Poznan University of Life Sciences, Poland, prepared four technologies of maize straw harvesting and storage as part of the project of the Ministry of Science and Higher Education N313 270938 'The Technology of Harvesting and Storage of Maize Straw as an Energy Biomass and Structural Substrate for Composting'. The aim of the study was to make an economic assessment of the technology of harvesting and storage of maize straw to be used for biogas production.

2 Material and methods

2.1 Maize straw harvesting technologies

In 2014 maize straw harvesting and storage technologies were investigated at two locations in Greater Poland Voivodeship. The main machinery used in the technologies included a field chaff-cutter, a self-loading wagon and around baler. Maize straw was stored in a field prism, in a flexible silo, in film-wrapped bales and in net-wrapped bales in a field prism (Fig. 1). The combine harvester which was used for harvesting maize grains was not taken into consideration as an element of any of the technological variants presented.

Technology A – Maize grains were harvested by a combine equipped with a harvesting head, which collected cobs but did not shred stalks. Maize straw was harvested by a field chopper equipped with a six-row unit for harvesting whole maize plants. The harvested and shredded biomass was transported by tractors with

* Corresponding author: dawid.wojcieszak@up.poznan.pl

high volume trailers. The straw was condensed with a device for compressing silo contentin a flexible silo.

Technology B – Maize grains were harvested by a combine equipped with a harvesting head, which collected cobs and shredded stalks. Maize straw rolls were made with a rotary swat her. Then the straw was collected by a collecting trailer equipped with a cutting system. The straw was stored in a field heap covered with a silage film.

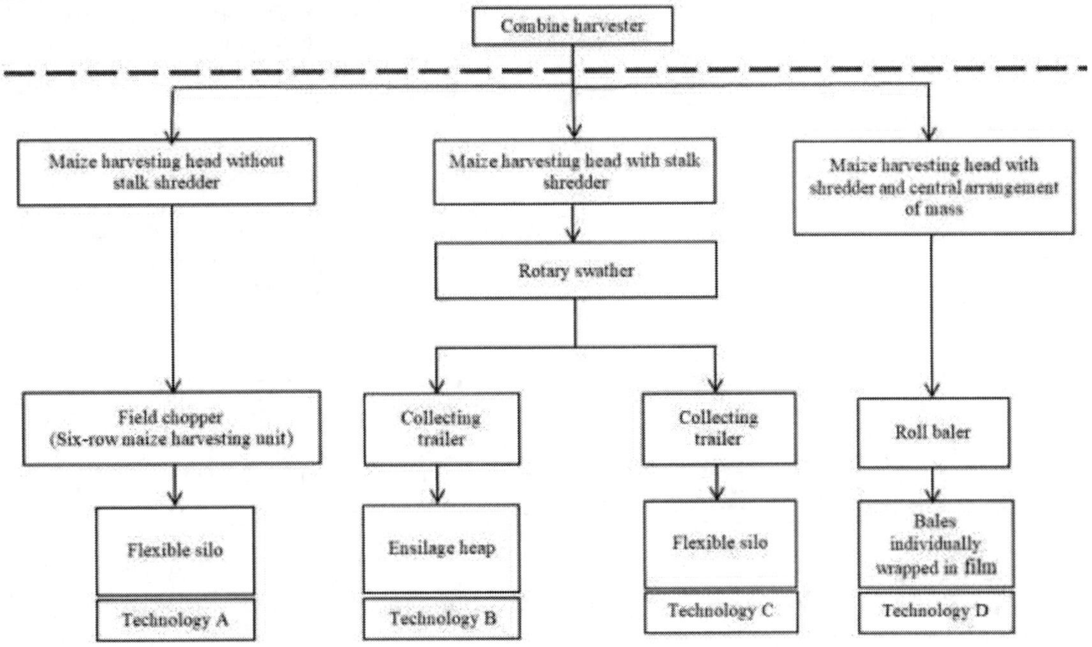

Fig. 1. A block diagram of maize straw harvesting and storage technologies.

Technology C – Maize grains were harvested by a combine equipped with a harvesting head, which collected cobs and shredded stalks. Maize straw rolls were made with a rotary swat her. Then the straw was collected by a collecting trailer equipped with a shredding system. The straw was placed in a flexible silo by means of a device for compressing silo content.

Technology D – Maize grains were harvested by a combine equipped with a harvesting head, which collected cobs, shredded stalks and made rolls of the shredded biomass. Next, net-wrapped cylindrical bales were formed. The bales were collected from the field and transported to the storage place, where they were wrapped in film.

2.2 Efficiency of machinery

The efficiency of the machinery and appliances used in the study was estimated by measuring the time and amount of work done. The time was measured to an accuracy of one second for the procedures which did not last longer than 20 minutes. The machinery uptime was measured at least during three work shifts.

The amount of fuel consumed by self-propelled machinery and tractors was measured with the 'full tank' method. The measurements were used to calculate the efficiency of the machinery and specific fuel consumption.

2.3 Machinery operating costs

The calculation method was used to estimate the costs of operation of technological means (machinery, tractors, appliances) and to analyse the costs of the technologies under study [11].

The costs of operation C_e of the machinery used in the maize straw harvesting and storage technologies were expressed as €/Mg d.m. and they were calculated according to the following formula:

$$C_e = \frac{C_m + C_u}{\left(\frac{W_{07}}{P_b}\right)} \quad (1)$$

where:
C_e – costs of operation of machinery, tractors, appliances [$€\cdot Mg^{-1}$ d.m.],
C_m – machinery maintenance costs [$€\cdot h^{-1}$],
C_u– costs of use [$€\cdot h^{-1}$],
W_{07}– efficiency of machinery [$ha\cdot h^{-1}$],
P_b– weight of maize straw harvested [Mg d.m.$\cdot ha^{-1}$].

The maintenance costs include depreciation, storage (garaging, maintenance) and insurance. The depreciation cost enables reconstruction of the value of machinery. According to the method, the cost of machinery storage amounts to 1% of the purchase price of a machine per year.The costs of maintenance of technological devices C_m were calculated according to the following equation:

$$C_m = \left(\frac{P_m}{U_y}\right) + \left(\frac{P_m S_f}{100}\right) + I_c \quad (2)$$

where:
C_m – costs of maintenance
P_m – price of machine [€],
U_y – machine uptime [h],
S_f – storage costs index [%],
I_c – insurance cost [€·year^{-1}].

Thecosts of use C_u include the costs of repairs, fuel, lubricants and accessory materials. The cost of repairs includes current repairs and overhauls:

$$C_u = \left[\frac{\left(\frac{P_m R_f}{U_y}\right)}{100}\right] + [1.06 F_u F_p] + [M_u M_p] \quad (3)$$

where:
C_u – costs of use
R_f – repair costs index [%],
F_u – fuel consumption [l·h^{-1}],
F_p – price of fuel [€·l^{-1}],
M_u – consumption of accessory materials [kg·h^{-1}],
M_p – price of accessory materials [€·kg^{-1}].

The costs of operation of machinery and tractors were calculated according to the current price lists provided by dealers. The prices of the models of machinery and tractors which are no longer manufactured were estimated according to the prices of their brand-new equivalents. The cost of compulsory liability insurance for tractors (€32.5 per year) and the price of diesel (€1.3 per litre) were included in the calculations. Table 1 lists the costs of repairs, uptime, specific fuel consumption and efficiency of machinery, which are necessary to calculate the costs of operation of technological devices.

Table 1. Data for the calculation of machinery operating costs.

Machinery	Power [kW]	Price [€]	Operating time [years]	Annual use [h]	Repair costs index [%]	Fuel consumption [l·h^{-1}]	W_{07} [ha·h^{-1}]
Technology A							
Field chopper	297	239.2		300	70	30.0	1.0
High-volume trailer	-	17.6		400	90	-	0.3
Tractor 1	114	51.7	25	600	90		-
Device for compressing silo content	-	95.7	15	300	100	-	1.0
Tractor 2	70	42.3		600	90	10.0	-
Technology B							
Rotary swather	-	33.2	20	460	80	-	2.5
Tractor 1	114	51.7	25	600	90	8.7	-
Collecting trailer	-	120.8	20	450	100	-	1.3
Tractor 2	188	196.6	25	1600	90	16.4	-
Telescopic loader	74	84.4	20	1200	70	8.0	3.0
Tractor 3	188	196.6	25	1600	90	16.4	-
Technology C							
Rotary swather	-	33.2	20	460	80	-	2.5
Tractor 1	114	51.7	25	600	90	8.7	-
Collecting trailer	-	120.8	20	450	100	-	2.3
Tractor 2	188	196.6	25	1.600	90	16.4	-
Device for compressing silo content	-	94.3	15	300	100	-	1.1
Tractor 3	93	76.5	30	1.350	600	10.0	-
Technology D							
Roll baler	-	29.4	20	300	80	-	0.8
Tractor 1	70	42.2	25	600	90	12.0	-
Trailer	-	12.9	25	200	90	-	3.0
Tractor 2	81	36.8	25	600	90	6.0	-
Front loader	-	4.2	20	320	100	-	3.0
Tractor 3	53	28.2	25	800	90	5.5	-
Bale wrapper	-	2.5	15	200	80	-	0.6
Tractor 4	37	19.6	25	800	90	4.5	-

3 Results

The maize straw harvesting and storage technology A, which involved the use of a field chopper, proved to be the most expensive. The total cost of this technology was €126.9 per Mg d.m. (Table 2).The transport of chaff from the field to the silo had the highest share in the cost structure of this technological solution. It amounted to 58% (€74.2 per Mg d.m.). Maize straw preservation in a flexible silo cost €34.9 per Mg d.m.(28%).The share of the cost of straw harvesting by a field chopper was the smallest as it amounted to 14% of the total cost of this technology (€17.7 per Mg d.m.).

The total cost of technology D was €25.6 per Mg d.m. Wrapping bales in film was the most expensive element in the cost structure of this technological solution as it amounted to 55% (€14.2 per Mg d.m.). It is possible to reduce the costs of this variant by using an inline wrapper, which needs about half of the amount of film that is necessary for an individual bale wrapper.

Table 2. Maize straw harvesting and storage – the cost structure.

Technological variant	**Costs**			
	Harvesting [€·Mg^{-1} d.m.]	**Transport [€·Mg^{-1} d.m.]**	**Storage [€·Mg^{-1} d.m.]**	**Total [€·Mg^{-1} d.m.]**
A	17.7	74.2	34.9	126.9
B	8.3		4.2	12.5
C	9.1		12.5	21.6
D	9.2	2.3	14.2	25.6

The costs of maize straw harvesting by a collecting trailer are very economical in comparison with the costs of the harvesting technology using a field chopper and a baler. The total cost of straw storage in a field prism was €12.5 per Mg d.m. When straw was ensilaged in a flexible silo, the cost was more expensive by €9.1 per Mg d.m. and it amounted to €21.6 per Mg d.m. A film-wrapped field prism was the most economical form of maize straw storage as it cost €4.2 per Mg d.m.

4 Conclusions

1. Maize straw can be effectively harvested with standard machinery for roughage and green forage harvesting.
2. It is the most economical to harvest maize straw with a collecting trailer equipped with a shredding system. The cost of this technology amounts to €12.5 per Mg d.m.
3. A film-wrapped field prism is the least expensive form of maize straw storage. The cost of storage in a film-wrapped prism amounts to €4.2 per Mg d.m.
4. It is possible to eliminate raking and thus reduce the costs of maize straw harvesting by using harvesting heads arranging the biomass into windrow centrally, between the wheels of a combine harvester.
5. The research findings can be used for estimating the profitability of harvesting maize straw for energy production and other industrial purposes.

References

1. S. Menardo, G. Airoldi, V. Cacciatore, P. Balsari. BiosystEng**129**, 352-359 (2015). http://dx.doi.org/10.1016/j.biosystemseng.2014.11.010
2. USDA. *Grain: World Markets and Trade*. Foreign Agricultural Service (2017)
3. CSO. *Agriculture in 2015*. (2016)
4. S. Sokhansanj, S. Mani, S. Tagore, A. F. Biomass Bioenerg**34**, 75-81 (2010). http://dx.doi.org/10.1016/j.biombioe.2009.10.001
5. A. Zych. *The vability of corn cobs as a bioenergy feedstock. A report of the West Central Research and Outreach Center* (2008)
6. Z. Zbytek, J. Dach, T. Pawłowski, A. Smurzyńska, W. Czekała, D. Janczak. MATEC Web Conferences**60,** 04008 (2016) doi: 10.1051/matecconf/20166004008
7. M. Cieślik, J., Dach, A. Lewicki, A. Smurzyńska, D. Janczak, J. Pawlicka-Kaczorowska, P. Boniecki, P. Cyplik, W. Czekała, K. Jóźwiakowski. Energy **115**, 495-1502 (2016) http://dx.doi.org/10.1016/j.energy.2016.06.070
8. J. Przybył, W. Kot, D. Wojcieszak, N. Mioduszewska, D. Karol. Agricultural Engineering **4(148),** 103-111 (2013)
9. M. Cieślik, J. Dach, P. Boniecki, D. Janczak, W. Czekała, A. Lewicki, P.C.R. Carmona, J. Pawlicka. *ICEE Internatonal Conference on Energy & Environmental* 300-306 (2015)
10. K. J. Shinners, B. N. Binversie. Biomass Bioenerg**31**, 576-584 (2007)
11. A. Muzalewski. *Koszty eksploatacji maszyn – wskaźniki eksploatacyjno-ekonomiczne nr 24* (Wydawnictwo IBMER) (2009)

The application of various software programs for mapping yields in precision agriculture

Mirosław Zagórda, *Maria* Walczykova*

University of Agriculture in Krakow,
Faculty of Production and Power Engineering

Abstract. Yields mapping can be done in GIS programs as well as in specialized software for agriculture. These can be programs designed for preparing informative and application maps and also software designed for integrated management of agricultural production that have a built-in GIS module. In the presented study, representatives of those three groups were taken into consideration – QGIS, Agro-Map, and Farm^{TM}Works Office (now Trimble Ag Software). The aim of the study was to evaluate the compatibility of the obtained yield maps in terms of basic statistical and geostatistical parameters, such as the average yield, the coefficient of variability, placement on the field zones with different yield ranges, and their percentage share in the area of the field surface. Calculations and visualization of the results were done for winter wheat, winter rape, and corn, cultivated in three-year rotation on one field. All three presented programs allow preparation of raw yield monitoring data so that yield maps produced with identical assumptions of interpolation parameters differ slightly. The difference in the use of these applications is mainly due to the further use of the resulting maps, the time required to learn the program, the task execution time, and the cost of purchase.

1 Introduction

In decision making and managing a farm, it is important to have knowledge about spatial variability of agricultural production and relationships between different phenomena. The basis of spatial dependence is the assumption that "everything is related to everything else, but near things are more related than distant things" [1], and this is fundamental for precision agriculture. One of the most widely used precision farming technologies is yield mapping [2]. Yield maps are a key component due to their usefulness in both the development and evaluation of precision management strategies [3]. These maps contain a wealth of information and can be an important tool for making informed decisions on paddock management [4]. Yield sensors in combined harvesters provide a cheap and readily available source of yield data for researchers, providing more data, and improving the effectiveness of yields as a measure for description of management zones [5]. Yield maps provide feedback information on how crops respond to certain soil and crop management practices and are used to determine recommendation rates for many inputs. Crop yield is used as an important integrator of varying crop and soil variables, such as nutrients, moisture, and pest problems. This establishes the grounds for managing farmlands based on crop and attributes [6]. Crop yield maps reflect stable yield patterns and annual random yield variation. Therefore, a single-year yield map is useful for the interpretation of possible causes of yield variation but may be of limited value for more strategic site-specific management decisions over medium- to long-term periods [7].

Yield monitor data contain systematic and random sources of measured yield variation, including more stable yield variability related to climate and soil-landscape features, variable management-induced yield variability, and measurement errors associated with the yield mapping process itself. Management induced variation includes random events that typically occur in small patches, such as planter skips, poor crop establishment, non-uniform fertilizer application, herbicide damage, lodging, or pest damage. The most significant measurement errors associated with raw yield datasets include grain flow and moisture sensor errors, inappropriate GPS recordings and combine movement, yield surges, operator errors, unknown crop width, other outlying values, and also data processing errors [2,4]. Based on the review of the literature, Arslan and Colvin [6] provided a classification of sources of errors in yield

*Corresponding author: rtwalczy@cyf-kr.edu.pl

data into four categories: sensor errors, errors due to operating conditions, operator errors, as well as data reduction and yield mapping errors. Researchers have reported that 10 to 30% [8] or even to 50% [3] of the observations in a given field contain significant errors and should be removed.

Methods for removing of these outliers from the raw yield data have not been standardized, although many different filtering techniques have been suggested to address specific error types. Sudduth and Drummond [3] describe several separate filtering techniques implemented in Yield Editor to remove measurements that are not consistent with the actual crop. Using these techniques, the following amendments can be done: correction for the transport time between the location where the crop is harvested and the location where flow rate is sensed; removing observations when entering and exiting each transect; removing data points collected at greater velocities and less than the specified limit; eliminating data where rapid velocity changes have occurred from one point to the next point along a transect; and deleting points with above and below the yield thresholds, removing data points with swath width readings lower than the minimum.

Dobermann et al. [7] processed raw data through a cleaning algorithm that deleted the following erroneous values: header status up, start and end pass delays (8 s) for both headlands and stop-and-go segments within the field, short segments (<12 data points), frequency distribution outliers (values outside the mean ± 3 SD), co-located yield records caused by global positioning system (GPS) drift, and local neighbourhood outliers. According to Gozdowski et al. [9], data cleaning of outliers is usually a method allowing the identification of erroneous observations for the whole set of data at once. Most often, the method based on the distribution presented as histogram is used. Nevertheless, the above cited authors found that, in order to get more reliable yield maps, the classical statistical method should be aided by the spatial statistics using Moran's I autocorrelation coefficient.

Yield maps can be processed in commercial or free GIS programs (e.g., QGIS, GRASS, or ArcView GIS), in commercial applications provided by the yield monitor producers for preparing not only yield maps but also informative and application maps for performing variable rate application of some agricultural supplies (for instance Agro-Map) or in programs designed to support the management of farm production processes like currently available 365FarmNet, Farm™Works Office (now Trimble Ag Software), Agrar Office, SGIS Agronomy, SMS™Software, etc. that have an in-built GIS module. As far as the GIS programs are concerned, a great deal of expertize is required, and these applications are typically used by companies providing services to agricultural producers in the area of precision farming. The other can be used directly by agricultural producers. Programs built as an integrated system of information on agricultural production space, having a built-in GIS module, usually contain several interconnected modules for field data, field operations, used machinery, supplies, machinery maintenance, employment, storage, business economics, soil mapping, yields, and proposals of field operations with possibility of processing application for different technological processes like fertilization, liming, planting etc. [10-11].

The aim of the study was to evaluate the compatibility of maps processed from the same raw yield data in three programs - QGIS, Agro-Map, and Farm™Works, in terms of basic statistical and geostatistical parameters and the fragmentation, i.e. the percentage share of zones with different yield ranges in the area of the field surface. The data were processed in the way as an ordinary user would do it, according to the method of processing indicated in each program. Methods of possible data cleaning and the choice of interpolation parameters in those three applications were presented as well as the use of created yield maps as a basis for preparing application maps for mineral fertilization. Calculations and visualization of the results were done for winter wheat, winter rape, and corn, cultivated in three-year rotation on one field.

2 Material and methods

The data for yield mapping were recorded for 3 years in a field of 19 hectares. The harvest was carried out with CLAAS Lexion 430 combine harvester equipped with an optical crop monitoring system. Data were recorded by the system at a frequency of 0.2 Hz at the average speed of 5.21 $km \cdot h^{-1}$, which resulted in the average distance between measurement points of 7.24 m. The spacing of points between successive runs resulted from the working width of the harvesting unit, and in the case of wheat and oilseed rape, it was, on average, 5.15 m (with a header width of B=5.2 m) and 4.46 m for corn (with a header width of B=4.5 m). The procedure for removing erroneous measurements was based on the assumption that the zero yield and the maximum yields, which exceed the potential yielding of the examined plant, would be discarded, taking into account the variety, the year of cultivation, and the site. The results of post-registration variety testing of COBORU (Research Center for Cultivar Testing) were applied and the following maximum limit values were assumed: for winter wheat, 12 $t \cdot ha^{-1}$ (for 14% m.c.) [14], winter oilseed rape 6 $t \cdot ha^{-1}$ (for 8 % m.c.) [15], and corn 16 $t \cdot ha^{-1}$ (for 14% m.c.) [16]. Filtration of other parameters such as working speed, mass flow, and grain moisture content depends on the capabilities of the used computer application. On the basis of the data prepared that way, basic statistics such as the average yield, standard deviation, coefficient of variation, and percentage share of zones with different yield ranges in the area of the field surface were determined.

3 Results and discussion

3.1. Processing the yield data in QGIS program

The yield data is first read from the monitoring device mounted on the combine by means of a simple view program that the manufacturer provides to the user free of charge. In case of the presented research, it was Agro-Map program, however, in a full version. In the QGIS, there is virtually no format on the list of spatial data formats that can be imported into the program in which the files are created in combine harvesters of different manufacturers, unless it is a file with a *.shp* extension. Therefore, it was necessary to convert the file and this was made in the MS Excel spreadsheet by adapting the data format to the requirements of the QGIS program. To remove the outliers, i.e. zero and unrealistic values of yields, histograms were made and the yields to the value of one standard deviation as well as exceeding the potential yield of the given crop were discarded (Fig. 1). For the data prepared that way, basic statistics was determined (Table 1). On the basis of this data, it can be stated that the most erroneous measurements took place during winter rape harvesting. Removal of the incorrect measurements resulted in a decrease in the yield variation coefficient, calculated from individual measurements, by approximately 46%, 44%, and 32%, respectively, for wheat, rape and maize. The Inverse Distance Weighted (IDW) interpolation method (Fig. 2) was used to prepare yield maps, which, besides kriging, is the most commonly used.

Five equal yield ranges that are considered optimum were chosen [12], and the values of interpolation parameters have been set. The interpolation parameters for modification are the power factor that determines the effect of the values and distances of the neighbourhood points on the interpolation results and the number of nearest points to the centre of the raster model cell [13]. Decisive in determination of these parameters is the combine working width, speed, and the frequency of yield data registration. The power factor was assumed to be 2, as a standard value [13]. On the basis of the prior information about the spacing of the measurements points, the pixel size of 10x10 m was used for interpolation, which corresponded to about 2 working widths of the harvester. At the surface of one pixel, there were about 4 measurement points. Based on the maps created (Fig. 2), the percentages of areas with different yield ranges (Table 2) were determined. As can be seen on the maps (Fig. 2), two yield areas with similar percentages were prevailing in winter wheat and oilseed rape, while in corn one yield range dominated with a high, almost 78% share, indicating very well-balanced yields.

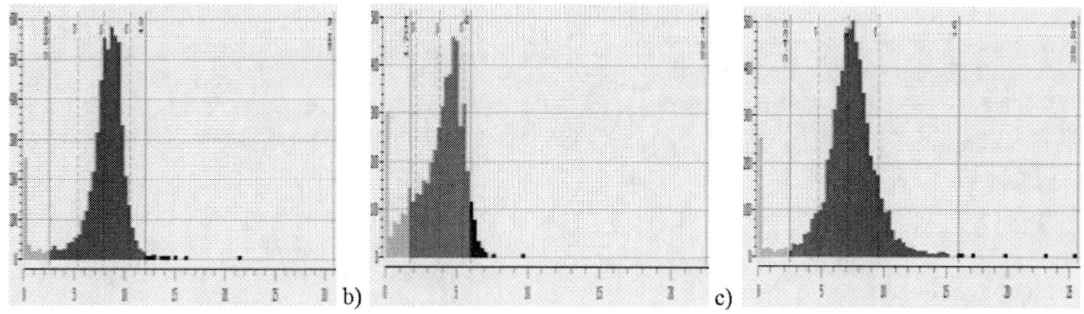
a) b) c)

Fig. 1. Histograms for: a) winter wheat, b) winter rape, c) corn.

Table 1. Descriptive statistics calculated from individual observations – results from QGIS program.

Specifications	Total number of observations	Removed observations %	Average yield $t \cdot ha^{-1}$ based on		Standard deviation SD $t \cdot ha^{-1}$ based on		CV %	
			total observ.	observ. cleaned from errors	total observ.	observ. cleaned from errors	total observ.	observ. cleaned from errors
Winter wheat	5563	7.55	7.852	8.300	2.589	1.455	32.97	17.53
Winter rape	5086	18.93	3.797	4.197	1.704	1.061	44.88	25.32
Corn	6437	5.75	7.126	7.453	2.432	1.743	34.13	23.39

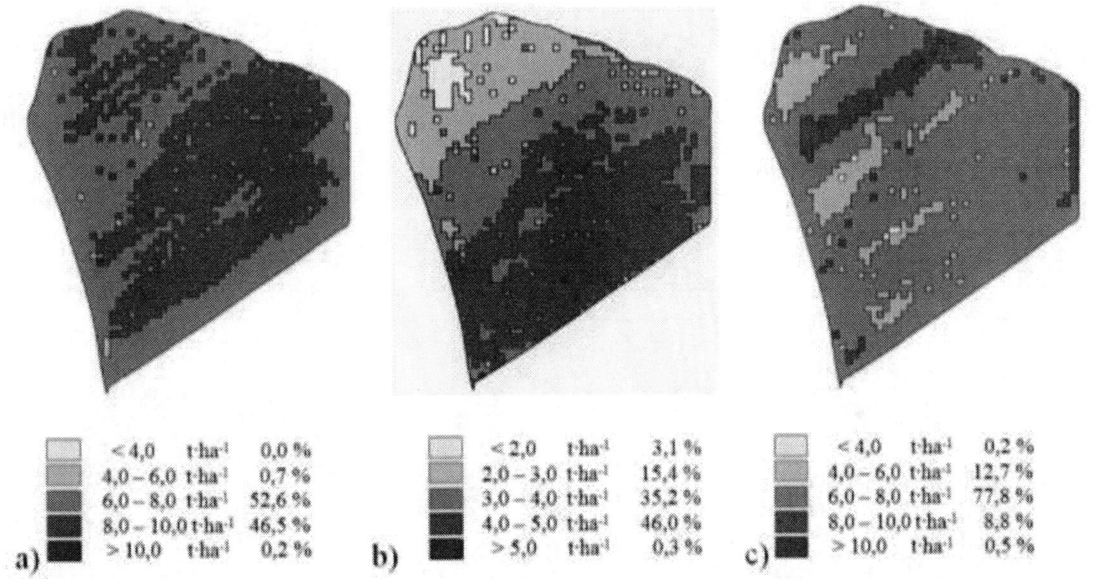

Fig. 2. Maps of the spatial distribution of yield prepared in QGIS: a) winter wheat, b) winter rape, c) corn.

Table 2. Percentage share of areas with different yield ranges – based on yield maps in Fig. 2, processed in QGIS program.

Specifications	Yield ranges (numbered according to the maps - from the smallest to the largest values)				
	1	2	3	4	5
Winter wheat	0.0	0.7	52.6	46.5	0.2
Winter rape	3.1	15.4	35.2	46.0	0.3
Corn	0.2	12.7	77.8	8.8	0.5

3.2. Processing yield data in Agro-Map

The raw yield data for this application may come from practically all currently used yield monitoring devices. Agro-Map program offers a kriging model for the creation of a contoured map, which is best for the cartographic representation of analyses. It provides a function that examines raw data after selection of the interpolation procedure. The yield values are organized into classes and the frequency limit, below which it is assumed that values are erroneous, is set at 3%, which proceeds invisibly [17]. When starting interpolation, the user assumes the resolution of the map (cell size), creates a legend, and introduces interpolation parameters (search radius, minimum, and maximum neighbours). Taking into account the recording parameters presented in the description of the work in the QGIS program, there were about 4 measurement points on the surface of 10 x 10 m. Consequently, it was reasonable to assume 20 m of the search radius and the number of adjacent points at the level of 16, i.e. interpolation was made for 3 runs on the distance of about 20 m. During map creation, the program proposes a range of yield values, without taking into account zero and unreasonable high yields. These suggested value limits can be altered at will if they do not appear sensible. In the dialog window, the moisture content of the harvested crop should also be corrected during the interpolation. Removing erroneous values caused by other factors (too small or too high working speed, zero moisture content of grain) before the interpolation is possible only manually in the source file, where the recorded data are in the form of a table. Table 3 presents the basic statistics computed for the two data cleaning variants - according to program hints and after manual removal of the measurements mentioned above. A method of removing erroneous measurements other than those used in QGIS application resulted in minor differences in mean yields as well as in the coefficient of variation. The maps in Figure 3 refer to the data cleaned from all measurements, which were considered to be incorrect in connection with the above-mentioned rules. Table 4 presents the percentages of areas with different yield ranges for both variants. As a consequence of the results included in Table 3, the differences between yields are insignificant in the variants, but this time, there is a greater fragmentation of the field than in the QGIS program (Table 4).

Table 3. Descriptive statistics calculated from individual observations – results from Agro-Map application.

Specifications	Removed observations %		Average yield $t \cdot ha^{-1}$		Standard deviation SD $t \cdot ha^{-1}$		Coefficient of variability CV %	
	(A) suggested by the program	(B) without all zero values	(A) suggested by the program	(B) without all zero values	(A) suggested by the program	(B) without all zero values	(A) suggested by the program	(B) without all zero values
Winter wheat	4.82	5.27	8.136	8.143	1.812	1.785	22.27	21.92
Winter rape	9.24	9.79	3.889	3.896	1.393	1.390	35.82	35.68
Corn	6.09	6.38	7.399	7.397	1.659	1.657	22.42	22.40

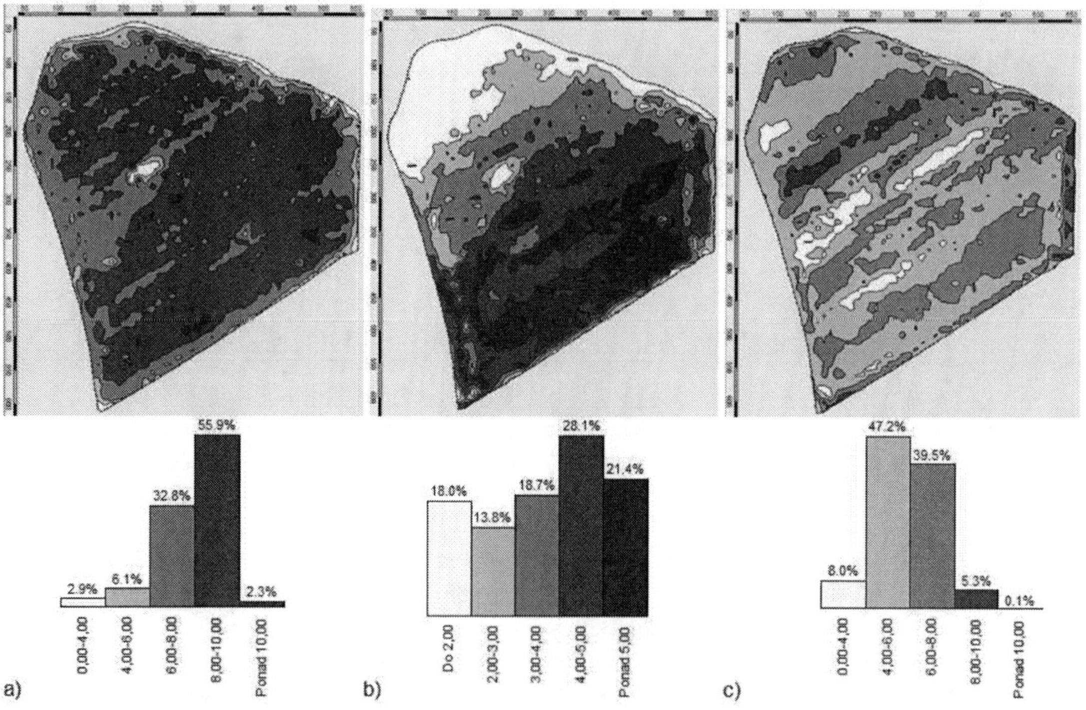

Fig. 3. Maps of the spatial distribution of yield with legend, prepared in Agro-Map: a) winter wheat, b) winter rape, c) corn.

Table 4. Percentage share of areas with different yield ranges – based on yield maps in Fig. 3, processed in Agro-Map application.

Specifications	Yield ranges (numbered according to the maps - from the smallest to the largest values)									
	1		2		3		4		5	
	(A)	(B)	(A)	(B)	(A)	(B)	(A)	(B)	(A)	(B)
Winter wheat	2.9	2.8	6.1	6.2	32.8	32.8	55.9	55.9	2.3	2.3
Winter rape	18.0	16.1	13.8	13.7	18.7	17.2	28.1	25.1	21.4	27.9
Corn	8.0	8.0	47.2	47.2	39.5	39.4	5.3	5.3	0.1	0.1

3.3. Processing data in Farm™Works Office

As in the Agro-Map application, raw yield data for mapping can come from practically all currently used crop monitoring devices (AGCO, AgLeader, Case IH, Claas Lexion, New Holland, Trimble, etc.). The first step in the mapping procedure is to discard the zero values of all parameters given in the filter by entering the minimum value of 0.1 and entering the maximum possible yield value for the given plant (Fig. 4a). After critical evaluation of the remaining parameters in the displayed window (Fig. 4b), further adjustments are made to the filter (Fig. 4c). As a minimum yield, the standard deviation is recommended. Table 5 includes values of the basic statistics obtained in this program, and they are similar to the results obtained in the QGIS program (Table 1). Both methods of interpolation proposed by the program, i.e. AVERAGE (AVG) and

IDW, were used for the mapping. The maps shown in Figure 5 were prepared by the IDW method, using interpolation parameters identical to those of QGIS. In order to show the impact of the interpolation technique on the fragmentation of the resulting yield map, i.e. the percentage share of areas with different yielding ranges, Table 6 also includes the results obtained by the AVERAGE method. It has been shown that the differences are minimal. In order to obtain more continuous boundary lines of homogeneous areas in the field, the edge smoothing function (possible in the range from 0 - 100%) and the function of removing small area polygons (size can be provided) were used. Figure 6 shows the results for yield maps at 50% smoothing, and removed areas less than 0.1 ha, and Table 7, the figures for both methods of interpolation.

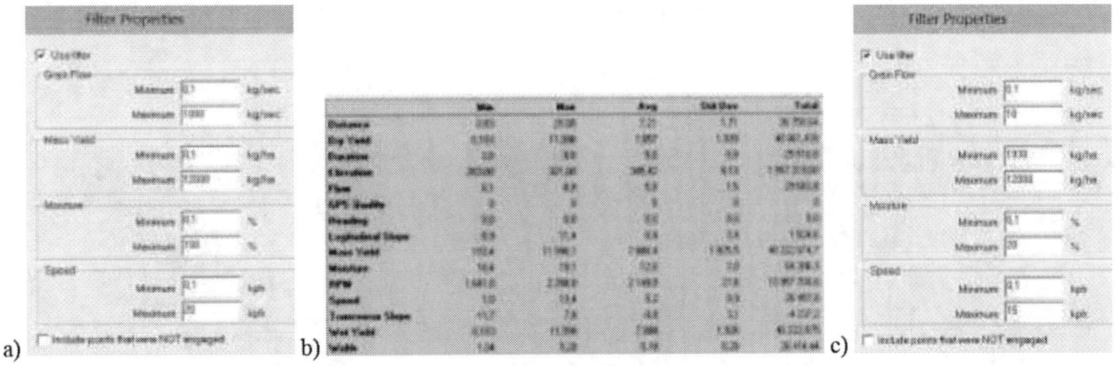

Fig. 4. Filtering data in tFarm™Works Office: a) filter pre-setting, b) statistics of the filtered data, c) correction of the filter settings.

Table 5. Descriptive statistics calculated from individual observations – results from Farm™Works Office.

Specifications	Removed observations %	Average yield t·ha^{-1}	Standard deviation SD t·ha^{-1}	Coefficient of variability CV %
Winter wheat	10.9	8.04	1.57	19.5
Winter rape	17.4	4.15	1.10	26.5
Corn	6.01	7.32	1.79	24.5

Fig. 5. Maps of the spatial distribution of yield with legend, prepared in Farm™Works Office: a) winter wheat, b) winter rape, c) corn.

Table 6. Percentage share of areas with different yield ranges – based on yield maps in Fig. 5, processed in Farm™Works Office.

Specifications	Yield ranges (numbered according to the maps - from the smallest to the largest values)									
	1		2		3		4		5	
	AVG	IDW	AVG	IDW	AVG	IDW	AVG	IDW	AVG	IDW
Winter wheat	0.3	0.3	4.5	5.3	36.7	36.6	57.4	56.5	1.1	1.3
Winter rape	7.8	7.4	13.2	13.4	20.8	21.5	37.3	37.0	20.9	20.7
Corn	0.5	0.8	14.5	15.9	63.2	60.9	19.2	19.1	2.6	3.3

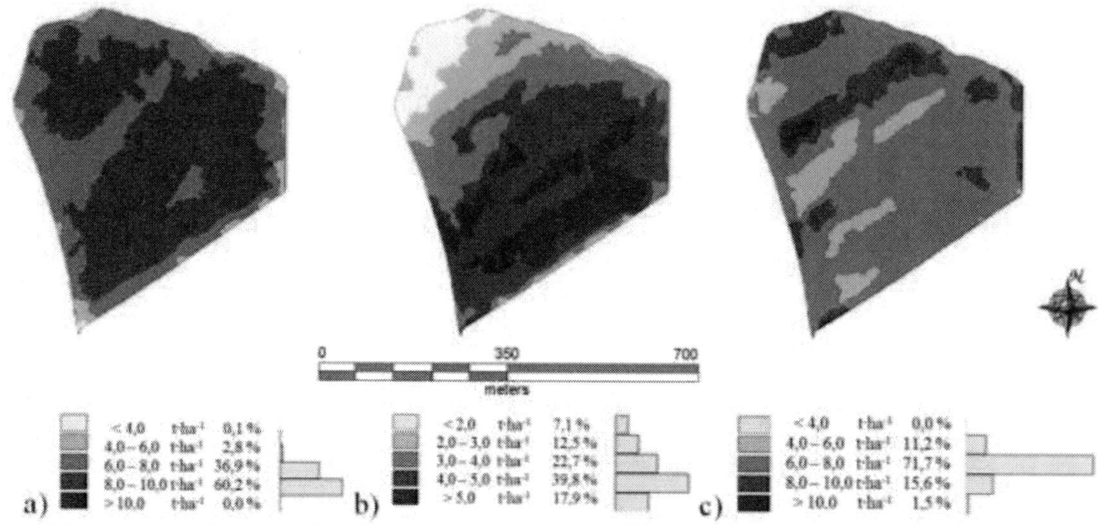

Fig. 6. Maps of spatial distribution of yield with legend, prepared in Farm™Works Office (smoothed 50%, with removed fragments smaller than 0.1 ha): a) winter wheat, b) winter rape, c) corn.

Table 7. Percentage share of areas with different yield ranges – based on yield maps in Fig. 6, processed in Farm™Works Office (smoothed 50%, with removed fragments smaller than 0.1 ha).

Specifications	Yield ranges (numbered according to the maps - from the smallest to the largest values)									
	1		2		3		4		5	
	AVG	IDW	AVG	IDW	AVG	IDW	AVG	IDW	AVG	IDW
Winter wheat	0.1	0.1	2.4	2.8	38.0	36.9	59.5	60.2	0.0	0.0
Winter rape	7.9	7.1	11.9	12.5	21.9	22.7	39.6	39.8	18.7	17.9
Corn	0.0	0.0	11.5	11.2	71.0	71.7	16.3	15.6	1.2	1.5

3.4. Use of yield maps for creating application maps for fertilization

The simplest method of preparing application maps is to prepare them on the basis of the previously made map of yield or soil nutrients content. On the basis of the prepared yield maps in all the discussed programs, it is possible to make application maps for mineral fertilization. One way to plan for mineral fertilization is to use the method of returning to the soil nutrients taken by crop that is based on information concerning yield (grain and straw) and the amount of mineral content contained therein [18] [19]. In the case of yield maps prepared in the presented programs, it is possible to assign the amount of taken nutrients calculated with the methods well-known in agronomy to the respective yield ranges. The yield map is converted into a fertilizer dosing (application) map that is exported to the spreader control panel. For technical reasons related to the application of fertilizers, application maps should be adapted to the equipment in terms of resolution (usually it is the spreader's working width), the format (file format supported by the dose control system), and the minimum size of polygons required for the variable rate application (e.g., the minimum size of the polygon in the field at the spreader with a working width of 20 m will be 20 m x 20 m, i.e. 0.04 ha). In Farm™Works Office and the QGIS, it is also possible to process data in such a way (smoothing out contours, removing small areas) so that the resulting yield map can be converted into the application map of fertilization by adding the appropriate fertilizer dose attribute to the particular yield ranges. The yield map prepared in Agro-Map with the use of the kriging method contains many small areas and requires additional processing by the user, which involves manually removing small areas on the application map. After that, the calculated fertilizer doses are assigned to the respective yielding ranges on the map.

5 Closing remarks

All three presented programs enable preparation of raw yield monitoring data so that yield maps produced with identical assumptions of interpolation parameters differ slightly. The difference in the use of these applications is mainly due to the further use of the resulting maps, the time required to learn the program, task execution time, and cost of acquisition. Of the three presented applications, it is only possible to add more layers of information on the field in order to determine the management zones as accurately as possible in the QGIS program. The Agro-Map application enables smart soil sampling on the soil nutrient content based on the created yield maps and preparing application maps for variable rate mineral fertilization as well as other supplies (seeds, herbicides, etc.). Farm^{TM}Works Office software also enables assessment of the economic efficiency of the areas with different yields. In each of these cases, a significant personal effort is required to learn how to handle these programs, but the most demanding is the QGIS application. The time of preparation of the yield map, including the upload of the original data, when carried out the way presented in the study, lasts from 15 to 70 minutes for the expert user in the program - the least in the Agro-Map program, the most in the QGIS program. As far as purchase costs are concerned, the QGIS is free and the highest acquisition cost is related to Farm^{TM}Works Office software, nevertheless, as mentioned previously, mapping is just one of the many farm management support modules that this program includes.

References

1. H. J. Miller, Annals of the Association of Am. Geogr. 94 (2), 284-89 (2004)
2. J.L. Ping, A. Doberman, Prec. Agric. 6, 193-212 (2005)
3. Sudduth K.A., Drummond S.T. Agron. J. 99:1471-1482 (2007)
4. T.P. Robinson, G. Metternicht, Agric. Syst. 85, 19-41 (2005)
5. S. Nawar, R. Corstanje, G. Halcro, D. Mulla, A.M. Mouazen, Adv. in Agron. 175-244 (2017)
6. S. Arslan, T. Colvin, Prec. Agric. 3, 135-154 (2002).
7. A. Dobermann, J.L. Ping, Y.I. Adamchuk, G.C. Simbahan, R.B. Ferguson, Agron. J. 95, 1105-1120 (2003)
8. Y. Rong-Cai, H. Zhinqui, ASA-CSAA-SSSA Annual Meeting, Long Beach, CA (2014)
9. D. Gozdowski, S. Samborski, E.S. Dobers, Colloq. Biometr. 40, 41-51 (2010)
10. V. Rataj, J. Galambošová, M. Macák, L. Nozdrovický, Profi Press SK (2014)
11. M. Walczykova, M. Zagórda, Polish Acad. of Arts and Sci. Commiss. on Agric., Forest. and Vet. Sc. 18, Monograph Vol. 22, 306-314 (2016)
12. R. Grisso, M. Alley, S. Phillips, Ph. McClellan, Virginia Coop. Exten., Precision Farming, Publication 442-500 (2002)
13. R. Pudełko, Pamiętnik Puławski (Polish Journal of Agronomy), z. 146: 99-111 (2007)
14. Zboża ozime (Winter grain) 2016. http://www.coboru.pl/DR/pdopublikacjecentralne.aspx
15. Rzepak ozimy (Winter rape) 2016. http://www.coboru.pl/DR/pdopublikacjecentralne.aspx
16. Kukurydza (Corn) 2016. http://www.coboru.pl/DR/pdopublikacjecentralne.aspx
17. Agro-Map Manual (2004)
18. S. Mercik, Wydawnictwo SGGW, Warszawa (Publishing House Warsaw University of Life Sciences), pp. 288. (2004)
19. Kopeć M., Gondek K., Materials for training in fertilizer advisory, Kraków (2002)

Root system architecture in winter varieties of spelt (*Triticumspelta* L.)

Andrzej Żabiński[1], *and Urszula* Sadowska[1]

[1]University of Agriculture in Krakow, Faculty of Production and Power Engineering Institute of Machinery Exploitation, Ergonomics and Production Processes, Poland

Abstract. The objective of the study was determination of the variability of morphometry and comparison of the morphological structure of the root system in winter cultivars of spelt. Four spelt cultivars were used in the study: Frankencorn, Oberkulmer Rotkorn, Schwabenkorn and Ostro. The material for the study originated from a field experiment. The roots were collected using the soil core method to the depth of 30 cm, from the rows and inter-rows, then the roots were separated using a semi-automatic hydropneumatic scrubber. The cleaned roots were manually separated and scanned, obtaining their digital images. Image analysis was performed using the Aphelion computer software. In order to characterize the root system of the spelt cultivars included in the study, values of the following indexes were determined: root dry mass (RDM), root length density (RLD), specific root length (SRL), mean root diameter (MD). Based on the obtained results it was determined that the RDM, MD and RLD indexes in all spelt cultivars attain the highest values in the row, at the depth 0–5 cm.The highest value of the RDM and MD indexes characterized the root system of the Ostro cultivar at the depth 0–5 cm. The Oberkulmerrotkorn spelt cultivar was distinguished among the tested objects by the highest value of the SRL index.

1 Introduction

Spelt (*Triticumaestivumsspspelta* L.) is a species included in the Poaceae family. It is one of the oldest wheat subspecies, which survived practically unchanged to the modern times. For centuries, it was the dominant cereal in Europe, cultivated in areas featuring cool climate: in Scandinavia, mountainous Germany, Switzerland and Poland. In the EU as well as Polish legislation spelt is considered to be a separate species (*Triticumspelta* L.) [1]. This cereal has gained increasing attention in recent years due to the possibility for cultivation under extensive conditions. In addition, its nutritional values are also appreciated. Spelt contains high amounts of essential nutrients such as proteins, unsaturated fatty acids, carbohydrates, vitamins and bioelements, as well as fiber, which supports body purification, distinguishing it from other cereals [2, 3]. Considering the increasing interest in spelt, primarily in the organic farming sector, a need has arisen for the study of natural characters of the existing cultivars of the cereal, determining the possibility to obtain satisfactory results of its cultivation in terms of qualitative as well as quantitative of the grain. The root system is the plant's organ with a highest impact on this issue.

Research on the root systems constitutes the basic method for the evaluation of plant growth. Twisting, elongation and branching are three basic cellular processes in roots, which constitute the basis for the architecture of the root system. Together, these elements determine the distribution of roots in the soil environment, thus determining the access to water and nutrients by the plant and its anchoring in the substrate [4]. Roots determine the proper course of the key plant physiological processes, such as photosynthesis and gas exchange [5]. Information on the root systems of crops are the foundations for the efficient water resources and nutrient management. Understanding of the development and architecture of a root system is the basis for the optimization of arable land use and increase in the efficiency of the plants cultivated on them. One of the indexes commonly considered in analyzes of root systems is the root length density (RLD). Its values are important for the modeling process of the movement of water and nutrients in the aeration zone and for the study of soil-root-stem-air interaction [6, 7, 8, 9].

2 Materials and methods

The study was conducted in a 2-year period as a field experiment, located in Mydlniki near Krakow (latitude 50°04'N longitude 19°51'E) at the Institute of Machinery Exploitation, Ergonomics and Production Processes of the University of Agriculture in Krakow. The climatic conditions of the area where the study was conducted are characterized by monthly accumulated rainfall and average monthly temperatures in years 2012–2013,

*Corresponding author: Andrzej.Zabinski@ur.krakow.pl

presented in Table 1. The presented data originate from the meteorological station located at a distance of about 3 km from the study area. The experimental site is characterized by a moderate-continental climate. The experiment was set up in individual years on a soil with the granulometric composition of clay sand, belonging to the good rye soil complex.

The study included 4 winter spelt cultivars. Frankenkorn, Schwabenkorn, Oberkulmer Rotkorn and Ostro. The forecrop for spelt ever year was white mustard, sown as an aftercrop intended as green manure. The surface area of the fields for individual objects was 4 m^2 (2m × 2m). The fields were separated with pathways, width 1.5 m. Mineral fertilization was used under plants at the following dosages of pure components: N–70 $kg \cdot ha^{-1}$, P_2O_5–60 $kg \cdot ha^{-1}$, K_2O–75 $kg \cdot ha^{-1}$. Phosphorus-potassium fertilizers were sown in autumn during forecrop cultivation, nitrogen was used during plant vegetation in spring. Spelt was sown in the first third of October. Sowing was performed manually into rows, at 12 cm spacing. The assumed plant density was 250 $pcs \cdot m^{-2}$.

The soil samples with roots were obtained from the experimental fields at the stage of mature waxy soft kernels (BBCH 85). They were collected separated from the rows and inter-rows in 3 repetitions within each cultivar, using the soil-core method [10]. The core diameter was 80 mm. The soil and root samples were obtained to the depth of 30 cm, then separated into layers: 0–5, 5–10, 10–15, 15–20, 20–25 and 25–30 cm. In order to remove the soil particles, roots were cleaned using hydropneumatic elutriation [11]. The air pressure utilized during cleaning was approximately 300 kPa. The cleaning time of each sample was 3 min. In order to remove excess water from the cleaned roots, they were transferred and placed on filter paper. Subsequently, the roots were manually separated, removing all remaining contaminants. The separated and cleaned roots were then placed in the scanner. Digital images were obtained using Epson Perfection 4870 Photo (scanner by Seiko Epson Corp., Suwa, Nuagano, Japan), recording them in TIFF format at the 600 dpi resolution. Example image of scanned roots is presented in figure 1.

Table 1. Monthly and annual temperatures and accumulated rainfall for the study period and long-term averages.

	1961-99	**2012**	**2013**
	Monthly average temperature [°C]		
January	-3,3	-0,1	-2,4
February	-1,6	-5,5	-0,5
March	2,4	5,8	-0,9
April	7,9	12,3	8,9
May	13,1	15,3	14,3
June	16,2	20,0	17,5
July	17,5	21,0	19,4
August	16,9	19,1	19,1
September	13,1	14,8	12,0
October	8,3	8,5	9,9
November	3,2	4,7	4,9
December	-1,0	-0,8	1,2
Annual mean	7,7	9,6	8,6
	Precipitation [mm]		
January	34	52	62
February	32	27	22
March	34	23	32
April	48	43	20
May	83	23	99
June	97	144	213
July	85	64	27
August	87	55	26
September	54	42	86
October	46	97	14
November	45	24	71
December	51	26	22
Annual sum	696	620	694

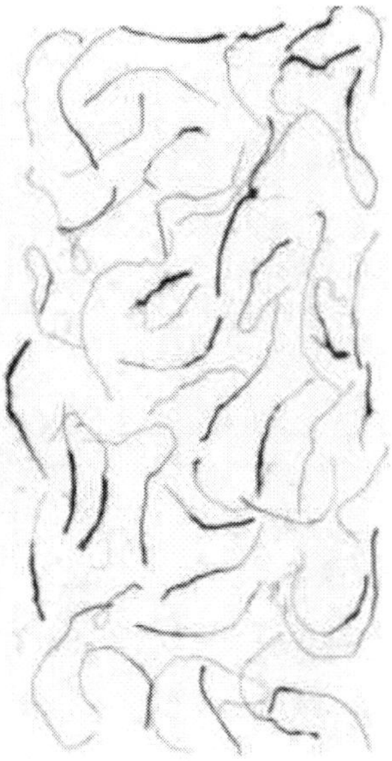

Fig.1. Example image of scanned spelt roots of the Oberkulmer Rotkorn cultivar, obtained from the depth of 0–5 cm.

The root images were subjected to analysis using the APHELION v3.2 software, following the methodology described by Bauhus and Messier [12]. The measured root length was separated into eight diameter classes: 0–0.02; 0.02–0.05; 0.05–0.1; 0.1–0.2; 0.2–0.5; 0.5–1.0; 1.0–2.0; >2.0. Root length density (RLD) was calculated dividing the total root length in the soil sample by the sample volume. Specific root length (SRL) was calculated by dividing the RLD by the dry mass of the roots. The mean root diameter (MD) was calculated as a weighted average of root lengths for individual diameter classes. After scanning, the roots were dried at temperature 70°C to determine their dry mass. Root dry mass (RDM) was calculated dividing the root dry mass in the soil sample by the sample volume.

In the period of obtaining soil-root samples the soil penetration resistance was also recorded using the STIBOKA penetrometer (EjkelkampAgrisearch Equipment, the Netherlands), with a conical base with the surface of 100 mm^2 and the apex angle 60° [13].

Results of the spelt root system study were statistically elaborated using the Statistica 12 software. In order to evaluate the significance of the differences between the means a Duncan test was used at the significance level of p=0.05, distinguishing homogeneous groups.

3 Results and discussion

Results of the measurement of soil penetration resistance on the experimental fields are provided in table 2. Analysis of the obtained values of the parameter demonstrated that on the depth of 25 and 30 cm the resistance for the penetrometer probe is highest and it does not exhibit variability in statistical terms. Its mean value for these depths was 3.494 MPa. In higher layers of soil, at depths of 10, 15 and 20 cm, the resistance was significantly lower and amounted to 2.685 MPa on average. The lowest value of penetration resistance was registered at the depth of 5 cm, where it was 1.345 MPa (Table 2).

Table 2. Soil penetration resistance at a given depth.

Depth (cm)	Soil penetration resistance (MPa)
5	1,354 d*
10	2,228 c
15	2,771 b
20	3,056 bc
25	3,254 a
30	3,735 a

*a, b, c, d – homogeneous groups according to the Duncan test

The performed analysis of variance of the RDM index describing the root dry mass exhibited interaction between individual cultivars and the vertical distribution of roots in the soil. The RDM index attained highest values at depth from 0 to 5 cm for the Ostro cultivar, where the mean value was 0.0023 g·cm^{-3} The remaining cultivars in the discussed soil layer were characterized by lower root dry mass amounting to the mean 0,0015 g·cm^{-3}. In the deeper soil layers (5–30 cm), the discussed index did not exhibit any statistically significant differences between the tested cultivars and its mean value was 0.0002 g·cm^{-3} (Table 3).

The root dry mass index further exhibited variability depending on the depth and sampling site. At the depth between 0 to 5 cm, higher root mass was determined in samples obtained from rows as compared to inter-rows (Table 3). This can be explained by the thicker initial and seminal roots with higher mass present in this zone.

The index for specific root length (SRL) is one of the most commonly determined morphological parameters of fine roots. It characterizes economic aspects of a root system, linked to the capability to absorb water and nutrients. The SRL index is negatively correlated to the root thickness. Higher value of the SRL index means thinner roots, thus their greater surface area. Root systems with a high value of the SRL index possess better contact with the soil, thus increasing their absorption potential [14, 15]. Under suboptimal concentrations of nutrients in soil, large root surface area is favorable, as it facilitates absorption of less mobile components, such as phosphorus (P) [16]. In the study conducted by Løes and Gahoonia [17], concerning root systems of spring wheat and barley, the obtained mean values of the SRL index under the conditions of soil cultivation were 30000 and 26300 cm·g^{-1}, respectively. Corneo et al., [18], who determined, among others, value of this index among 23 genotypes of wheat, belonging to different breeding groups observed that in the layer 0–30 cm it ranges from 2895 to 8962 cm·g^{-1}. For the tested spelt cultivars the mean value of specific root length in the 0–30 cm layer remained at the level of 13291.9 cm·g^{-1}, not exhibiting variation on specific depth levels (Table 4). But, significant differences were observed for the values of this index between individual cultivars. The highest specific root length characterized the Oberkulmer Rotkorn cultivar, for which the mean value of the parameter was 14873.6 cm·g^{-1}. The cultivars: Ostro and Schwabenkorn were characterized by similar values of the SRL index, whose mean value for these cultivars was 13114.7 cm·g^{-1}. The lowest specific root length with the value of 12064.674 cm·g^{-1} characterized the Franckenkorn cultivar. The SRL index also exhibited statistically significant variation depending on the depth and sample collection site. For the depth 0–5 cm and samples obtained from row, the specific root length was lowest, attaining the value of 4357.6 cm·g^{-1} (Table 4).

Table 3. Index values describing the root dry mass (RDM).

Cultivar	Collection site	RDM (g·cm^{-3}) for the depth					
		0-5 cm	5-10 cm	10-15 cm	15-20 cm	20-25 cm	25-30 cm
Franckenkorn		0,00146 bc*	0,00033 d	0,00024 d	0,00022 d	0,00018 d	0,00013d
Schwabenkorn		0,00114 c	0,00026 d	0,00023 d	0,00025 d	0,00021 d	0,00015 d
Oberkulmer Rotkorn		0,00179 b	0,00028 d	0,00020 d	0,00021 d	0,00020 d	0,00011 d
Ostro		0,00235 a	0,00033 d	0,00027 d	0,00025 d	0,00022 d	0,00014 d
	Row	0,00288 A**	0,00033 BC	0,00021 C	0,00022 BC	0,00020 C	0,00015 C
	Inter-row	0,00049 B	0,00027 BC	0,00026 BC	0,00025 BC	0,00021 C	0,00012 C

*a, b, c, d – homogeneous groups for the cultivar x depth interaction based on the Duncan test
**A, B, C – homogeneous groups for the collection site × depth interaction based on the Duncan test

Table 4. Index values of specific root length (SRL).

Cultivar	Collection site	SRL ($cm \cdot g^{-1}$) for the depth						Mean for cultivar ($cm \cdot g^{-1}$)
		0-5 cm	5-10 cm	10-15 cm	15-20 cm	20-25 cm	25-30 cm	
Franckenkorn		9127,4	13028,8	13204,8	12303,9	12825,9	11897,2	12064,7 c*
Schwabenkorn		10708,6	15197,5	13097,9	13846,5	14242,0	12653,3	13291,0 b
Oberkulmer Rotkorn		9299,6	16086,6	16044,6	16324,0	15493,2	15993,8	14873,6 a
Ostro		7669,18	14368,4	13129,3	14612,5	14777,4	13074,2	12938,5 b
	Row	4357, 6 D**	14109,9 ABC	14405,8 ABC	14866,7 AB	14259,0 ABC	13759,7 ABC	
	Inter-row	13759,7 ABC	15230,7 A	13332,5 BC	13676,7 ABC	14410,3 ABC	13049,6 C	

*a, b, c, d – homogeneous groups for cultivars based on the Duncan test
*A, B, C – homogeneous groups for the collection site × depth interaction based on the Duncan test

Analysis of the architecture of a root system indicates the importance of root diameter. It is directly linked to the level of biomass accumulation in the roots [19]. The diameter influences the possible root length and number of roots in the system [20]. Root systems with particularly high number of very fine roots, formed by numerous plant species, especially monocotyledons, result from optimization of the root length to mass (nutrient reservoir) ratio, determining the absorption surface. Thin roots posses limited capability to penetrate soil due to their lower length [21]. The root diameter is likely to have an influence on the efficiency of water transport to the stem via the vascular system. Longitudinal flow resistance increases in fine roots.

The mean root diameter index (MD) for the tested spelt cultivars exhibited variability depending on the cultivar and depth. Its greatest value, amounting to 0.189 mm was determined for the Ostro cultivar at the depth 0–5 cm. For the remaining cultivars, in the same layer of soil, the MD index attained lower and similar values, remaining at the average level of 0.022 mm. All tested objects formed thinnest roots in the layer 5–10 cm. The mean spelt root diameter also depended on the sample collection site. Considering both depth and root sample collection site, its greatest diameter was recorded in row at depth 0–5 cm (Table 5). Greater root diameter in the surface layer and in the row is likely linked to the greatest share of thicker initial and seminal roots. Corneo et al., [18], who tested root systems of different genotypes of wheat determined that in the soil layer 0–30 cm, the MD values remain in the range from 0.25 to 0.40 mm. The spelt root diameter is lower, in the arable layer to the depth of 30 cm it remained in the range from to 0.18 mm. As mentioned earlier, the root system characterized by a low diameter of the basic roots creating the system, has a greater absorbing surface and absorption capabilities. But, fine roots have a limited range in the vertical soil profile, limiting the capacity to absorb water from deeper soil layers in deeper layers of soil during prolonged droughts. In the case of the tested spelt cultivars, this phenomenon is exhibited by the values of the RLD and RDM indexes. Their considerably decrease was marked already from the depth of 5 cm.

Table 5. Values of the mean rood diameter (MD) index.

Cultivar	Collection site	MD (mm) for the depth					
		0-5 cm	5-10 cm	10-15 cm	15-20 cm	20-25 cm	25-30 cm
Franckenkorn		0,169 bc*	0,127 ijkl	0,129 hijk	0,135 fghijk	0,137 fghij	0,143 efgh
Schwabenkorn		0,159 cd	0,124 ijkl	0,138 fghi	0,138 fghi	0,146 defg	0,155 de
Oberkulmer Rotkorn		0,174 b	0,115 l	0,121 kl	0,125 ijkl	0,135 fghijk	0,133 ghijk
Ostro		0,189 a	0,122 jkl	0,126 ijkl	0,131 hijk	0,135 fghijk	0,148 def
	Row	0,221 A	0,124 FG	0,127 EFG	0,130 DEF	0,139 D	0,143 BC
	Inter-row	0,125 EFG	0,120 G	0,130 DEF	0,134 CDE	0,139 D	0,146 B

*a, b, c, d, e, f, g, h, i, j, k, l – homogeneous groups for the cultivar x depth interaction based on the Duncan test
**A, B, C, D, E, F,G – homogeneous groups for the collection site × depth interaction based on the Duncan test

RLD is an important character of plant root system, changing with water availability. If it is unlimited, the RLD values may be higher in the surface layer of soil, yet during prolonged droughts it may also attain higher values in the deeper layers of the soil profile [22]. Values of the index constitute a basic information on the capability of a root system to utilize the water and nutritional resources of soil. The literature data shows that the critical value of a RLD, sufficient for the use of these resources is 1 cm cm^{-3} [23, 24, 25]. High variability of RLD depending on the moisture conditions of soil results in different publications, concerning the same species a wide range of the values can be found. According to Gregory et al., [26], Barraclough et al., [24, 27], the root length density (RLD) for the winter wheat cultivars in soil to depth of 20 cm is from 5.6 to 10 cm cm^{-3} and from 0.4 to 1.5 cm cm^{-3} at the depth from 80–100 cm. Pillinger et al., [28] report that in the surface layer of soil RLD, also in the case of winter wheat may attain values at the level of 7 cm cm^{-3} and 0.2 cm cm^{-3} at depth 80–100 cm. Hoad et al., [29] state that in the arable layer of soil, at good root development of winter heat, the RLD amounts from 3 to 6 cm cm^{-3} and below 1 cm cm^{-3} at depth exceeding 40 cm. In the case of inferior root development, RLD may amount to less than 1 cm cm^{-2}, also in the arable layer. In Australia, Atta et al., [30], who published results of study concerning 15 genotypes, report that the values of the RLD index in the upper 15 cm soil layer remain in the range from 0.6 to 2 cm cm^{-3}, whereas Narayanan et al., [31], who examined 30 spring wheat genotypes state that in the soil layer between 0 to 30 cm the values of the index range from 0.86 to 1.83 cm cm^{-3}.

In the case of the tested spelt cultivars in the layer 0–30 cm, values of the tested index remained in the range from 0.179 cm·cm^{-3} in the inter-row at depth 25–30 cm to 1.388 for row and surface layer 0–5 cm (Table 6).

A statistical analysis of the values of the root length density (RLD) index did not exhibit significant differences in this field between the tested spelt cultivars at individual depths. The mean value of the index was 0.467 cm·cm^{-3}. The root length density exhibited variation depending on the depth and site of sample collection. The highest values of the index were obtained in row at depth 0–5 cm, where its mean value was 1.388 cm·cm^{-3}. On the other hand, the lowest values were recorded at the depth from 20 to 30 cm, both in the row as well as in the inter-row (Table 6). These results demonstrate that to the depth of 5 cm the root systems of the tested cultivars may exhibit satisfactory water and nutrient absorption capacity, yet it solely applies to a thin layer located near the plant, as in the inter-row the RLD index assumes values below 1 cm·cm^{-3} for the same depth.

Table 6. Values of the root length density (RLD) index.

Cultivar	Collection site	RLD (cm·cm^{-3}) for the depth					
		0-5 cm	5-10 cm	10-15 cm	15-20 cm	20-25 cm	25-30 cm
Franckenkorn		1,023	0,504	0,370	0,342	0,289	0,192
Schwabenkorn		0,862	0,477	0,353	0,361	0,320	0,204
Oberkulmer Rotkorn		1,047	0,552	0,416	0,421	0,334	0,214
Ostro		1,069	0,574	0,393	0,388	0,328	0,182
	Row	1,388 a	0,546 bc	0,365 def	0,364 def	0,309 efg	0,217 fg
	Inter-row	0,612 b	0,508 bcd	0,401 cde	0,391 cdef	0,327 efg	0,179 g

*a, b, c–homogeneous groups for the collection site x depth interaction based on the Duncan test

4 Conclusions

Analysis of the obtained study results allows to formulate the following conclusions:

1. The highest value of the RDM and MD indexes characterized the root system of the Ostro cultivar at the depth 0–5 cm.
2. Spelt of the OberkulmerRotkorn cultivar is distinguished among the remaining objects the greatest value of the specific root length (SRL) index, amounting to 14873.6 cm·g^{-1}.
3. The root length density (RLD) index in the layer 0–30 cm assumes similar values for all tested spelt cultivars. Its mean value remains at the level of 0.467 cm·cm^{-3}.
4. For all tested spelt cultivars, the RDM, MD, RLD indexes assume greatest values in rows at depth 0–5 cm.

References

1. S. Stankowski, G. Hury, A. Makrewicz, G. Jurgiel-Małecka, M. Gibczyńska. Inż. Ekolog., **49** (2016)
2. H. Sulewska, W. Koziara, K. Panasiewicz, G. Ptaszyńska, M. J. Morozowska. Res. Appl. Agric. Engng. **53**, 4 (2008)
3. J. Tyburski, M. Babalski. (CDR w Brwinowie Oddział w Radomiu, Radom, 2006)
4. S. M. Rich, M. Watt. J. Exp. Bot. **64**, 5 (2013)
5. A. Elazab, G. Molero, M. Dolores Serret, J. Luis Araus. Funct. Plant Biol., **39** (2012)
6. M. A. Hajabbasi.. J. Agric. Sci. Technol., **3** (2001)
7. K.G Mandal, K.M. Hati, A.K. Misra, P.K. Ghosh, K.K. Bandyopadhyay. Jour. Agric. Physics. **3,** (1, 2) (2003)
8. G. Den Herder, G. Van Isterdael, T. Beeckman, I. De Smet. Trends Plant Sci. **15,** 11 (2010)
9. B. M. Atta, T. Mahmood, R. M. Trethowan. Aust. J. Crop. Sci. **7**, 13 (2013)
10. W. Bohm. Springer-Verlag (Germany, Berlin 1979)
11. A.J.M. Smucker, S.L. Mc Burney, A.K. Srivastova. Agron J., **74** (1982)

 https://doi.org/10.1051/bioconf/20181001019

12. J. Bauhus, Ch. Messier. Agron. J., **91** (1999)
13. American Society of Agricultural Engineers. Soil cone penetrometer. ASAE Standard S313.3, (2002)
14. W. Larcher. Physiological plant ecology, **3** (1995)
15. D.M. Eissenstat, C.E. Wells, R.D. Yanai, J.L. Whitbeck. New Phytol.**147**, 1 (2000)
16. A. Fitter. Plant Roots, **3** (2002)
17. A. K. Løes, T. S. Gahoonia. Euphytica., **137** (2004)
18. P. E. Corneo, C. Keitel, M. A. Kertesz, F. A. Dijkstra. AgrEcosyst Environ., **246** (2017)
19. D.M. Eissenstat. J. Plant Nutr. **15**, (6-7) (1992)
20. L.P.R. Bidel, L. Pagès, L.M. Rivière, G. Pelloux, J.Y. Lorendeau. Ann. Bot. **85**, 6 (2000)
21. L.J. Clark, A.H. Price, K.A. Steele, W.R. Whalley. Funct. Plant. Biol. **35**, 11 (2008)
22. A. Blum. Aust. J. Agric. Res. **56**, 11 (2005)
23. M. van Noordwijk Proceedings of the third ISRR (symposium Austria, 1983)
24. P.B. Barraclough, H. Kuhlmann, A.H. Weir. J. Agron. Crop. Sci., **163** (1989)
25. C. White, R. Sylvester-Bradley, P.M. Berry. J. Exp. Bot. **66**, 8 (2015)
26. P.J. Gregory, M. McGowan, P.V. Biscoe, B. Hunter. J AgrSci-Cambridge, **91** (1978)
27. P.B. Barraclough, A.H. Weir, H. Kuhlann. Developments in agricultural and managed forest ecology, **24** (1991)
28. C. Pillinger, N. Paveley, M.J. Foulkes, J. Spink. Plant Pathology **54**, 4 (2005)
29. S.P. Hoad, G. Russell, P.S. Kettlewell, M. Belshaw. HGCA Project Report, **351** (2004)
30. B.M. Atta, T. Mahmood, R.M.Trethowan. Aust. J. Crop. Sci. **7**, 13 (2013)
31. S. Narayanan, A. Mohan, K.S. Gill, P. V. V. Prasad. PLoS ONE **9**, 6 (2014)

The effect of conditions and storage time on course of moisture and temperature of maize grains

Marek Angelovič[1], *Koloman* Krištof[1], *Ján* Jobbágy[1], *Pavol* Findura[1], and *Milan* Križan[1]

[1]Faculty of Engineering, Slovak University of Agriculture, 949 76 Nitra, Tr. A. Hlinku 2, Slovakia

Abstract. Characteristics of stored cereals significantly affect the technical solution of individual storage facilities, as well as the technical appliances used to provide storage technologies, i.e. storing aeration and off-loading. The objective of the study was to monitor the influence of atmospheric temperature and extraneous moisture on the course of temperature and moisture of maize grain during the storage. An average value of maize grain moisture was 13.6% and an average value of temperature inside of the silo was 12.3°C within a defined timeframe. An average moisture value of maize grain stored in the floored warehouse was 13.7%, but an average temperature inside the warehouse was 21.9°C, however extraneous conditions during the storage were unchanged. Average value of extraneous moisture was 86.5 % and an average value of atmospheric temperature was 9.17°C. It was not recorded any occurrence of animal pests in stored maize grain during the storage. Grain water content differs with the change of air moisture and temperature. Reduction of grain water content is most frequently realized by residual drying (moisture is less than 15%); the storage stability is provided at the moisture of 12%.

1 Introduction

Maize is an important crop which is the essential for stock rising as energy food. Drying is vital for maize safe storage. However, the unsuitable drying could decrease its feeding and processing quality [1, 2].

Field grain harvesting is followed by post-harvest treatment, which is essential for storage. During the post-harvest treatment, grain is cleaned from substantial part of impurities and grain is dried to the moisture of 14 to 14.5% of its weight, if necessary. The purpose of the storage is to preserve and improve the grain quality. Technological properties can be improved by proper storage. Warehouses should be steady enough to protect grain against weight reduction, caused by atmospheric conditions and animal pests. We cannot avoid grain weight reduction during the storage, which is caused by respiration, diffusion and mechanical damage [3-4].

Moreover, [5] found out that previous storage history (moisture content and duration) affected subsequent storability. Storage for longer periods, given the same moisture content, resulted in lower subsequent storability, and storage at higher moisture content, given the same time period, also resulted in lower subsequent storability [5].

Even [6] stated that corn seed (Zea mays L.) very often is harvested on the ear at high moisture so as to reduce the risk of frost, insect, and disease damage. The harvest is started at seed moisture as high as 40%, and ears are subsequently dried mechanically with heated air before shelling. The drying process has been a frequent cause of seed injury and great economic loss. Damage is most frequently encountered when the drying temperature is elevated or the loss of moisture is delayed. Dehumidification has long been used to safely dry perishable products and to maintain their quality during storage. Utilizing a dehumidification system which was based on heat pump technology allowed acceleration of the drying process without elevating the temperature. The heat pump system is also less expensive to operate than traditional natural gas fired systems and produces high quality seed [6].

Subsequently, [7] concluded that ear maize up to 30-33% seed moisture content can be successfully quicker dried at 50°C. Selecting a combination of high and low temperature degrees can reduce the drying time as compared with a low temperature drying. The best combination maintains standard germination above 90% soon after artificial drying was at 50°C for 20 h before completed drying at 38°C for 12 h. But after 6 months storage, the best one was 50°C for 16 h before completed drying at 38°C for 17.30 h which maintains standard germination above 88%. These combinations result in saving of about 21 and 16.25% drying time, respectively with maintaining adequate levels of seed and seedling vigor [7].

Furthermore, [8] suggest that drying maize on their stem was the traditional technique frequently used; this technique must be improved to avoid contaminations and to increase the quality of drying. However, the method of storage is accountable for the most significant losses after harvest, because mildew develops when the conditions of storage (too high temperature and moisture of the air) do not tally with the final content of the dried

*Corresponding author: pavol.findura@uniag.sk

product. Sorption isotherms of products are most important to model moisture uptake during storage and distribution. The monolayer moisture content was found to decrease with increasing temperature. These results can be used to predict the potential changes in the stability of maize grains and later for the development of a system of suitable drying [8].

Moreover, [9] found out that there were no significant changes in the content of individual carotenoids within genotypes during kernel development from 45 days after pollination through the time of harvest. Carotenoid losses through traditional grain drying were also minimal (<9%). However, the stability of carotenoids in maize kernels over storage time after harvest was found to be dependent on both temperature and humidity. Therefore, these results support the notion that effective control of moisture content and temperature of the kernels during storage conditions is essential to reduce the speed of degradative reactions [9].

Temperature and moisture are basic indicators affecting the realization of successful long-term cereals storage. The temperature of the stored grain is an important parameter, which is influenced by the oxygen availability. Grain storage in closed areas causes a formation of carbon dioxide and oxygen reduction, because of permanent respiration. Therefore, respiration intensity of grain is reduced gradually. Aeration should provide air supply with low air moisture; even the air temperature should be at least by 5°C lower compared to the temperature of ventilated grain. Hygroscopicity, an ability to absorb and to exclude water vapor, has a significant impact on the grain quality and the storage ability. Grain water content differs with the change of air moisture and temperature. Reduction of grain water content is most frequently realized by residual drying (moisture is less than 15%); the storage stability is provided at the moisture of 12% [2, 4, 10, 11].

In addition, [12] suggested the need to determine sorption isotherms for tropical hybrid maize genotypes in relationship with seed morphology. Meanwhile, generalized prediction of seed moisture during drying or dry-storage of these varieties in tropical seed stores can be done using the Henderson-Thompson and Chung-Pfost models [12].

Furthermore, [13] suggested that alternative grain storage systems have been adopted when traditional bin storage capacity is exceeded or unavailable. One method is grain bag (silo bag) storage systems. This system has been used for silage or high-moisture grain storage. It was concluded that researchers, producers, and grain managers could use a similar system to monitor grains stored using this alternative storage system and to make effective management decisions [13].

2 Materials and methods

2.1 Characteristics of grain warehouse

Currently, most frequently used grain warehouses are steel silos (Fig. 1). They are high buildings with shaft chambers, which are called silo cells, with circular or hexagonal shape of floor projection. They are use to store only dry grain with the moisture content up to 14%. Silo cells are made of ferroconcrete or metal. These height containers have different air distributions used for ventilation and aeration of stored grain in relation to the type and structure [2, 4].

Hangar-type warehouse are generally used for grain storage at post-harvest lines, for seed storage, as well as for temporal storage of damaged stock (Fig. 2). They are formed by a closed external roofed building structure. These buildings have quite complicated built-in ventilation system, which enables complete drying and aeration of stored grain [2, 14-15].

Fig. 1. Steel silos.

Fig. 2. Floored warehouse.

2.2 Monitoring of the effect of storage conditions on product quality

Characteristics of stored cereals significantly affect the technical solution of individual storage facilities, as well as the technical appliances used to provide storage technologies, i.e. storing, aeration and off-loading.

While, [1] concluded that Pasting properties of maize dried in hot air showed a decrease of peak when maize drying temperatures increased (≥60°C), while final viscosity rose significantly. The construction of maize was also modified after hot air treatment. All these changes result in the moisture content of maize died at 105°C added by conditioning was just 2%. When the conditioning temperature was above 75°C, bulk density was increased significantly with the increasing temperature, but the moisture content added by conditioning and the final moisture content of the product were not changed. When the conditioning temperature below 75°C, the quantity of feedstock significantly decreased, indicating the relative amount of steam was small, a bulk density decreased significantly, the state was like dry extruded [1].

The storage temperature is one of the main factors that affect the quality of storage, and considering the fact that limited studies have been conducted with maize. [16] conducted a study aimed to evaluate the quality of maize grains stored at 5, 15, 25 and 35°C for 12 months. The results of water content, thousand grain weight, germination, electrical conductivity, and fatty acid profile indicated that major changes were observed in grain storage at higher temperatures, especially at 25 and 35°C. Therefore, it was concluded that the safe time for storage of grains under these conditions is less compared to storage at lower temperatures [16].

Complicated chemical processes occur during cereal storage. These can ultimately cause the loss of nutritional value. Part of the loss is caused by a natural reaction (breathing kernels); another part of the loss is caused by chemical reactions that are a result of the violence of technological discipline at post-harvest treatment and storage of cereals [2, 4, 17].

Main characteristics of the grain in relation to grain storage are:

- grain volume weight and void fraction,
- pourability of cereals,
- temperature of cereals,
- balanced water content of cereals,
- unfavourable impurities content.

We monitored:

- the effect of crop temperature at the beginning of storage on ventilation frequency,
- the effect of grain moisture on storage period,
- occurring of animal pests in relation on the ambient temperature and the storage period.

2.3 Machines, devices and instruments

We used machines, devices and equipment on the agricultural farm AgroDivízia, Ltd. in Selice for our experimental measurements. We used specialized devices and instruments belonging to the laboratory equipment for agro-physical characteristics detection, i.e. hygrometer, thermometers, digital scales, measuring device for volume weight etc.

2.4 Experiment evaluation methods

We used verified statistical calculation procedures for measured values evaluation. We used Microsoft Excel software for graphical processing and mathematical-statistical results evaluation.

3 Results and discussion

Technology and methods of storage affect the temperature and moisture of the material, which could have an impact on the storage period and ton animal pests evolution.

We realized our experimental measurements of maize grain storage at steel silos and floored warehouses considering equipment opportunities of farm AgroDivízia, Ltd. in Selice.

3.1 Monitoring of parameters of moisture and temperature of maize grain stored in steel silos

Measured values of temperature and moisture of maize grain stored in monitored steel silo are represented graphically in Figure 3.

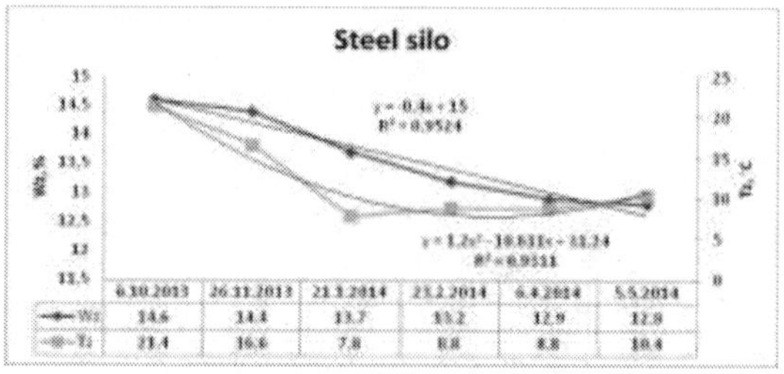

Fig. 3. Progress of moisture and temperature of maize grain stored in steel silo in dependance on storage period: W_z - moisture of maize grain ,T_z - temperature of maize grain.

From the measured values we can state, that an average value of maize grain moisture is 13.6% within the timeframe, with the maximum value of 14.6% and minimal value of 12.8%. We can state, that that an average value of temperature inside of the silo is 12.3°C within the timeframe, with the maximum value of 21.4°C and minimal value of 7.8°C. The moisture of stored grain correlates positively with storage temperature of the grain.

Simultaneously, with the measurement realized inside of the steel silo, we monitored external conditions during the storage, i.e. air moisture and temperature, which have a significant impact on storage conditions.

Moreover, [18] stated that grains stored at 35°C during the 12 months period showed the greatest decrease in grain yellowness, pH, protein solubility and breakdown viscosity. An increase in disulfide bonds within the protein structure and interaction between starch and non-starch components seems to be responsible for the changes in protein solubility and pasting properties determined in maize during the storage period. Fat acidity and the percentage of grains infected by visible molds were concluded to be very dependent of moisture content. The result of this study demonstrated that the minimum temperature of 5°C was able to maintain the quality of maize stored for up 12 months [18].

Measured values of atmospheric temperature and extraneous moisture in the area of monitored steel silo are represented graphically in Figure 4.

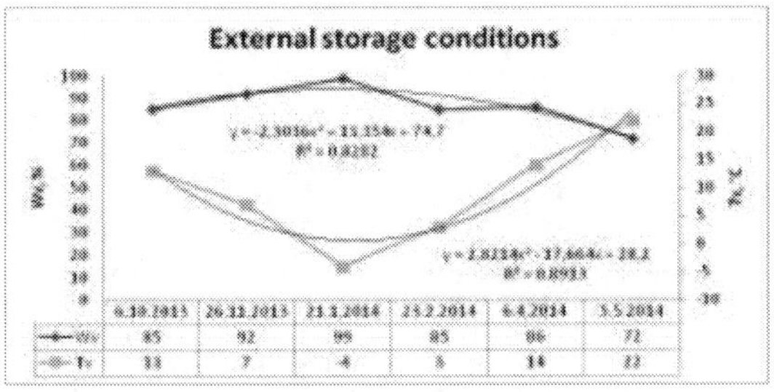

Fig. 4. Progress of air moisture and temperature during the storage in the area of monitored steel silo in dependance on storage period: W_v - air moisture, T_v – air temperature.

From the measured values we can state, that an average value of air moisture is 86.5%, within the timeframe, with the maximum value of 99%, and minimal value of 72%. We can state, that that an average value of air temperature in the area of silo was 9.17°C within the timeframe, with the maximum value of 22°C and minimal value of -4°C. Similarly, air moisture correlates with air temperature. A reduction of atmospheric temperature is increasing extraneous moisture, resulting in a direct impact on the course of moisture and temperature of stored maize grain in steel silo (Figure 3).

[19] conducted a study to determine influence of the changes in relative humidity and temperature in storage facilities on the moisture in grain of stored soybean and maize. Grains were stored during 34 days at the temperatures of 0°C and 20°C, and relative humidity of 55%, 73%, 80% and 98%. It was observed that at the temperature of 0°C and relative humidity of 55% and 73%, moisture in soybean grain decreased, 2.4% and 1.9%, while at the relative humidity of 80% and 98% it increased, 0.2% and 0.6% after 34 days storage. At the temperature of 20°C and relative humidity of 55% and 73%, moisture in soybean grain decreased 4.0% and 0.7%, while at the relative humidity of 80% and 98% it increased 0.8% and 2.3%, as following. During 34 days storage at 0°C and relative humidity of 55%, moisture in maize grain decreased 0.2%, while at the relative humidity of 73%, 80% and 98% it increased 0.4%, 1% and 1.5%. At the temperature of 20°C and relative humidity of 55%, moisture in maize grain decreased 1.5% and at the relative humidity of 73%, 80% and 98% increased, 0.2%, 0.9% and 1.7% [19].

3.2 Monitoring of parameters of moisture and temperature of maize grain stored in floored warehouse

Even [20] conducted a study where moisture content (m.c.) and temperature were measured in two 1300 t semi-underground hermetic concrete silos filled with maize which was stored for nearly three years in an upland equatorial climate. Gradual cooling of the maize took place in the first year of storage and the temperature remained constant thereafter. This could be explained by calculations of the effects of heat loss by conduction and air convection and heat gain through grain respiration. Considerable m.c. increases occurred at the maize top surface and calculations indicate that this was mainly due to air convection currents at night when the silo structure above ground was cool compared with the grain interior [20].

Measured values of temperature and moisture of maize grain stored in monitored floored warehouse are represented graphically in Figure 5.

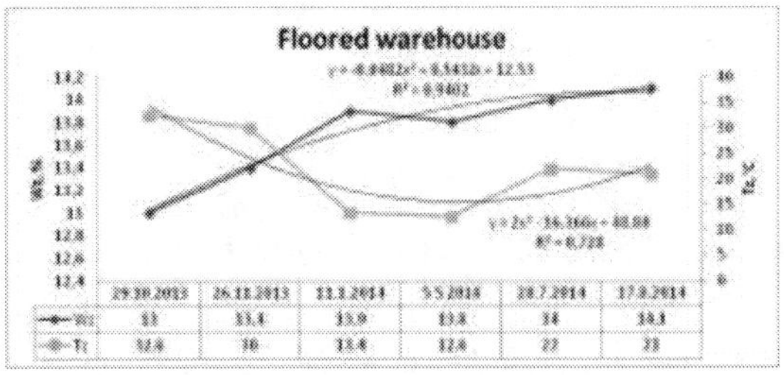

Fig. 5. Progress of moisture and temperature of maize grain stored in floored warehouse in dependance on storage period :W_z - moisture of maize grain ,T_z - temperature of maize grain.

From the measured values, we can state that, that an average value of maize grain moisture is 13.7% within the timeframe, with the maximum value of 14.1% and minimal value of 13%. We can state, that that an average value of temperature inside of the floored warehouse is 21.9°C within the timeframe, with the maximum value of 32.6°C and minimal value of 12.6°C. The moisture of stored grain correlates with storage temperature of the grain, but with different course than in the steel silo.

As an alternative, [21] used a simulation model to quantify the effect of the AERO controller on dry matter loss, moisture content, grain temperature and required aeration time. Decision making of the AERO controller is based on simulation of the aeration process and on real time data acquisition. It proved to be an effective strategy and showed its significant potential as a non-chemical preventative practice for safe storage [21].

Moreover, nutritional changes in maize grains stored at 10, 25 and 45°C for 6 months were studied by [22]. Significant decrease in pH and increase in titratable acidity was observed during storage of maize grains at 25 and 45°C. Moisture contents of maize grains decreased by 25% at 25°C and 38% at 45°C after six months of storage. Total soluble sugars increased by 10.7% at 10°C and 17.3% at 25°C, whereas a 39.5% decrease was observed after 6 months storage at 45°C. Total available lysine and thiamine contents in maize grains decreased by 13 and 9.26% at 25°C, 16 and 20.4% at 45°C, respectively, after 6 months of storage. Protein digestibility decreased by 5.19 and 9.0% at 25 and 45°C, respectively, whereas decrease in starch digestibility was 9.86% at 25°C and 15.1% at 45°C on storage of maize grains for 6 months. However, no significant nutritional changes occurred during storage of maize grains at 10°C [22].

Moreover, [23] conducted a study where maize was brought to three moisture contents (m.c.) within the range typically observed in farm and commercial storage, and was inoculated to simulate the amount of storage mold contamination typical of this situation. Certain of the experimental units were packed in

https://doi.org/10.1051/bioconf/20181002001

insulation so that heat could accumulate within the grain masses to simulate hot spots. The wettest grain heated rapidly and became semi-anaerobic. The hot grain then dried rapidly, with the amount of moisture loss influenced by the ratio of water vapor pressures inside and outside the grain. The hot grain cooled and became more aerobic over time. New infections by storage molds, disappearance of viable field molds, development of kernel damage, and changes in atmospheric gases within the grain masses were correlated with the grain moisture or temperature and the rate at which the moisture and temperature changed. The rate of increase in new kernel damage was as high as 3.3% per week [23]. Both the rate of respiration and the estimated ratio of starch to fat consumed were associated with the mean grain m.c. during the trial, and the estimated ratio of starch to fat consumed was in the range of 2.2/1 - 2.6/1 in the grain containing 16.6 - 18.2% m.c. That both fat and starch were consumed calls into question loss estimates based on starch metabolism alone. The fat content of the grain decreased more than 10% in some experimental units, but increased less than 5% in others. The protein content generally increased as other grain constituents were consumed [23].

Simultaneously, with the measurement realized inside of the floored warehouse, we monitored external conditions during the storage, i.e. air moisture and temperature, which have a significant impact on storage conditions.

Measured values of atmospheric temperature and extraneous moisture in the area of monitored floored warehouse are represented graphically in Figure 6.

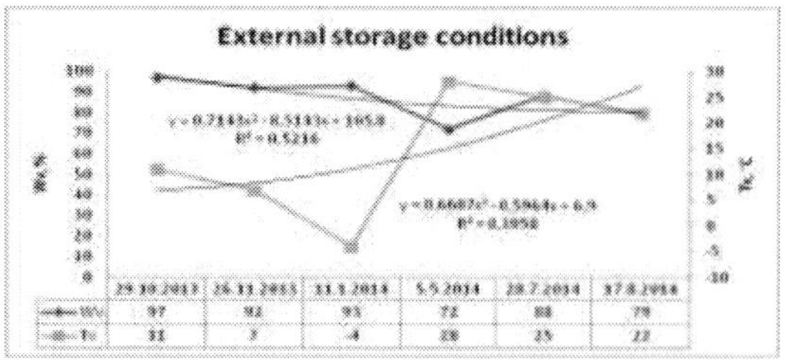

Fig. 6. Progress of air moisture and temperature during the storage in the area of monitored floored warehouse in dependance on storage period :W_v - air moisture, T_v – air temperature.

From the measured values we can state, that an average value of air moisture is 86.8% within the timeframe, with the maximum value of 97%, and minimal value of 72%. We can state, that that an average value of air temperature in the area of warehouse is 14.8°C, with a maximum value of 28°C and minimal value of -4°C. Similarly, air moisture correlates with air temperature. A reduction of atmospheric temperature is increasing extraneous moisture, resulting in a direct impact on the course of moisture and temperature of stored maize grain in floored warehouse (Figure 5).

However, [24, 26] conducted a study to assess whether maize can be safely stored in PICS bags without loss of quality, we carried out laboratory studies of maize grain infested with Sitophilus zeamais (Motshulsky) and stored in PICS triple bags or in woven polypropylene bags. In this study, over an eight month observation period, temperatures in the bags correlated with ambient temperature for all treatments. Relative humidity inside PICS bags remained constant over this period despite the large changes that occurred in the surrounding environment. Relative humidity in the woven bags followed ambient humidity closely. PICS bags containing S. zeamais-infested grain saw a significant decline in oxygen compared to the other treatments. Grain moisture content declined in woven bags, but remained high in PICS bags. Seed germination was not significantly affected over the first six months in all treatments, but declined after eight months of storage when infested grain was held in woven bags [24]. Relative damage was low across treatments and not significantly different between treatments. It was concluded that overall, maize showed no signs of deterioration in PICS bags versus the woven bags and PICS bags were superior to woven bags in terms of specific metrics of grain quality [24].

In addition, [25] concluded that PICS bags are effective at preserving the dryness of maize seed in storage during high relative humidity conditions, which leads to maintenance of seed quality [25].

4 Conclusions

Based on experimental measured results and their evaluation can be stated, that extraneous conditions during the storage, i.e. atmospheric temperature and extraneous moisture have a significant impact on the conditions of maize grain storage.

Considering conditions, better results were observed for steel silo storage with a possibility of extra-ventilation stored layers. An average value of maize grain moisture was 13.6% and an average value of temperature inside of the silo was 12.3°C within a defined timeframe.

An average moisture value of maize grain stored in the floored warehouse was 13.7%, but an average temperature inside the warehouse was 21.9°C, however extraneous conditions during the storage were unchanged. Average value of extraneous moisture was 86.5% and an average value of atmospheric temperature was 9.17°C.

We did not record any occurrence of animal pests in stored maize grain during the storage.

Grain water content differs with the change of air moisture and temperature. Reduction of grain water content is most frequently realized by residual drying (moisture is less than 15%); the storage stability is provided at the moisture of 12%.

The topic of the cereals storage in relation to machines, technological and economical point of view poorly understood in the Slovak Republic, hence this topic will be the subject of further research by the Department of Machinery and Production Systems, Faculty of Engineering, SUA.

Acknowledgements
The research leading to these results has received funding from the European Community under project no 26220220180: Building of Research Centre „AgroBioTech".

References

1. J. Yang, H. Wang, Y. Gao, J. Li, B. Hu, Influence of hot air and natural drying on extrusion properties of *maize. American Society of Agricultural and Biological Engineers Annual International Meeting,* **3**: 2238-2249, (2011)
2. M. Angelovič, J. Jobbágy, P. Findura, S. Fiantoková, Savremena poljoprivredna tehnika, **41**(1): 1-8 (2015)
3. P. Doležal et al. Krmovinářství, **4**: 28 – 31 (2014)
4. M. Angelovič, P. Findura, S. Fiantoková, A. Krakowiak, U. Ziemanczyk, K. Rutkowski, Komunální technika. **8**: 20-26 (2014)
5. B. P. Marks, R. L. Stroshine, Journal of Stored Products Research, **31**(4): 343-354 (1995)
6. J. S. Burris, Postharvest Biology and Technology, **3**(2), 155-164 (1993)
7. M. I. El-Abady, Research Journal of Seed Science, **7**: 75-86 (2014)
8. A. Talla, Predicting sorption isotherms and net isosteric heats of sorption of maize grains at different temperatures. *International Journal of Food Engineering*, **10**(3): 393-401 (2014)
9. D. Ortiz, T. Rocheford, M. G. Ferruzzi, Journal of Agricultural and Food Chemistry, **64**(13), 2727-2736 (2016)
10. J. Jech, et al., Research of the effects of technical and technological conditions on efficiency of post-harvest treatment lines for cereals [In Slovak: Výskum vplyvu technických a technologických podmienok na účinnosť liniek pozberovej úpravy obilnín]. In: *Záverečná správa čiastkovej úlohy 01-VTP.č. 27-23, KSVS.* Nitra: SUA, 140 p (2002)
11. M. Angelovič, S. Fiantoková, M. Angelovič, Scientific papers Animal Science and Biotechnologies, **46**(2): 431-436 (2013)
12. I. O. Daniel, K. O. Oyekale, M. O. Ajala, L. O. Sanni, M. A. Okelana, J. A. Adetumbi, A. C. Akintobi, A. Adebisi, Research Journal of Seed Science, **5**(1): 32-37 (2012)
13. J. K. Ward, J. D. Davis, Transactions of the ASABE, **56**(4): 1503-1509 (2013)
14. J. Skalický et al. 2008. *The treatment and storage of cereals in steel silos and floored warehouses – Methodology MZe ČR [In Czech: Ošetřování a skladování zrnin ve věžových zásobnících a halových skladech (Metodická příručka MZe ČR).* Prague: Research institute of agricultural machinery (2008)
15. P. Findura, J. Turan, M. Angelovič, P. Zemánek, M. Prístavka, O. Kuruc, Acta technologica agriculturae, **14**: 5-7 (2011)
16. Niemiec M, Sikora J., Szeląg-Sikora A., Kuboń M., Olech E., Marczuk A. Przemysł chemiczny 96(**3**),685-688 (2017)
17. R. T. Paraginski, B. A. Rockenbach, R. F. Dos Santos, M. C. Elias, M. De Oliveira, Revista Brasileira de Engenharia Agricola e Ambiental, **19**(4): 358-363 (2015)
18. J. Mareček, H. Frančáková, The effect of storage in large-capacity warehouses on the qualitz of selected varieties of malting barley [In Slovak: Vplyv veľkokapacitného skladovania na kvalitu vybraných odrôd sladovníckeho jačmeňa]. In: Safety and conrol of foodstuff (Proccedings of VIII. international conference, Nitra, 30-31 March 2011)[In Slovak: *Bezpečnosť a kontrola potravín (Zborník prác z VIII. medzinárodnej vedeckej konferencie, Nitra 30. – 31. marec 2011).* Nitra: SUA, 2011, pp. 250 – 252. ISBN 978-80-552-0559-5 (2011)
19. R. T. Paraginski, N. L. Vanier, J. D. J. Berrios, M. de Oliveira, M. C. Elias, Journal of Stored Products Research, **59**: 209-214 (2014)
20. M.Volenik, V. Rozman, I. Kalinovic, A. Liska, D. Kiš, B. Šimić, Agriculturae Conspectus Scientificus, **72**(3): 215-219 (2007)
21. M. C. Gough, Journal of Agricultural Engineering Research, **31**(1): 55-65 (1985)
22. D. C. Lopes, J. H. Martins, A. J. Steidle Neto, A. F. Lacerda Filho, E. C. Melo, P. M. B. Monteiro, Biosystems Engineering, **101**(3): 325-330 (2008)
23. Z.U. Rehman, F. Habib, S. I. Zafar, Food Chemistry, **77**(2): 197-201 (2002)
24. C. Reed, S. Doyungan, B. Ioerger, A. Getchell, Journal of Stored Products Research, **43**(4): 443-458 (2007)
25. S. B. Williams, L. L. Murdock, D. Baributsa, Storage of maize in Purdue Improved Crop Storage (PICS) bags. *PLoS ONE*, **12**(1): e0168624 (2017)
26. I. Afzal, M. A. Bakhtavar, M. Ishfaq, M. Sagheer, D. Baributsa, *Journal of Stored Products Research*, **72**: 49-53 (2017)

Computer-aided fertilisation using the Nawozy-5 (Fertiliser-5) software

Michał Cupiał[1*], and *Zbigniew* Kowalczyk[1]

[1]University of Agriculture in Krakow, Faculty of Production and Power Engineering, ul. Balicka 116B, 30-149 Krakow, Poland

Abstract. The original program Nawozy-5 is an application supporting mineral fertilisation. It is intended for small farms that are not equipped with facilities enabling application of precision agriculture and do not use any commercial software for farm management support. Computational algorithms in the program Nawozy-5 make use of the nutrient balance method. The application calculates requirement for nitrogen, phosphorus, potassium, magnesium, calcium and sulphur and then selects particular fertilisers and calculates their quantities. The program is free of charge and available online on the author's website www.cupial.mcpk.net.

1 Introduction

Mineral fertilisation is one of the most important factors that play an essential role in producing high crop yields in agriculture. Wrong dosage results in decreased plant yields and increased costs of cultivation, and has adverse effect on the environment. This aspect is becoming more and more important as the mineral nutrients that are not used by the plants pose a considerable threat to the environment [1].

Precision agriculture presupposes accurate dosage of mineral fertilisers in quantities adjusted to the nutritional needs of the plants. Modern spreaders can dose fertilisers changing the dose in different parts of the field taking advantage of GIS maps and GPS navigation [3,4]. This solutions allow to utilise fertilisers in an effective manner; however, due to high prices of the machines and devices, they are, as a rule, not used in small farms. Farmers who only have several hectares of land calculate the dose of fertilisers in a simplified way, which may imply that plants will not receive the desired amount of nutrients or that too much fertiliser will be applied.

The correct fertiliser dose selection should be based on investigations into the soil content of phosphorus, potassium and magnesium, as well as the soil pH and also the need for liming. These as well as other specialist analyses can be performed at the Regional Chemical-Agricultural Stations or by companies offering such services. The analyses can be supplemented with fertilising recommendations.

Functions enabling calculation of the correct dose of mineral fertilisers are available in specialist software dedicated exclusively to this purpose and in some selected programs intended for full farm management. Such programs include NawSald, Naw-3 and InfoPlant offered by the Institute of Soil Science and Plant Cultivation [5] and also by other suppliers [6].

Farmers who have small farms typically do not use commercial software for a number of reasons. The price of the application, which does not seem high for a farmer who earns a substantial income, is often unacceptable for a small farm. An agricultural producer who owns just several hectares as a rule does not need any complex application with a lot of functions and consequently complicated to use. An essential factor is also the conviction that the benefits from applying such a program will not outweigh the cost of its purchase.

The original computer program Nawozy-5 has been developed in order to enable selection of a correct dose of fertilisers. It successor the program Nawozy-2 [7,8]. In opposite to the previous version of the application in in described program, fertiliser needs are computed based on soil richness analysis using the nutrient balance method. The software has a user-friendly interface, thanks to which it can be used by people with only limited IT skills. The paper presents a description of the software as well as the algorithms and computational techniques applied in it.

2 Input assumptions

The fundamental assumption in the creating of the software was to achieve maximum simplicity of use. The software is supposed to fulfil its functions and be friendly to the user. It was assumed that the operation of the application (in its basic version) would be intuitive and would not require any additional training. The interface and the mode of operation are similar to those in the previous version of the application (Nawozy-2), the former functions have been retained, and new ones have been added.

Calculation of the dose of pure component is done by means of balancing the needs of the plant and the available components. Apart from the availability of the

* Corresponding author: michal.cupial@ur.krakow.pl

components resulting from soil analysis, the balance is affected by other factors, such as e.g. forecrop, remains found in crop residues, organic fertilisation, green manure, etc. In the program, it is possible (albeit not recommended) to make calculations by means of the tabulation method used in the previous version of the program.

The program has been written in the Embarcadero RAD Studio environment and is intended for PCs with Windows operating system. It also runs on Linux. The application is completely free and available online. The language of the application is Polish, but the interface has been partially translated into English. It was assumed that the databases used would most of all concern conditions found in Poland; thus, full translation of the interface will only be expedient if the data are supplemented.

3 Description of the system

The interface of the Nawozy-5 program is built in the form of tabs. Calculations are performed automatically after the user has entered the required data.

3.1 Computational methods and algorithms

Calculation of the required dose of a pure component is made using the balance method. Fig. 1 shows a simplified calculation algorithm. The application calculates the balance using the data entered by the user as well as the data contained in the built-in database. The values are recalculated each time when the input parameters are changed. If the user has not entered any data, default values are taken for the calculations.

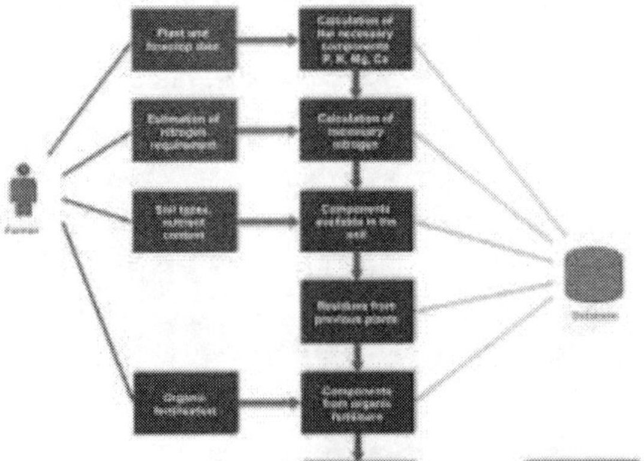

Fig. 1. Simplified calculation algorithm in the Nawozy-5 program.

The calculations start after the plant for which fertilisation is to be calculated has been selected. Data necessary for the calculations, contained in text files included with the program (marked as database in the diagram) are: information on mineral and organic fertilisers, fertilising needs of the plants, plant data as well as conversion tables.

The basis for the calculations are data obtained from soil richness analysis, such the type of soil (agronomic category), phosphorus (K_2O), potassium (P_2O_5) and magnesium (Mg) contents as well as the soil pH.

In order for the plant to be capable of producing proper crops, it has to absorb a certain quantity of nutrients. Nutritional requirements, i.e. the number of kilograms of each component per tonne of product or per hectare of land are calculated on the basis of the anticipated main and secondary crop yields. The amount of the necessary components is calculated by the program using the data contained in the tables specifying the amount of nutrients needed to produce 1 t of crop. The calculated value is adjusted by means of appropriate conversion coefficients that depend on soil richness, the type of component and organic fertilisation.

The mineral components that the plant can absorb from the soil are subtracted from the requirement calculated in this manner. They are found in organic fertilisers, in crop residues ploughed into the soil and in forecrop residues. All these parameters can be entered by the user. After the balancing is complete, the program calculates the quantity of pure component that should be supplied to the plant in mineral fertilisers.

Nitrogen (N) fertilisation is affected by other factors; therefore, additional procedures have been applied for this component in the program. In this case, the calculations require some extra parameters to be entered. As a result, a calculation is made of the number of points that enable adjustment of the calculated values. However, like in the case of the other mineral components, after the balancing the end result is the quantity of the component that should be supplied to the plant.

Based on the calculated dose of the pure component, the program allows to determine the types and quantities of the particular fertilisers that should be applied. This can be done automatically in the application or with the use of a calculation table.

In the case of automatic calculations, the program iteratively repeats the computational loop to calculate successively the quantities of fertilisers that ensure coverage of the pure component requirement. In the case of compound fertilisers, the fertiliser cover the needs with regard to at least one component while it does not exceed the needs for the other ones. In this manner, up to 5 fertilisers can be selected (automatically). There is also an option of dividing the dose of fertilisers in set proportions, which is particularly essential in the case of nitrogen fertilisers. Due to the fact that the program calculates the cost of fertilisation, the user can select the most economic variant of fertilisation. The built-in database contains most of the commercially available products along with current prices.

The application also provides access to a table (a spreadsheet) that is helpful in calculating fertiliser doses.

3.2 Layout and operation of the program

The interface of the Nawozy-5 program has been built in the form of tabs. It was assumed that such a form of application would be most intuitive for less advanced users. This form of application allows to enter data that are necessary for performing the calculations while maintaining the default values of the remaining parameters.

When the program is launched, it opens the tab that enables the user to enter the plant and forecrop. Entering a cultivated plant is the first parameter that is needed to start the calculations. This is the only required parameter; the other ones can be left as default.
Fig.2 shows the tab with the content of mineral nutrients in the soil.

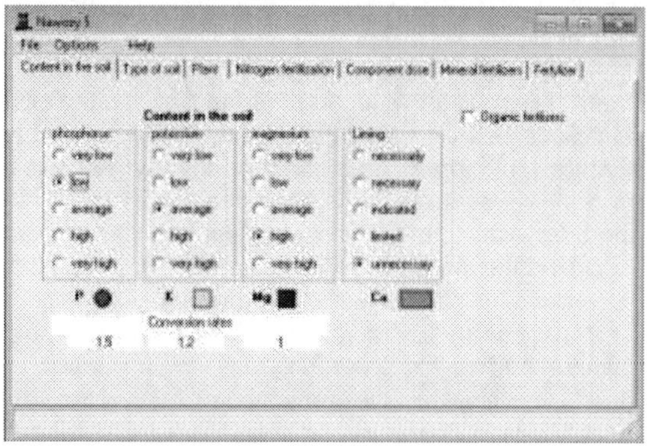

Fig. 2. Example of a screenshot of the Nawozy-5 program (data entry window).

The colours intuitively correspond to soil richness. These data are the results that were obtained in the course of soil testing at the Regional Chemical-Agricultural Stations. In this tab it is also possible to specify organic fertilisation (type and dose).

A separate tab concerns nitrogen fertilisation. In it, the user can specify the factors that affect fertilisation with this component. The point score obtained has an impact on the balance of this component.

Enabling visibility of the calculation tables in the software options enables checking the individual components of the balance.

In the program, there are two tabs that contain results of the final calculations: the dose of the component and mineral fertilisers (Fig. 3).

The table with the calculated dose of the individual components contains values in kilograms of the pure component per hectare as well as the total dose for the whole field. The final calculations are contained in the tab that enables selection of mineral fertilisers. The user can change the types of fertilisers until the desired effect is achieved. If the selected fertilisers do not allow for the components to be balanced, then the program will signal this with a proper message. The automatic selection of fertilisers can be replaced with a table allowing the quantities of fertilisers to be entered manually.

The last tab is used for calculations of the operational parameters of the fertiliser spreader.

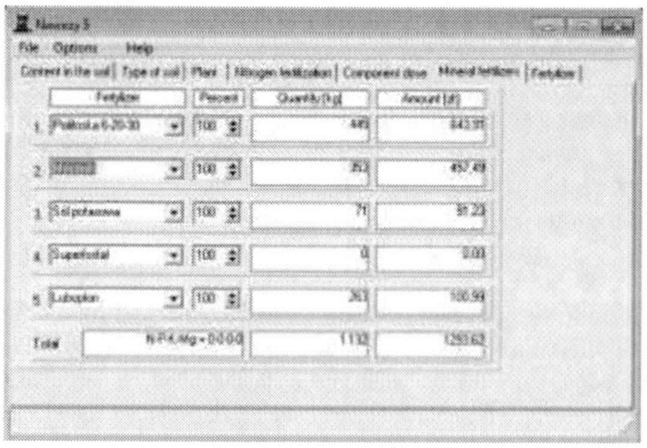

Fig. 3. Example of a screenshot of the Nawozy-5 program (calculated dose of fertiliser).

Calculations made for each field can be saved in a file for future reference. The program is provided with files containing data that are necessary for the calculations. These files can be modified by the user with any text editor with the structure of the file retained. Data concerning fertilisers and cultivated plants can be edited from within the application. This way, the fertiliser prices can be updated.

3.3 Availability of the application

The program Nawozy-5 is generally available for both private users and companies and institutions. It can be downloaded from the author's website and it is free from any charges. Even though the program is free, its registration is required in order to obtain access to some functions (Fig.4).

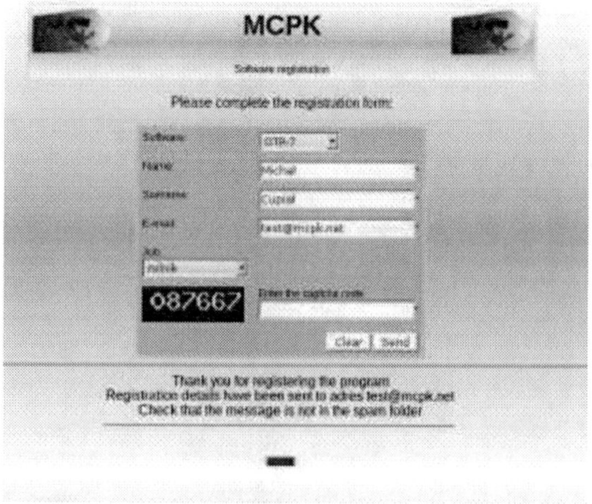

Fig. 4. Software registration website.

Registration of the program is free of charge. It was introduced in order to determine the number of users. Even though the application was not additionally promoted, 326 downloads of the program were recorded (including 27 downloads from the English version of the website) within 6 months of commencing the statistics. During this time, the software website was visited by 579 users. Not many people decided to register the program. In the period being analysed, there were 64 registrations, wherein only 32 by farmers and 6 by consultants.

It turns out that a lot of users prefer to use the less functional version of the program without having to register it. One of the reasons for this state of matters is the reluctance to provide personal data, even in such a limited scope. Another major reason is the fear that providing the email address online will result in receiving spam in later time.

4 Summary and conclusions

Calculation of the optimal dose of fertilisers, although seemingly easy, is a complicated task. Even with the knowledge of appropriate algorithms and calculation formulas, it is still necessary to bear in mind that biological material is characterised by great variability. In the case of field calculations, an additional factor is soil variability or the changing weather conditions. All these factors affect the correct choice of fertilisation. As a consequence, calculations are made for averaged conditions. While the impact of the soil variability on the surface of the field can be taken into account in precision agriculture by means of maps, then it is difficult to foresee the weather conditions, especially in the long term. Also the tables of fertilisation needs averaged for the given species may be inadequate for the new varieties.

It should be noted that there are two additional factors that have impact on the balancing of nutrients for plants. On the one hand, plants can take advantage of the components that are left behind after the preceding plants and on the other hand, they are not capable to fully utilise the fertilisers supplied to them in the particular year. Also, the balancing of nitrogen is difficult even though it is only to a small degree accumulated in the soil and its action is largely limited to one vegetation season. But unused nitrogen undergoes constant changes. Accordingly, it disappears from the soil being either washed away or released into the atmosphere [9].

The program Nawozy-5 is not intended for modern farms that apply precision agriculture using expensive and modern machines. Being intuitive and easy to use as well as free of charge, it has the task to improve the efficiency of fertilising on small farms which so far have not been applying rational fertilisation. The effect of its application should be not only increased crop yields but also elimination of unnecessary costs associated with over-fertilisation. This last factor is essential not only for economic reasons. Of great importance is also the pro-environmental action as the excess of unneeded fertilisers is a threat to the natural environment. The program Nawozy-5 also works with the OTR-7 software (Organizer of Agricultural Technology) which designed for optimization of the machinery park equipment of farm [10, 11].

References

1. M. Niemiec, J. Sikora, A. Szelag-Sikora, M. Kubon, E. Olech, A. Marczuk, Przem. Chem., **96(3)**, 685-688, (2017)
2. S. Fountas, et al., Comput. Electron. Agric., **115**, 40-50, (2015)
3. A. Kaloxylos, et al., Comput. Electron. Agric., **89**, 130-144, (2012)
4. R. Nikkila, I. Seilonen, K. Koskinenet, Comput. Electron. Agric., **70(2)**, 328-336, (2010)
5. A. Zaliwski, C. Pietruch, Agric.Eng., **2(90)**, (2007)
6. F. Terribile, et al., Solid Earth, **6(3)**, 903, (2015)
7. M. Cupiał, Agric.Eng., **9(14)**, 65-68, (2005)
8. M. Cupiał, A. Szeląg-Sikora, M. Niemiec, Proc. ECOpole, **9**, (2015)
9. W. Wu, M. Baoluo, Sc.T. Env., **512**, 415-427, (2015)
10. M. Cupiał, A. Szeląg-Sikora, M. Niemiec, Agric.& Agric. Sc. Proc., **7**, 64-69, (2015)
11. Kuźnia M., Jerzaka W., Łykob P., Sikora J. Journal of Power Technologies. **95(2)**,158-165, (2015)

Effect of hydrothermal treated corn flour addition on the quality of corn-field bean gluten-free pasta

Ahlem Dib[1], *Agnieszka* Wójtowicz [2,*] *Leila* Benatallah[1], *Abdallah* Bouasla[1], and *Mohammed Nasreddine* Zidoune[1]

[1] Laboratoire de Nutrition et Technologie Alimentaire, Institut de la Nutrition, de l'Alimentation et des Technologies Agro-Alimentaires, Université des Frères Mentouri, Constantine, Algeria
[2] Department of Food Process Engineering, University of Life Sciences, Lublin, Poland

Abstract. Corn semolina supplemented by field bean semolina in ratio of 2/1 (w/w) were used for obtaining protein and fiber enriched gluten-free pasta. The effect of hydrothermal treatment of corn flour on its applicability as gluten-free pasta improver was tested. A central composite design involving water hydration level and the amount of hydrothermal treated corn flour were used. Instrumental analyses of pasta (cooking loss, water absorption capacity, hydration and pasting properties, textural parameters and microstructure) were carried out to assess the impact of experimental factors. Results showed that hydrothermal treatment of corn flour affected in different extent on pasta properties, improving cooking and textural characteristics of pasta. The optimum formulation of corn-field bean contained 7.41 g of treated corn flour and 77.26 mL of water was selected on the base of desirability function approach with value of 0.825 which showed the best pasta properties. Obtained results showed also that addition of treated flour induced significant differences ($p < 0.05$) in all parameters in comparison with control pasta.

1 Introduction

Pasta is a traditional cereal-based food product that is becoming increasingly popular worldwide because of its convenience, palatability, and nutritional quality. High quality cooked pasta from durum wheat semolina has pleasant taste and firm texture, low surface stickiness and does not release an excessive amount of organic matter into the cooking water [1]. This is clearly related to the fact that gluten is the main structure-forming protein and it is considered as a fundamental factor of wheat pasta quality during cooking [2]. However, wheat pasta is not suitable for people with celiac disease because their immune system generates antibodies against gluten, causing damage to the tiny hair-like projections in the small intestine [3]. It is known that the treatment of celiac disease is based on life-long excluding of gluten from the diet [4, 5].

Numerous studies have focused on improving the sensory and nutritional qualities of gluten-free products especially bakery and pasta products. This has been done by using other ingredients to solve the issue of the lack of nutrients in these products and to prevent deficiency diseases such as celiac disease. Moreover, supplementation of cereals based pasta with leguminous is envisaged in order to improve the nutritional value of formula. Field bean constitutes an interesting source of fiber and protein enabling to supplement protein intake from cereals and tubers for the better amino acids balance [6] and it could enhanced nutritional value of the final product [7, 8].

The replacement of gluten presents a major technological challenge leading to the search for alternatives to gluten in the manufacture of gluten-free (GF) pasta products [2]. Currently, there have been extensive researches for the development of GF dough with appropriate viscoelastic properties, involving diverse approaches. As regards pasta, several ingredients (modified starches, gluten-free flours, and additives) have been used as an alternative to gluten in order to create a starchy network that can withstand the physical stress of cooking and impart firmness to the cooked product [9, 10].

One of the usual methods to produce GF pasta is to obtain pre-gelatinized starch through heat and cool stages, thus forming a rigid network based on the retrograded starch [11, 12]. Pre-gelatinization can help to improve functional properties and give body and texture to the product. In the literature, it has also been reported that substances that swell in water could replace gluten in the dough [13]. The pre-gelatinized starch is composed of small gel particles, that when dissolved, yield solutions of high viscosity [14]. Many foods have include pre-gelatinized starch to provide stable texture at room and refrigeration temperatures, among these are instant puddings, gravy mixes, pie fillings and glazes [15].

The various reasons, like the low availability and diversity, the high costs of GF products and the needs of

* Corresponding author: agnieszka.wojtowicz@up.lublin.pl

the celiac disease population, are considered to find appropriate products on the market. These reasons constitute our fundamental arguments to propose formula for GF laminated pasta.

The aim of this study was to evaluate the effect of hydrothermal treated corn flour addition as an improver in manufacturing of GF laminated pasta based on corn and supplemented with field bean. The evaluation was carried out using central composite design (CCD) and response surface methodology (RSM) to define the levels of water and treated corn flour needed to ensure improvement of various properties of GF pasta products. Moreover, optimum formula composition was selected on the base of desirability function approach (DFA) with best characteristics.

2 Materials and Methods

2.1 Raw materials

As Corn (*Zea mays* L) semolina (particle size below 500 µm) was provided by PZZ Lubella Sp. z o. o. Sp. K. (Lublin, Poland), containing g per 100 g of moisture - 14.5, ash - 0.1, fat 3.1, protein 6.5, fiber 2.2. Field bean semolina (*Vicia faba minor*) (particles smaller than 500 µm, composition in g per 100 g of moisture 10.15, ash - 2.37, fat - 1.29, protein - 31.2, fiber - 11.03) was obtained after grinding the dehulled bean seeds purchased from Alamir Company (Al Behera, Egypt).

2.2 Chemical composition analysis

The standard procedures of AACC methods [16] were used for analyzing fat, ash and protein content in the pasta samples making. Soxhlet extraction with n-hexane (AACC 30-10) was used for determining the fat content. The amount of ash was determined by incinerating in a muffle furnace at 550°C (AACC 08-01) and the Kjeldahl procedure (AACC 46-10) was used for determining the protein content in three replications. The procedure of AOAC [17] was used for determining fiber content with Scharrer-Kürschner method. Measurements were performed in double.

2.3 Hydrothermal treatment of corn flour

Hydrothermal treatment of corn flour (HTCF) was carried out according to Bourekoua et al. [18]. Corn flour (PZZ Lubella Sp. z o. o. Sp. K., Lublin, Poland) with particles smaller than 200-µm was used for produced treated flour. Flour was mixed with distilled water (21 ± 1°C) at the ratio 1/5 (w/w). The mixture was heated for 6-7 min with continuous stirring until the inner temperature reaches 65°C. Afterwards, the obtained paste was cooled at room temperature for 1 h, and then kept at 4°C for 24 h. Before being used as pasta ingredients, the treated corn flour was maintained at room temperature for 1 h.

2.4 Experimental Design

A CCD with 2 factors was employed to determine the effect of two variables: hydration level (X_1) and hydrothermal treated corn level (X_2) on cooking and textural properties of pasta followed by optimisation of the process using RSM. The CCD included two sections: factorial section was a 2^2 tests; the star section included four tests (2^2), and five replicated center-points were also added, for a total of $2^2+2^2+5=13$ runs (Table 1). For each response, model selection appeared to be quadratic. Responses of each variable were subjected to statistical analysis in order to define the optimal recipe using DFA. The DFA is a multi-criteria optimization method useful to find the best compromise between several responses. $D = d_i1\times d_i2 \times d_i3 \ldots\ldots \times d_in)^{1/n}$, were d_i was the desirability indices for each response (if $d_i = 0$ least desirable; if $d_i = 1$ most desirable) and n was the number of response in the measure [1].

2.5 Pasta-making process

Corn-field bean semolina mixture (CFBS) in a ratio of 2/1 (w/w) cereal-leguminous was used in GF recipe, aim to offering a better nutritional balance in amino-acids [6]. Also, supplementation corn formula based with field bean in order to achieve technological approach for processing and given a good machinability of laminated dough. The hydration level was determined by preliminary experiments. The resulting ranges of water addition were 57.49 to 97.03 mL for 100 g of recipe. Treated corn flour was added in amounts ranging from 0 to 14.82 g for 100 g of recipe. Hydration and HTCF levels were expressed based on the CFBS blend. The basic pasta recipe consisted of 66.66 g of corn semolina, 33.33 g of field bean semolina, 2 g of salt and the amount of distilled water defined in the experimental design.

Pasta produced without addition of HTCF (run 1) was considered as control pasta (Table 1), add optimal pasta preparation.

Table 1. Factors, levels and code values used in the Central Composite Design (CCD) for gluten-free CFBS pasta.

Run	Hydration X_1 (mL)		HTCF X_2 (g)	
	Code	Value	Code	Value
1	0	77.26	-1.414	0
2	+1.414	97.03	0	7.41
3	+1	91.23	-1	2.17
4	0	77.26	0	7.41
5	0	77.26	0	7.41
6	+1	91.23	+1	12.64
7	0	77.26	0	7.41
8	0	77.26	+1.414	14.82
9	0	77.26	0	7.41
10	-1.414	57.49	0	7.41
11	-1	63.28	+1	12.64
12	-1	63.28	-1	12.17
13	0	77.26	0	7.41

The pasta-making process involved premixing of dry ingredients and then distilled water was added. When HTCF was added, the corresponding amount of corn semolina and distilled water were replaced. It must be kept in mind that slurries were contained corn flour and distilled water at the ratio of 1/5 (w/w).

In the first step, the mixture was mixed for 15 min at 25°C using a kitchen Aid model kPM5 (St. Joseph., Michigan, USA). The resulting dough was rounded, divided into balls of 50 g and covered with sealed plastic wrap and then subjected to rest at 25°C for 1 h. The dough was molded and passed through reduction rolls of a pasta machine Marcato Ampia type 150 (Campodarsego, Italy) and the roller gap was maintained at 1.5 mm for the last sheeting. Finally, pasta samples were dried in oven with air circulation at 40°C for 4 h, such that it reached final moisture content below 12% and were stored in sealed plastic bags at room temperature.

2.6 Cooking properties

The optimal cooking time (OCT) of GF pasta products was determined in triplicate according to the method described by Chillo et al. [19]: 10 g of pasta was cooked in 300 mL of boiling distilled water. The OCT corresponded to the disappearance of the white core inside pasta after being compressed between two Plexiglas plates.

Cooking losses (CL) were determined by entirely evaporating the combined cooking and rinsing water in an Erlenmeyer glass beaker in an air oven (SLW 53 STD POL-EKO-APARATURA, Wodzisław Śląski, Poland) at 105°C. The residue was weighed and reported as percentage of dry sample weight before cooking [20]. Water absorption capacity (WAC) was measured as the weight increase of pasta after cooking, and was expressed as percent weight gain with respect to the weight of uncooked pasta [1].

2.7 Hydration properties

The hydration properties of optimum pasta and control sample were determined at ambient temperature designating the water absorption index (WAI) and water solubility index (WSI). Ground pasta samples (0.7 g) were mixed in centrifugal tubes with 7 mL of distilled water. After 5 min of rest the samples were hydrated for 10 min and mixed every minute for uniform reconstitution, followed by centrifugation (15000 rpm, 10 min, 21°C) in a T24 Centrifuge (VEB MLW, Leipzig, Germany). The supernatant was dried to constant weight at 105°C. Tests were carried out in four replications. Calculations of WAI and WSI were made using the equations (1) and (2), respectively, according to Wójtowicz and Mościcki [20].

$$\mathrm{WAI\ (g/g)} = \frac{\text{Wet sediment weight}}{\text{Dry sample weight}} \tag{1}$$

$$\mathrm{WSI\ (\%)} = \frac{\text{Dry supernatant weight}}{\text{Dry sample weight}} \times 100 \tag{2}$$

2.8. Pasting properties

Pasting properties of ground dry optimum and control pasta were performed according to Bouasla et al. [21] using a Brabender Micro-Visco-AmyloGraph (Brabender OHG, Duisburg, Germany) operated under constant conditions of measuring range (speed: 250 rpm; sensitivity: 235 cmg). The following time-temperature profile was used: heating from 30 up to 93°C, holding at 93°C for 5 min, cooling from 93 to 50°C, and holding at 50°C for 1 min. A heating/cooling rate of 7.5°C·min^{-1} was applied. The following indices were considered: Pasting temperature (PT) (°C), Initial viscosity (IV) (mPas), Peak viscosity (PV) (mPas), Final viscosity (FV) (mPas), Setback (Set) (mPas), and Breakdown (BD) (mPas). Measurements were performed in duplicate.

2.9 Texture measurements

Selected texture properties of GF pasta were evaluated. Texture measurements of pasta were performed on Zwick Roell BDO-FB0.5 TH (Zwick GmbH& Co., Ulm, Germany) equipped with Warner-Bratzler's knife (3 mm thick and 60 mm long, double face truncated at angle 45 °) for cutting tests of cooked optimal products and control pasta, and with OTMS Ottawa cell for firmness, stickiness and cohesiveness test of cooked pasta samples.

The hardness (H) of single cooked strand of optimal pasta samples and control pasta were measured in five replications as cutting force (N) as described by Wójtowicz and Mościcki [20]. Measurements were carried out using the 0.5 kN working head with a speed of 8.33 mm·s^{-1}. The values and curves of cutting forces were recorded and analyzed with *testXpert®* 10.11 software.

For firmness (F) (N), stickiness (S) (mJ) and cohesiveness (CH) (mJ) of cooked pasta products, 50 g of cooked and drained pasta was placed in the testing OTMS chamber, and double-compressed with a head speed 3.3 mm·s^{-1}. The software *testXpert®* 10.11 was used to record data during the tests as means of triple measurements [22].

2.10 Microstructure of dry pasta

Scanning electron microscope (SEM) pictures of optimal dry pasta as well as control pasta were taken for microstructure analysis. The surface and the cross-sections of dry samples were observed [23]. Samples of dry pasta products were cut into 5 mm pieces for surface samples and specimens of 2-3 mm in length for cross-sections were applied on carbon discs using a silver tape and sprayed with gold in a vacuum sublimator k-550X (Emitec, RC, England). The microstructure of samples was observed by a VEGA LMU microscope (Tescan, USA) at the accelerating voltage of 30 kV. Pictures were taken with magnifications × 200, × 600 for surface and × 200, × 600, × 1500 for cross-sections.

2.11 Statistical analysis

Multiple regression analysis was performed to fit second order model to dependent variables by using JMP software version 7 (SAS, USA). The models were used to determine responses surfaces in Statistica software version 10 (StatSoft. Inc., USA).

One way analysis of variance (ANOVA) was applied to compare the effect of water (X_1) and treated corn flour (X_2) on the dependent variables (Y). A coefficient of determination (R^2) was computed and the adequacy of models was tested by lack-of-fit. The model proposed for each response was:

$$Y = b_0 + b_1X_1 + b_2X_2 + b_{11}X_1X_1 + b_{22}X_2X_2 + b_{12}X_1X_2 \quad (3)$$

where: b_0 is the value of the fitted response at the center point of design, that is point (0,0); b_1 and b_2 are the linear regression terms; b_{11}, $b_{22}$are the quadratic regression terms; and b_{12} is the cross-product regression term (interaction coefficients).

Optimization was performed with JMP software version 7 to find the best formula of GF corn-field bean pasta with DFA method. For comparison of measured parameters (cooking, textural parameters, hydration index and pasting properties) between optimized recipe and control pasta results were analyzed with ANOVA test at $p \leq 0.05$ level of significance followed by the Fisher's LSD using Statistica 10 software.

3 Results and discussion

3.1 Chemical composition

The results of chemical composition of pasta samples presented in Table 2 showed that the supplementation of corn semolina with field bean semolina increased significantly ($p < 0.05$) the amount of protein and fiber level. The enrichment of formula studied based on corn semolina with legume causes an increase of 220% of protein content and 231% of fiber level. This crop could be used as good ingredients for enrichment recipe because of its great nutritional value benefit in the human diet.

In addition, celiac consumers have low intake of dietary fiber [21], so proposed formula could be beneficial as highly valuable with increased fiber content. Field bean provide an important amount of protein with good amino acids balance and supplementation of cereals with legumes could enhanced nutritional value of the final product [7, 8, 24].

Table 2. Proximate composition of gluten free pasta samples (g per 100 g^{-1} db).

Component	Corn pasta	Corn-field bean pasta
Protein	6.55 ± 0.11^b	14.47 ± 0.06^a
Fat	3.15 ± 0.06^a	0.82 ± 0.02^b
Ash	0.16 ± 0.03^b	2.44 ± 0.09^a
Fiber	2.21 ± 0.01^b	5.12 ± 0.12^a

$^{a-b}$ Means with the same superscript within row are not significantly different ($p > 0.05$).

3.2 Improving effect of hydrothermal treated corn flour on laminated GF pasta quality

3.2.1 Model fitting

The analysis of variance induced by the hydration level and the amount of treated corn flour on the quality parameters of corn-field bean semolina pasta is shown in the Table 3. The statistical analysis indicated that the fitting models were adequate because they gave satisfactory values of the R^2 and the lack-of-fit test was not significant for all responses indicating a significance and adequacy of the models.

Table 3. Regression coefficients for the predictive models for cooking loss, water absorption capacity, firmness and stickiness.

Model term	CL (%)	WAC (%)	F (N)	S (mJ)
X_1	+0.523	+0.979	-0.016*	+0.023*
X_2	+0.580	+0.012*	+0.396	-0.037*
X_1X_1	+0.025*	-0.135	+0.000*	+0.556
X_2X_2	+0.014*	-0.000*	+0.301	+0.001*
X_1X_2	+0.341	-0.001*	+0.984	-0.870
Lack of fit	ND	ND	ND	ND
F	3.66^{NS}	17.54^{NS}	12.88^{NS}	8.37^{NS}
R^2	0.72	0.93	0.90	0.86

CL: cooking loss, WAC: water absorption capacity, F: firmness, S: stickiness, X_1: water, X_2: HTCF, X_1X_1, X_2X_2: quadratic coefficients, X_1X_2: interaction between coefficients, *F*: variance Ficher- Snedecor, R^2: coefficient of determination, NS: not significant ($p > 0.05$),* significant at $p \leq 0.05$.

3.2.2 GF pasta quality characteristics

The effect of different levels of water and treated corn flour on cooking loss (CL) and water absorption capacity (WAC) of pasta is shown respectively on the Fig. 1 and Fig. 2. The CL of pasta samples varied from 11.00 to 16.06% (Fig. 1). CL decreased with increasing addition of HTCF up to an incorporation level of 8%, which increased with the maximum incorporation as its quadratic effect was positive ($p < 0.05$, Table 3). In addition, CL decreased with increasing level of water and the data in Table 3 showed the presence of positive quadratic effect of water on the value of CL. Heat-treatment of corn flour seems to led to the creation of less soluble components, responsible for the low CL.

In addition, starch is often considered as the main structural network in GF pasta due to its functional properties [25]. The formation of a strong network of retrograded starch around the gelatinized starch triggers less leaching of starchy components from the surface of GF pasta, therefore, less cooking loss [26].

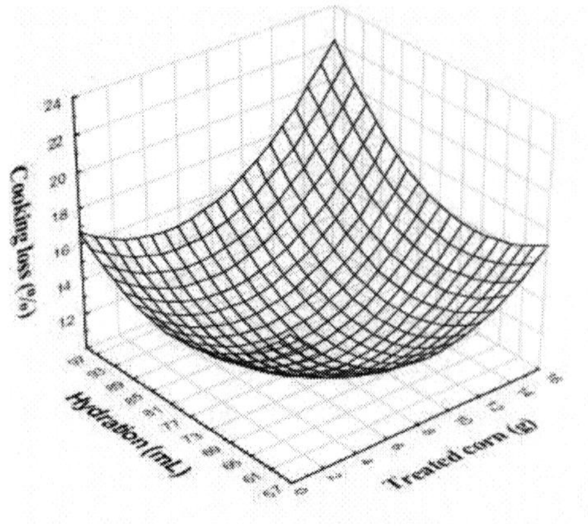

Fig. 1. Response surface of CL of gluten-free CFBS pasta depend on hydration level and HTCF amount.

The water absorption capacity (WAC) of pasta from all 13 runs ranged from 160.10 to 237.00% (Fig. 2). WAC increased with increasing addition of HTCF up to an incorporation level of 14% as its positive linear effect and decreased with maximum incorporation as its negative quadratic effect ($p < 0.05$, Table 3). Moreover, the chart (Fig. 2) showed a negative interaction effect between two variables (water and HTCF) ($p < 0.05$, Table 3).

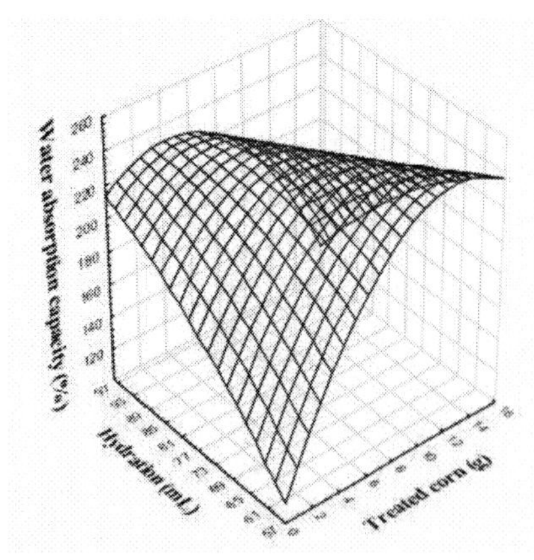

Fig. 2. Response surface of WAC of gluten-free CFBS pasta depend on hydration level and HTCF amount.

WAC is an important measured index, which depends on the amount of damaged starch and the weakness of its granules [21]. This behavior suggests that conditions of hydrothermal treatment of corn flour seemed led to the formation of more hydrophilic structure causing higher water absorption. Moreover, low WAC indicates poor quality of cooked pasta due to the chewy texture [27]. Many authors confirmed the dependence of GF pasta cooking properties with water amount and indicated also its dependence on the type of raw material [3].

Regarding texture, response surface for all 13 runs showed that the firmness ranged from 352.90 to 516.00 N (Fig. 3).

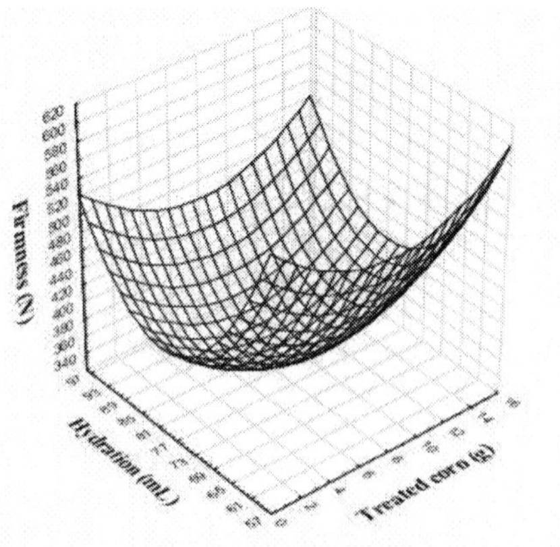

Fig. 3. Response surface of F of gluten-free CFBS pasta depend on hydration level and HTCF amount.

The graph showed that the firmness of samples decreased slightly with the addition of HTCF up to an incorporation level of 6%, which increased with the maximum incorporation. The chart (Fig. 3) showed that the firmness decreased with increasing level of water in the recipe up to 82 mL as its linear effect was negative which increased with the maximum hydration as its quadratic effect was positive ($p < 0.05$, Table 3). Response surface obtained for the stickiness measurement showed that this parameter ranged from 22.14 to 47.5 mJ (Fig. 4). Stickiness was much higher if higher amount of water was used during pasta preparation as its linear effect was positive ($p < 0.05$, Table 3). Moreover, increasing amount of treated corn flour affected higher pasta stickiness. the chart (Fig. 4) showed that the stickiness decreased with increasing level of up HTCF up to an incorporation level of 8% as its linear effect was negative which increased with the maximum addition as its quadratic effect was positive ($p < 0.05$, Table 3).

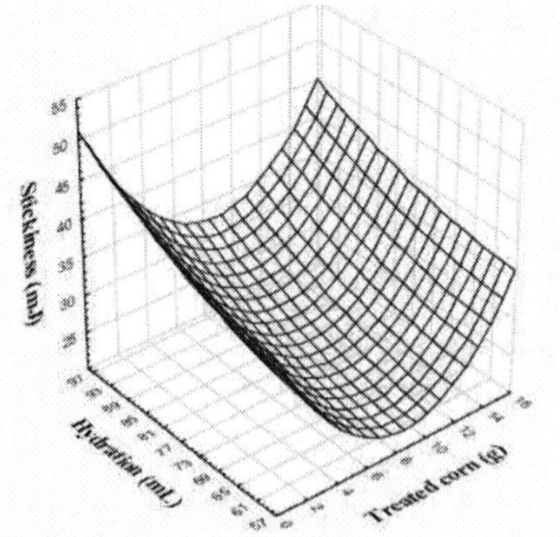

Fig. 4. Response surface of S of gluten-free CFBS pasta depend on hydration level and HTCF amount.

Generally, the addition of treated corn may improve the textural quality of GF pasta particularly in term of stickiness, because the partial gelatinization of starch increases the dough consistency. Heat treatment of corn flour seem to favor the formation of a strengthened starchy network, involving the majority of starch macromolecules (as exhibited by its low cooking loss) with a positive effect on the texture of cooked pasta in terms of its high firmness and low stickiness. These modifications of starch organization are responsible for decreasing stickiness and increasing hardness of pasta. In addition, Raina et al. [28] reported that the textural quality of both uncooked and cooked rice pasta improved significantly when treated rice flour was used. Moreover, the intensity of flour pre-gelatinization plays a very important role in imparting a desirable pasta texture [2]. Also, a similar behavior was found by Hormdok et al. [29] when used rice from different treatment (annealing and treatment at low moisture and high temperature) in order to improve of rice noodles quality. These particular hydrothermal treatment suppress granule swelling, retard gelatinization, and increase starch paste stability [29, 30], thus improving cooking behavior and texture properties of rice noodles [2].

3.3 Characteristics of optimum GF pasta

3.3.1 Cooking and textural quality

The selection of improver content and water level were obtained by applying the DFA to obtain optimum recipe with low CL and stickiness, higher WAC and firmness (Table 4).

Table 4. Criteria used to obtain the desirability functions, predicted responses.

Parameter	Objective	Predicted response	Experimental response
CL (%)	minimum	11.05[a]	11.03[a]
WAC (%)	maximum	236.47[a]	236.42[a]
F (N)	maximum	353[a]	353.03[a]
S (mJ)	minimum	16.10[a]	26.12[a]

CL: cooking loss, WAC: water absorption capacity,
F: firmness, S: stickiness,
[a]Experimental responses with the same superscript than predictive response indicate there are not significant differences ($p > 0.05$).

The optimal amounts of 100 g of CFBS were 7.41 g of treated corn and 77.26 mL of water with a desirability value of 0.825. These levels of improver and water resulted in GF pasta with good quality characteristics compared to the pasta without improver. The cooking behavior, textural properties (cutting force, firmness, stickiness and cohesiveness) of optimum and control GF pasta are shown in the Table 5.

The optimum pasta had significantly lower CL (11.03%) than control pasta (16.12%). As regard WAC, optimum recipe showed higher values (236.46%) than control pasta (160.17%). In addition, optimum sample presented the higher and better values of hardness of cooked pasta (1.43 N), firmness (353.03 N) and cohesiveness (24.13 mJ) with lower stickiness (26.12 mJ) than the control pasta (0.51 N, 324.07 N, 15.59 mJ, 47.74 mJ, respectively).

Table 5. Characteristics of optimal and control gluten-free CFBS.

Parameters	Optimum pasta	Control pasta
	Cooking quality	
CL (%)	11.03 ± 0.03[a]	16.12 ± 0.12[b]
WAC (%)	236.46 ± 0.16[a]	160.17 ± 0.19[b]
	Textural parameters	
H (N)	1.43 ± 0.03[a]	0.51 ± 0.03[b]
F (N)	353.03 ± 0.07[a]	324.07 ± 1.63[b]
S (mJ)	26.12 ± 0.05[a]	47.74 ± 0.22[b]
Ch (mJ)	24.13 ± 0.07[a]	15.59 ± 0.08[b]
	Pasting properties	
PT (°C)	76.73 ± 0.20[a]	73.33 ± 0.49[b]
IV (mPas)	17.33 ± 0.57[a]	13.58 ± 0.52[b]
PV (mPas)	175.33 ± 3.51[a]	261.33 ± 3.21[b]
FV (mPas)	305 ± 6.80[a]	461 ± 2.64[b]
BD (mPas)	19.66 ± 2.51[a]	234.33 ± 3.05[b]
Set (mPas)	149.66 ± 2.51[a]	34.33 ± 2.08[b]
	Hydration properties	
WAI ($g \cdot g^{-1}$)	2.66 ± 0.07[a]	1.15 ± 0.11[b]
WSI (%)	9.2 ± 0.04[a]	13.13 ± 0.04[b]

[a-b] Means with the same superscript within row are not significantly different ($p > 0.05$).

Hydrothermal treatment of corn flour which has partially gelatinized starch improved the quality of the corn-field bean pasta, in terms of cooking loss and water absorption. The treated corn flour may have act as a binder, re-polymerizing into a network around the starch granules because of the different gelatinization temperatures, thereby increasing their tolerance to cooking stress, as suggested by Pagani [31] and Marti et al. [2]. Moreover, the best cooking and textural properties in optimum formula might be related to the pasting properties derived from pre-gelatinized corn flour. Hydrothermal treatment of corn flour might stimulate the variation in the physical and chemical properties of starch. These modifications of starch organization are responsible for stronger structure than control pasta resulting reduced stickiness and cooking loss and improved pasta texture.

3.3.2 Hydration properties

The results of hydration properties of pasta sample are presented in Table 5 and showed significances differences in WAI and WSI between optimum and control pasta ($p < 0.05$). The significant increase of WAI in optimum sample as compared to control results probably from greater possibility of exposed hydrophilic groups to create bonds with water molecules while gel formation. Moreover, the reduction in WSI was observed compare to the control sample without HTCF.

Hydrothermal treatment applied in our work probably leads to re-organization of the starch granules, what entails other specific interactions between the starch molecules [32] such as formation of more ordered double helical amylopectin side-chain clusters and amylose–lipid complexes located within the granules. The similar phenomenon of decrease of solubility index was reported in other studies [33, 34].

3.3.3 Pasting properties

The pasting properties of both optimum and control recipe were investigated by the measurements of viscoamylograph test. The results of pasting parameters presented in Table 5 showed some significances differences in pasting properties of optimized formula (PT, IV, PV, FV, BD and Set) in comparison with the control pasta without HTCF ($p < 0.05$). These differences can be due to the modification and changes of rigidity of starch granules after treatment. The Brabender viscograms showed increasing the IV of HTCF (17.33 mPas) as confirmation of partial gelatinization of starch under hydrothermal treatment comparing to control pasta (13.58 mPas). The high IV for HTCF was attributed to the modification in the molecular organization of the starch during treatment, which resulted in loss of granule integrity and breaking the order of crystallinity of the starch [32, 33]. Hydrothermal treatment of corn flour exhibited a significant lowering of PV in optimum recipe (175.33 mPas) comparing to control pasta (261.33 mPas). The decrease in PV might be relating with the limited swelling capacity and had an effect on increasing PT (76.73°C) in comparison with control formula with HTCF (73.33°C). This corresponds with the increased temperature needed to gelatinize undamaged starch granules, as found by Hormdok and Noomhorm [29]. Observations of the increase in pasting temperature, decrease in PV, FV and BD, might be elucidated by alterations of pasting properties in treated flour possibly due to production of the bonds between the chains in the amorphous regions of starch granules as well as modification of crystallinity during a hydrothermal treatment, as reported previously by other authors [34, 35]. Moreover, reduction in BD after heat treatment indicates higher stability of starch subjected to continuous heating, as reported by Adebowale et al. [36].

3.3.4 Microstructure of dry pasta

Surface of traditional laminated GF pasta is presented on the Fig. 5. Structure of pasta fabricated from CFBS recipe without improver (Fig. 5A) was compact with small amount of empty spaces but lack of continuity due to the absence of gluten and starch granules were not well agglomerated and separate starch granules were visible on pasta surface, especially at high magnification (Fig. 5B).

Some structures of proteins matrix can be observed around starch granules which affects low amount of components leaching from the surface during cooking. More compacted and uniform structure was observed on the surface of pasta laminated from optimum recipe with addition of treated corn flour (Fig. 5C). Agglomerates of starch and proteins are surrounded by pre-gelatinized starch which forms continuous phase responsible for better results of CL and lower amount of starchy components passing into cooking water. Pre-gelatinized corn flour forms aggregates linked all pasta components what is clearly visible at high magnification (Fig. 5D).

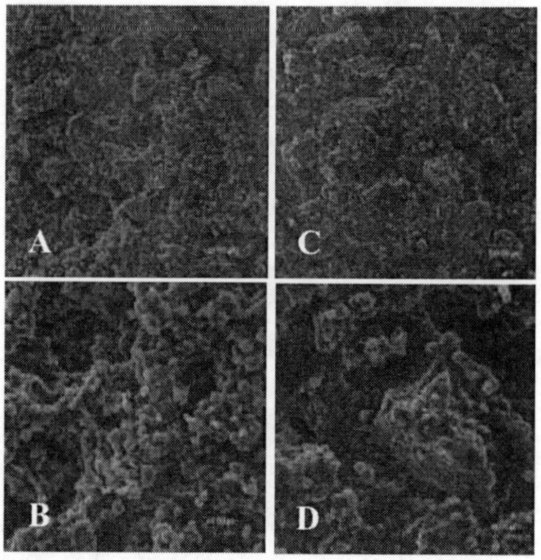

Fig. 5. SEM surface of laminated pasta from control and optimized recipe with different magnifications; control pasta: A x 200, B x 600, optimum pasta: C x 200, D x 600.

Analyzing cross-sectional microstructure of pasta presented on the Fig. 6, it could be concluded similar observations.

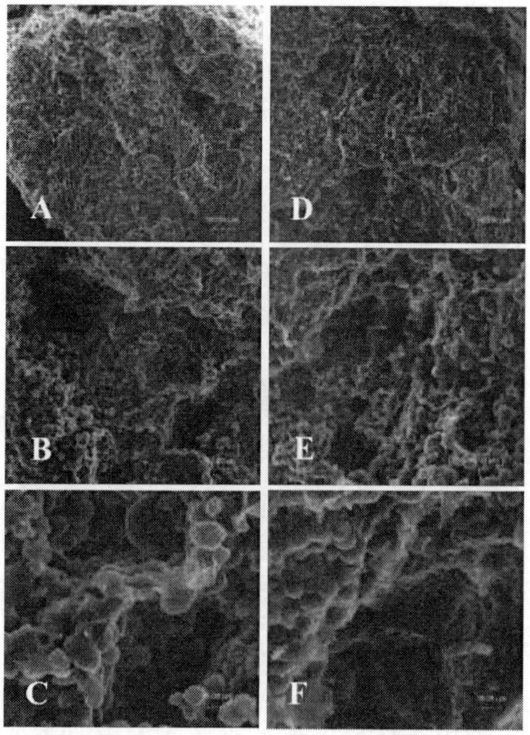

Fig. 6. Cross-section with various magnifications of control CFBS pasta (A x 200, B x 600, C x 1500) and optimum pasta with the addition of HTCF (D x 200, E x 600, F x 1500) observed with SEM.

Fig. 6 B shows the loose and discontinuous internal structure of CFBS pasta with small amount of swollen starch granules with visible higher dimensions and only few places are linked by protein fibrous fraction comes from field bean flour (Fig. 6C).

More compact and homogenous structure can be observed if pre-gelatinized corn flour was used as pasta improver (Fig. 6D). Comparing the size and amount of visible starch granules pasta with addition of treated corn flour shows lower amount of free starch and more uniform distribution of pasta components inside the pasta tread. Starch granules are surrounded by continuous phase of gelatinized starch and packed tightly next to each other, which limits leaching of unbounded components during cooking and guarantee better firmness of improved pasta (Fig. 6E and 6F).

4 Conclusion

A recipe based on corn-field bean semolina was used with hydrothermal treated corn flour added as pasta improver. Enrichment of formula with field bean semolina provide an important amount of dietary fiber and protein with good amino acids balance and could improved nutritional value of the final product. An experimental design, used to optimize the level of water addition and the amount of improver, revealed the impact of those factors on the quality characteristics of GF pasta and allowed determining the optimal levels for improving the cooking quality and textural parameters. The optimal GF corn-field bean pasta was produced by incorporating 7.41 g of treated corn flour and 77.26 mL of water (based on 100 g of semolina blends). According to the results of cooking and texture parameters, pasting and hydrations properties and microstructure observations, it can be concluded that optimized corn-field bean pasta was obtained using hydrothermal treated corn as pasta improver, leading to produce GF laminated pasta for celiac consumers.

References

1. V. Larrosa, G. Lorenzo, N. Zaritzky, A. Califano, LWT-Food Sci. Technol. **70,** 96 (2016)
2. A. Marti, R. Caramanico, G. Bottega, M.A. Pagani, LWT-Food Sci. Technol. **54,** 229 (2013)
3. V. Larrosa, G. Lorenzo, N. Zaritzky, A. Califano, J. Cereal Sci. **57,** 520 (2013)
4. K.A. Bascuñán, M.C. Vespa, M. Araya, Eur. J. Nutr. in Press (2016)
5. G. Vici, L. Belli, M. Biondi, V. Polzonetti, Clin. Nutr. **35(6),** 1236 (2016)
6. L. Benatallah, M.N. Zidoune, C. Michon, Int. Rev. Chem. Eng. **4,** 474 (2012)
7. M.A. Gimenez, S.R. Drago, D. De Greef, R.J. Gonzalez, M.O. Lobo, Food Chem. **134,** 200 (2012)
8. G. Giuberti, A. Gallo, C. Cerioli, P. Fortunati, F. Masoero, Food Chem. **175,** 43 (2015)
9. N. Sozer, Food Hydrocoll. **23,** 849 (2009)
10. M.A. Loubes, S.K. Flores, M.P. Tolaba, LWT-Food Sci. Technol. **69,** 280 (2016)
11. F. Cabrera-Chávez, A.M. Calderón de la Barca, A.R. Islas-Rubio, A. Marti, M. Marengo, M.A. Pagani, S. Iametti, LWT-Food Sci. Technol. **47(2),** 421 (2012)
12. A. Marti, K. Seetharaman, M.A. Pagani, J. Cereal Sci. **52,** 404 (2010)
13. S. Chillo, V. Civica, M. Iannetti, N. Suriano, M. Mastromatteo, M.A. Del Nobile, Carbohydr. Polym. **78,** 932 (2009)
14. J. BeMiller, R. Whistler, Carbohydrates. In: Fennema O. (Ed.), *Food Chemistry*. Taylor & Francis (1996)
15. C. Moore, J. Tuschhoff, C. Hastings, R. Schanefelt, applications of starches in foods. In: Whistler, R. BeMiller, J. E. Paschall (Eds.), *Starch: Chemistry and Technology*. Academic Press (1984)
16. AACC, Approved methods of the American Association of Cereal Chemists (9th ed.) St. Paul, USA (1995)
17. AOAC, Official methods of AOAC International (17th ed.). AOAC International (2002)
18. H. Bourekoua, L. Benatallah, M.N. Zidoune, C.M. Rosell, LWT-Food Sci. Technol. **73**, 342 (2016)
19. S. Chillo, J. Laverse,P.M. Falcone, M.A. Del Nobile, J. Food Eng. **83,** 492 (2007)
20. A. Wójtowicz, L. Mościcki, LWT-Food Sci. Technol. **59,** 1175 (2014)
21. A. Bouasla, A. Wójtowicz, M.N. Zidoune, LWT-Food Sci. Technol. **75,** 569 (2017)
22. C. Martinez, P. Ribotta, A. Leon, C. Anon, J. Texture Stud. **38,** 666 (2007)
23. M.A. Sadeghi, S. Bhagya, J. Food Sci. **73,** 229 (2008)
24. A. Kocira, S. Kocira, M. Świeca, U. Złotek, A. Jakubczyk, K. Kapela, Sci. Hortic. **76**, 214 (2017)
25. M. Lucisano, C. Cappa, L. Fongaro, M. Mariotti, J. Cereal Sci. **56,** 667 (2012)
26. A. Bouasla, A. Wójtowicz, M.N. Zidoune, M. Olech, R. Nowak, M. Mitrus, A. Oniszczuk, J. Food Sci. **81(5)**, C1070 (2016)
27. N. Wang, L. Maximiuk, R. Toews, Food Chem. **133,** 742 (2012)
28. C.S. Raina, S. Singh, A.S. Bawa, D.C. Saxena, J. Texture Stud. **36,** 402 (2005)
29. R. Hormdok, A. Noomhorm, Food Sci. Technol. **40,** 1723 (2007)
30. R. Hoover, T. Vasanthan, Carbohydr. Res. **252,** 33 (1994)
31. M.A. Pagani, Pasta products from non-conventional raw materials. In: Mercier, C. Cantarelli, C. (Eds) *Pasta and extruded products*. Elsevier Applied Science (1986)
32. H.M. Lai , H.H. Cheng, Int. J. Food Sci. Technol. **39,** 201 (2004)

33. A. Marti, M.A. Pagani, Trends Food Sci. Technol. **31,** 63 (2013)
34. O.O. Olayinka, K.O. Adebowale, B.I. Olu-Owolabi, Food Hydrocoll. **22,** 225 (2008)
35. E.R. Zavareze, A.R.G. Dias, Carbohydr. Polym. **83,** 317 (2011)
36. K.O. Adebowale, T.A. Afolabi, B.I. Olu-Owolabi, Food Hydrocoll. **19,** 974 (2005)

Models of adaptation of the milking machines systems

Vasyl Dmytriv[1,*], *Ihor* Dmytriv[1], *Yurij* Lavryk[1], and *Ivan* Horodeckyy[1]

[1] Lviv National Agrarian University, Faculty of Mechanic and Power Engineering, Lviv-Dubliany, Ukraine

Abstract. Two systems of milking machines were considered - biotechnical and vacuum. Methodology for estimation of efficiency of the "machine-animal" biotechnical system was worked out. The dependences of the efficiency parameters of the technical system operation were analyzed. The K_{MMO} operator load factor of the milking machine was proposed. The factor characterizes the technological process of the machine milking. The analytical dependences were worked out for the simulation of the productivity of milking machines and oscillation of the vacuum-gage pressure. As to the simulation results, when the vacuum pipeline diameter was increased the oscillation of the vacuum-gage pressure decreased. The results of analysis and theoretical researches on technological process of the cow machine milking gave a possibility to define the requirements to the improvement of technological process and technical equipment, which will provide the increase of efficiency of the milking systems. Usage of the developed cyber-physical system of the machine milking of cows will increase the productivity of the milking machine in 1.26...1.85 times. At the vacuum gage pressure oscillation of Δp_{vp} = 2500 Pa the suction ability of milking machine will be E=4.093 m/s accordingly. The defined index of efficiency of the adapted systems functioning of the milking machine is K_{BTS2} = 55.3.
Key words: milking machine, mathematical model, vacuum, amplitude, load factor.

1 Introduction

We will consider the process of cows milking as a biotechnical system of the "person-machine-animal", that does possible the functional ensuring of realization of the genetic potential of the cow productivity with the interaction of objects of this system. Efficiency of functioning of the biotechnical system depends on the parameters of its objects that ensure the quality and efficiency of technological functions.

As a basic performer of the technological process of milking, the milking machine provides adaptation of the technical system to physiology of milk ejection of the cow through vacuum gage pressure with preset parameter and possibility of its adjusting. The milking machine provides also the parameters control and maintenance at the set level during milking of cows. Pulsating air system of milking machine is the peculiarity of unit arrangement that changes the configuration and functional possibilities. Pulsating air system gives an opportunity also to work autonomically, or in composition of the automatic control system of technological process (Industrial Control – IC) of the machine milking. To operate in the composition of IC the milking machine was additionally completed by the measuring device of milk ejection intensity, microprocessor control unit and other elements of electronics [1, 2]. However, the vacuum gage pressure forming is still one of important factors of the parameters forming directly that does possible adaptation of the "machine-cow" system. The question of parameters adaptation of the milking machine systems was examined by Dmytriv V.T. and other [3-5], where influence of the technological and construction parameters on work mode of systems is analyzed. Also, the stability of vacuum gage pressure was estimated directly by the group of researchers consisting of Pazzona A., Murgia L., Zanini L. et al., depending on the method of adjusting in the vacuum pipeline or in a milking machine. They set that stabilizing of vacuum by the regulator of gravitational type is more dynamic, and the permanent to time is twice more lower from the computer-based system [6]. An analogical conclusion is also in accordance with research results of the regulators of pulsating vacuum gage pressure [7]. In particular Reinemann D.J., Schuring N. and Badel R.D. investigated the vacuum gage pressure oscillation depending on: a) configuration of the vacuum and milk hose systems (length of pipelines, hose diameter and other parameters influencing the losses of pressure); b) flowrates of milk in the milk pipeline; c) rates of air movement in the vacuum pipeline. They set that the vacuum gage pressure was raised up as in a vacuum- and milk pipeline, so in the under-teat chamber of the teat cup in the process of diminishing of the intensity of milk ejection [8].

The analysis of research works shows that the grounding of parameters of the effective functioning of the cows milking bioengineering system requires the development of conception and methodology of

* Corresponding author: Dmytriv_V@ukr.net

optimization of the mentioned system parameters and the same for the technical constituent of this system. Optimization of the bioengineering system parameters is also important for the increase of efficiency of milking machines.

2 Materials and methods

2.1. Effectiveness of the "operator-machine-animal" bioengineering system

The effectiveness of work of the machine milking operator (MMO) was determined by the productivity for the maximum system load. This parameter depends on the n_{mu} quantity of the milking units and the t_m duration of the milking of one cow (with consideration of the preparation and final operations $t_m = t_{mm} + t_{p\text{-}f}$), and also the Z_{li} labour inputs:

$$W_{mm} = f(n_{mu}, t_m, Z_{li}), \quad (1)$$

where: W_{mm} – the productivity of the milking machine, cows/h., n_{mu} – quantity of the milking units, pcs., t_m – duration of the milking of one cow, h., Z_{li} – labour inputs, MJ.

Effectiveness of functioning of the technical system of milking was estimated by the quality parameter of functioning of the vacuum system, namely the Δp_{VP} amplitude of pressure oscillation of the vacuum system of milking machine. The vacuum gage pressure oscillation depends on the Π_{es} parameters of the mentioned technical system elements, the K_{fes} quality of its interaction and the Z_{TS} energy expense for the system functioning:

$$\Delta p_{VP} = f(\Pi_{es}, K_{fes}, Z_{TS}), \quad (2)$$

where Δp_{VP} – the amplitude of vacuum gage pressure oscillation, Pa, Π_{es} – parameters of technical system elements, K_{fes} – the quality coefficient of the interaction of technical system elements, Z_{TS} – the energy expense for the technical system functioning, MJ.

Effectiveness of the bioengineering system depends on the cow milk ejection that is limited by the pressure oscillation in the vacuum system of the milking machine. The milk yield decrease can be described by the function as a result of the vacuum gage pressure oscillation [9]:

$$\Delta Q_M = Q_p \cdot \left[\frac{\varepsilon}{\varepsilon+1} \cdot \left| 1 - \frac{2 \cdot E_1}{3 \cdot J} \right|^{3/2} + \frac{1}{\varepsilon+1} \cdot \left(\left| 1 - \frac{2 \cdot E_2}{3 \cdot J} \right|^{3/2} - \left| 1 - \frac{2 \cdot E}{3 \cdot J} \right|^{3/2} \right) \right], \quad (3)$$

where ΔQ_M – the loss of cow milk yield as a result of the pressure oscillation, MJ, Q_p – the potential of the cow milk yield, MJ, ε – the ratio of the pressure oscillation duration to the total milking duration, E_1, E_2 – the suction ability of the milking machine with the modes accordingly $p_1 = p + \Delta p_{VP}/2$ i $p_1 = p - \Delta p_{VP}/2$, p – the vacuum gage pressure in the under teat chamber of the teat cup, Pa, E – the suction ability of the milking machine for the vacuum gage pressure p in the under teat chamber of the teat cup, $J = E \cdot 2/3$ – parameter of the milk ejection.

The suction ability of the milking machine was determined according to the described procedure [9]:

$$E = K_f \cdot K_\mu \cdot (2 \cdot p / \rho_M)^{1/2} / (K_f + 1), \quad (4)$$

where K_f – the proportion of phases of the milking machine, K_μ – flow coefficient, ρ_M – milk density, kg/м^3.

On the basis of the above mentioned parameter the effectiveness of the milking machine system is expressed by the formula:

$$K_{BTS} = \frac{1}{Z_{BTS}} \sum_{n=1}^{i} (Q_p - \Delta Q_M), \quad (5)$$

where Z_{BTS} – the energy expense for the functioning of the milking machine system, $Z_{BTS} = Z_{li} + Z_{TS}$, MJ.

The demonstrated formula (5) presents that effectiveness depends of the parameters of the technical system. The mentioned technical system has to ensure the conditions of the milk ejection from the cow udder according to the lactogenesis physiology demand with the reduction of the MMO labour inputs and also total energy expense lowering for the functioning of the milking machine systems.

Taking into account the order of the preparation and final operations of machine milking one can determine the K_{on} load factor of the milking machine operator by the formula [10]:

$$K_{MMO} = \frac{t_m + (n_{mu} - 1) \cdot t_f}{n_{mu} \cdot t_{p\text{-}f}}, \quad (6)$$

where t_m – duration of the one cow milking, $t_m = t_{mm} + t_{p\text{-}f}$, s [11], t_f – duration of the final operations, s.

If the $K_{MMO} > 1$ – operator is underloaded, accordingly the milking machine is not overdone on the udder teats, if the $K_{MMO} < 1$ – the machine milking operator is overloaded and the schedule of preparation and final operations is not kept, or the milking machine is overdone on the udder teats of cow.

From the equation (6) we shall derive the formula for determining the optimal quantity of the milking machines, taking into account that $t_{p\text{-}f} = t_p + t_f$, $t_m = t_p + t_{mm} + t_f$:

$$n_{mu} = \left[K_{MMO} + \frac{t_f}{t_p} (K_{MMO} - 1) \right]^{-1} \cdot \left[\frac{t_f}{t_p} + 1 \right], \quad (7)$$

where t_p – the duration of the preparation operations by the MMO, s.

Taking into account the average distribution of the duration of preparation and final operations ($t_w = K_{MMO} \cdot n_{mu} \cdot [\bar{t}_p \pm S(t_p) + \bar{t}_f \pm S(t_f)]$) the equation of the productivity of milking machine will have the appearance:

$$W_{mm} = \frac{3600 \cdot n_{mu}}{K_{MMO} \cdot n_{mu} \cdot [\bar{t}_p \pm S(t_p) + \bar{t}_f \pm S(t_f)] + (\bar{t}_f \pm S(t_f)) \cdot (1 - n_{mu})}, \quad (8)$$

where $\bar{t}_p, \bar{t}_f$ – the average value of the duration of preparation and final operations of the cow machine milking accordingly, s, $S(t_p)$, $S(t_f)$ – the root-mean-square deviation of the duration of preparation and final operations of the cow machine milking accordingly, s.

2.2 Amplitude of the pressure oscillation of vacuum system of the milking machine

The construction of the milking machine can be presented as the functionally finished elements, which are the constituents of the vacuum system of the milking machine (Fig. 1) [12].

Taking into account the law of mass conservation for the gas in the controlled volume by way of equation of the air motion mechanical energy and work for overcoming of the frictional force, the second airflow during the pumping-out of the vacuum system will be [13]:

$$m = S_{csa} \cdot \sqrt{\frac{2 \cdot g}{1+\xi} \cdot p_1 \cdot \rho_1 \cdot \left(x^{\frac{2}{n}} - x^{\frac{n+1}{n}} \right)}. \tag{9}$$

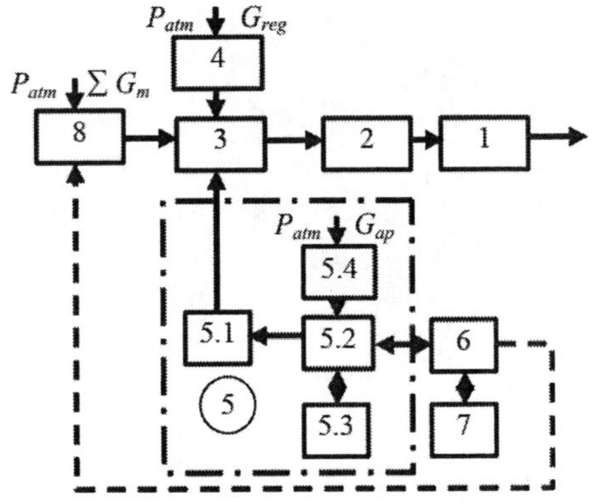

Fig. 1. The structure and functionally scheme of the vacuum system of milking machine: 1 – vacuum pump; 2 – vacuum reservoir; 3 – vacuum hose; 4 – vacuum gage pressure regulator; 5 – pulsator of the milking machine; 5.1 – chamber of the constant vacuum gage pressure; 5.2, 5.3 – chambers of the variable pressure; 5.4 – chamber of the atmospheric pressure; 6 – controller of the milking machine; 7 – interwall chamber of the teet cup; 8 – milk pipeline.

Differential equation of the V air volume pumping-out at the condition of M air mass by the τ time equals to $dM = V \cdot dp$ the mass increase [14]:

$$dM = S_{csa} \cdot \psi \cdot \sqrt{p_i \cdot \rho_i} \cdot d\tau, \tag{10}$$

where $\psi = \sqrt{\frac{2 \cdot g}{1+\xi} \cdot \left(x^{\frac{2}{n}} - x^{\frac{n+1}{n}} \right)}$ – velocity aspect ratio defines the pressure relation at the air pumping-out, x – ratio of the current value of the vacuum gage pressure to the atmospheric pressure, $x = p_i/p_A$, p_i і ρ_i – current value of the pressure and the density in the volume V.

Taking into account that $\rho_i = \rho_A \cdot \left(\frac{p_i}{p_A} \right)^{\frac{1}{n}}$, the equation (10) will have the appearance:

$$dM = S_{csa} \cdot \psi \cdot \sqrt{p_i \cdot \rho_A \cdot \left(\frac{p_i}{p_A} \right)^{\frac{1}{n}}} \cdot d\tau = \\ = S_{csa} \cdot \psi \cdot \sqrt{p_A \cdot \rho_A} \cdot \sqrt{\left(\frac{p_i}{p_A} \right)^{\frac{n+1}{n}}} \cdot d\tau \tag{11}$$

Also taking into account that $dM = Vd\rho_i$, we will have:

$$dM = V \cdot \rho_A \cdot d\left[\left(\frac{p_i}{p_A} \right)^{\frac{1}{n}} \right] = \frac{V}{n} \cdot \rho_A \cdot \left(\frac{p_i}{p_A} \right)^{\frac{1-n}{n}} \cdot d\left(\frac{p_i}{p_A} \right). \tag{12}$$

After the dependences of (11) and (12) were equated, the sign was accounted and formula was reduced by $(p_i/p_A)^{1/2}$, the differential equation of air pumping out from the V volume will have the appearance:

$$\rho_A \cdot \frac{V}{n} \cdot \left(\frac{p_i}{p_A} \right)^{-1} \cdot d\left(\frac{p_i}{p_A} \right) = \\ = -S_{csa} \cdot \psi \cdot \sqrt{p_A \cdot \rho_A} \cdot \sqrt{\left(\frac{p_i}{p_A} \right)^{\frac{n-1}{n}}} \cdot d\tau \tag{13}$$

The (13) formula is transformed and integrated with the limitations from 1 to p_i/p_A and from 0 to τ:

$$\frac{1}{n} \cdot \int_1^{\frac{p_i}{p_A}} \left(\frac{p_i}{p_A} \right)^{\frac{1-3n}{2n}} \cdot d\left(\frac{p_i}{p_A} \right) = S_{csa} \cdot \psi \cdot \frac{1}{V} \cdot \sqrt{\frac{p_A}{\rho_A}} \cdot \int_0^t d\tau. \tag{14}$$

The integration results define the duration of air pumping out of the V volume from the pressure p_A to the vacuum gage pressure p_i and will be determined in series by the formula:

$$\frac{1}{n} \cdot \frac{\left. \left(\frac{p_i}{p_A} \right)^{\frac{1-n}{2n}} \right|_1^{\frac{p_i}{p_A}}}{\frac{1-n}{2n}} = S_{csa} \cdot \psi \cdot \frac{1}{V} \cdot \sqrt{\frac{p_A}{\rho_A}} \cdot \tau \Big|_0^{\tau}, \tag{15}$$

or

$$\tau = \frac{2}{1-n} \cdot \frac{V}{S_{csa} \cdot \psi} \cdot \sqrt{\frac{\rho_A}{p_A}} \cdot \sqrt{\left(\frac{p_i}{p_A} \right)^{\frac{1-n}{n}}} - 1. \tag{16}$$

To simulate the characteristic of the vacuum gage pressure changes in the vacuum system of the milking machine one has to know the time parameters of the vacuum pressure – the time interval of the air pumping out from the volumes of the variable vacuum gage pressure. From the equation (16) the dependences of pressure changes in time have been derived:

$$p_i = p_A \cdot \sqrt[1-n]{\frac{(\tau+1)^{2\cdot n}\cdot(1-n)^{2\cdot n}\cdot S_{csa}^{2\cdot n}\cdot\psi^{2\cdot n}\cdot p_A}{4^n\cdot V^{2\cdot n}\cdot\rho^n}}, \quad (17)$$

where V – the volume of chambers of the variable vacuum gage pressure of the milking machine, м^3, S_{csa} – the cross-sectional area of the pulsator drain port, м^2, ψ – velocity aspect ratio defines the pressure relation, m$^{1/2}$/s, ρ_A – the air density in the V volume for the atmospheric pressure, kg/м^3, p_A – the air atmospheric pressure in the V volume, kg/м^2, p_i – the value of the vacuum gage pressure in the i-th point of time, kg/м^2, n – polytropic coefficient, n = 1,41.

Thus, the quantity of pumped out air will be:

$$G_i = \frac{p_i \cdot V}{R_n \cdot \Theta}, \quad (18)$$

where V – the volume of chambers of the variable pressure, м^3, R_n – gas constant of the air, R_n = 287,72 J/(kg·°C), Θ – air temperature, °C.

For connection of the i chambers with the given volume one can determine the volume, pressure and the air quantity accordingly [12]:

$$V = \sum_{i=1}^{n} V_i, \quad p_V = \frac{1}{n}\sum_{i=1}^{n} p_i, \quad G = \sum_{\tau=0}^{1} G_i. \quad (19)$$

The amplitude of the vacuum gage pressure oscillation in the vacuum pipeline is determined by the formula:

$$\Delta p_{VP} = \frac{R_n \cdot \Theta \cdot \left(G_{VP} + G + G \cdot \left(N^+_{nmu} - 1\right) \cdot P(N_{nmu})\right)}{V_{VP}}, \quad (20)$$

where G_{VP}, G – the air quantity in vacuum pipeline and volumes of the variable vacuum gage pressure accordingly, kg, V_{VP} – the volume of the vacuum pipeline of the specific milking machine, м^3, $P(N_{nmu})$ – the probability of simultaneity and coincidence of the phase and work times of the specific milking machines:

$$P(N_{nmu}) = P(N^+_{nmu}) \cdot P(N^-_{nmu}), \quad (21)$$

where $P(N^+_{nmu}) = \frac{a_1^{N_{nmu}}}{N^+_{nmu}\downarrow} \cdot e^{-a1}$ – the probability of simultaneity and coincidence of the phase and work times of the milking machines according to the Poisson limit theorem [15], $P(N^-_{nmu}) = \frac{a_2^{N_{nmu}}}{N^-_{nmu}\downarrow} \cdot e^{-a2}$ – the probability of unsimultaneity and uncoincidence of the phase and work times of the milking machines according to the Poisson limit theorem [15], $a_1 = n_{mu} \cdot p(^+_{nmu})$, $a_2 = n_{mu} \cdot p(^-_{nmu})$, n_{mu} – the quantity of the simultaneously working milking machines, pcs., N^+_{nmu} – the quantity of the simultaneously working milking machines with the coincidence in time of the phase and work times of pulsator, pcs., N^-_{nmu} – the quantity of the simultaneously working milking machines with the uncoincidence in time of the phase and work times of pulsator, pcs., $p(^+_{nmu})$ – probability of the coincidence in time of the phases and work times of pulsator for N^+_{nmu} – simultaneously working milking machines, $p(^+_{nmu}) = N^+_{nmu}/n_{mu}$, $p(^-_{nmu})$ – probability of the uncoincidence in time of the phases and work times of pulsator for N^-_{nmu} – simultaneously working milking machines, $p(^-_{nmu}) = N^-_{nmu}/n_{mu}$.

3 Results

The productivity of the milking machine (equation 8) depends on the duration of preparation t_p and final t_f operations, duration of machine milking t_{mm} and the cow productivity. Also the productivity of the milking machine depends of the quantity of the milking units, which are operated by the MMO with the regulated load factor K_{MMO}. The results of the modelling are shown in Fig. 2 and 3.

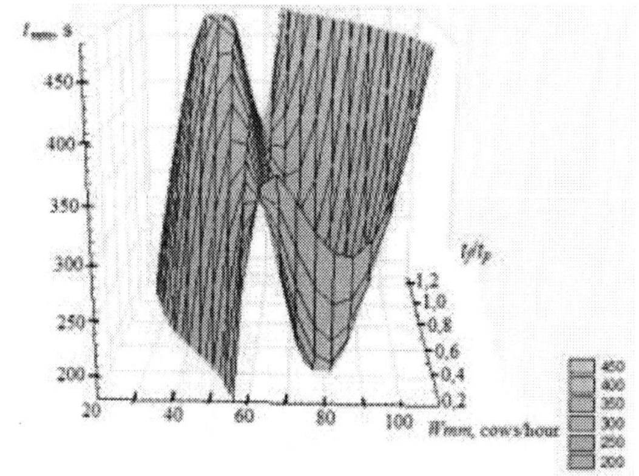

Fig. 2. Dependence of the productivity of the milking machine from the duration t_{mm} of machine milking of cow and ratio of t_f/t_p, by the $t_f \rightarrow max$, $t_p \rightarrow max$.

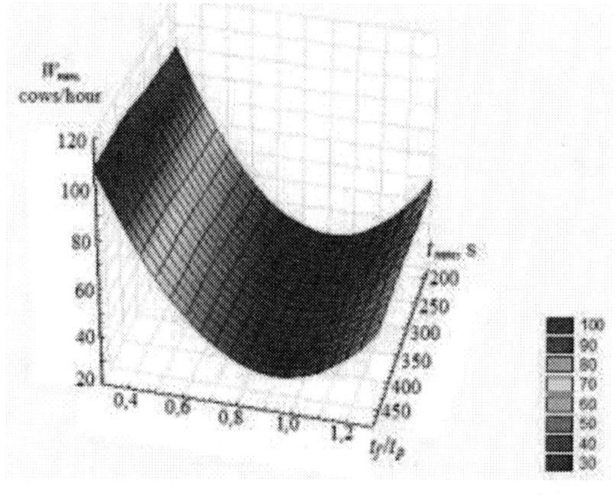

Fig. 3. Dependence of the productivity of the milking machine from the time t_{mm} of machine milking of cow and ratio of t_f/t_p, by the $t_f \rightarrow min$, $t_p \rightarrow min$

The analysis of dependence of the productivity of the milking machine from the duration t_{mm} of machine milking of cow and ratio of t_f/t_p (Fig.3) demonstrates when $t_f \to max$, $t_p \to max$, the productivity of the MMO work is not submitted to the specific law. The area on the 3-D graph of the 26...34 cows/hour productivity fits with the milking machine for the bucket milking, 34...50 cows/hour – pipeline milking in the barn, from 51 cows/hour – parlor milking.

The analysis of dependence of the productivity of the milking machine from the time t_{mm} of machine milking of cow and ratio of t_f/t_p (Fig.2) demonstrates when $t_f \to min$, $t_p \to min$, the productivity of the MMO work is submitted to the second-order equation.

The area on the 3-D graph of the 29...36 cows/hour productivity fits with the milking machine for the bucket milking, 37...50 cows/hour with the pipeline milking in the barn, from the 51 cows/hour and high with the parlor milking.

Modelling of the pressure oscillation amplitude in the vacuum pipeline is demonstrated in Fig. 4.

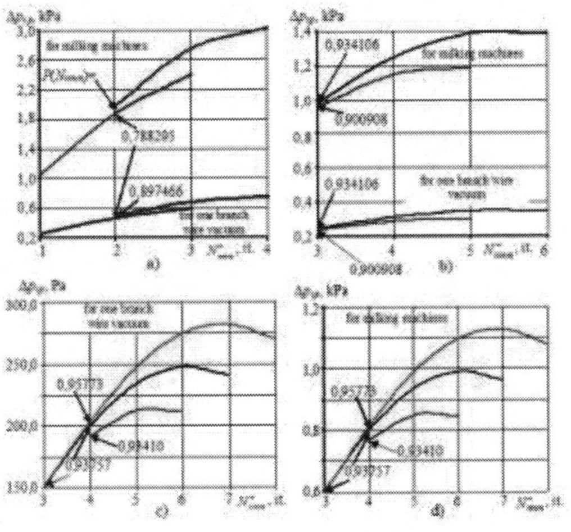

Fig. 4. Graphs of modelling of the vacuum gage pressure oscillation Δp_{vp} in the vacuum pipeline of milking machine in dependence of quantity of the milking units N^+_{nmu} with the coincidence in time of the phase and work times of pulsator with probability of $P(N_{nmu})$ for diameter of vacuum pipeline D: a – D = 25,4 mm, milking unit with pneumatic pulsator; b – D = 40 mm, milking unit with pneumatic and electric pulsator of the pair-wise action; c, d – D = 50 mm, milking unit with pneumatic and electric pulsator of the pair-wise action.

Analysis of results of the vacuum pressure oscillations modelling shows that the maximum oscillations amplitude of vacuum pressure at a frequency of 1 Hz does not exceed Δp_{vp} = 1.4 kPa for vacuum pipeline of D = 40 mm diameter with a probability of $P(N_{nmu})$ = 0.76212. To vacuum pipeline with diameter of D = 25.4 mm the maximal oscillations amplitude of vacuum pressure is Δp_{vp} = 3.03 kPa with probability of $P(N_{nmu})$ = 0.628837.

For the number of milking machines which are working synchronous with the maximum probability of coincidence in time cycles and phase of pulsators work the vacuum pressure oscillations does not exceed Δp_{vp} = 2 kPa. For the vacuum pipeline of D = 50 mm diameter the vacuum pressure oscillations are Δp_{vp} = 800...812 Pa with the maximum probability of $P(N_{nmu})$ = 0.93757...0.95773.

Graphic dependence of the vacuum pressure oscillation amplitude changes Δp_{vp} from the number of simultaneously working milking machines N^+_{nmu} with coincidence in time of the cycles and phases of pulsator work with probability $P(N_{nmu})$ and the vacuum pipeline D diameter was shown in Fig. 5.

The analysis of simulation results shows that with an increase of the number of milking machines the vacuum pressure amplitude is increased. From $N^+_{nmu} = 8$ with the n_{mu} = 8 probability of coincidence in time of the pulsator cycles and phases of $P(N_{nmu})$ = 0.592547 the oscillations amplitude of vacuum pressure will be Δp_{vp} = 4.72 kPa for the diameter of milking machines vacuum pipeline of D = 25.4 mm. For the vacuum pipeline diameter of D = 50 mm and the same number of milking machines the amplitude of vacuum pressure oscillations will be Δp_{vp} = 1.08 kPa.

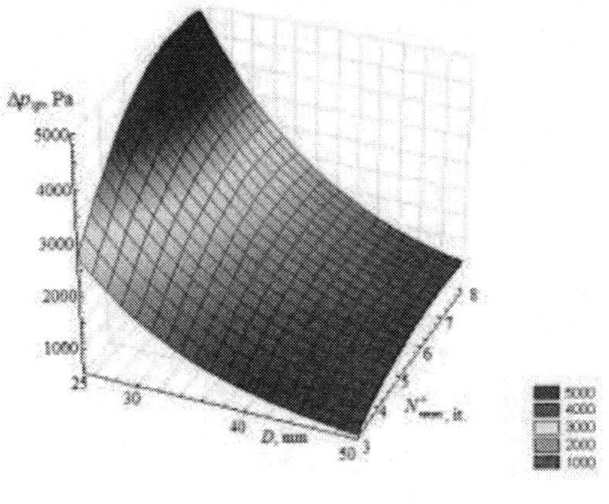

Fig. 5. The dependence of change of the Δp_{vp} vacuum pressure amplitude oscillation from the number of simultaneous working milking machines N^+_{nmu} with coincident in time of pulsator cycles and phases and D machine vacuum pipeline diameter .

The maximum probability of coincidence in time of the pulsators cycles and phases is $P(N_{nmu})$ = 0.95773 for $N^+_{nmu} = 4$ machines, then the amplitude of vacuum pressure oscillations will be equal to Δp_{vp} = 3.55 kPa per one cycle of pulsator work (frequency of 1.0 Hz) for milking machines vacuum pipeline diameter of D = 25.4 mm and Δp_{vp} = 0.81 kPa for D = 50 mm.

3 Discussion and conclusion

The dependences (1-5) show that efficiency depends on the parameters of the technical system which have to provide the conditions for the milk take out from the cow udder in accordance with the physiology of milk lactogenesis and ejection with reduction of both the MMO labor inputs and energy expense in the process of milking machine system functioning.

The requirements of the machine milking process were fulfilled for the $K_{MMO} \geq 1$. The milking duration in the ranging ratio of $t_f/t_p = 0.5...0.8$ are reduced due to the optimal number of milking machines which are operated by MMO with the $K_{MMO} \approx 1$. At the conditions of $K_{MMO} < 1$ the MMO has no enough time to fulfil all operations by the requirements of the machine milking process due to unreasonable number of operated milking machines.

It can be noted that the maximum vacuum oscillation in milking machines vacuum pipeline is not more for the permissible parameters ($\Delta p_{vp} \leq 2.5$ kPa) at the milking in the cattle barn with a frequency of 1 Hz. With the diameter of vacuum pipeline increasing the amplitude of vacuum gage pressure oscillation is decreased.

The analysis and the theoretical studies of technological process of the cow machine milking make possible to define the requirements for the improvement of technological process and technical means with the efficiency rise of milking systems in the following way:

- to provide a flexible relationship amongst the intensity of milk ejection and constructive and technological parameters of milking machine systems, to provide the adaptation of the technical system into the function of milk ejection of each individual cow;
- to create of cyber-physical milking machine that would conform with the physiological characteristics of cows as to the milk ejection;
- to optimize of the organizational and technological and structural energetic parameters of technological process and milking machines for adaptation both the technical system and the physiological state of the cow organism.

Using of cyber-physical systems of cows machine milking increases the milking machines productivity by the 1.26...1.85 times.

For the vacuum pressure oscillations of $\Delta p_{vp} = 2500$ Pa the suction ability of milking machines for modes will be accordingly of $E = 4.093$, $E_1 = 4.177$, $E_2 = 4.007$ m/s. For the milking machine phase proportions were taken the standard value. Accordingly the reducing of milk yields per cow in the energy equivalent is 65.1 MJ (3.08 MJ/kg). The energy equivalent of milk production will be 11450 MJ accordingly. The energy expenses of milking machine for the year in energetic equivalent will be 16556.4 MJ. The efficiency index of conventional milking machine will be $K_{BTS1} = 46.6$. According to the results of studies of the technological process of milking, the 450 W (1.62 MJ) reducing of the vacuum machine drive power will enable annual saving of 1773.9 MJ. Accordingly in practice one can use the 1 MMO less with saving in the energy equivalent of 1.971 MJ. Then the efficiency index of adapted systems functioning of the milking machines will be $K_{BTS2} = 55.3$. Accordingly, the efficiency index of the milking machines systems functioning will increase by 18.7%.

References

1. V. T. Dmytriv, New direction of development of the technologies and technical means in dairy cattle husbandry: Materials of the 13th International Symposium on Machine Milking of Farm Animals, 27-29 of June, 2006. *Milking machine with microprocessor control*, 161-165 (2006).
2. V. T. Dmytriv, Mechanization and Electrification of Agriculture, Interdepartmental special scientific coll. Hlevakha, *Automated milking machine*, **84**, 134-136 (2010)
3. V. T. Dmytriv, Mechanization in Agriculture. International Scientific, Scientific Applied and Informational Journal, Sofia, *Adaptive machine milking system*, **10**, 15-18 (2015)
4. V. T. Dmytriv, Mechanics and Technology Bases of the Milking Machine Systems. Theory and Practice: Monograph (2017)
5. V. Dmytriv, V. Adamchuk, I. Dmytriv, MOTROL. Commission of Motorization and Energetics in Agriculture, *Adaptive machine milking with the pneumatic electric magnetic pulsator*, **17(9)**, 83-87 (2015)
6. A. Pazzona, L. Murgia, L. Zanini, et all. ASAE Annual International Meeting, Riviera Hotel and Convention Center, Las Vegas, Nevada, USA, *Dry tests of vacuum stability in milking machines with conventional regulators and adjustable speed vacuum pump controllers* (2003)
7. V. Dmytriv, MOTROL. Commission of Motorization and Energetics in Agriculture, *Analytical study of the pressure regulator for pulsating vacuum systems*, **18/8**, 69-76 (2016)
8. D. J. Reinemann, N. Schuring, R. D. Badel, The Sixth International ASABE Dairy Housing Conference, Minneapolis, Minnesota, USA, *Methods for Measuring and Interpreting of the Milking Vacuum* (2007)
9. L.P. Kartashov, *Methods for calculating of the biological and technical parameters of the "human-machine-animal" system*, Orenburg: OSAU Publishing Center (2007)
10. V. T. Dmytriv, Research Journal of National University of Life and Environmental Sciences of Ukraine: Technics and Energy of Agriindustrial Complex, *Modeling of the number of milking machines at the optimum workload of the milking machine operator*, **212/2**, 167-173 (2015)
11. V. T. Dmytriv, *Bases of the Theory of the Machinery Usage in the Animal Husbandry*, Lviv : Afisha (2008)
12. V. Dmytriv, Journal of Lviv National Agrarian University : Agrarian Engineer Research, *Model of the air flow by the elements of the milking machine*, **10**, 483-488 (2006)
13. I. V. Dmytriv, V. T. Dmytriv, Research Journal of the Kharkiv Petro Vasylenko National Technical University of Agriculture "Technical Systems and Technologies of the Animal Husbandry", *Model of air pumping from the chambers of variable vacuum gage pressure of the milking machine*, **132**, 207-212 (2013)

https://doi.org/10.1051/bioconf/20181002004

14. V. Dmytriv, I. Dmytriv, MOTROL. Commission of Motorization and Energetics in Agriculture, *Modeling of the time of air outflow from the bounded space,* **15/4**, 193-197 (2013)
15. P. F. Ovchinnikov, B. M. Lisitsin, V. M. Mikhailenko, Higher Mathematics, Kyiv : Higher school, (1989)

A Multistate Model of Reliability of Farming Machinery

Karol Durczak[1,*], *Piotr* Jurek[1], *Jan* Beba[1], *Adam* Ekielski[2] *and Tomasz* Żelaziński[2]

[1]Institute of Biosystems Engineering, Poznań University of Life Sciences, Poland
[2] Department of Organisation and Production Engineering, Warsaw University of Life Sciences, Poland

Abstract. The article describes a multistate model of reliability of farming machinery as a deductive stochastic model of the process of changes in the technical conditions observed during operation. These conditions determine the capacity of machinery to fulfil functions, simultaneously keeping safety and maintaining acceptable costs of possible repairs. The theory of semi-Markov processes was used to solve the problem. After detailed analysis of the symptoms of damage to exemplary groups of farming machinery (rotary mowers, rotary harrows and harvesting presses) we obligatorily and arbitrarily proposed an optimal four-state reliability model to describe changes in technical conditions. In contrast to the classic reliability theory, which allows only two states of technical usability (either a machine is fit to function or not), we also allowed intermediate states, because not all types of damage affect the functionality of machinery. This approach increases the probability of technical usability of machinery and rationally delays the moment of premature repair.

1 Introduction

Reliability is the ability of a technical device to perform its required functions under stated conditions for a specified time. According to the classic reliability theory, there are only two states of usability of a technical device, i.e. the state of being usable and the state of being unusable. A device is technically usable if it is able to perform its required functions at a specified moment or interval of time. The state of unusability is the opposite. The two-value approach to reliability, without intermediate states, simplifies reality, because there are repairable devices, composed of numerous parts, which form systems. Farming machinery is an example of such devices [1, 2, 3 and 4].

Contemporary machines, which are used at each farm, must be reliable due to the specific conditions of their work. Their periods of operation are usually long (ranging from a dozen years to several decades). They work together with other machines in simple or complex units (e.g. a tractor and plough unit, a sowing and cultivation unit) during short agrotechnical terms (ranging from a few hours while spraying orchards to a few weeks during harvesting) and under difficult field and weather conditions. Due to the specific character of their work an agrotechnical procedure can be conducted even if the machine is not fully technically usable. Therefore, the traditional reliability theory should be modified by introducing additional, acceptable states of reliability.

A four-state stochastic model was proposed to model the reliability of cars [5]. Apart from the state of full (complete) usability and the state of full (complete) unusability, the authors also assumed two intermediate states, i.e. the state of partial (incomplete) usability and the state of incomplete task-related usability. This approach seems to be justified and it can also be applied for the quantification of reliability of tractors and farming machinery. Hypothetically, there could be a different number of intermediate states between the state of usability and the state of unusability, so it requires an individual approach, depending on the group of machinery under consideration.

The aim of the study was to make methodological assumptions of a multistate probabilistic reliability model, which will allow for actual changes in the technical condition of farming machinery during field work. The theory of semi-Markov processes was used to solve the problem, as it is successfully used in many branches of business activity and fields of science [6, 7, 8, 9, 10, 11, 12, 13 and 14].

2 Material and methods

Many methods, which differ in precision and range of application, were developed to calculate indicators of the reliability of technical devices with a complex structure. There are analytical methods, simulations and mixed methods. Analytical methods of calculation enable detailed analysis of random events and processes. The Markov process method and, increasingly often, the semi-Markov process method are methods specialising in the analysis of random processes [15, 16, 17 and 18]. If the Markov process method is to be applied, we need an exponential distribution of the probability of damage, which causes some limitations. This distribution is characterised by constant intensity of damage $\lambda = \text{const}$.

* Corresponding author: karol.durczak@up.poznan.pl

In practice, semi-Markov processes are proposed for the modelling of actual operation processes. They are characterised by the possibility to use various distributions of probability freely, which are generalisations of the exponential distribution, e.g. the Weibull distribution and the gamma distribution.

At the beginning it is necessary to make a very precise definition of complete usability of farming machinery. Only a brand-new machine which is released from a factory and has not worked even one hour ($t = 0$) is in the state of complete usability (CU) and its reliability amounts to $R(0) = 1$. As the machinery uptime increases, its reliability decreases. If the state of technical usability is not restored regularly, the state of complete unusability (CUn) will result. In practice, this situation should not occur and the state of usability should be restored regularly upon the observation of symptoms (organoleptic tests and diagnostic systems). A used machine will never achieve the 100% usability of a brand-new product. However, we made some simplifications of material reality in this study and the machine will achieve the CU state after being repaired.

According to the definition of reliability presented in the Introduction, a farming machine is in the state of usability only if is able to perform its required functions. The basic utility functions of farming machinery are listed in Polish standards – General Requirements and Tests, in the chapter: Operating Requirements. For example, the basic utility functions of tractor-drawn mouldboard ploughs are: stubble-field skimming and shallow pre-sowing ploughing on light and medium soils for skimming ploughs as well as medium and deep ploughing (pre-sowing and pre-winter ploughing) on all soil types for medium and deep ploughs (PN-89/R-36598). These requirements are also specified for other groups of farming machinery, e.g. rotary mowers (PN-90/R-36506). Their basic utility function is to mow meadow grass and low-stem green forage and to arrange it into windrows. A harvesting press is a tractor-drawn or self-propelled machine for collecting hay and straw in windrows or raked into rolls. Other utility functions are pressing (piston press) and baling (baler) (PN-91/R-55026).

According to the Lorenz curve, even the simplest machines lose their properties during operation and pass into intermediate states (IN). If the symptoms indicating partial usability of a machine are ignored or overlooked, the usability of the whole machine may be reduced. In consequence, it may generate costs, which are particularly significant in the case of expensive farming machinery. Delays in sowing, spraying or harvesting due to an unexpected downtime and a major repair cause serious economic problems in agricultural enterprises.

The authors of this study proposed the graph method and the theory of semi-Markov processes to solve the problem of farming machinery passing from the CU through the IN to the CUn state.

The graph theory enables pictorial presentation of the functioning scheme of a particular process. Semi-Markov processes are stochastic processes with specific properties. In order to model the process of change in the technical condition of farming machinery and the states of its reliability the semi-Markov process $\{W(t){:}\ t \geq 0\}$ can be defined by means of the homogeneous Markov renewal process [19]. The process is fully defined if its function matrix is known:

$$Q(t) = [q_{ij}(t)] \tag{1}$$

This function matrix is an adjacency matrix whose elements are the probabilities of transition from state s_i to state s_j over a period of time which is not longer than t ($i \neq j$; $i, j = 1, 2,..., k$).

Another condition is to know the initial distribution of this process:

$$p_i = P\{W(0) = s_i\},\ s_i \in S;\ i = 1, 2,\ldots, k \tag{2}$$

The construction of the semi-Markov model $\{W(t){:}\ t \geq 0\}$ of the actual process of change in the technical condition of farming machinery as the states of its reliability is the sine qua non to apply the theory of semi-Markov processes. These models require that the following conditions be met:

- the Markov condition, i.e. the future evolution of the state of any technical device must only depend on its present condition;
- random variables: T_i (the duration of state s_i, no matter which state follows) and T_{ij} (the duration of state s_i on condition that it is followed by state s_j) must have non-exponential distributions.

3 Results

The operation of farming machinery under natural conditions enables distinction of a few characteristic changes in the technical condition. However, not all of these states make further effective and high quality operation of the machinery impossible. The long-term experience gained during research on operation let us observe several characteristic states of partial usability of farming machinery. Table 1 lists typical symptoms of the states of unusability of exemplary groups of farming machinery, which can be categorised as four universal classes.

For further considerations concerning the reliability of farming machinery a set of four states was arbitrarily assumed $S = \{s_1, s_2, s_3, s_4\}$, as is shown in Figure 1.

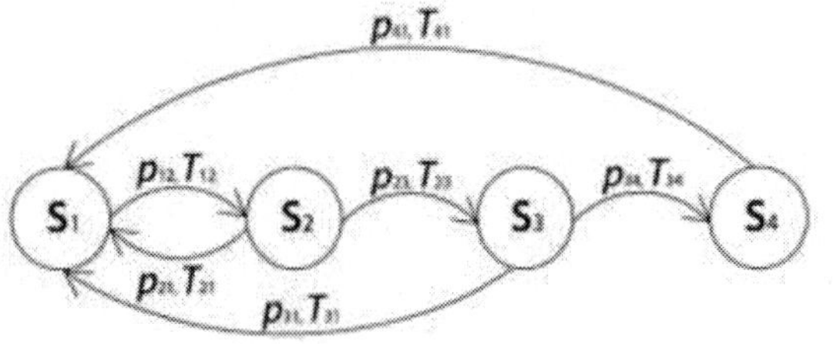

Fig. 1. A directed graph of the four-state model of the actual process of farming machinery operation.

Table 1. The states of reliability of selected groups of farming machinery and possible symptoms of damage

Machine	Technical condition	Symptoms of unusability in particular state
Rotary mower	s_1 – state of complete usability (CU)	The machine is operated within the entire range of its capacity according to specifications, without any symptoms of damage.
	s_2 – state of partial usability	The machine can be operated like in state s_1, but the quality of work is worse (blades are worn out or minimally damaged – lower quality of work, but the machine is still functional)
	s_3 – state of incomplete task-related usability	Blades are damaged considerably – lesser weight of the blade or one blade is missing because the fixing element has been broken. The machine is still basically functional, but the quality of work is very low, because it does not mow evenly. Apart from that, the missing blade causes uneven load of the rotating parts and possible damage to the propulsion system.
	s_4 – state of complete unusability (CUn)	The device does not function adequately (blades are missing, the propulsion system is broken and it is impossible to operate the device).
Rotary harrow (active)	s_1 – state of complete usability (CU)	The machine is operated within the entire range of its capacity according to specifications, without any symptoms of damage.
	s_2 – state of partial usability	The coupling in the propulsion system is not adjusted. Higher load at high revolutions.
	s_3 – state of incomplete task-related usability	The coupling in the propulsion system does not work. It may be maladjusted or broken. If the machine is overloaded, the working elements and racks and pinions in the rotor propulsion system may be badly damaged.
	s_4 – state of complete unusability (CUn)	The propulsion system is broken – it is impossible to operate the machine, especially at high revolutions.
Harvesting press (baler)	s_1 – state of complete usability (CU)	The machine is operated within the entire range of its capacity according to specifications, without any symptoms of damage.
	s_2 – state of partial usability	The terrain copy system is broken or the pick-up is maladjusted. The machine collects biomass, so it is functional, but the malfunctioning system of pick-up level adjustment (the work level is too low) may cause premature wear of working parts, such as pick-up fingers, the propulsion system and pick-up cover. They may also let foreign bodies, such as stones enter the chamber. When pick-up fingers are too close to the ground, they wear out and become damaged soon. They may be broken off. In consequence, there are gaps, which may cause inadequate picking up of the mass. The operator makes a decision when to qualify a particular degree of wear as terminal. When the pick-up working position is too low, soil may enter the press chamber and cause premature wear of rotating metal elements of the pick-up, the rotor, blades and press chamber.
	s_3 – state of incomplete task-related usability	The working elements of the pick-up, the rotor, blades and press chamber may become badly damaged. In consequence, the cutting system may become damaged (the feeding rotor and blades). The machine still makes bales although the system which should shred the collected material is out of order.
	s_4 – state of complete unusability (CUn)	A foreign body enters the machine and causes permanent damage to the working elements and the propulsion system, which renders the machine unusable.

Compiled by the authors

The states are marked with nodes. The arcs with arrows show the relations, i.e. possible changes in the technical condition. Two values were allocated to the arcs, i.e. probability (p) and the time (T) of transition between individual states.

This arrangement comprises the use of a machine in state s_1, when it is fully usable. When the machine is in the state of partial usability s_2, it should be used until a certain agrotechnical process is finished and it should undergo renewal immediately afterwards. If it does not undergo renewal, it will pass into the state of incomplete task-related usability s_3, which will increase costs and reduce safety and quality of work. When the machine is in this state, it should undergo renewal to state s1 with the least delay possible. States s_2 and s_3 should be as short as possible and the machine should undergo complete rather than partial renewal. Initial symptoms might signal a breakdown which will be expensive to repair. The transition from state s_2 or s_3 to state s_1 does not require so much workload as the transition from state s_4 to state s_1. The final state s_4 is the result of a random event known as total damage. A random breakdown event may cause this damage and disqualify the device from further use. The machine will lose its functionality.

All the states belonging to set S are values of a semi-Markov process. The elements of this set are consecutive states of reliability in a causal relationship.. The transition graph showing changes in the technical condition of farming machinery during operation (Fig. 1) was used to make the following square function matrix:

$$\boldsymbol{Q}(t)=\begin{bmatrix} 0 & q_{12}(t) & 0 & 0 \\ q_{21}(t) & 0 & q_{23}(t) & 0 \\ q_{31}(t) & 0 & 0 & q_{34}(t) \\ q_{41}(t) & 0 & 0 & 0 \end{bmatrix} \tag{3}$$

According to the farming machinery operation schedule and the state of usability, a device can fulfil its tasks in two variants $\{W(t)\colon t \geq 0\}$ [20]. These tasks must be performed while keeping adequate safety, efficiency and profitability, where the initial distribution $\{W(t)\colon t \geq 0\}$ is written as follows [21]:

$$p_1=P\{W(0)=s_1\}=1,\ p_i=P\{W(0)=s_i\}=0 \text{ for } i=2,3,4 \tag{4}$$

The probabilities of change in states of reliability of farming machinery $P = [p_{ij};\ i,j = 1, 2, 3, 4]$ make the following transition probability matrix:

$$\boldsymbol{P}=\begin{bmatrix} 0 & 1 & 0 & 0 \\ p_{21} & 0 & p_{23} & 0 \\ p_{31} & 0 & 0 & p_{34} \\ 1 & 0 & 0 & 0 \end{bmatrix} \tag{5}$$

As results from the matrix, only two transitions into states of the technical condition of farming machinery are certain in the process of their operation: transition from state s_1 to s_2 ($p_{12} = 1$) and the necessary transition from state s_4 to s_1 ($p_{41} = 1$) to make a repair.

In our example of the four-state model the limiting distribution π_j ($j = 1, 2, 3, 4$) of the process $\{W(t)\colon t \geq 0\}$ is determined by the following dependence [22]:

$$P_j=\frac{\pi_j E(T_j)}{\sum_{k=1}^{4}\pi_k E(T_k)};\ i=1,2,3,4 \tag{6}$$

where the limiting distribution is in accordance with the following equations:

$$\left.\begin{aligned} &[\pi_1, \pi_2, \pi_3, \pi_4]\cdot\begin{bmatrix} 0 & 1 & 0 & 0 \\ p_{21} & 0 & p_{23} & 0 \\ p_{31} & 0 & 0 & p_{34} \\ 1 & 0 & 0 & 0 \end{bmatrix}=[\pi_1, \pi_2, \pi_3, \pi_4] \\ &\pi_1+\pi_2+\pi_3+\pi_4=1 \end{aligned}\right\} \tag{7}$$

When we solve the system of equations from the matrix (7) and use the theorem (6), we will receive the following equations:

$P_1=E(T_1)M^{-1}$

$P_2=E(T_2)M^{-1}$

$P_3=p_{23}E(T_3)M^{-1}$

$P_4=p_{23}p_{34}E(T_4)M^{-1}$

for $M=E(T_1)+E(T_2)+p_{23}E(T_3)+p_{23}\,p_{34}E(T_4)$

where:

$E(T_j)$ – the expected value of duration of state $s_j \in S (j = 1, 2, 3, 4)$; i.e. the average durability of a technical device

p_{ij} – the probability of the process $\{W(t)\colon t \geq 0\}$ transition from state s_i to state s_j (s_i, $s_j \in S$; $i, j = 1, 2, 3, 4$; $i \neq j$)

P_j – the probability of the process $\{W(t)\colon t \geq 0\}$ remaining in state s_j ($j = 1, 2, 3, 4$).

The variant of changes in the states of reliability discussed in this article presents the situations in which the user may deliberately take the risk to continue work, remembering that it will reduce its quality and safety and it will involve higher costs. As results from experience, it is recommended to operate farming machinery in the state of complete usability s_1 and possibly in the state of partial usability s_2, having made a deliberate decision to do so. The reliability of this system increases to the value P_1+P_2. States s_3 and s_4 should be treated as states of damaged machinery, which make further operation impossible.

4 Conclusion

We used the system method of constructing deductive models and the theory of semi-Markov processes to make a four-state model of reliability of farming machinery. The process is continuous in time and discreet in states. The mathematical and logical analysis of the process as well as the examples given in the study led to the following conclusions:

1. The four-state model of reliability correctly describes the processes occurring during the operation of farming machinery. It presents the actual changes in the technical condition better than the classic reliability theory.
2. The model allows for the state of partial technical usability of machinery (state s_2). Under normal conditions of use there is damage which does not influence effective operation of the machinery.
3. When we know intermediate states, in theory it is possible to restore complete usability of machinery repeatedly until it is morally obsolete. Therefore, the indicators of costs of repair of farming machinery might be as high as 100-150% of its price [23].

4. The multistate approach is a reliable reflection of the material reality. The article provides the method which can be adapted and used according to one's heuristic knowledge by making individual adjustments concerning intermediate states and the probable consequences of ignoring them. It is the user that makes an arbitrary decision about the acceptable quality of work, safety and costs of repair of machinery.

References

1. W. Adamkiewicz i in., *Badania i ocena niezawodności maszyn w systemie transportowym.* (Wydawnictwa Komunikacji i Łączności, 1983)
2. L. Dwaliński, *Niezawodność maszyn rolniczych.* (Wema, 1988)
3. T. Otmianowski, *Procesy odnowy maszyn i ciągników rolniczych.* (Państwowe Wydawnictwo Rolnicze i Leśne, 1983)
4. C. Rzeźnik, *Podstawy obsługi technicznej maszyn rolniczych.* (Wydawnictwo Akademii Rolniczej w Poznaniu, 2008)
5. J. Girtler, M. Ślęzak, Eksploatacja i Niezawodność – Maintenance and reliability **15(2)**, 156–160 (2013)
6. J. Paska, *Niezawodność systemów elektroenergetycznych.* (Oficyna Wydawnicza PW, 2005)
7. S. McClean, A. Papadopoulou, G. Tsaklidis, Communication in Statistics – Theory and Methods **33(6),** 623-638 (2004)
8. M. Matalytski, O. Tikhonenko, *Procesy stochastyczne.* (Akademicka Oficyna Wydawnicza EXIT, 2011)
9. R. Magiera, *Statystyczne funkcje decyzyjne,* (Oficyna Wydawnicza GiS 2016)
10. M. Sobczyk, *Statystyka,* (Wydawnictwo Naukowe PWN 2017)
11. J. Jiang, G. G. Roussas, F. J. Samaniego, *Nonparametric Statistical Methods and Related Topics,* (World Scientific Publishing Company, 2012)
12. D. J. Smith, *Reliability maintainability and risk. Practical Methods for Engineers,* (Butterworth-Heinemann, 2011)
13. E. Zio, *Introduction to the Basics of Reliability and Risk Analysis,* (World Scientific Publishing Company, 2007)
14. S. Keller-McNulty, *Modern Statistical And Mathematical Methods in Reliability,* (World Scientific Publishing Company, 2005)
15. P. Lev'y, *Proceesus semi-markoviens.* (Proc. Int. Cong. Math, Amsterdam, 416–426, 1954)
16. R. Pyke, Ann. of Math. Statist., **32**, 1243–1259 (1961)
17. R. Pyke, R. Schaufele, Ann. of Math. Statist., **32**, 1746–1764 (1964)
18. M. Rosenblatt, *Procesy stochastyczne.* (PWN, 1967)
19. J. Girtler, Zagadnienia Eksploatacji Maszyn **3**,419-428 (1996)
20. F. Grabski, *Teoria semi-markowskich procesów eksploatacji obiektów technicznych,* (AMW, 1982)
21. J. Girtler, The Archives of Automotive Engineering (Archiwum Motoryzacji) **3**,165-196 (2004)
22. F. Grabski, *Semi-markowskie modele niezawodności i eksploatacji.* (IBSPAN, 2002)
23. A. Muzalewski, *Koszty eksploatacji maszyn.* (Wyd. IBMER, 2006)

Influence of raw material properties on energy consumption during briquetting process

Tatiana Ivanova[1], *Bohumil* Havrland[1,*], *Radek* Novotny[1], *Alexandru* Muntean[1], and *Petr* Hutla[2]

[1]Czech University of Life Sciences Prague, Faculty of Tropical AgriSciences, Department of Sustainable Technologies, Kamýcká 129, 165 21 Prague 6, Czech Republic
[2]Research Institute of Agricultural Engineering, Drnovská 507, 161 01 Prague 6, Czech Republic

Abstract. Biomass is doubtless a very significant source of renewable energy being worldwide abundant with high energy potential. This paper deals with assessment energy consumption at especially grinding and briquetting processes, which should result in essential economy of energy at solid biofuel production. Various types of raw materials were used at the experiment such as hemp (*Cannabis sativa* L.) biomass, two species of Miscanthus (*Miscanthus sinensis, Miscanthus x gigantheus*) and apple wood biomass. These materials were dried, grinded and pressed by piston press having pressing chamber diameter of 65 mm. Materials were grinded into three fractions (4 mm, 8 mm and 12 mm). Material throughput (kg.h^{-1}) and energy consumption (kWh.t^{-1}) were registered. As to results: the highest throughput at both grinding cases as well as briquetting was found at apple wood biomass; however the energy consumption during briquetting of apple wood was relatively high. The worst results concerning throughput and energy consumption (especially at briquetting) were found at hemp biomass. Nevertheless, briquettes made of hemp had the best mechanical durability. Both Miscunthus species (herbaceous biomass) have very similar parameters and showed quite good relation between throughput and energy consumption at the used machines.

1 Introduction

In recent years, there has been a growing interest at the global level in the use of energy biomass [1]. Biomass is the third largest alternative energy source in the world, after coal and oil. Half of the world uses biomass energy as primary energy source [2]. Biomass can be divided into waste biomass, agricultural waste and deliberately grown biomass (energy crops). Biomass cycle can help to reduce CO_2 concentrations [3] and it is the only renewable source that can be directly converted into solid, liquid or gas forms/biofuels [4].

Briquettes are produced by pressing equipment (piston press, screw press etc.). Briquetting increases energy density and significantly reduces volume of solid biofuel [5]. The big advantage is that briquetting process is pressing loose raw material into briquettes with a particular shape and reduces transportation and storage costs [1]. On the other hand, briquetting process is expensive; purchase price of briquetting lines and energy inputs are generally high [6].

There are several parameters which play an important role in briquetting process and affect the quality of briquettes as well as costs. Particle size of input material is very important for densification. According to [7] the optimum particle size is 6-8 mm with 10-20% powdery component (i.e. particles less than 4 mm). Large particles (more than 12 mm) are not suitable for briquette production, because final briquettes are not compacted. Input material has to contain smaller and also larger particles (random distribution of particle size is an optimum) as it contributes to better mechanical properties of produced briquettes [8]. According to [2] particle size and shape have also important effect on flow characteristics. The granular (preferably 6–8 mm in size) homogeneous materials which can flow easily in conveyors, bunkers and storage silos, are suitable for briquetting. Use of material with moisture content in a definite range (between 10-15%) is very important for briquettes' quality and final product properties /both thermal and mechanical, too. For example, briquettes with high moisture content have worse combustion properties as high moisture leads to loss of heat and releasing large amount of water vapour, which can then cool and damage the heater [9]. Biomass drying process and its operation is expensive. Nevertheless, modern methods are improving drying technologies and efficiency of the dryers [10, 11]

In order to produce good quality briquettes, briquetting press parameters (technological parameters) are also important [7]. As for example, compacting pressure influences strength of final briquettes. The quality of briquettes is increasing with increasing pressure in briquetting chamber [12]. Compacting

* Corresponding author: bohumilhavrland@gmail.com

pressure depends on type of briquetting press. Compacting pressure can be higher, which is related to higher energy consumption. Compacting pressure can be lower, but the pressing temperature has to be higher. That means, when the temperature is higher, final briquettes will have similar quality as high compressed briquettes [13]. Also holding time is connected with compacting pressure. [14] made a research about effect of pressure and holding time by using hydraulic press with maximal pressure 138 MPa and different holding times ranging between 0 to 60 sec. Experiment showed that holding pressure was more effected at lower pressure than at higher pressure. It follows from the study that compacting pressure is more important than holding time and the holding time at high pressure is negligible [14].

The present research paper is focused on analysis of energy consumption and material throughput, which were measured and calculated on examples of different biomass materials during both briquettes' production processes (grinding and briquetting) in order to evaluate the influence of raw material properties on above mentioned important processing characteristics. The following research hypotheses were formulated: 1) Hammer mill and piston press throughput and energy consumption differs depending on raw materials; 2) The particle size influences briquetting press throughput and energy consumption; 3) The grinding process is more energy intensive than briquetting process.

2 Material and methods

In total four different biomass materials were used for briquettes production: energy crops *Mischanthus x giganteus* and *Mischanthus sinensis* – herbaceous species, industrial hemp (*Cannabis sativa* L.) – fibrous textile plant and apple wood biomass/apple tree trimming residues. All the materials were dried naturally.

2.1. Biomass grinding and briquetting

The grinding process was carried out using Hammer mill ŠV-15 (Stoza, Czech Republic) with energy input 15 kW. The materials were crushed in three fractions (screening holes diameters: 4 mm, 8 mm and 12 mm). Hammer mills widely used for material grinding as they are relatively cheap, easy to operate and produce wide range of particle sizes [15].

Briquetting was performed on hydraulic piston briquetting press Brikstar 30-12 (Briklis, Czech Republic) with working pressure 18 MPa, power intake 4.4 kW, operation temperature 60°C, control voltage 400 V Briklis and diameter of pressing chamber 65 mm [16]. Briquettes were produced without any additional binders.

2.2 Determination of throughput and energy consumption

Material throughput during grinding as well as briquetting process was determined from weight of used material and total grinding time, and recalculated per one hour.

In order to determine energy consumption (the energy spent on unit mass per time) during both processes, the measurements of electric energy consumption were conducted using Prowatt-3 with 16 A, 380 V, Chauvin arnoux, France). Prowat-3 (see Figure 1) can measure three phases current and it is more accurate than normal electric meter. For each material and fraction the readings were taken every 10 minutes. The time is necessary for calculation, but also the mass of material is important. The energy consumption was then calculated for 60 min and energy consumption for one tonne of input material.

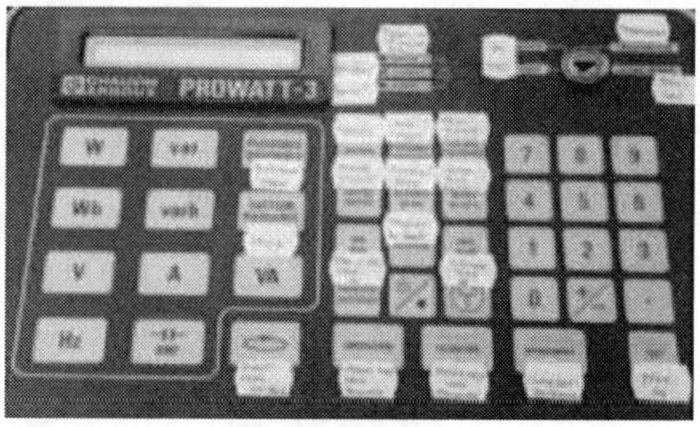

Fig. 1. Prowatt-3.

2.3. Determination of biomass moisture content

Determination of moisture content of biomass before briquetting process was conducted according to the applicable standard EN ISO 18134–2:2015-1: method of drying at a temperature of 105 °C in laboratory drier UFE 500 – MEMMERT. For calculation of moisture content the following equation was used:

$$w = \frac{m_w - m_d}{m_w} \times 100, \% \qquad (1)$$

where: w – moisture content (%), m_w – wet material weight (g), m_d – dry material weight (g).

2.4. Determination of mechanical durability of briquettes

Mechanical durability determination of produced briquettes was done according to the standard EN ISO 17831–2:2015 in briquette durability drum of cylindrical shape and volume of 160. Each sample portion of about 2 kg was rotated at 21 ± 0.1 revs. min^{-1} for 5 minutes. The mechanical durability was calculated by following equation:

$$DU = \frac{m_A}{m_E} \times 100, \% \qquad (2)$$

where: DU – mechanical durability (%), m_A – weight of input briquettes (kg), m_E – weight of output briquettes (kg).

3 Results and discussion

3.1. Evaluation of hammer mill throughput

The results of material consumption of the hammer mill for each tested material and each fraction are presented at the Figure 2. From the Fig. 2 is evident that the material throughput varies (differs) by used material and there is significant difference between apple wood and other materials, especially hemp. It is visible that the apple wood had the highest (the best) throughput. From the result of apple wood is also clear that the fraction 12 mm has higher throughput than the others fractions, followed by 8 mm and then 4 mm. Hammer mill throughput showed rising trend with increase of screen size during grinding of this wood material. *Miscanthus x gigantheus* (149.65, 140.29 and 148.64 kg.h^{-1}) and *Miscanthus sinensis* (129.48, 136.00 and 149.65 kg.h^{-1}) have almost similar throughput. During the grinding of these straw materials were not found differences between fractions. Hemp have different (very low) values (80.40, 41.50 and 51.71 kg.h^{-1}) comparing to other materials. During the measurements were observed complications with grinding of hemp due to the rolling of hemp fibres on hammers of the grinding machine.

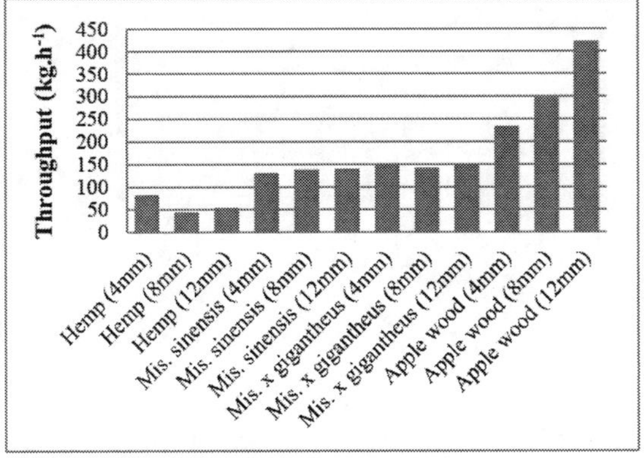

Fig. 2. Hammer mill throughput.

[17] determined material throughput for switchgrass, wheat straw and corn stover. The throughput was similar for each material and the value was150 kg.h^{-1}. [18] used chopping machine in the research and corn stalk as tested material. The particle size of the corn stalk from the chopping machine was 4 mm. The authors has recorded that the chopping capacity of the machine ranges from 300 to 500 kg.t^{-1}. It could be said that corn stalk has structure similar to *Miscantus x gigantheus*, however, in the present research *Miscanthus x gigantheus* with fraction 4 mm had throughput for hammer mill at least two times smaller.

3.2. Evaluation of hammer mill energy consumption

Figure 3 illustrates the average energy consumption for grinding of selected biomass using the hammer mill with three different screen sizes. Energy consumption to reduce hemp to particle sizes 4, 8 and 12 mm was 65.97, 117.47, 66.85 kWh.t^{-1}, respectively. Hemp had medium and non-trended energy consumption in comparison with other materials. From the Fig. 3 is evident that the *Miscunthus sinensis* and *Miscunthus x gigantheus* had the lowest energy consumption, and apple wood had the highest energy consumption (141.25, 135.29 and 129.52 kWh.t^{-1}), that can be explained as wood is hard material for grinding.

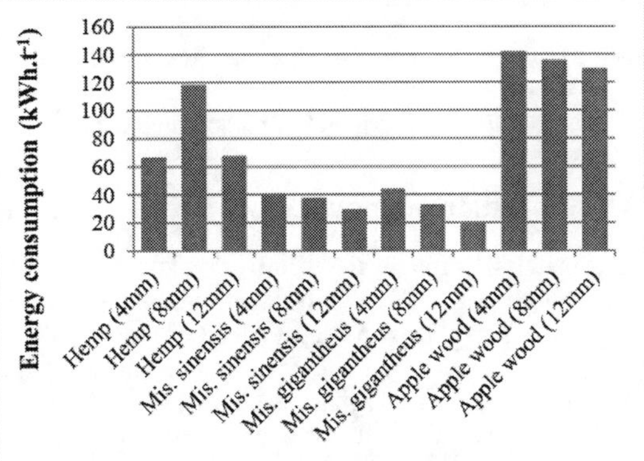

Fig. 3. Hammer mill energy consumption.

Fig. 3 also shows differences between fractions. In case of both Miscanthus and apple wood energy consumption was rising with decrease of screen size, i.e. fractions 12 mm and 8 mm had less energy consumption than fraction 4 mm. [19] measured energy consumption of hammer mill using switchgrass, which has stalk structure more similar to *Miscanthus sinensis*, and screen sizes 0.8, 1.6 and 3.2 mm. The results confirmed the difference between fractions. The smallest fraction 0.8 mm had the highest energy consumption of 62.55 kWh.t^{-1}, followed by fraction 1.6 mm - 51.76 kWh.t^{-1} and 3.2 mm - 23.84 kWh.t^{-1}. The energy consumption of hammer mill was also comparable with the work of [20], where the same relation between screen size and energy consumption was found. In contrast, [18] has measured energy consumption of chopping machine using corn stalk and screen size 4 mm, which was equal to 20 kWh.t^{-1}. The hammer mill showed higher energy consumption and only *Miscanthus x gigantheus* (fraction 12 mm) has energy consumption 19.08 kWh.h^{-1}. [19] has also reported a relation between energy consumption and moisture content of input material, i.e. the higher the moisture, the higher the energy consumption is.

3.3. Evaluation of moisture content of tasted materials

The results of material moisture contents after grinding before briquetting process are presented on the Figure 4. It is visual that the moisture content is not more than 15%. This value is a maximum border for briquettes production by Brikstar press 30-12 [16]. According to [16] the lowest moisture content of input material is 8%, but in the present research the materials with even lower moisture contents were tested for densification as well.

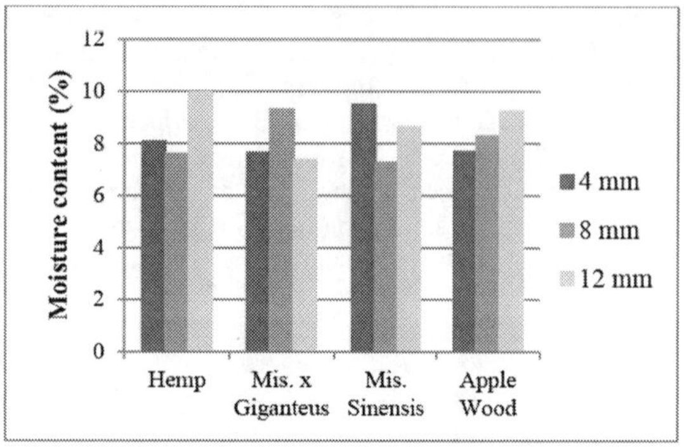

Fig. 4. Biomass material moisture content.

3.4. Evaluation of briquetting press throughput

The material throughput during briquetting is evident from the Figure 5 and it shows that throughput depends on particle size. The same dependence was mentioned by [2]. Particle size 4 mm had higher material throughput than other fractions within each tested material, so material with smaller particle size was easily pressed.

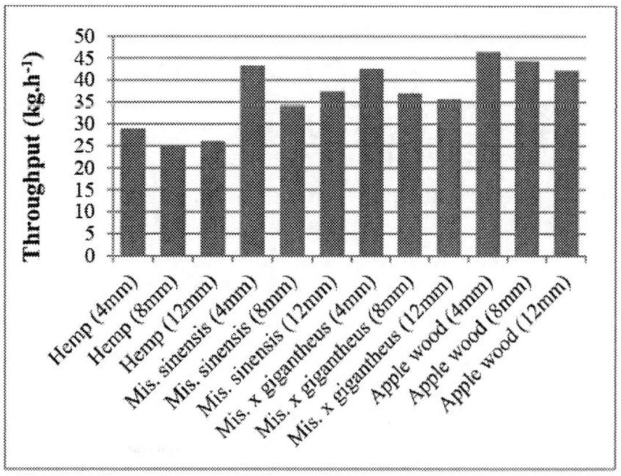

Fig. 5. Briquetting press throughput.

The throughput for fractions 4 mm was: hemp - 28.71 kg.h^{-1}, *Miscanthus sinensis* - 43.16 kg.h^{-1}, *Miscanthus x gigantheus* - 42.40 kg.h^{-1} and apple wood - 46.30 kg.h^{-1}. Both Miscanthus species showed overall similar throughput during briquetting. Briquetting throughput for apple wood was in general the highest in comparison with other materials as well as it corresponds to the best throughput of wood biomass during grinding, however, the difference between throughput of wood and Miscanthus materials within briquetting is not such significant like during the grinding. As in case of all the materials the best briquetting throughput showed smaller fraction 4 mm, but in case of hammer mill throughput it was different (opposite) and the fraction 12 mm had mainly highest throughput there. The lowest throughput was again detected for hemp (Fig. 5).

According [16] the piston press Brikstar 30-12 has material throughput 20 kg.h^{-1} to 40 kg.h^{-1}. From the Fig. 5 is evident that the measured throughputs are basically in the range, nevertheless all fractions of apple wood showed higher values (46.30 kg.h^{-1}, 44.20 kg.h^{-1} and 42.00 kg.h^{-1}). From the Fig. 4 and Fig. 5 is obvious that hemp with moisture content of 7.60%, 8.08% and 9.97% has the lowest briquetting throughput while apple wood with moisture content of 7.69% or 8.29% has higher throughput, so impact of moisture is now visible. It seems that throughput depends on material type and particle size. Additionally, it should be mentioned that hammer mill throughput was in general notably higher than throughput of briquetting.

3.5. Evaluation of briquetting press energy consumption

The energy required for compaction of materials into briquettes at different particle sizes is shown in the Figure 6. The energy consumption was in the range of 43.91 kWh.t^{-1} to 115.24 kWh.t^{-1}. From the energy consumption is clear which material is more suitable for briquettes production. The highest energy consumption had hemp fractions 4 mm and 8 mm (115.24 kWh.t^{-1} and 107 kWh.t^{-1}).

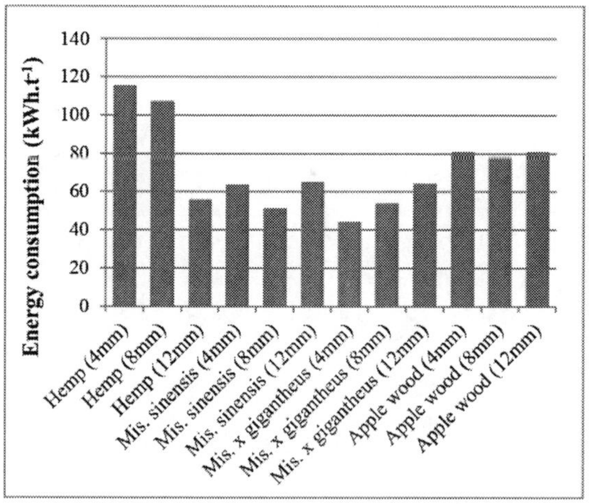

Fig. 6. Briquetting press energy consumption.

Positive results, i.e. relatively low consumption and generally the best values comparing to apple wood and hemp, had both Miscanthus for all fractions. Energy consumption for these materials was in the range of 43.91 kWh.t^{-1} to 64.67 kWh.t^{-1} and it was visible that briquetting process is more energy intensive than grinding process in general. Apple wood with energy consumption 4mm - 80.65 kWh.t^{-1}, 8mm - 78.00 kWh.t^{-1} and 12 mm - 80.67 kWh.t^{-1} is characterized by higher energy consumption then Miscanthus, but lower then hemp (see Fig. 6). [21] published that the specific energy consumption of briquetting press is in the range of 44 kW.t^{-1} to 70 kW.t^{-1}. The energy consumption of Miscanthus biomass is in this range.

3.6. Evaluation of mechanical durability of produced briquettes

From the point of view of the mechanical durability (Figure 7) the briquettes from apple wood showed just average values (and even the lowest durability), while hemp briquettes showed the best one. In research of [22] sawdust and shavings (woody materials) had also mechanical durability of 91% (sawdust) and 92% (shavings).

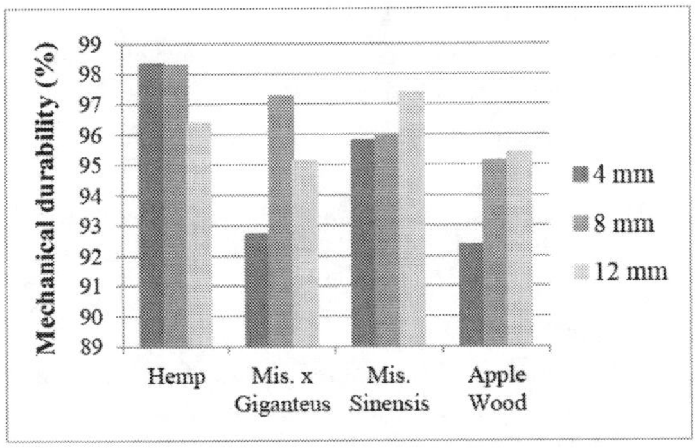

Fig. 7. Mechanical durability of produced briquettes.

The interesting fact is in contrast between briquettes made from hemp and other materials; it can be probably explained by hemp biomass structure - by better connection of fibrous particles.

[23] shows relevance between briquetting temperature and briquettes density. It was stated that the briquettes density grows proportionally to increasing temperature; when the temperature is higher it is possible to use lower pressure for briquetting. [2] has also conducted a research about temperature influences and has found that the briquettes made with temperature more than 90°C have higher density then other briquettes made at room temperature. According to [12] temperature at the pressing process is significant for lignin excretion from cellular structure. When the temperature rises to 80°C lignin is secreted from the structure of the cellulose skeleton and intercellular space [5]. [24] indicated that lignin present in the biomass is fluidized and acts as glue when the temperature increases. The temperatures between 85-105°C would result in much stronger briquettes. Relatively low operational temperature of briquetting press Brikstar 30-12 used in the present research may explain the generally lower mechanical properties of produced briquettes.

3.7. Summarization of grinding and briquetting processes of used materials

The total throughput for both processes grinding and briquetting as well as the total energy consumption during grinding plus briquetting are presented in the Figure 8. From the results of throughput is evident that the most suitable material for processing is apple wood, while hemp is the less suitable. The total throughputs corresponding to different fractions of apple wood were: 4 mm - 277.80 kg.h^{-1}, 8 mm - 339.87 kg.h^{-1} and 12 mm 462.00 kg.h^{-1}, and for hemp: 4 mm - 109.11 kg.h^{-1}, 8 mm - 66.40 kg.h^{-1} and 8 mm - 77.57 kg.h^{-1}. *Miscanthus sinensis* and *Miscanthus x gigantheus* had similar material throughput, which ranged from 171.00 kg.h^{-1} to 192.05 kg.h^{-1}.

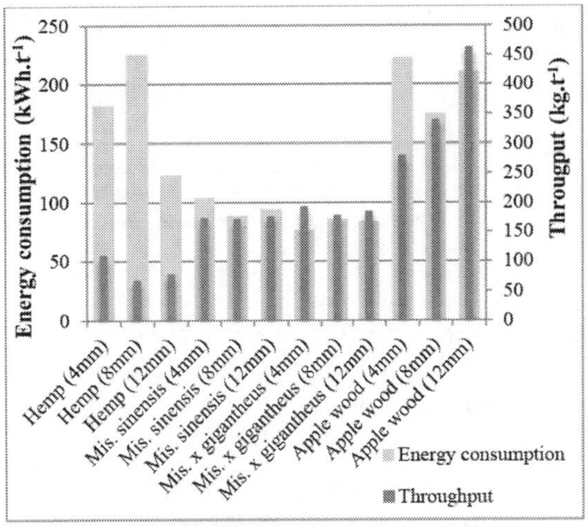

Fig. 8. Total throughput and energy consumption of the whole production process.

Research revealed that the apple wood had similar (high) energy consumption as hemp, and the total energy required for briquettes production from Miscanthus biomass was the lowest. Taking into account all the experimental results, it could be stated that wood material is applicable for briquettes' production, but processing of bigger screen sizes is recommended, as well as herbaceous Miscanthus biomass showed relatively good relation between throughput and consumed energy.

3 Conclusions

The research confirmed the first hypothesis. During the evaluation of throughput of different materials, in both processes (grinding and briquetting) the best results showed apple wood, very similar throughput was between two types of Miscanthus (herbaceous biomass) a significantly worse throughput had the hemp (fibrous material). According to the used material, energy consumption during grinding and briquetting can also differ. The lowest consumption had the herbaceous biomass. The highest energy consumption was recorded during grinding of hard wood and during briquetting of hemp.

Research partly confirmed the second hypothesis. During the evaluation of the influence of fractions on throughput, the best throughput during briquetting had all materials with the smallest fraction of 4 mm. During grinding process the significant difference between fractions was visible only in case of wood, where on contrast the bigger fraction(s) caused better throughput. From the energy consumption point of view, the finer the fraction was during grinding, the higher was the energy consumption – as confirmed form wood and both Miscanthus species. During briquetting any typical trend was not visible.

Research partly confirmed the third hypothesis. In case of wood the energy consumption during grinding was significantly higher than during briquetting. On the contrary, during briquetting of two types of Miscanthus, the energy consumption was higher (almost double). Average energy consumption during grinding and briquetting of hemp was similar, slightly higher during briquetting.

The study was supported by the Internal Grant Agency of the Faculty of Tropical AgriSciences, Czech University of Life Sciences Prague, research grant number 20165012 and 20175011 as well as by Internal project in the framework of Institutional support of RIAE development (decision number RO0614) – project title "New technologies of targeted biomass processing into raw materials and advanced biofuels".

References

1. Z. Wang, T. Lei, M. Yang, Z. Li, T. Qi, X. Xin, X. Yan, Applied Energy **192**, 83-94 (2017)
2. L. Chen, L. Xing, L. Han, Renew Sust Energ Rev **13**, 224-232 (2009)
3. B. Havrland, T. Ivanova, B. Lapczynska-Kordon, M. Kolarikova, 12th International scientific conference Engineering for rural development, Jelgava, 541-544 (2013)
4. H. Kambo H., A. Dutta, Applied Energy **135**, 182-191 (2014)
5. A. Muntean, T. Ivanova, B. Havrland, V. Pobedinsky, 11th International scientific conference Engineering for rural development, Jelgava, 496-499 (2012)
6. P. Král, J. Hrázný, *Utilization of wood and other lignocellulosic wastes* (Mendel University, Brno, 1999)
7. D. Grover, K. Mishra, *Biomass briquetting: technology and practices (*FAO, 1996)
8. J. Oladeji, C. Enweremadu, International Journal of Energy Engineering **2**, 22-27 (2012)
9. P. Jevič, P. Hutla, Z. Šedivá, Sustainable production and quality control of solid fuels based on agricultural bioproducts (Research Institute of Agricultural Engineering, Prague, 2008)
10. B. Havrland, P. Hutla, D. Adamovský, Agricultura tropica et subtropica **43**, 19-28 (2010)
11. T. Ivanova, *Research of energy plants processing to solid biofuels* (Czech University of Life Sciences Prague, Prague, 2012)
12. P. Križan, L. Šooš, M. Matúš, M. Svátek, D. Vukelic, 9th International Conference APLIMAT, Bratislava, 124-136 (2009)
13. J. Kers, P. Kulu, A. Aruniit, V. Laurmaa, P. Križan, L. Šooš, Ü. Kask Estonian Journal of Engineering **16**, 07–316 (2010)
14. Y. Li, H. Liu, Biomass Bioenergy **19**, 177-186 (2000)
15. P. Lopo, Feed Management **53**, 23-26 (2002)
16. Briklis, Available at http://www.briklis.cz/briketovaci-lis/cs/#product-tabs=obecne_informace (2011)
17. V.S. Bitra, A.R. Womac, N. Chevanan, P.I. Miu, C. Igathinathane, S. Sokhansanj, D.R. Smith, Powder Technol **193**, 32–45 (2009)
18. J. Hu, T. Lei, Z. Wang, X. Yan, X. Shi, Z. Li, Energy **61**, 557-566 (2014)
19. S. Mani, L.G. Tabil, S. Sokhansanj, Biomass Bioenergy **27**, 339-352 (2001)
20. L. Cadoche, G. López, Biological Wastes **30**, 153-157 (1989)
21. T. Ivanova, A. Muntean, B. Havrland, V. Pobedinsky, Agronomy Research **11,** 47-52 (2013)
22. M. Brožek, A. Nováková, M. Kolářová, Research in Agricultural Engineering **58**, 30-35 (2012)
23. A. Demirbasa, Physical properties of briquettes from waste paper and wheat straw mixtures, Energ Convers and Manage **40**, 437-445 (1999)
24. A. Tripathi, P. Iyer, T. Kandpal, Biomass Bioenergy **14**, 479-488 (1998)

Quality assessment of solid biofuel made of sweet sorghum biomass

Tatiana Ivanova[1,*],*Alexandru* Muntean[1], *Bohumi l*Havrland[1], and *Petr* Hutla[2]

[1]Czech University of Life Sciences Prague, Faculty of Tropical AgriSciences, Department of Sustainable Technologies, Kamýcká 129, 165 21 Prague 6, Czech Republic
[2]Research Institute of Agricultural Engineering, Drnovská 507, 161 01 Prague 6, Czech Republic

Abstract. The present article relates to assessment of energy utilization of sweet sorghum waste biomass as solid biofuel (briquettes). The briquettes were produced from biomass of pure sweet sorghum after juice extraction, mixture of sorghum with wood sawdust (ratio 1:1) and mixture of sorghum with wood shavings (ratio 1:1). Chemical, physical and mechanical properties of produced briquettes were measured in accordance with appropriate standards. The research results showed that the mixed sorghum briquettes with wood shavings have the highest mechanical durability and the lowest ash content; on the other hand, briquettes made of sweet sorghum and wood sawdust havethe best values of all other parameters, including higher calorific values, density, etc. Although addition of residual wood biomass improved the general quality of sorghum based briquettes, it was stated that the briquettes made of pure processed sorghum stalks belong to the category of high quality agricultural solid biofuels. It can be concluded that sweet sorghum is of very good prospects and thus it is a promising biomass feedstock for solid biofuels production (not only for the production of liquid biofuel as it has been used by today and has been known before).

1Introduction

Bioenergy, i.e. energy produced in the field or gained from waste biomass is one of promising trends in agricultural engineering research that merits higher attention of researchers and investments into application of its results. This is because the increasing energy consumption leads at present to a fast shrinking fossil fuel deposits and strong environment pollution probably reflecting in warming up the Earth's atmosphere. Therefore, there is a need to utilizing alternative energy resources including biomass and biomass-based biofuels [1]. Biomass of energy crops has been gaining a considerably wide energy application during recent years [2]. Due to high biomass yields with a high sugar content (fermentable saccharides) as compared to other crops yielding sugar or starch like sugarcane, wheat or corn, the sweet sorghum gives a very attractive raw material for the liquid biofuel - bioethanol production [3]. According to [4] and [5] sweet sorghum has on average 85–90% sugar fermentation efficiency for bioethanol.

Sorghum (*Sorghum bicolor* (L).Moench), including sweet sorghum, is one of the longest growing crops, it is widely adapted to diverse environments and grows almost everywhere in the world [6]. Sorghum is multipurpose plant, also called 4F crop, because it is used for fuel, food, fodder and feed [5]. In the world cereal's rank sorghum is in the fifth place following corn, rice, wheat and barley as useful to human daily need [7]. Sweet sorghum is annual or short perennial plant and according to growing conditions and varieties, it reaches a height of about 120–400 cm [8]. Sorghum is C4 crop with high photosynthetic efficiency. It accumulates carbon dioxide during the night and then has a lower photorespiration rate during the day that reduces waste of carbon and water [9]. In most of tropical and temperate countries, where sugar cane or corn cannot grow, sweet sorghum is growing well and tolerates salinity and drought [7, 10]. As confirmed by [5] that sweet sorghum has low water requirements for high yield, as comparing to sugarcane, which requires three times more. Sweet sorghum resists to biotic and abiotic factors. It has low requirement for fertilizers, one third of sugar cane or corn need in the same conditions [11]. Sweet sorghum is concerned as low inputs and low costs crop [12, 13], which has many advantages in contrast to other energy crops.

Sweet sorghum has been considered as a particular energy crop for production of bioethanol since 1970s. It has high yield of biomass (54–69 t ha^{-1}), i.e. stalks with high sugar content [14]. Amount of the main substances that compose sweet sorghum stalk as carbohydrates (glucose, sucrose and fructose) often depends and changes according to the maturity stage, cultivar and harvesting period [15]. Nevertheless, growing production and consumption of bioethanol generates large quantities of waste biomass remaining from sorghum processing – sugar juice extraction [10]. In

* Corresponding author: ivanova@ftz.czu.cz

accordance with [16] residual biomass from sweet sorghum contents mainly cellulose 34–44%, hemicellulose 25–27% and lignin 18–20%. Crushed residual stalks were burnt by [17] to generate electricity or steam. But for a more efficient combustion residue obtained from sorghum processing should be used in a form of compacted solid biofuel. Sweet sorghum is important for refined sugar and bioethanol production. Effective use of waste biomass can contribute to additional lowering of sugar production costs and good waste to energy management.

The main objective of this research is the analysis of potential and suitability of sweet sorghum biomass after juice extraction to be used for energy purposes in the form of solid biofuels, particularly as briquettes.

2 Material and methods

For experimental purposes of this research the briquettes (three kinds of briquettes) were produced from the following feedstock materials:

1)Briquettes from sweet sorghum without any blender addition or any mixing with other crops. This type was made of sweet sorghum stalks in dry and grinded form after sugar juice extraction. For briquetting 10 kg of biomass with moisture content 12% was used, abbreviated as sample S_0.

2)Mixed briquettes made of 50 % of sweet sorghum stalk and 50 % of wood sawdust, named as SW_1.

3)Mixed briquettes made of 50 % sweet sorghum stalk and 50 % of wood shavings, assigned as SW_2.

Sweet sorghum biomass used for the research purposes was shipped from Moldova. The briquettes were produced by hydraulic piston briquetting press BrikStar 50-12 (Briklis, Czech Republic) with working pressure 18 MPa and a cylindrical matrix of 65 mm in diameter.

Fractional composition (particle size distribution) of sweet sorghum biomass used for briquetting was done by sieve analysis method according to EN 15149-1:2011and it is presented in the Figure 1.

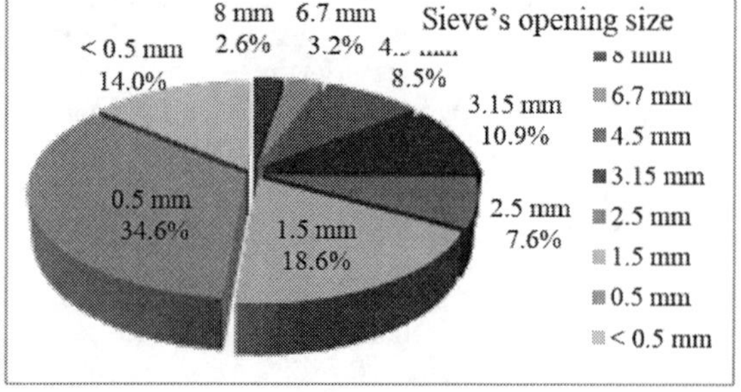

Fig. 1. Particle size distribution of sweet sorghum biomass used in experiment

All required briquettes' parameters were measured in the laboratory according to the methodology given by EN and ISO standards for solid biofuels.

Determination of moisture content (w) of briquettes as well as initial raw biomass was performed according to EN ISO 18134-2:2015 and EN ISO 18134-3:2015,respectively. Tested samples were dried out at 105 °C in laboratory oven Memmert 100-800 and the following equation was used for calculation:

$$w = \frac{m_2 - m_3}{m_2 - m_1} \times 100, \% \quad (1)$$

where: m_1– mass of empty crucible, g; m_2– mass of crucible with sample before drying, g; m_3– mass of crucible with sample after drying, g.

The gross calorific value (GCV) for dry basis (d.b.) and as received (a.r.) samples was measured by bomb calorimeter Laget MS-10A according to EN 14918:2009and was calculated by simplified equation [18, 19]:

$$GCV = \frac{dTk \times Tk - c}{m}, \text{J g}^{-1} \quad (2)$$

where: dTk – temperature jump, °C; Tk– mean value of the effective heat capacity of the calorimeter as determined in the calibrations (9,161 J °C^{-1}); c– total repair (repair on benzoic acid plus repair on burning spark fine wire), J; m – weight of material sample, g.

For calculation of net calorific value (NCV) from gross calorific value was used equation applied by [20]:

$$NCV = GCV - 24.53 \times (w + 9H), \text{J g}^{-1} \quad (3)$$

where: 24.53 – heat of water evaporation; w – moisture content of the sample, %; 9 – coefficient for conversion of hydrogen to water, H – hydrogen content in the sample, %.

It should be mentioned that according to [21] moisture content has significant impact on a net calorific value of solid biofuels and the quality of biofuel largely depends on the moisture content of initial raw material. Calorific value of briquettes also depends on the origin and treatment of the feedstock material, especially on material's particle size. Other authors [22] have stated that solid biofuel with high density have a higher calorific value.

Briquettes' average dimensions (length and diameter) were measured by digital calliper and their density (ρ) was simply calculated by the equation (4):

$$\rho = \frac{m}{\pi \times r^2 \times l}, \text{kg m}^{-3} \quad (4)$$

where: m – weigh of the briquette, kg; r – D/2 half of the diameter, m; l – length of the briquette, m.

Mechanical durability (DU) of produced briquettes was determined using a special rotary drum in accordance with standard EN ISO 17831-2:2015and was calculated as follows:

$$DU = \frac{m_A}{m_E} \times 100, \% \quad (5)$$

where: m_A – mass of sieved briquettes after the drum treatment, g; m_E – mass of pre-sieved briquettes before the drum treatment, g.

Proximate analysis (ash content and volatile matter) and ultimate analysis (C, H, O, N, S and Cl) of tested briquettes were performed as well.

In case of ash content measurements, the standard EN ISO 18122:2015was respected and ash content (A) of dry samples was calculated by the equation (6):

$$A = \frac{(m_3 - m_1)}{(m_2 - m_1)} \times 100, \% \quad (6)$$

where: m_1 – mass of empty dish, g; m_2 – mass of dish plus test sample, g; m_3 – mass of dish plus ash, g.

Determination of volatile matter content (V_d) was done in accordance with standard EN ISO 18123:2015and the following equation was used for calculation (7):

$$V_d = \left[\frac{100 \times (m_2 - m_3)}{(m_2 - m_1)} - w\right] \times \left(\frac{100}{100 - w}\right), \% \quad (7)$$

where: m_1 – weight of empty vessel, g; m_2 – weight of vessel with sample before heating, g; m_3 – weight of vessel with sample after heating, g; w – moisture content, %.

Chemical elements were determined according to the respective standards: EN ISO 16948:2015for carbon, hydrogen, nitrogen and EN ISO 16994:2015 for chlorine and sulfur. Oxygen content was then calculated.

3 Results and discussion

Quality of solid biofuels depends on number of parameters. These parameters could be divided into two groups: chemical properties and physic-mechanical properties.

In the Table 1, all parameters assessed in this research are presented and compared to the approximate values given by solid biofuels standards.

From the Table 1 is visible that all produced briquettes showed very good quality, i.e. briquettes made of pure sweet sorghum biomass fulfil standard requirements for graded non-wooden briquettes and according to the stated parameters belong to A quality class (the best one). It was found that S_0 briquettes have relatively high gross and net calorific values, low ash content and the contents of all measured chemical elements below limits. Gross calorific values (w.b. and a.r.) of pure sweet sorghum stalk briquettes correspond to gross calorific values of sorghum biomass published by other authors, for example 17.9 MJ kg^{-1} [23] or 17–19 MJ kg^{-1}[14]. Compared to other sources of biomass: $GCV_{d.b.}$ of sorghum is higher than $GCV_{d.b.}$ of hay (grass in general) – 18 MJ kg^{-1} [21], switchgrass or corn stover – 17.3 MJ kg^{-1} [24] and it is almost equal to $GCV_{d.b.}$ of famous energy crop Miscanthus – 19 MJ kg^{-1}, however lower than $GCV_{d.b.}$ of wood logging residues – 19.7 MJ kg^{-1} (broad-leaf wood) and 20.5 MJ kg^{-1} (coniferous wood) (EN ISO 17225-1:2014).

The research results showed very good (low) value of ash content in S_0 briquettes (see Table 1). According to [25] ash content of herbaceous biomass reaches values up to 10%, for example ash content of hay found by [26] is 6.3%, which is almost twice higher in comparison with sorghum. On the other hand, ash content of graded wood briquettes should not exceed maximum 3% (EN ISO 17225-3:2014). As [27] reported that low ash content leads to better suitability of fuels for thermal utilization in contrast to a fuel with high ash content. High ash content causes high dust emissions and negatively affects combustion efficiency.

Another fuel parameter, which has an impact of combustion process, is volatile matter. According to [28] high volatile matter (from 70 to 86%) improves the combustion rate of the biomass during the devitalization phase. On the contrary, low volatile matter causes high smoke within incomplete combustion and release toxic gases. [29]published that the values of carbon content in biomass, which can be used as a fuel vary between 30–60%, for hydrogen 5–6%, for oxygen 30–40% and for the rest elements less than 1%. By [21] carbon, hydrogen and oxygen are the main components of solid fuels, which directly contribute to the fuel combustion (affecting calorific value). Chlorine, nitrogen and sulphur have an effect on ash formation as well as significantly impact on the formation of harmful emissions, mainly mono-nitrogen oxides (NO_x) and sulphur oxide (SO_x) and highly corrosive hydrochloric acid (HCl) [21, 30]. Taking into account all above mentioned facts, it can be concluded that briquettes from sweet sorghum biomass seem to be very suitable for a clean and efficient combustion.

From the Table 1 is also visible that addition of wood residual biomass, i.e. production of mixed briquettes improves the total quality of sweet sorghum based briquettes, including chemical and physic-mechanical properties. The mixed briquettes are more compact as confirms briquettes dimensions and density. The addition of wood sawdust showed slightly better results, except for mechanical durability and ash content, which are insignificantly better in case of briquettes enriched with wood shavings.

The differences between mechanical durability of SW_1 and SW_2 can be caused by different fraction composition, i.e. particle size distribution of the initial raw materials and joining mechanisms between different sized particles during mechanical compaction (briquetting). [31]reported that the compacted briquettes made of raw materials with larger particles (> 6 mm) have usually better hardness and briquettes densification in general. Sweet sorghum biomass used for the experimental purposes contains of particles smaller than 6.7 mm from more than 97%, where almost 70% of the particles are smaller than 1.5 mm (see Fig. 1). Hence, the addition of larger particles into the mixtures or crushing the sorghum through bigger screens can have a positive impact on mechanical durability. Mechanical durability is not presented as a normative parameter in the solid biofuels standards; nevertheless, high mechanical durability is preferable as it is a measure of resistance of densified fuels towards shocks and/or abrasion in consequence of transport and handling processes [32].

Table 1. Determined properties of briquettes based on sweet sorghum biomass and typical standards' requirements.

Parameter	Briquette type			Standard requirements*		
	S_0	SW_1	SW_2	Typical variations[1]	Property class[2]	
					A	B
Moisture content, a.r., %	7.7	7.3	7.9	–	≤ 12	≤ 15
Ash content, d.b, %	3.9	2.2	2.1	2–10	≤ 6	≤ 10
Volatile matter, d.b. %	70.8	76.3	75.5	–	–	–
GCV d.b., MJ kg^{-1}	18.9	19.7	19.5	16.6–20.1	–	–
GCV a.r., MJ kg^{-1}	17.7	18.6	18.2	–	–	–
NCV d.b., MJ kg^{-1}	17.7	18.4	18.2	15.8–19.1	–	–
NCV a.r., MJ kg^{-1}	16.3	17.1	16.7	–	≥ 14.5	≥ 14.5
Length, mm	77.4	49.4	51.7	< 50 to >400	–	–
Diameter, mm	66.9	66.1	66.8	25 to >125	–	–
Mechanical durability, %	90.5	94.1	95.1	<90 to ≥ 95	–	–
Briquettes density (kg m^{-3})	617.5	745.9	645.5	–	–	–
C, d.b. %	43.1	45.7	45.4	41–50	–	–
H, d.b. %	5.27	6.02	6.08	5.4–6.5	–	–
O, d.b. %	37.3	39.0	38.9	36–45	–	–
N, d.b. %	0.61	0.32	0.34	0.2–1.5	≤ 1.5	≤ 2.0
S, d.b. %	0.04	0.03	0.03	< 0.05 to 0.2	≤ 0.20	≤ 0.30
Cl, d.b. %	0.09	0.06	0.08	< 0.1 to 1.2	≤ 0.10	≤ 0.30

*[1]Typical variations were obtained from European standard for solid biofuels EN 14961-1 (2010) - for virgin straw materials, with or without insignificant amounts of grains;
[2]Specification of graded non-woody briquettes for property classes A and B according to EN ISO 17225-7 (2014);
a.r. – as received; d.b. – dry basis; S_0 – pure sweet sorghum briquettes, SW_1 – mixed briquettes from sweet sorghum and wood sawdust (1:1); SW_2 – mixed briquettes from sweet sorghum and wood shavings (1:1).

According to EN 14961-1 (2010) S_0 and SW_1 briquettes have a medium category mechanical durability DU90.0 and SW_2 briquettes reached the highest category durability DU95.0 (see Table 1). Besides the material fraction, compaction pressure is another factor which affects the mechanical durability, i.e. with increasing pressure durability of the briquettes increase [32]. Low pressure used during the briquettes production had probably impact on final briquette density as well, which is relatively low for all briquettes' types. [33] stated that too low pressure produces poor quality of briquettes. For example, screw presses have higher density of briquettes. On the other hand, therefore higher pressure used for briquetting produces briquettes with better mechanical properties (dense and durable), but much energy is spent [32].

4 Conclusions

Usage of suitable waste biomass in the form of biofuels for energy purposes contributes to the support of renewable energy development, replacement of pollutant and limited fossil fuels, protection of forest reserves as well as it has positive social/energy/environmental impact and leads to sustainable development. Sweet sorghum is almost worldwide spread crop with high biomass yield. The big advantage of sorghum biomass utilization for solid biofuels' production is in an effective recycling of waste from the bioethanol production, which reflects on good waste management and makes the sweet sorghum processing waste-free technology.

Based on references review and research results, the sweet sorghum waste biomass can be evaluated as very promising feedstock for solid biofuel (briquettes) production. All three types of briquettes produced from sweet sorghum and tested during this study showed very good quality, which fulfil the standard requirements. Briquettes made of pure sorghum processed stalks are characterized by high calorific value, low ash content and low content of chlorine, nitrogen and sulfur. The study also concludes that the addition of woody biomass, i.e. wood sawdust and wood shavings improve overall quality of sorghum based briquettes. Two weakest parameters of sorghum briquettes which were found by the research measurements and are recommended to be improved, i.e. mechanical durability and briquettes' density (mechanical parameters), are not such decisive in biofuel's quality assessment as they can be technologically improved and do not depend on the biomass properties like other parameters.

The study was supported by the Internal Grant Agency of the Faculty of Tropical AgriSciences, Czech University of Life Sciences Prague, research grant number 20165012 and 20175011. Acknowledgements also belong to Ing. TsotsoAkofa Mensah for her appreciable help with laboratory works.

References

1. R.P. Srinivasa, S.S. Rao, N. Seetharama, A.V. Umakanth, P.S. Reddy, B.V.S. Reddy, Ch.L. Gowda, *Sweet sorghum for biofuel and strategies*

for its improvement,Info. Bull. No. 77 (International Crops Research Institute for the Semi-Arid Tropics, Andhra Pradesh, 2009)

2. C. Zhang, G. Xie, S. Li, L. Ge, T. He, Appl. Energy **87**, 2360-2368 (2010)
3. Reddy, B.V.S., S. Ramesh, P.S. Reddy, B. Ramaiah, P.M. Salimath, R. Kachapur, ISMN **4**,79-86 (2005)
4. S. Prasad, A. Singh, N. Jain, H.C. Joshi, Energy Fuels **21**, 2415-2420 (2007)
5. A. Almodares, M.R. Hadi, J Agric Res **4**, 772-780 (2009)
6. F.S.K.N. Rai, B.V.S. Reddy, B. Diwakar, *Development of cultivars and seed production techniques in sorghum and pearl millet*(International Crops Research Institute for the Semi-Arid Tropics, Andhra Pradesh, 1997)
7. C.W. Smith, R.A.Frederiksen, *Sorghum: origin, history, technology and production*(John Wiley & Sons, Inc., New York, 2000)
8. Q. Lu, Song, R., Lu, F. 1997. Rainfed Crops **4**, 19-23 (1997)
9. J.E. Keeley, P.W. Rundel, Int. J. Plant Sci. **164**, 55-77 (2003)
10. J. Woods, *Integrating sweet sorghum and sugarcane for bioenergy: modelling the potential for electricity and ethanol production in SE Zimbabwe* (King's College, London, 2000)
11. M. Kim, D.F. Day, J IndMicrobiolBiotechnol **38**, 803-807 (2011)
12. EIA, *International Energy Outlook 2011. Report No. DOE/EIA-0484*(Energy Information Administration, Washington, 2011)
13. M. Calvino, J. Messing, CurrOpinBiotechnol **23**, 323-329 (2012)
14. A. Almodares, M. Jafarinia, M.R. Hadi, American-Eurasian J. Agric. & Environ. Sci. **6**, 441-446 (2009)
15. D. Mamma, D. Koullas, G. Fountoukidis, D. Kekos, B.J. Macris, E. Koukios, Process Biochem. **31**, 377-381 (1996).
16. B. Sipos, J. Réczey, Z. Somorai, Z. Kádár, D. Dienes, K. Réczey, Appl. Biochem. Biotechnol. **153**,151-162 (2009)
17. W. Vermerris, J. Erickson, D. Wright, Y. Newman, C. Rainbolt, IFAS, SS AGR **293**: 1–4 (2008)
18. B. Havrland, T. Ivanova, B. Lapczynska-Kordon, M. Kolarikova, 12th International scientific conference Engineering for rural development, Jelgava, 541-544 (2013)
19. T. Ivanova, A. Muntean, V. Titei, B. Havrland, M. Kolarikova, Agron. Res. **13**, 311-317 (2015)
20. T. Dlouhý, *Calculations of boilers and flue gas heat exchangers*(Czech Technical University in Prague, Prague, 2007)
21. P. Jevič, P. Hutla, Z. Šedivá, *Sustainable production and quality control of solid fuels based on agricultural bioproducts*(Research Institute of Agricultural Engineering, Prague, 2008)
22. J.S. Tumuluru, C.T. Wright, J.R. Hess, K.L. Kenney, Biofuel BioprodBior **5**, 683–707 (2011)
23. J. Kára, Z. Strašil, P. Hutla, S. Usťak, *Energy crops. Technology for growing and use*(Research Institute of Agricultural Engineering, Prague, 2005)
24. J.S. Tumuluru, Front. Energy Res. **3**, 1–7 (2015)
25. A. Kotlánová, Biom.cz, Available at http://biom.cz/cz/odborne-clanky/metody-zkouseni-fyzikalne-chemickych-vlastnosti-tuhych-biopaliv (2010).
26. P. Hutla, Biom.cz, Available at http://biom.cz/cz/odborne-clanky/tuha-biopaliva-z-mistnich-zdroju (2010)
27. H.J. Kim, G.Q. Lu, I. Naruse, J. Yuan, K. Ohtake, J. EnergyResour. Technoll **23**, 27-31 (2000)
28. S.V. Loo, J. Koppejan. *The handbook of biomass combustion and co-firing*. Earthscan, London, Sterling, VA, 2008)
29. J. Chaney, *Combustion characteristics of biomass briquettes* (University of Nottingham, Nottingham, 2010)
30. J.S. Tumuluru, J.R. Hess, R.D. Boardman, C.T. Wright, T.L. Westover, IndBiotechnol **8**, 113–132 (2012)
31. Y. Song, J.S. Tumuluru, K.L. Iroba, L.G. Tabil, M. Xin, V. Meda, XVIIth World congress of the International Commission of Agricultural and Biosystems Engineering, Quebec, 1-14 (2010)
32. T. Ivanova, M. Kolarikova, B. Havrland, L. Passian, 13th International scientific conference Engineering for rural development, Jelgava, 131-136 (2014)
33. B. Havrland, V. Pobedinschi, V. Vrancean, J. Pecen, T. Ivanova, A. Muntean, A. Kandakov, *Biomass processing to biofuel*(Powerprint, Prague – Chisinau, 2011)

Application of FTIR spectroscopy for analysis of the quality of honey

Monika Kędzierska-Matysek[1,*], *Arkadiusz* Matwijczuk[2,**], *Mariusz* Florek[1], *Joanna* Barłowska[1], *Anna* Wolanciuk[1], *Alicja* Matwijczuk[2], *Edyta* Chruściel[2], *Radosław* Walkowiak[2], *Dariusz* Karcz[3], and *Bożena* Gładyszewska[2]

[1]University of Life Sciences in Lublin, Department of Commodity Science and Processing of Raw Animal Materials, 20-950 Lublin, ul. Akademicka 13, Poland
[2] University of Life Sciences in Lublin, Department of Physics, 20-950 Lublin, ul. Akademicka 13, Poland
[3]Krakow Technical University, Department of Analytical Chemistry (C1)

Abstract. Every kind of honey is a very precious natural product which is made by Mellifera bees species. The chemical composition of honey depends on its origin or mode of production. Honey consists essentially of different sugars, predominantly fructose and glucose. There are also non – sugar ingredients like proteins and amino acids, as well as some kind of enzymes, such as: invertase, amylase, glucose oxidase, catalase and phosphatase. The fact that honey is one of the oldest medicine known worldwide is remarkable. Scientists all over the world have been trying to improve analytical methods as well as to implement new ones in order to reaffirm the high quality of honey the benefits of which may be distracted or disturbed. There are many methods and popular analytical techniques, including as follows: mass spectroscopy and molecular spectroscopy (especially FTIR spectroscopy). The infrared spectroscopy technique is one of the most common analytical methods which are used to analyse honey nowadays. The main aim of the task was to use ATR-FTIR infrared spectroscopy to compare selected honey samples as well as typical sequences coming out from certain functional groups in the analysed samples.

1 Introduction

Honey is a natural product produced by bees of *Apis mellifera* species from nectar or honeydew. It has a distinctive sweet taste due to a high concentration of sugar i.e. glucose and fructose.

During storage in the vast majority of honeys, a spontaneous crystallization process which determines their consistency takes place [1]. On one hand, the solid phase of honey contains glucose, which is less soluble than fructose. Moreover, glucose can crystallize spontaneously at the room temperature and can be in the form of

* Corresponding author: monikakedzierskamatysek@poczta.onet.pl
** Corresponding author: arkadiusz.matwijczuk@up.lublin.pl

α-D-glucose, monohydrate and α-D-glucose and anhydrous β [2-3]. On the other hand, the liquid phase makes up other ingredients, and the most important is the aqueous solution of glucose. The crystallization process of honey continues until equilibrium, i.e. when glucose monohydrate crystals stop growing and do not dissolve [4]. It should be noted that many factors affect honey crystallization. The chemical composition of honey, primarily the type and concentration of sugars, and water content play an important role in this process. Honey during crystallization, in which the ratio of glucose to water (G/W) is less than 1.7, often tends to delaminate. Crystallization of honey accelerates due to the presence of crystallization germs (i.e. air bubbles, dust, pollen or previously crystallized honey) [5]. The optimum temperature for the crystallization process of honey ranges from 10 °C to 14 °C. Lupano in his works [6] showed that in honeys stored at 20 °C large crystals were formed with the melting temperature (T_m) ranging from 45 to 65 °C, while in honey stored in temp. - 20 °C small crystals were formed with the T_m ranging from 25 to 45 °C.

The crystal structure of honey also depends on the nectar origin. Rapeseed honey is characterized by small and irregular crystals. Crystals in buckwheat honey can be of a considerable size and relatively regular shapes [7]. At this point it is worth to emphasise that the crystallization phenomenon does not adversely affect the quality characteristics of honey [8].

An important tool for determining the crystalline structure of natural honey and its purity is ATR-FTIR IR spectroscopy [9]. This spectroscopy can be successfully applied for the classification of the honey type by replacing the long and time-consuming pollen analysis. Analysis of honey with the infrared spectroscopy can be used to determine the crystalline structure as well as the purity [2]. It also enables quick determination of the quantitative and qualitative components which define the specific quality of natural honey. The spectra allow distinguishing honeydew honey, artificial, or nectar honey, and the use of specialized software gives the ability to detect even the varieties of the nectars' honey.

FTIR spectrums of natural honeys may significantly differ with some wavelength ranges. It also should be emphasised that very often different bands (in samples) overlap and the sample may not be sufficiently characterized, which can be due to the presence of substances like polysaccharides, water and many other pollutants resulting in a wide band in analytically important spectral ranges. The objective of this paper was to specify certain properties of the selected 5 types of honey using the ATR-FTIR analysis.

2 Materials and Methods

Honeys were purchased in the retail network of the city of Lublin. Five types were analyzed, including buckwheat, multifloral, sunflower, linden honey and honeydew with honeydew leaf. According to the declaration of the manufacturers on the labels, honeys were harvested both in the European Community (EC) and non-EC countries. Before the analysis, the samples were stored in a dark place at the room temperature. Honeys have been naturally crystallized to form on the surface of a thin liquid layer. Physical and chemical properties of honey were determined in mixed samples and presented in Table 1. Water and extract content in honeys was determined by measuring the refractive index by Abbe refractometer Carl Zeiss (Jena, Germany) [10].

PH, free acidity and electrical conductivity of honey were determined as described earlier [10] using pIONneer 65 Meter (Radiometer Analytical, Villeurbanne, CEDEX-France) with a combined pH electrode (E16M340) and a 4-pole conductivity cell (CDC 30T) with a built-in temperature sensor.

2.1 FTIR measurements

The measurements of ATR-FTIR background-corrected spectra were carried out in solvents using a HATR Ge through (45° cut, yielding 10 internal reflections) a crystal plate for liquids and were recorded with 670-IR spectrometer (Varian, USA). Typically, 25 scans were collected, Fourier-transformed, and averaged for each measurement. The IR absorption spectra at the resolution of one data point per 1 cm^{-1} were obtained in the region between 4000 and 400 cm^{-1}. The instrument was purged with argon for 40 min before and during the measurements. The Ge crystal was cleaned with ultra-pure organic solvents (Sigma-Aldrich Co.) All measurements were taken at the room temperature (T = 23 °C). The spectra was measured both in the mixed and non-mixed samples. All experiments were carried at 20 °C.

3 Results and discussion

Table 1 presents physicochemical properties of the analysed honeys. All evaluated honeys met the limits required by the national legislation [11] (The Resolution of the Minister of Agriculture and Rural Development of 18 February 2004 amending the regulation on specific requirements for the commercial quality of honey, Journal of Laws of 2004, no. 40, Item 370), in terms of water content, acidity and conductivity (no more than 20%, 50 mEq/kg and ≥ 0. 8 mS/cm for honeydew honeys and ≤ 0. 8 mS/cm for all other subjects). All honeys should be considered as acidic, with a pH ranging from 3.96 (sunflower) to 4.73 (multifloral). In addition, the moisture content in the analysed honeys ranged from 17.2 (honeydew honey) to 19.4 (linden honey). The content of extract (%) varied from 79.2, in the case of linden honey to 81.3 for honeydew honey. Free acid values varied from 4.5 (multifloral honey) to 22 (honeydew honey). Electrical conductivity values ranged between 0.361 (sunflower) and 0.991 mS/cm (honeydew honey).

Table 1. Physicochemical characteristics of analysed honeys.

Honey type	Moisture %	Extract %	pH	Free acid mEq/kg	Electrical conductivity mS/cm
Buckwheat	18.6	79.9	4.65	16.5	0.537
Multifloral	18.4	80.1	4.73	4.5	0.371
Sunflower	17.6	80.9	3.96	18.5	0.361
Linden	19.4	79.2	4.13	14.5	0.579
Honeydew	17.2	81.3	4.60	22.0	0.991

2.2 FTIR spectroscopic studies

Only naturally obtained honey has been used to characterize the spectroscopic properties. In order to compare the samples FTIR spectroscopy was used as an efficient method. Typically, the characteristic differences in the FTIR spectral analysis for natural honeys were observed. It was probably related to the content of carboxylic acids in different types of honey resulting from the floral origin, geographic location and possible environmental pollution [2, 12-17]. Similarly as for alcohols, the O–H stretching vibration band in carboxylic acids is very broad and occurs in the field of 3300-2500 cm^{-1} [18-22] with the maximum at 3000 cm^{-1}. This is the same area as the stretching vibration region for carbon and aromatic C–H groups [23]. Therefore, carboxylic acids are characterized by an irregular character of absorption in 3300-2500 cm^{-1} with a wide range of the O-H stretching vibration band for C-H [12]. Strong hydrogen bonds present in dimeric carboxylic acids are the reason for the significant extension of the band stretching vibration of the OH group [24].

The stretching vibration band of the C=O carboxylic acids group occurs between 1760-1690 cm^{-1} [13, 25]. The exact position of the band depends on whether the acid is saturated or unsaturated, dimerized or associated, etc. The C-H band can appear in 1320-1210 cm^{-1}, and the deformation band of the O–H group approx. in 1440-1395 cm^{-1} and 950-910 cm^{-1} [26]. However, the band under 1440-1395 cm^{-1} cannot be distinguished from C-H bending bands, which also occur in the same frequency region. In general: we observe O–H stretching in 3300-2500 cm^{-1} [27]/ C=O stretching in 1760-1690 cm^{-1} [26]/ C-O stretching in 1320-1210 cm^{-1} [28]/ O–H bending in 1440-1395 and 950-910 cm^{-1} [29-30]. To facilitate the characterization of bands and comparison of honeys selected for testing Figure 1 shows an ATR-FTIR spectrum of the tested honeys with major high bands. Table 2 shows the exact position of the bands together with the assignment of relevant vibration to specific functional groups. Table 3 presents a comparative position of the maximum vibration of multifloral and linden honey made from mixing and after thorough mixing of the selected samples for testing.

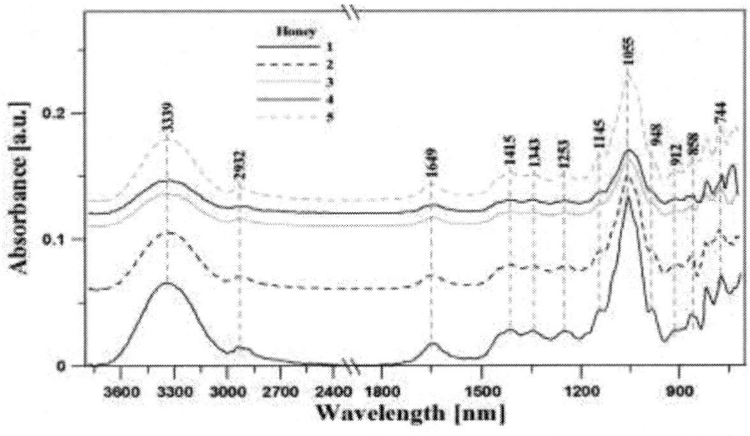

Fig. 1. ATR-FTIR absorption spectra obtained for honeys: 1 - buckwheat honey, 2 - multifloral, 3 - sunflower, 4 - linden and 5 - honeydew honey. Detailed data on the locations of specific bands are shown in Table 1.

ATR-FTIR spectra of 5 samples of pure quality honey were obtained in the same conditions, as was described in the materials and methods section. The spectra of the selected honeys observed between 3800 and 700 cm^{-1} shows four typical absorption zones dominated by two water bands at 3339 cm^{-1} (ν(O-H) and 1649 cm^{-1} (OH deformation). In the same figure (Fig. 2, Table 2 and Table 3) a band at 2932 cm^{-1} was observed, which according to Anios et. al. (2015) also corresponds to C–H stretching of carboxylic acids and NH_3 stretching band of free amino acids. The band from approximately 1500–750 cm^{-1} corresponds to the most sensitive absorption region of the major components of honey, particularly the most suitable region to quantify honey sugar (about 60–75 %) and organic acids. The contribution of sucrose, glucose, fructose, show characteristic bands in the region between

1500 and 900 cm^{-1}. Another important spectral region located at 900–750 cm^{-1} is characteristic for the saccharide configuration. Authors in many publications define the following spectral peaks as important to sugar characterization.

The vibrations with the maximum bandwidth of about 1450 cm^{-1}, cm^{-1} 1418 and 1347 cm^{-1} are characteristic for bending vibration of O-CH and C-C-H in the carbohydrate structure or bending vibration coming from OH in the C-OH grouping. The band range from 1250 to 1140 cm^{-1} is, in turn, a vibration characteristic for the stretching vibration of the C-H in carbohydrates or stretching vibration of C-O in carbohydrates. The vibrations with the maximum at approximately 1100 and 1080 cm^{-1} constitute a band that can originate in the vibration of C-O in the C-O-C group. Then the area of the bands extending from 1050 to 970 cm^{-1} of the C-O stretching vibrations can be assigned to the C-OH group or to the stretching in the carbohydrate structure of C-C. Band with the maximum at approximately 920 cm^{-1} are bending vibrations bands of the C-H group. The last region which is very distinctive in the evaluation and description of honey is the band from 890 to 810 cm^{-1} characteristic for the vibration anomeric region of carbohydrates or C–H deformation.

To sum up the above considerations, distinct differences in the intensity of the bands which are first of all characteristic for vibration of the OH groups should be also emphasised. These changes inform clearly about varied water content in honey samples selected for testing. The greatest intensity of these bands with the maximum at 3340 cm^{-1} was in case of buckwheat honey, while the lowest for linden honey and sunflower honeys. The area of bands with the intense maximum at approximately 1055 cm^{-1} coming from stretching vibrations of both C-O in C-OH and C-C group in the carbohydrate structure also seems to be very interesting. It clearly indicates varied sugar composition of tested samples.

Fructose is the essential sugar in honey with the average value of 36% ranging from 30% to 50% in typical samples. Glucose is the second most representative monosaccharide in honey and varies from 16% to 34% with the mean value of 25%. Very significant changes were found also within the area of 900-750 cm^{-1} which confirms the variable saccharide configuration.

In addition, it is noteworthy to present the differences in the band with the maximum at approximately 1440, 2880, 1410 and 1250 and 1190 cm^{-1}. The above mentioned bands vary with regard to location and this greatly changes the crystals, which have been made in the test samples of honey. These effects will be considered in the subsequent preparation of the topic presented in this article using methods of both spectroscopic and thermographic measurement with the use of differential scanning calorimetry (DSC) and measurements of Raman spectra.

Table 2. Location of maxima of absorption bands FTIR [12-13, 29, 31-34], together with the assignment of appropriate vibration for selected honeys: buckwheat, multifloral, sunflower, linden, and honeydew honey made in terms of spectral 3700-900 cm^{-1}.

FTIR					**Type and origin of vibrations**
Position of bands [cm^{-1}]					
Buckwheat	Multifloral	Sunflower	Linden	Honeydew	
3340	3327	3335	3325	3343	ν (O-H) in H_2O
2937	2941	2932	2921	2931	ν (C-H) or/and ν (NH_3) of free amino acids
2888	2879	2875	2882	2887	

1646	1643	1640	1644	1646	δ (O-H) from H_2O
1455	1456	1451	1453	1454	δ (O-CH) and δ (C-C-H)
1415	1420	1408	1418	1417	δ (O-H) in C-OH group + δ (C-H) in the alkenes
1347	1345	1344	1347	1342	δ (-OH) in C-OH group
1249	1256	1253	1253	1259	ν (C–H) in carbohydrates or/and ν (C–O) in carbohydrates
1189	1186	1193	1189	1189	
1143	1145	1145	1153	1148	ν (C–H) in carbohydrates or/and ν (C–O) in carbohydrates
1101	1101	1102	1103	1105	ν (C-O) in C-O-C group
1078	1080	1076	1080	1079	
1055	1056	1056	1056	1055	ν (C-O) in C-OH group or ν (C-C) in the carbohydrate structure
1033	1032	1032	1033	1033	
1014	1011	1012	1017	1014	
981	981	978	985	982	
966	966	964	976	964	
917	917	923	933	920	δ (C–H)
895	897	894	897	892	anomeric region of carbohydrates or δ (C–H)
866	866	866	868	861	
844	830	839	846	839	
819	815	814	816	820	

ν – stretching vibrations, δ – deformation vibrations, s – symmetric, as – asymmetric, st – strong, * denotes frequencies that may originate from the solvent used

Table 3. Location of maxima of FTIR absorption bands with arrangement of appropriate vibration for selected honeys: multifloral, linden (mixed and non-mixed) in terms of spectral 3700-900 cm^{-1}.

FTIR				**Type and origin of vibrations**
Position of bands [cm^{-1}]				
Multifloral		Linden		
mixed	non-mixed	mixed	non-mixed	
3337	3327	3333	3325	ν (O-H) in H_2O
2934	2941	2929	2921	ν (C-H) or/and ν (NH_3) of free amino acids
2888	2879	2888	2882	
1644	1643	1647	1644	δ (O-H) from H_2O
1446	1456	1453	1453	δ (O-CH) and δ (C-C-H)
1412	1420	1417	1418	δ (O-H) in C-OH

				group + δ (C-H) in the alkenes
1346	1345	1348	1347	δ (-OH) in C-OH group
1253	1256	1247	1253	ν (C–H) in carbohydrates or/and ν (C–O) in carbohydrates
1189	1186	1182	1189	
1147	1145	1149	1153	ν (C–H) in carbohydrates or/and ν (C–O) in carbohydrates
1100	1101	1103	1103	ν (C-O) in C-O-C group
1081	1080	1080	1080	
1056	1056	1055	1056	ν (C-O) in C-OH group or ν (C-C) in the carbohydrate structure
1031	1032	1038	1033	
1014	1011	1012	1017	
981	981	981	985	
965	966	966	976	
918	917	923	933	δ (C–H)
896	897	893	897	anomeric region of carbohydrates or δ (C–H)
867	866	862	868	
835	830	844	846	
815	815	817	816	

ν – stretching vibrations, δ – deformation vibrations, s – symmetric, as – asymmetric, st – strong, * denotes frequencies that may originate from the solvent used

4 Conclusions

In conclusion this paper referred to some physicochemical properties of the selected types of honey. Evaluated honeys met the limits required by the national legislation in terms of water content, free acid and electrical conductivity. The moisture content in all samples differed significantly as shown by both the physicochemical and spectroscopic measurement. Application of analysis of ATR-FTIR indicated key areas that serve both determination of the composition of the honey samples selected for testing and may also be used to determine the region of origin of these samples. The infrared spectroscopy analysis results confirm significant variation of the water content in the samples as well as the variable composition of the sugar in the selected samples.

The samples that before the measurements of the spectrum have not been mixed were also the location of the maximum bands assigned to the structures that we associate with the process of crystallization in honeys. FTIR-ATR showed to be a reliable method to quantify the majority of the sugar content in honey and is easily adapted to routine analysis of this product.

References

1. A. Oryan, E. Alemzadeh, A. Moshiri J Tissue Viability. **25**, 98-118 (2016)
2. I. Martos, F. Ferreres, F.A. Tomás-Barberán J Agric Food Chem. **48**, 1498-1502 (2000)
3. F. E. Young J. Phys. Chem. **61**, 616-619 (1957)
4. S. Bakier Pszczelarstwo **4**, 2-4 (2007)

5. Y.W.Chen C.H. Lin, F.Y. Wu, H.H. Chen J Food Process Eng **32**, 512-527 (2009)
6. C.E. Lupano Food Res Int **30**, 683-688 (1997)
7. S. Bakier Postępy Techniki Przetwórstwa Spożywczego **16**, 30-34 (2006)
8. R. Subramanian, H. Umesh Hebbar, N. Rastogi Int J Food Prop **10**, 127-143 (2007)
9. S.Sivakesava, J. Irudayaraj J. Sci. Food Agric **81**, 683-690 (2001)
10. Bogdanov, S. The honey book (2009)
11. J.S. Zegar Z badań nad rolnictwem społecznie zrównoważonym (10). Raport końcowy, synteza i rekomendacje, Instytut Ekonomiki Rolnictwa i Gospodarki Żywnościowej-Państwowy Instytut Badawczy, 2009.
12. S. Gok, M. Severcan, E. Goormaghtigh, I. Kandemir, F. Severcan Food Chem. **170**, 234-240 (2015)
13. O. Anjos, M.G. Campos, P.C. Ruiz, P. Antunes Food Chem. **169**, 218-223 (2015)
14. B. Özbalci, İ.H. Boyaci, A. Topcu, C. Kadılar, U. Tamer Food Chem. **136**, 1444-1452 (2013)
15. L. Lenhardt, R., Bro, I. Zeković, T. Dramićanin, M.D. Dramićanin Food Chem. **175**, 284-291 (2015)
16. X. Yang, Y. Zhuo, S. Zhu, Y. Luo, Y., Feng, Y., Dou Biosens. Bioelectron. **60**, 292-298 (2014)
17. H. Madupalli, B. Pavan, M.M. Tecklenburg J. Solid State Chem. **255, 27-35** (2017)
18. Z. Gan, Y., Yang, J. Li, X. Wen, M. Zhu, Y. Jiang, Y. Ni J Food Eng **178**, 151-158 (2016)
19. U. Turaga, V. Singh, A. Gibson, S. Maharubin, C. Korzeniewski, S. Presley, E. Smith, R.J. Kendall, S. Ramkumar AATCC Rev **3**, 25-31 (2016)
20. E.M. Nickless, S.E. Holroyd, G. Hamilton, K.C. Gordon, J.J. Wargent Vibrational Spectroscopy **84**, 38-43 (2016)
21. N.I. Zakaria, A.R. Yacob eProceedings Chemistry **2**, (2017)
22. E. Mellado-Mojica, N.P. Seeram, M.G. López J. Food Comp. Anal. **52**, 1-8 (2016)
23. L. Svečnjak, S. Prđun, J. Rogina, D. Bubalo, I. Jerković Food Chem. **232**, 286-294 (2017)
24. H.E. Tahir, Z. Xiaobo, L. Zhihua, S. Jiyong, X. Zhai, S. Wang, A.A. Mariod Food Chem. **226**, 202-211 (2017)
25. I. Kasprzyk, J. Depciuch, D. Grabek-Lejko, M. Parlinska-Wojtan Food Control (2017)
26. S. Li, X. Zhang, Y. Shan, D. Su, Q. Ma, R. Wen, J. Li Food Chem. **218**, 231-236 (2017)
27. N. Cebi, M.Z. Durak, O.S. Toker, O. Sagdic, Arici, M. Food Chem. **190**, 1109-1115 (2016)
28. G. Bázár, R. Romvári, A. Szabó, T. Somogyi, V. Éles, R. Tsenkova Food Chem. **194**, 873-880 (2016)
29. K. Ruoff, W. Luginbühl, R. Künzli, M.T. Iglesias, S. Bogdanov, J.O. Bosset, K. von der Ohe, W. von der Ohe, R. Amadò J Agric Food Chem. **54**, 6873-6880 (2006)
30. F. Anguebes, L. Pat, B. Ali, A. Guerrero, A.V. Córdova, M. Abatal, J.P. Garduza J Anal Methods Chem **2016**, 1-14 (2016)
31. N.F.M. Sabri, H.H. See, H. H. eProceedings Chemistry **1**, 22-26 (2016)
32. P. Mühlschlegel, A. Hauk, U. Walter, R. Sieber Food Addit Contam 1-8 (2017)
33. H. Dong, D. Luo, Y. Xian, H. Luo, X. Guo, C. Li, M. Zhao J. Agric. Food Chem **64**, 3258-3265 (2016)
34. C. Das, S. Chakraborty, K. Acharya, N.K., Bera, D. Chattopadhyay, A. Karmakar, S. Chattopadhyay Talanta **171**, 327-334 (2017)

The concept of algorithm supporting the process of scheduling production tasks

Adrian Knapczyk[1,*], *Sławomir* Francik[1], *Zbigniew* Ślipek[1]

[1]University of Agriculture, Faculty o Production and Power Engineering, Department of Mechanical Engineering and Agrophysics, Balicka 120, 30-149 Kraków, Poland

Abstract. The purpose of the work was to develop a concept of an algorithm which supports the process of scheduling production tasks. The designed algorithm performs a task queue creation process in six steps: selecting a schedule type, declaring a queue length, selecting orders for queue development, loading data, automatic job analysis, and queue and Gantt chart generation, taking a decision on the acceptance of the designed production program. The designed system will use single-attribute priority rules (defined by experts) and multi-attribute rules that can be created by combining single-attribute rules. The concept of an algorithm that supports the process of scheduling production tasks, developed and tested by conducting simulations, can be the basis for developing a computer application supporting decision-making process in a manufacturing or service company.
Keywords: task scheduling, priority rules, task queuing algorithm, optimization, scheduling problem

1 Introduction

Production management in a modern company combines production engineering, organizational technology and staff experience. The aim is to minimize costs, while preserving the assumed level of production, while ensuring the quality of the offered products / services.

One of the key but at the same time the most difficult tasks in manufacturing management is task scheduling. A well-prepared schedule ensures manufacturing stability and minimizes decision-making uncertainty. Scheduling, as part of organizational techniques used in the enterprise, combined with other resources (e.g. staff experience, innovative technologies) is one of the conditions for a successful enterprise development. The schedule must take into account a number of factors (e.g. production costs, inventory values, storage costs, customer satisfaction, meeting delivery deadlines, minimizing delays, use of machine stock, minimizing rearming time, maximizing use of labor force).

The high complexity of production processes presents a serious challenge when scheduling production. Companies produce a wide range of products. There is also a multitude of production technologies. Different technologies use machines of different capacities, and this directly affects the manufacturing times. The additional difficulty is that not all production tasks are known at the beginning of the planning process. They may appear after the schedule has been prepared, which necessitates corrections. All this makes it difficult to coordinate production in time [1–5].

The scheduling methods can be divided into [6]:

• static - the priority value is not changed while waiting for production,

• dynamic - the value of the priority changes while waiting for production.

Most often, production scheduling requires a job scheduling / queuing process.

The problem of scheduling, outside manufacturing, occurs also in many other industries, such as information technology [7], transportation ([8], medicine [9], scheduling classes (as well as project management) [10].

Agricultural engineering uses queuing quite broadly. For the case of agricultural field operations, according to ASABE Standards, the concept of scheduling is defined as ''determining the time, when various operations are to be performed. Availability of time, labour and machinery supply, job priorities and crop requirements are some important factors'' [11,12]. In the agri-food and biomass supply chain, four main functional areas can be identified, namely, production, harvest, storage, and distribution [13].

Problem of scheduling multiple machines, executing multiple operations at multiple fields, is difficult to find an optimal solution [11, 14, 15]. Many factors must be taken into account [15], like available labour and machinery, timeliness functions and workability [12].

The most common objectives in the machinery planning and scheduling problems are the minimisation of the total travelled distance or operational time [13] completion time (makes plan) or by assessing the trade-off between time and cost [11] In operations research, the makes plan of a project is the total time that elapses

* Corresponding author:adrian.knapczyk91@gmail.com

from the beginning to the end. The term commonly appears in the context of scheduling.

Developing an optimal schedule requires decision making support tools [11, 15]. The scheduling of field tasks problem is a critical aspect in the design of decision support systems (DSSs) for complex farm advising [16]

Currently, the planning of the biomass collection operations is performed relying on the contractor's experience without the use of any dedicated planning tool [17]. In the supply chain planning, one-attribute methods such as FIFO, LIFO, and FEFO are used [18]. Biological products require additionally the transport within set deadlines. An example of such products may be biomass intended for energy-generating purposes [19].

The essence of the prioritising method is performing tasks in the order from the highest to lowest priority (Figure 1). This method is justified in situations where it is difficult to determine which task is to be implemented first. Specifying an order queue can reduce production costs [20]. In the case of tasks with the same priority, an additional decision-making criterion (e.g. another rule) or random ordering is required [21].

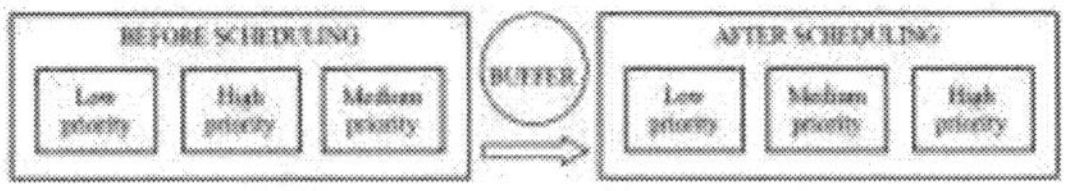

Fig. 1. Task prioritising diagram.

2 Purpose and scope of work

The purpose of the work was to develop a concept of an algorithm which supports the process of scheduling production tasks.

In order to achieve the purpose of the research, existing scheduling methods were analysed. This was necessary for a rational selection of the prioritising method to be used in the proposed system. It was also necessary to carry out identification of elements of the system, to describe them, and to formulate requirements. The last step was to develop an algorithm for creating order queues.

In order to verify the correctness of the developed algorithm operation a simulation was carried out. The simulation was carried out for real products.

3 System design

In this publication, the system (task scheduling support system) was designed based on the analysis of scheduling problems (described in the literature and present in a real production plant).

Task scheduling methods fall into two main groups: single-attribute and multi-attribute priority rules. The one-attribute priority rules include [22–24]:

- FIFO (First in first out)
- LIFO (Last in first out)
- FEFO (First expired first out)
- SPT (Shortest processing time)
- LPT (Longest processing time)
- MOR (Most operations remaining)
- LOR (Least operations remaining)
- EDD (Earliest due date)
- EMODD (Earlies modified operational due date)

The multi-attribute priority rules include [24]:

- CR+SPT (Critical ratio + the shortest process time)
- S/OPN (minimum Slack time per remaining Operation)
- S/RPT+SPT (Slack per remaining process time + the shortest process time)
- PT+WINQ (Process time + work in the next queue)
- PT+PW (Process time + wait time)

A large number of existing priority rules makes it necessary to analyze their suitability for solving the problem and to choose the best rules for the given problem.

Vinod and Sridharan [24] conducted a comparative study of single- and multi-attribute priority rules. Simulation studies focused on optimizing payment deadlines, using dynamic priority rules. The authors compared the effectiveness of three single attribute priority rules (FIFO, SPT, EMODD) and four multi-attribute priority rules (CR + SPT, S / RPT, PT + WINQ and PT + WINQ + SL). The worst effects have been shown by FIFO and EMODD rules. Good results were obtained for the SPT and PT + WINQ methods.

Similar conclusions were reached by Wiśniewski [20], [25]. Studies conducted in the printing industry showed that the use of single-attribute priority rules does not produce the best results. Three products (leaflets, brochures, books) and eleven priority rules (FIFO, LOR, MOR, LPT, SPT, EDD, EMODD, CR + SPT, S / OPN, S / RPT+SPT, PT+WINQ+SL) were used. The results of the simulation show that there is no significant difference in waiting time between the applied priority rules for lower machine utilization. However, at the highest level of machine utilization (> 85%), the FIFO, LOR, SPT, EDD rules were the least effective. Single-attribute priority rules showed little flexibility, which directly affected product waiting time. The application of multi-attribute rules resulted in much better effects.

Apart from the analysis of the results presented in the literature on the subject, this study examined the opinions of experts-practitioners who daily work on scheduling production task queues in multi-product production systems. As a result, it was found out that the best would be the application of single rules with the possibility of combining them into multi-attribute rules. According to experts, scheduling tasks should include rules such as FIFO, SPT, EDD and LOR. In their opinion, for simple cases it is possible to use single-attribute priority rules, while in more complex cases, it is advisable to use multi-attribute priority rules.

Thus, it was assumed that in the designed system, one-attribute priority rules (defined by experts) and multi-attribute rules obtained by combining one-attribute rules would be used. The user will be able to freely connect them and use them for selected application. Multithreaded priority rules that can be achieved are: FIFO + SPT, FIFO + EDD, SPT + EDD, SPT + LOR, EDD + LOR, FIFO + SPT + EDD.

Table 1. Description of the system elements.

SYSTEM ELENETS	DESCRIPTION	FEATURES
Products /services	They are a collection of information about manufactured products / services. The product type directly affects the production time.	- product / service type - group of products / services - production / services times at each stage
Posts /machines	A collection of all the posts / machines used in the production /service process. It is important to specify groups of posts / machines and their efficiency.	- type of posts / machines - group of posts / machines - efficiency
Rearmament	They provide information on the rearmament times for each type of post / machine.	- type of rearmament - time to rearm
Malfuntions	A collection of information on the types and times of malfunctions occurring in the production / service processes	- type of malfunction - time
Status	Information on the status of orders. The following statuses were identified during the implementation; finished, frozen	- type of status
Scheduling period	Information about how many hours and days a task queue will be created	- working time per shift (8-hours, 10-hours , 12- hours) - number of days
List of sequences	A list of posts / machines used in the manufacturing of given products	- List of posts / machinesincluded in the sequence
List of sequence steps	Determines the sequence of performing tasks / activities on individual posts / machines	- Sequence steps for given products
Orders	Detailed information on orders. The client states what products are ordered, the quantity and order delivery deadline. The production route is now automatically set for given types of products.	- name (order code) - type of ordered products - quantity of ordered products - order delivery deadline - production route - production time
Scheduling methods	Information on the scheduling methods available in the system. The user may select an optimal method in given applications.	- groups of methods (single-attribute, multi-attribute) - declaration of the selected method

As a result of the conducted research which included a detailed analysis of the production processes and a series of brainstorming sessions with the expert panel, detailed requirements were formulated with regard to the structure of the databases. Various elements of the system and necessary features describing these elements were identified (Table 1).

4 The scheduling algorithm

The tool (computer program) that supports the task scheduling process consists of two essential elements: the operation algorithm and the application interface (the outer part of the application). The external part of the application is designed and created mainly for the user. The interface allows the user to enter and edit data, and to generate report results.

The algorithm determines the order in which individual actions are performed, the input data required at each stage and the system- generated responses. Correct design of the algorithm is a key issue in developing a system concept that supports the scheduling of production tasks.

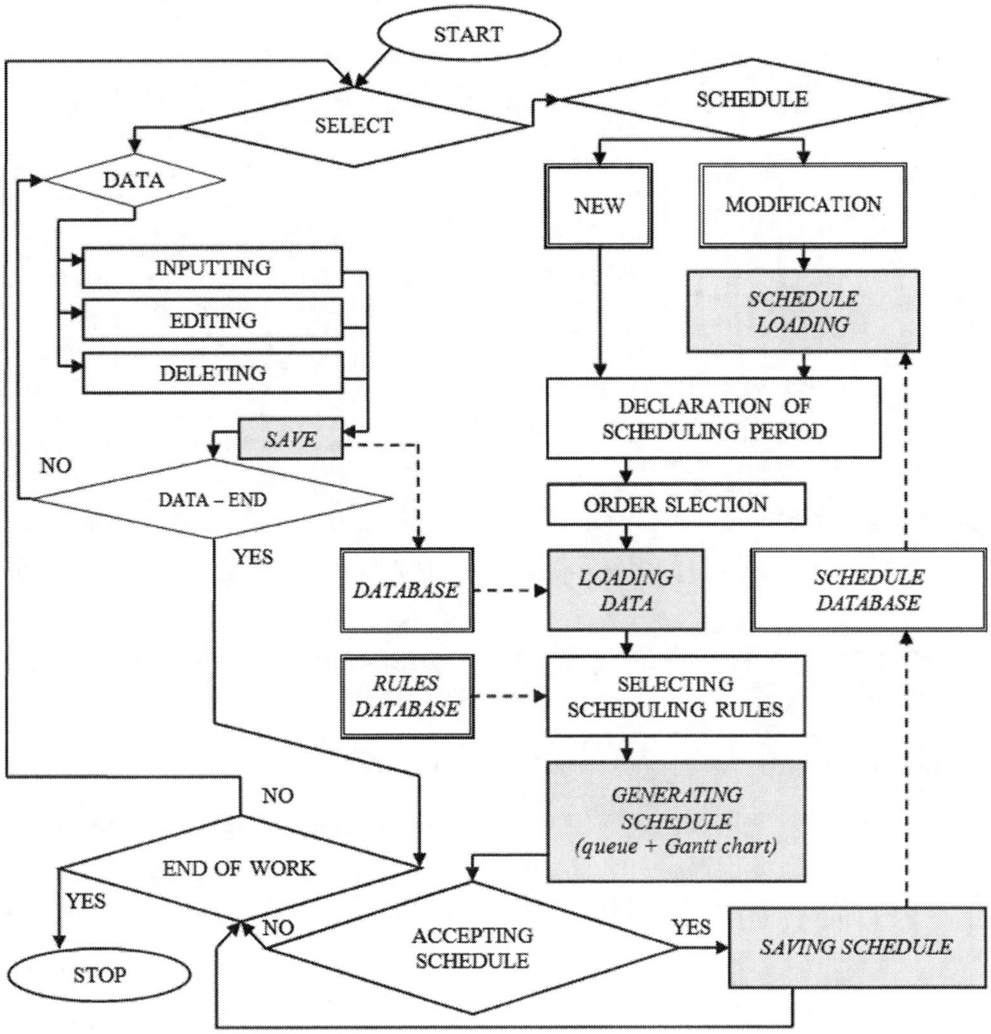

Fig. 2. Task queuing algorithm scheme.

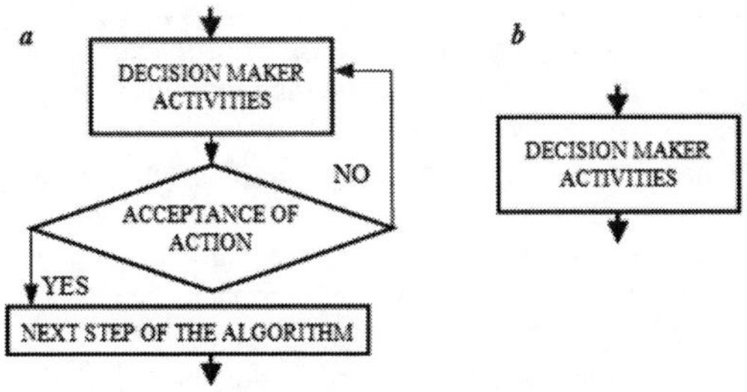

Fig. 3. Actual activity scheme (a) and simplified block (b) of the decision maker activities in generating the schedule.

Once the system elements have been defined, a scheduling algorithm (order queue) has been designed. Its scheme is shown in Figure 2.

In order to increase the clarity of the scheme (Figure 2) actions taken by the decision maker (the scheduler) were simplified: actual activities of the decision-maker (Figure 3a). were replaced with a simplified block diagram (Figure 3b). At each stage of the operation, the decision maker has the opportunity to improve his choices before moving on to the next step in the algorithm.

Before starting to create a queue, it is necessary to update the database in the following categories: products, jobs, sequence list, list of sequence steps, orders, buffers, failures, rearmaments, statuses. The decision maker can make an update after entering the "DATA" option. The decision maker has the possibility to enter new data or edit and delete existing data. When entering data, the system pre-checks their correctness by verifying the type and completeness of the data.

When the input procedure is complete, the data is stored in the database, from where it is later loaded into the production task scheduling procedure.

In addition to the production data database, we have also designed:

• rule database - used for storing priority (one-attribute and multi-attribute) rules which the decision maker can select in the task scheduling procedure,

• schedule database - used for storing the generated schedules. It is necessary if they are later modified.

The main part of the program is the production tasks scheduling procedure. It starts when the user selects the option "SCHEDULE". It consists of six steps:

1. Selection of a schedule type. The decision maker selects an option to create a new schedule or to modify an existing one. In case of modifications, the previously generated and saved schedule must be loaded.

2. Declaration of length of time for the queue. After choosing the option "scheduling", the user should declare the period for which the task / order queue is to be worked out (number of queue hours, number of queue days).

3. Selection of orders to be taken into account when preparing the queue. The decision maker chooses from the previously declared tasks those that are to be taken into account when creating the queue.

4. Loading data. In this step, the necessary data from the previously updated database is automatically loaded.

5. Generating a queue and a Gantt chart. Predefined variables are used when analyzing. At this stage, combined single-attribute rules are used. The queue is presented graphically to the decision maker as a chart, which facilitates the analysis of the resulting production program.

6. Making a decision. After analyzing the generated queue, the user has the possibility to accept the given production program. In case of acceptance, it is possible to create a final report. If the decision is not accepted by the decision maker, he / she can change the selected parameters and repeat the procedure.

Despite the automation of the task scheduling procedure in the proposed system, ultimately it is the decision maker who is responsible for the implementing the generated schedule. By modifying the existing queue, the user can make corrections that occurred during the production process. The goal is to protect the system from errors resulting from unforeseen events.

5 Simulation test

The simulation of the developed algorithm was carried out using Microsoft Excel. Real data from the production process was used in the simulation. In the simulation real data from the production process was used.

Data required for the simulation were obtained from a company manufacturing electronic parts. The company did not agree to disclose its name or to use real product names. To present the results, commercial information was coded.

The company manufactures three types of products (marked as A, B and C) which belong to 5 groups of products (G1÷G5). The total production offer covers 255 different products but the number of products in various groups differs G1 = {A1 ÷ A48}, G2 = {A49 ÷ A99}, G3 = {A100 ÷ A174}, G4 = {B1 ÷ B57}, G5 = {C1 ÷ C24}). Individual products are manufactured using a different number of posts – the total number of posts is 49.

A change of the manufactured product requires the rearmament of posts. Time of post rearmament within a group is 7 minutes and in case of the change of the group, it is 20 minutes.

Series of simulations were performed, with the type and size of order based on the interpretation of random numbers. Different types of orders and different sizes of orders (from 100 to 500 units) were compared. An assumption was made that the number of products to manufacture did not change; it was a set of 10 products (2 most diversified products from each of the five groups): G1 = A1 and A48, G2 = A49 and A99, G3 = A100 and A174, G4 = B1 and B57, G5 = C1 and C24. The number of orders was 10, 50, 100 and 255 from the 10 selected products.

After selecting at random a set of production orders, the developed algorithm was applied to generate production plans with the use of different priority rules (8 rules: LOR, MOR, SPT, LPT, EDD+LOR, EDD+MOR, EDD+ SPT, EDD+ LPT).

Fig. 4 presents the results of scheduling, obtained during the simulations for the 10 selected products.

As the combined production times were the same for all methods, the factor differentiating the product manufacturing time was the rearming time.

After the analysis of the simulation results, it can be observed that the most crucial factor is the number of actions to be performed. All 4 methods using that factor (EDD+MOR, EDD+LOR, MOR, LOR) achieved the best results. It can also be noted that the effectiveness of using scheduling rules increases in line with increasing number of products, and that multi-attribute rules give better results than the single-attribute rules.

Another observation is that the validity of using priority rules increases with the increase of products to be manufactured. At the level of 10 products most of the priority rules had the same average rearming time. When the number of products increased (to 50, 100, 255), the dispersion of the average rearming time increased, too. In consequence, lower and lower values of the average rearming time were obtained. The simulation results also show the importance of the adequate selection of the priority rule.

A group of the company experts involved in production scheduling positively evaluated the results of the analyses. They stated that the results were in line with their expectations resulting from professional experience and that the use of priority rules allows to speed up the development of optimal production schedules.

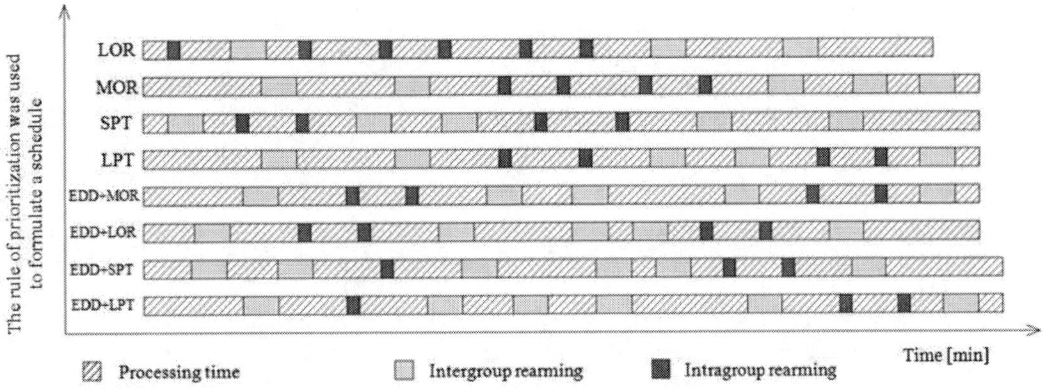

Fig. 4. Comparison of schedules developed using different scheduling rules for an example order (10 products)

Table 2. Simulation results – average rearming time for different number of products.

10 products		50 products		100 products		255 products	
Methods	**Average rearming time [min]**	**Methods**	**Average rearming time [min]**	**Methods**	**Average rearming time [min]**	**Methods**	**Average rearming time [min]**
LOR	10.2	MOR	9.2	MOR	9.14	LOR	8.15
MOR	12.8	LOR	9.2	LOR	9.14	MOR	8.20
EDD+MOR	12.8	EDD+MOR	11.8	EDD+MOR	10.44	EDD+LOR	9.22
EDD+LOR	12.8	EDD+LOR	11.8	EDD+LOR	10.44	EDD+MO R	9.27
LPT	12.8	EDD+SPT	15.18	LPT	16.16	SPT	16.10
SPT	12.8	SPT	15.96	SPT	16.42	LPT	16.15
EDD+SPT	14.1	LPT	15.96	EDD+SPT	16.94	EDD+SPT	16.61
EDD+LPT	14.1	EDD+LPT	16.22	EDD+LPT	17.07	EDD+LPT	16.61

6 Summary

The developed concept of order-queuing algorithm as an IT tool supporting the process of scheduling production tasks is the basis for the development of a computer application to support a decision-making process in manufacturing or service company

The designed algorithm performs a task queue creation process in six steps: selecting a schedule type, declaring a time length for the queue, selecting orders considered in queue development, loading data, automatic job analysis and queue generation, and Gantt charting as well as making a decision with regard to the acceptance of the generated production schedule.

In the designed system, one-attribute priority rules (determined by experts) and multi-attribute priority rules (obtained by combining single-attribute rules) were used.

The effectiveness of a particular priority rule may be related to the characteristics of a particular product group.

Simulation experiments have confirmed the assumption that the use of priority rules is valid. Selecting the right priority rule allows to significantly improve results – to shorten average rearming time.

Based on the performed simulations, it was shown that the use of an appropriate priority rule is particularly important for a large number of products (over 50). For example, for simulating the production of 255 products, the average rearming time for certain rules differed twice (8.15 min for LOR and 16.61 min for EDD + SPT and EDD + LPT).

Acknowledgement

This research was financed by the Ministry of Science and Higher Education of the Republic of Poland (statutory activities DS-3600/WIPiE/2017, Faculty of Production and Power Engineering, University of Agriculture in Krakow).

References

1. A. Bozek and M. Wysocki, Management and Production Engineering Review, vol. 7, no. 1, 21–32 (2016)
2. R. Dylewski and A. Jardzioch, Management and Production Engineering Review, vol. 5, no. 3, 3–8 (2014)
3. W. Zhong and Z. L. Chen, Journal of Scheduling, vol. 18, no. 4, 411–422 (2015)
4. B. Scholz-Reiter, J. Heger, and T. Hildebrandt, 'Analysis And Comparison Of Dispatching Rule-Based Scheduling In Dual-Resource Constrained Shop-Floor Scenarios', in WCECS 2009: World Congress on Engineering and Computer Science, 2009, vol. II, 921–927
5. R. Tavakkoli-Moghaddam, F. Taheri, and M. Bazzazi, IJE Transactions A: Basics, vol. 21, no. 3, 269–278 (2008)
6. J. Vaario and K. Ueda, Journal of Intelligent Manufacturing, vol. 9, 129–140 (1998)
7. Y. Xiong, S. Wan, J. She, M. Wu, Y. He, and K. Jiang, Journal of Network and Computer Applications, vol. 59, 63–73 (2016)
8. A. Kwasiborska and J. Skorupski, Prace Naukowe Politechniki Warszawskiej. Transport, vol. 101, 55–62 (2014)
9. S. Topaloglu,, Computers & Industrial Engineering, vol. 51, no. 3, 375–388 (2006)
10. I. Pisz and Z. Banaszak, Applied Computer Science, vol. 4, no. 1, 114–123 (2008)
11. A. Orfanou, P. Busato, D. D. Bochtis, G. Edwards, D. Pavlou, C. G. Sorensen, and R. Berruto, Computers and Electronics in Agriculture, vol. 94, 12–19 (2013)
12. C. G. Sorensen, L. Pesonen, S. Fountas, P. Suomi, D. Bochtis, P. Bildsøe, and S. M. Pedersen, Computers and Electronics in Agriculture, vol. 73, no. 1, 44–55 (2010)
13. D. D. Bochtis, Agricultural Engineering International: CIGR Journal, vol. 12, no. 2, 55–63 (2010)
14. C. B. Basnet, L. R. Foulds, and J. M. Wilson, International Transactions in Operational Research, vol. 13, no. 1, 1–15 (2006)
15. G. Edwards, C. G. Sorensen, D. D. Bochtis, and L. J. Munkholm, Computers and Electronics in Agriculture, vol. 117, 102–113 (2015)
16. B. Recio, F. Rubio, and J. A. Criado, Decision Support Systems, vol. 36, no. 2. 189–203 (2003)
17. D. D. Bochtis, P. Dogoulis, P. Busato, C. G. Sorensen, R. Berruto, and T. Gemtos, Computers and Electronics in Agriculture, vol. 91, 49–56 (2013)
18. A. R. East, Biosystems Engineering, vol. 109, no. 1, 1–9 (2011)
19. P. P. Ravula, R. D. Grisso, and J. S. Cundiff, Biomass and Bioenergy, vol. 32, no. 4, 314–325 (2008)
20. T. Wiśniewski, *Zastosowanie priorytetów dynamicznych do optymalizacji wieloproduktowych systemów produkcyjnych w poligrafi*, West Pomeranian University of Technology in Szczecin, (2013)
21. P. Szymczyk and M. Szymczyk, Automatyka, vol. 13, no. 2, 607–614 (2009)
22. B. Chen and T. I. Matis, International Journal of Production Economics, vol. 141, no. 1, 360–365 (2013)
23. A. Jardzioch and K. Bulwan, Advances in Manufacturing Science and Technology, vol. 37, no. 4, 49–59 (2013)
24. V. Vinod and R. Sridharan, International Journal of Production Economics, vol. 129, no. 1, 127–146 (2011)
25. T. Wiśniewski, S. Rymaszewski, and P. Korytkowski, Postępy Nauki i Techniki, vol. 13, 14–24 (2012)

Farm size and volume of external transportation services at selected farm businesses

Stanisław Kokoszka[1,*], and *Zbigniew* Daniel[1]

[1]Department of Production Management and Applied Computer Science, Balicka 116B, 30-149 Kraków, Poland

Abstract. The present survey and data analysis confirmed the impact of the farm size (based on the Utilized Agricultural Area) on the volume of external transportation services despite the overall improvement of owned means of transport. The main transport characteristic, i.e. shipment distance increases with the farm size. Hence, the largest farm businesses characterized by higher commercial outputs use external transportation services to a greater extent although they own better quality and higher number of farm vehicles. At the average freight volume transported by external services of 4.37 tons per hectare of the Utilized Agricultural Area, the share of contracted transportation services in the total freight transport increases from 13.95 % in the smallest farms to 28.42 % in the largest farm businesses. This analysis was confirmed by the correlation analysis between the farm size and volume of external transportation services.

1 Introduction

Due to continuous growth of commercial agricultural production, farm businesses are forced to unwillingly become transportation businesses. Thus, transportation costs decisively contribute to profitability of production [1, 3, 10]. Minimization of transportation costs depends, among other things, on transport provision methods, including the use of transportation services. In fact, production services for agriculture are a component of the modern agriculture sector and belong to efficient means of overcoming the opposition between consequences of scientific and technical progress and costs of its implementation [2, 5, 7, 11]. According to Tabor et al. [6] farms better equipped in technical means of production show a limited need for technical services. At the same time, often the lack of necessary equipment, time saving and in many cases, according to farmers, lower costs induce to contract external services [2, 4].

According to our previous studies [4, 8, 9], in southern Poland, 59.04% of farm businesses engaged in commercial production responded that transportation services were needed, of which 48.80 – 68.18% already contracted those services while 29.51 – 72.73% provided transportation services as a way of increasing their income.

2 Scope and aim of the study

The aim of the study was to assess the volume of external services contracted by farm businesses in relation to the farm size (based on the Utilized Agricultural Area (UAA)). The studies and analyses were undertaken to determine whether transportation services for agriculture (farm businesses) are needed and to estimate their level.

The survey was conducted in 653 farm businesses located in 21 municipalities of southern voivodeships (Małopolskie, Podkarpackie, Świętokrzyskie, Śląskie). The survey included all transport activities carried out in these farms. Farms contracting transportation services either to transport businesses or to neighbors were selected for the study. The analyzed group of farm businesses represented 35.22% of the whole population of farms surveyed. It should be mentioned that of the whole sample of farms contracting services, 37% also provided services, however, 100% of them provided services to their neighbors either as the "payment in work" or as an additional source of income.

3 Methodology

The survey was carried out via a guided survey regarding most of all transport activities. The first part of the survey was related to general characteristics of the farm (the structure of land use, structure of sowing, stocking density, structure and volume of production), the second part regarded farm vehicles and their use and the third part was related to working conditions of transport means and the use of transportation services (transport volume, service provider, motivation of contracting transportation services).

Based on the obtained data, the total freight transport volumes and share of external transportation services were determined and related to the farm size. The surveyed farms were divided into 3 groups depending on UAA:

- <10 UAA hectares – 50.43%, in UAA – 20.67%

* Corresponding author: Stanislaw.Kokoszka@urk.edu.pl

- 10.01 do 20.00 UAA hectares – 30.87%, in UAA – 32.98%
- > 20.01 UAA hectares – 18.70%, in UAA – 46.35%

4 Sample and study results

The survey sample comprised 230 farm businesses located in southern voivodeships. The surveyed farms represented different segments of agricultural sector from general non-specialized agricultural production to specialized agribusinesses in potato, vegetable, milk or fruit segments. Thus, cargo types, freight volumes and payload per trip vary widely. The freight volume averages 32.42 tons per hectare of UAA ranging from 4.73 do 221.75.

The second major transport characteristic, i.e. shipment distance is equally variable. Basic characteristics of the studied farms relevant to transportation are presented in Table 1.

Table 1. Selected characteristics of surveyed farms.

Characteristics		Whole sample	UAA groups		
			<10 ha UAA	10.1-20 ha UAA	>20 ha UAA
Utilized Agricultural Area (ha)	AVG.	14,05	5,76	15,01	34.84
	min.	0.80	0.80	10.15	20.01
	max.	155.00	10.00	20.00	155.00
Average shipment distance (km)	AVG.	5.70	4.32	7.05	7.20
	min.	0.06	0.06	0.52	0.78
	max.	41.57	31.71	41.57	31.56
Number of vehicles per farm	AVG	3.02	2.31	3.39	4.33
	min.	1.00	1.00	1.00	4.00
	max.	10.00	6.00	10.00	8.00
Average vehicle payload (tons)	AVG	3.06	2.57	3.57	3.54
	min.	0.50	0.50	1.00	2.00
	max.	9.00	8.00	8.70	9.00
Share of cars in equipment (%)	AVG.	4.94	3.18	5.52	8.75
	min.	0.00	0.00	0.00	0.00
	max.	50.00	50.00	33.33	40.00
Share of transportation services (%)	AVG	29.49	23.89	31.73	40.91
	min.	1.20	1.20	2.36	4.12
	max.	87.82	59.83	87.82	83.59
Sales (tons per UAA hectare)	AVG	7.13	4.73	8.38	11.53
	min.	0.00	0.00	0.00	7.74
	max.	47.33	45.82	32.88	47.33

At the average farm area of 14.05 UAA hectares, variability in the studied group is very high and ranges from the minimum of 0.80 hectares to the maximum of 155.0 hectares which means that the largest farm has 193 times greater UAA than the smallest one. The largest farms classified into the group >20 UAA hectares are on the average 2.3 times bigger compared to the medium group (10.1-20 UAA hectares) and almost 6 times bigger than the smallest ones (<10 UAA hectares). The average shipment distance is linked to the farm size. The shipment distance averages 5.70 km for the whole sample, ranging from 4.32 for the smallest farms to 7.20 km for the largest farm businesses, which means almost 167% rise. It results not only from the increased shipment distance in internal transport (due to a larger farm area) but also from the external transport. Larger farms generate a greater freight volume resulting from higher output, and they search for more price-efficient markets which is most often connected with the increasing shipment distance. At the same time, variability of the basic transport characteristic, i. e. shipment distance is very wide also in all UAA groups. The average variability range is from the minimum of 0.06 km to the maximum of 41.57 km, thus, the difference in the shipment distance between the farm with the most advantageous location in relation to fields and upstream and downstream markets and the farm with the least advantageous location is 6928%. This fact in connection with the freight volume indicates a diverse burden of transport in farm businesses. Likewise, equipment of farm business in transport means, decisive for efficient transportation, vary both within and between the UAA groups On average, the surveyed farm businesses own and operate 3.02 vehicles ranging from 1 to 10 vehicles. The number of owned vehicles increases with the farm size. Specifically, the increase in the farm size between UAA groups by almost 600% resulted in a rise in the number of vehicles per farm by 187%. It indicates that considerable differences in the above parameters are most probably the effect of increased efficiency of farm vehicles in the biggest farm businesses. This is confirmed by a much better quality of transport means in the largest farm businesses as estimated based on the average payload of a vehicle which averages 2.57 t in the smallest farms businesses and 3.54 t in the largest ones. In this aspect, it is also beneficial that the share of vehicles with higher efficiency (i.e. trucks) grows with the farm size. The largest farms own over 2 times heavier vehicles than the smallest ones. The next factor that can significantly contribute to contracting transportation services is related to the share of outer transport in the total transport volume. Transportation services are more often contracted for outer transport over longer distances in much better conditions frequently during the off season. Their share in total transport volumes is also significantly influenced by the level of commercial production related, among other things, to the farm size. This view is confirmed by the increase in the share of transportation services with the farm size. Specifically, the share of transportation services averages 29.49% of the total transport volumes for the whole study sample and increases from 23.89% in the smallest farms to 40.91% in the largest UAA group. However, variability within the UAA groups is very high. The share of external transport in the total transport volume is the effect of sold production (tons) and impacts on transport volume but also on demand for transportation services which sometimes, depending on the cargo type, require specialized equipment. The freight volume related to the sold production clearly increases with the farm size. It averages 7.13 tons per hectare of the UAA and increases by ca. 244% between the smallest and the largest farms. The results of this analysis are presented in Tab. 2 showing the volume of transportation services and their share in the total transport volume.

Table 2. Volume of transportation services.

Characteristics		Whole sample	UAA group		
			<10 ha UAA	10.1- 20 ha UAA	>20 ha UAA
Freight volume (tons per hectare of UAA)	AVG.	4.37	3.20	5.01	6.51
	min.	0.10	0.10	0.19	0.29
	max.	14.94	14.43	14.94	8.69
Share of transportation services (%)	AVG	18.02	13.96	18.35	28.42
	min.	0.11	0.11	0.64	0.66
	max.	68.22	54.25	51.79	68.22
Shipment distance (km)	AVG	14.18	10.47	18.52	17.00
	min.	1.40	1.00	0.60	7.10
	max.	100.00	80.00	120.00	100.00

Data presented in Table 2 confirm the view that the largest farm businesses characterized by the highest commercial output to a greater extent use external services although they are equipped with better quality farm vehicles and their higher number.

The average freight volume transported by external transportation services is 4.37 tons per hectare of UAA and ranges in the UAA groups from 3.20 to 6.51 tons per hectare of the UAA. Despite the increase in the total transport volumes in farm businesses, resulting from a larger area, the share of transportation services in the total transport volumes rises from 13.96% in the smallest farms to 28.42% in the largest farm businesses. However, it should be emphasized that the variability of the share of transportation services in the total transport volumes is extensive both when the whole sample is concerned and when the UAA groups are analyzed separately. It appears that the main cause of contracting transportation services by a farm business is related to the shipment distance resulting from locations of upstream and downstream markets. This can be corroborated by the difference between the average shipment distance in all farm businesses surveyed, i.e. 5.70 km and the shipment distance for transportation services averaging 14.18 km, thus the difference is almost 250%. In addition, the maximum shipment distance for external transportation services reaches 100 -120 km while the respective value for own farm vehicles (without services) is ca. 40 km.

At the final stage of analysis, the effect of farm size on the volume of contracted transportation services was estimated. A relationship between the independent variable (farm size) and dependent variable (the volume of contracted services) is presented in Figure 1.

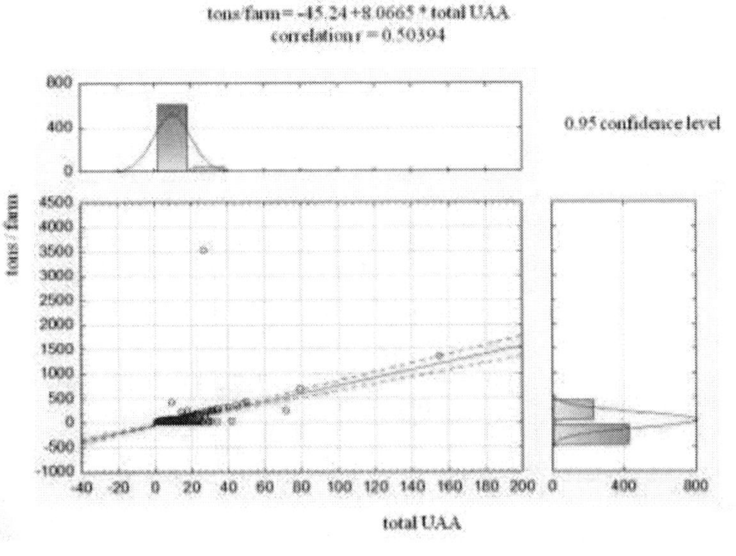

Fig.1 Effect of farm size (UAA hectares) on volume of transportation services (tons) per one hectare of UAA.

At the correlation coefficient of 0.50394 and significance level of 0.05, the equation of the relationship between the farm size in the UAA and volume of transportation services (freight tons) assumes the form:

$$Utr = -45,24 + 8,0665 \cdot P_{UR} \quad (1)$$

where:
Utr – volume of transportation services in tons
P_{UR} – utilized agricultural area (UAA hectares))

The above equation can be used for forecasting the demand for transportation services based on the freight volume per hectare of the UAA in farm businesses with the area ranging from 0.8 to 155 UAA hectares. According to the equation (1), transportation services will be contracted by farm businesses larger than 5.61 UAA hectares and they will amount to 0.013 tons. Farm business with 155 UAA hectares will use external services to transport 1205 tons.

5 Summary and conclusions

Owned farm transport means, considerably influencing the use of transportation services, show improvement with the increasing farm size, especially in terms of the vehicle quality (payload of a vehicle). However, the basic transport characteristic i.e. shipment distance increases with the farm size which is disadvantageous. Thus, the largest farm businesses characterized by higher commercial production to a greater extent use external transportation services although they are equipped with better quality and higher number of farm vehicles owned. Specifically, the average transport volume carried by transportation services is 4.37 tons per UAA hectare ranging from 3.20 to 6.51 tons per UAA hectare. The share of transportation services in the total transport volumes increases from 13.96% in the smallest farms to 28.42% in the largest farm businesses. The presented values can result from the difference in the average shipment distance of 5.70 km and shipment distance for external services averaging 14.18 km, thus these values differ by 250%. The presented analyses confirm that the demand for transportation services increases with the farm size.

References

1. J. Bielejec, *Transport rolniczy*, (IBMER, Warszawa, s.1-220, 2001)
2. R. Jabłonka, K Kapela, Inżynieria Rolnicza, **1(119)** (2010)
3. S. Kokoszka, M. Kuboń, S. Roczkowska-Chmaj, J Borcz, *Analiza i uwarunkowania rynku usług i urządzeń w transporcie wiejskim* (PTIR Kraków, 2001)
4. S. Kokoszka, Journal of Reseaarch and Applications in Agricultural Engineering, Vol. **58(2)** (2013)
5. S. Tabor, Inżynieria Rolnicza, **10(85)** (2006)
6. S. Tabor, M Kuboń, Inżynieria Rolnicza, **7(125)** (2010)
7. T. Szuk, Inżynieria Rolnicza, **8(117)** (2009)
8. S. Kokoszka, S. Tabor, Journal of Reseaarch and Applications in Agricultural Engineering, Vol. **61(1)** (2016)
9. S. Kokoszka, S. Tabor, Journal of Reseaarch and Applications in Agricultural Engineering, Vol. **62(1)** (2017)
10. D. Gollin, R. Rogerson, Journal of Development Economics, **107** (2014)
11. S. Kallas, Launch of the Green Freight Europe Initiative - Speech **12/230** (2012)

Research on power consumption of screw press for pressing of oil from rape seed

Stepan Kovalyshyn, and *Vasyl* Tomyuk

Lviv National Agrarian University, 80381, str. Dublyany, Volodymyra Velykoho 1, Ukraine

Abstract. The work highlights the analysis of the influence of technological and structural parameters of the screw press, namely the step of the turns of the screw shaft of its speed and the area of openings for the removal of cake on the energy parameters of the process of oil expression.
Based on the results of the multifactorial experiment, regression dependence allows us to estimate the influence of the parameters and operating modes of the screw press on the power consumption of the screw press and can be used in the process of synthesis and modeling of machines intended for pressing of vegetable oils from oil-containing raw materials. The surfaces of the response of the dependence of the power consumption of the screw press from the area of the openings for the removal of the cake as well as the step of the turns and the speed of the screw shaft have been constructed.
Providing the following parameters of the pressing process: turn pitch of the auger shaft $X_{10} = 21.8$ mm; area of openings for oil cake withdrawal $X_{20} = 107.99$ mm^2; rotation frequency of the auger shaft $X_{30} = 5.12$ min^{-1}, the power consumption of the screw press will be 170 W, ensuring the maximum yield of oil q = 36%.

1 Analysis of recent research works and publications

The process of oil expression from seed of oil-bearing crops by pressing is difficult and energy-consuming. The results of the research of this process indicate that energy costs consist of three main components: the capacity to consolidate the oilseed material, the power to overcome the friction forces between the raw material, the screw shaft and the extracting box, as well as the power required to squeeze oil [1, 5, 12, 13, 16, 21]. It is possible to improve the energy indexes of a screw press by optimizing its technological and constructive parameters. The first ones include the speed of the screw shaft and the area of the openings for the removal of cake, and to the other - the length of the screw shaft, its inner and outer diameters, the step and shape of the screw channel. In addition, the energy indicators are influenced by the physical and mechanical properties of oil-bearing raw materials - the geometric dimensions of the seeds, the coefficient of friction slip, etc.

The results of research on the influence of technological parameters of the screw oil press have shown that with increasing the rotational speed of the shaft, the power of oil extraction is linearly increased, and along with this, the energy parameters of the process deteriorate [9]. In addition, they also depend on the geometric features of the screw channel [21] and the length of the extracting box [1], and, correspondingly, the screw shaft. In works [5, 16] it is noted that it is expedient to increase the length of the shaft, since in this case more pressure is developed in the pressure chamber [3, 6] and there is an increase in the time of the latter's action on the pressed material, which leads to deeper oil expression. The disadvantage of long shafts is that they are quickly overloaded. This causes loss of strength that is why it is necessary to use high-quality steels, which work well for complex deformation. In addition, the size of the press increases, and operation and maintenance are complicated.

In this connection, it is necessary to study in more detail the influence of the step of the turns of the screw shaft of its rotational speed and the area of the openings for the removal of cake on the energy parameters of the pressing process.

2 Setting objectives

The aim of the work was to increase the efficiency of the process of pressing the seeds of oil-bearing crops in the screw oil press by reducing energy costs by substantiating the steps of the spindle shaft of its rotational speed and the area of openings for the removal of cake.

3 The main material

The achievement of research has been conducted by multifactor experiment as for the determination of the influence of some parameters of pressing process - turn pitch of auger shaft (h), area of openings for oil cake

* Corresponding author: tomyuk.lnau@gmail.com

withdrawal (s) and rotation frequency of auger shaft (n) on the completeness of power consumption (p). Experiments have been conducted using the improved construction of auger press Auger Oil Press-1, the constructive form of which is given in Fig. 1.

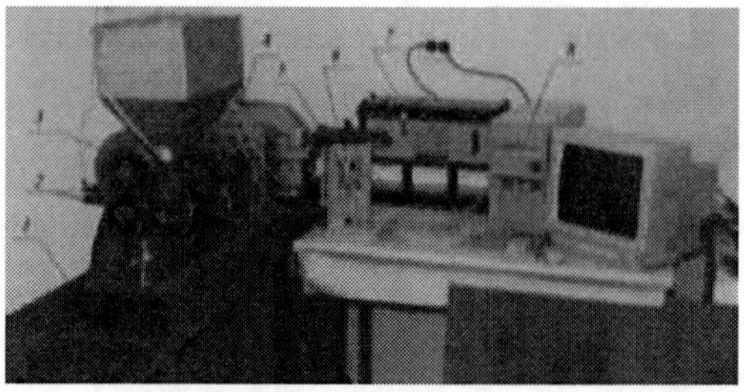

Fig. 1. General view of the measuring equipment for determining the pressure in the pressing chamber of the oil extraction press:

1 - bed; 2 - pressure sensors from the firm Keller PA-6T; 3 - screw press; 4 - cylindrical gearbox; 5 - frequency converter ATX-3.0; 6 - strain gauge amplifier; 7 - measuring station (KI - 505); 8 - PC with built-in board L-154.

It is known that the generalized result of multifactor experiment is equation of regression and when we research it as for the extremum one can optimize the meaning of researched parameters. In our case for developing an equation and determination of its coefficients, one can use the three-layered plan of the second range of Box-Benkin [8]. According to it one should determine zero levels of researched parameters. Taking into account the methodology of experimental data [2, 8] and the construction of the improved sample of experimental press, it is determined: turn pitch of auger shaft X_{10} = 22 mm; area of openings for oil cake withdrawal X_{20} = 109 mm^2 ; rotation frequency of auger shaft X_{30} = 45 min^{-1} . For every parameter the intervals for varying have been determined. They have been determining the pressing regimes or in other words the conditions of experiment conducting (Table 1).

Table 1. Results of coding of investigated factors.

Factors and units of measurement	**The usual sign**	**The encoded sign**	**Interval of varying**	**Levels of varying**					
				usual			**encoded**		
				Uppers(+1)	**Zero (0)**	**Lower (-1)**	**upper (+1)**	**Zero (0)**	**Lower (-1)**
Turn pitch of auger shaft(h), mm	X_1	x_1	2	24	22	20	+1	0	-1
Area of openings for oil cake withdrawal (s), mm^2	X_2	x_2	20	129	109	89	+1	0	-1
Rotation frequency of auger press, (n), min.$^{-1}$	X_3	x_3	30	75	45	15	+1	0	-1

Experiments have been conducted using seeds of winter rape of Danhal sort with the moisture 7%. Single auger shafts (Fig. 2) with the changing turn pitch have been set up on press. The form of their riffle has been done in the form of circle segment. Rotation frequency of auger shaft has been changed with the help of frequency converter of current power of electric engine.

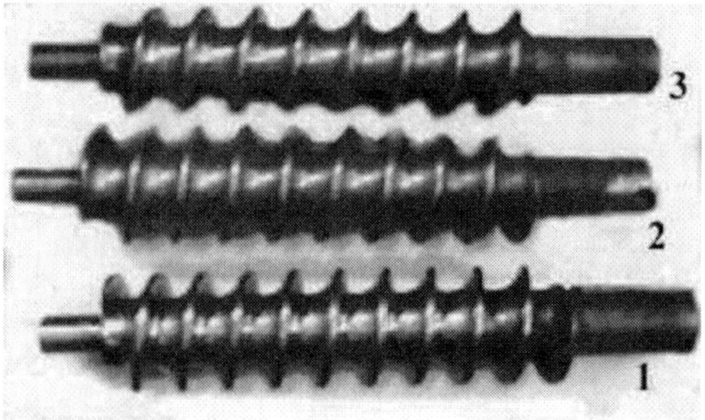

Fig. 2. Constructions of auger shafts: 1, 2, 3 – turn pitch of shafts, according to 20 mm, 22 mm, 24 mm.

Table 2. Matrix plan and results of multi-factor experiment.

№	Value of coded factors			Power of the press drive, W			Average value	S_u^2
	x_1	x_2	x_3	Sample 1	Sample 2	Sample 3		
1	-1	-1	0	594.0	594.0	594.1	594.03	0.003
2	1	-1	0	375.2	375.3	375.1	375.20	0.010
3	-1	1	0	318.2	318.4	317.9	318.17	0.063
4	1	1	0	169.3	169.2	168.8	169.10	0.070
5	-1	0	-1	293.2	293.1	293.0	293.10	0.010
6	1	0	-1	90.9	90.0	91.2	90.70	0.390
7	-1	0	1	733.5	734,2	733.6	733.77	0.143
8	1	0	1	568.2	568.7	568,9	568.60	0.130
9	0	-1	-1	395.3	395.8	396.5	395.87	0.363
10	0	1	-1	70.1	75.4	78.5	74.3	0.163
11	0	-1	1	716.2	716,5	717.9	716.87	0.823
12	0	1	1	612.3	612.5	612.9	612.57	0.093
13	0	0	0	389.0	389.5	390.1	389.53	0.303
14	0	0	0	389.0	389.5	390.1	389.53	0.303
15	0	0	0	389.0	389.5	390.1	389.53	0.303

As a result of processing of the obtained data Table 2 of the multivariate experiment and the use of software Statistica 8.0, the regression equation for power consumption is:

$$y_e=390-92x_1-120.3x_2+229.2x_3+17.5x_1x_2+9.5x_1x_3-68.5x_2x_3-19.5x_1^2-6.15x_2^2+51.19x_3^2. \quad (1)$$

As a result of calculations according to Table 2, the calculated Cochran criterion made up – $G_{calc}=\frac{0.823}{3.170}=0.260$. The table value of the Cochran criterion at $n = 15$ and $f = 3-1 = 2$ is 0.335. Therefore, the process is considered reproducible, since $G_{calc}=0.260<G_{table}=0.335$.

The hypothesis about the adequacy of the description of the equation (1) of the experimental results can be considered correct with a 95% probability, since Fisher's calculation criterion is less tabular:

$$F_{calc}=1.90<F_{table}=2.12.$$

Let's represent the regression equation (1) in the following way:

$$y_e=1052.7+1136x_1-17.42x_2-13.4x_3+0.43x_1x_2+0.15x_1x_3+0.01142x_2x_3-4.87x_1^2-0.0154x_2^2+0.0569x_3^2. \quad (2)$$

On the basis of equation (1), the surface of the response of the power consumption of the screw press from the area of the openings for the removal of the cake, the step of the turns and the speed of the screw shaft is constructed. During the construction of the surface of the response varied only two factors, and the third remained equal to zero. For this we used the software Statistica 8.0.

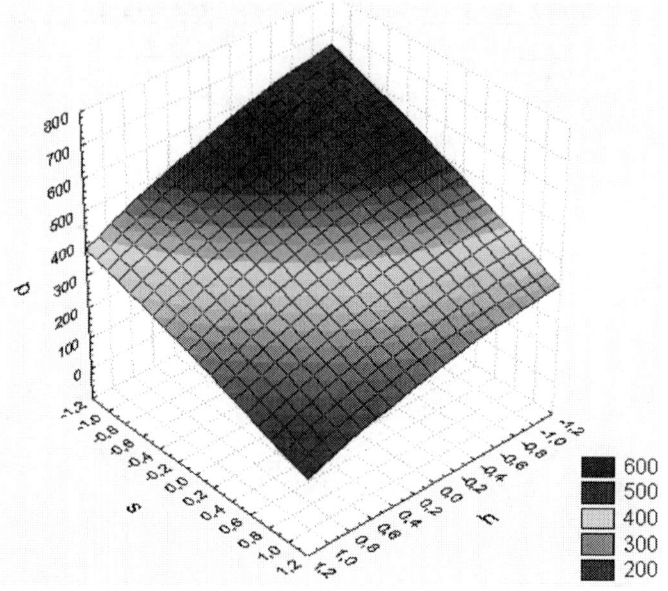

Fig.3. The equilibrium surface $p = f(h, s)$.

When analyzing the response surface, it should be noted that the minimum power, less than 200 W, is provided for maximal opening of openings and maximum step of turns.

Dependence $p = f(h, s)$ (Fig. 3) shows that at the turn pitch of the auger shaft X_{10} = 22 mm; area of openings for oil cake withdrawal X_{20} = 129 mm^2; rotation frequency of the auger shaft X_{30} = 45 min $^{-1}$, the lowest power consumption is achieved. For such structural parameters, the power of the screw press is 169.10 Watts.

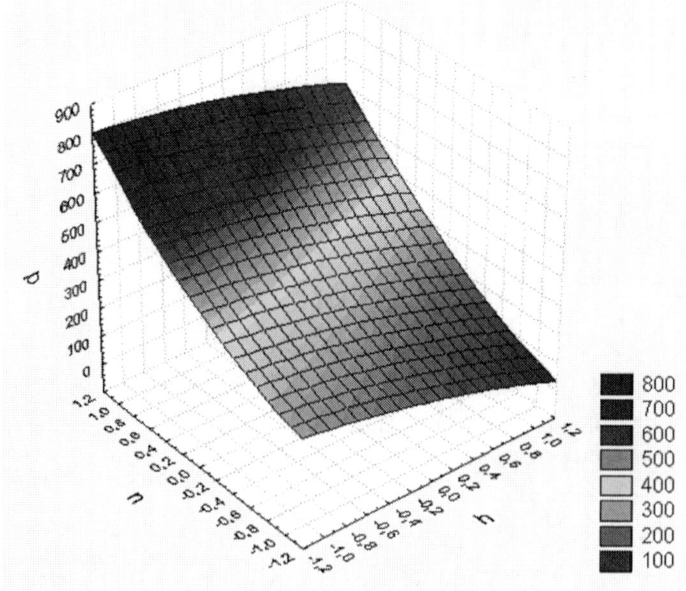

Fig.4. The equilibrium surface $p = f(n, h)$.

Dependence $p = f(n, h)$ (Fig. 4) shows that at the turn pitch of the auger shaft X_{10} = 20 mm; area of openings for oil cake withdrawal X_{20} = 129 mm^2; rotation frequency of the auger shaft X_{30} = 15 min $^{-1}$, the lowest power consumption is achieved. With such design parameters, the power of the screw press is 90.1 Watts.

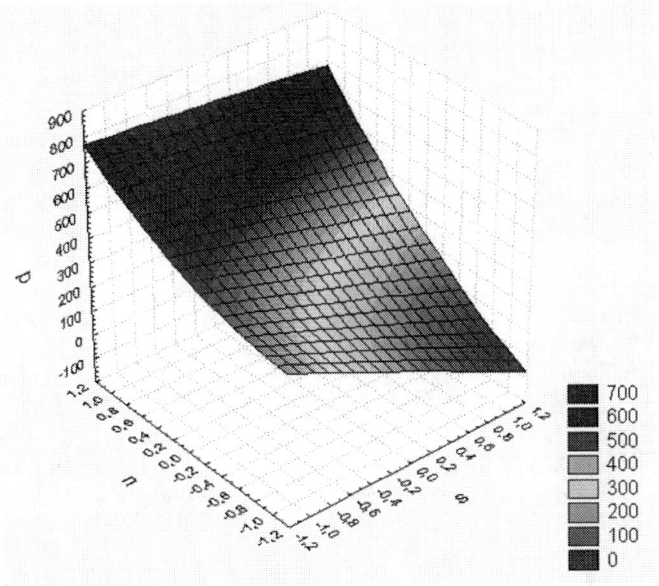

Fig.5. The equilibrium surface $p = f(n, s)$.

Dependence $p = f(n, s)$ (Fig. 5) shows, that at turn pitch of auger shaft X_{10} = 22 mm; area of openings for oil cake withdrawal X_{20} = 109 mm^2 ; rotation frequency of auger shaft X_{30} = 15 min^{-1} the lowest power consumption is achieved. With such design parameters, the power of the screw press is 170 Watts.

It is possible to optimize the parameters of the power consumption by comparing the obtained results with the results of the influence of the structural and technological parameters of this press on the output of oil. Since the maximum oil extraction q = 36% [14] is ensured at the turn pitch of auger shaft X_{10} = 21.8 mm; area of openings for oil cake withdrawal X_{20} = 107.99 mm^2; rotation frequency of auger shaft X_{30} = 5.12 min $^{-1}$ then the power consumption of the screw press at such parameters will be 170 W.

4 Conclusions

1. Use of the proposed design of oil auger press in the process of pressing seeds of oil-bearing crops with reasonable technological parameters allows for obtaining high-quality at low energy costs, with maximum output of oil.
2. The results of investigations of the influence of technological parameters of the screw oil press have shown that with an increase in the rotational speed of the shaft, the power of the oil press is linearly increased, and at the same time, the energy parameters of the process deteriorate.
3. Providing the following parameters of the pressing process as turn pitch of auger shaft X_{10} = 21.8 mm; area of openings for oil cake withdrawal X_{20} = 107.99 mm^2 ; rotation frequency of auger shaft X_{30} = 5.12 min^{-1} the consumed power of the screw press will be 170 W, providing the maximum yield of oil q = 36%.

References

1. Bartashevich N.A. Intensification of oil extraction by pressing of Far Eastern varieties of soybean seeds: *Author's abstract. Dis. Cand. Tech. Sciences.* Vladivostok. 24 p. (2006)
2. Vedenyapin G.V. The total experimental methods and data processing experienced data. Moscow: *Kolos.* 159p. (1976)
3. Bratsikhin E.A., Mikulin S.S., Streltsov N.N. Processing of plastic in the product. Moscow: *Chemistry.* 399p. (1966)
4. Voznesenskyy V.A. Statistical methods for planning experiment in techno-economic studies [Text].Moscow: *Statistics.* 175 p. (1974)
5. Holdovskyy A.M. Theoretical base production of vegetable oils. Moscow: *Pischepromizdat.* 496 p. (1958)
6. Gruznov G.F. Machines for the processing of plastics. Moscow: *Mechanical Engineering.* 238 p. (1970)
7. Gritsenko V.T., U. Duryn. Technology and oil seed processing line using the extruding method of extraction of oil. *Scientific and Technical Bulletin of the Institute of oilseed agriculture. - Berlin: IEC Agrarian Sciences* – Vol. 7. P. 181-183. (2002)
8. Declarations. Pat. a useful model number 31 622 U Ukraine, IPC V30V 9 / 12 (V30V 9 / 18). Screw press for oil extraction / R. Shevchuk, V. Tom'yuk; *applicant and patentee Lviv State Agrarian University.* Bul. № 7. 4 p. (2008)
9. Didyk T.A. Decrease in energy intensity of press extruders // Vavilov Readings - 2004: *Proceedings of the All-Russian Scientific and Practical Conference, dedicated to the 117th anniversary of*

the birth of Academician N.I. Vavilov. Saratov: FGOU VPO "Saratov State Automated Information System". P. 49-52. (2004)

10. Zarembo G.V. Study managering pressed vegetable oils in screw presses: *Thesis. Dis. candidate. Engineering. Science.* Krasnodar. 36 p. (1962)
11. Kychyhyn V.P. Technology and production control vegetable oils. Moscow: *Food processing industry*. 359 p. (1976)
12. Kovalyshyn S., Tomyuk V. Impact factors pressing for release of rapeseed oil: *Bulletin of Sumy NAU "Mechanization and automation of production processes*. Vol. 2. (18). (2008)
13. Kopeykovskyy V.M. Technology production of vegetable oils. Moscow: *Lehkay apyshevaya industry*.415 p. (1982)
14. Kovalyshyn S., Tomyuk V. Optimization of oil press parameters. *Motrol.* Lublin. Vol. 13. P. 173 – 182. (2011)
15. Markman A.L., Hlushenkova A. N. Basics of fat-and-oil productions. Moscow: *Food processing industry*. 279 p. (1965)
16. Maslykov V.A. Technological equipment production of vegetable oils. Moscow: *Food processing industry*.439p. (1966)
17. Melnikov S. Aleshkyn V., Roshchin P. Experiment planning of the agricultural studies the processes [Text]. Leningrad: *Kolos*. 199 p. (1972)
18. Sergeyev A.G. Guide to obtain technology and processing of vegetable oils and fats, ed. Leningrad: *VNIIZH*. Vol. I. - Book 2. 727 p. (1975)
19. Rumshynskyy L.Z. Mathematics, Processing of results experiment. Moscow: Nauka, 192p. (1971)
20. Tomyuk V. Shevchuk R. Screw compression press for oil. *Kinematics Lugansk NAU "Engineering"*. № 91. P. 253-254. (2008)
21. Tomyuk V. Investigation of pressure distribution along the shaft of oil press. *Bull. Kharkiv National Technical University of Agriculture P. Vasilenko*. № 78, P. 242-249. (2009)`

Modified gelatine structures as packaging material for frozen agricultural products

Katarzyna Kozłowicz[1*], *Franciszek* Kluza[1], *Dariusz* Góral[1], *Piotr* Nakonieczny[1], and *Maciej* Combrzyński[2]

[1]University of Life Sciences in Lublin, Department of Refrigeration and Food Industry Energetics, ul. Doświadczalna 44, 20-280 Lublin, Poland
[2] University of Life Sciences in Lublin, Department of Process Engineering, ul. Doświadczalna 44, 20-280 Lublin, Poland

Abstract. Gelatine gels modified by hydrated paper pulp (PP), ground extruded starch (ES) and hydrogel balls (HGB) were formed so that after the crosslinking process, the plates of suitable thickness could be obtained. Then the material was freeze-dried and the obtained porous gelatine structures as well as styrofoam (XPS) as the reference material were tested for thermal conductivity and bending. The temperature fields were examined during the thawing process of products placed inside the test packagings with modified gelatine structures and polystyrene as well as temperature field profile on outer surface of these packs. The studied materials resemble styrofoam in terms of thermal insulation and exhibited medium thermal conductivity ranging between 0.047 and 0.081 [$W \cdot (mK)^{-1}$]. The bending strength of the materials under investigation proved higher than that of styrofoam and this fact supports their applicability as an alternative for frozen agricultural products packages.

1 Introduction

The primary function of packaging material is to maintain freshness and quality of raw materials and agricultural products, especially those chilled and frozen. Packaging needs to possess suitable barrier properties, a degree of thermal insulation and mechanical strength. The material for food packaging should meet a number of essential technical and legislative regulations [1-2]. It is also expected to be biodegradable, namely, the most eco-friendly possible. Biodegradable packaging material can be produced from animal-derived substances (collagen, gelatine, casein, albumin, etc) or plant-based (cellulose and its derivatives, starch). The choice of suitable packaging material and its form also relies on the factors associated directly with the physicochemical characteristics of packed products, i.e. chemical state, texture, porosity as well as the time and storage conditions [3]. In response to increased awareness of consumers and high demand for quality, new packaging materials have been introduced [4-8].

The objective of the study was working out, development and characterization of the basic thermophysical properties of the material based on modified gelatine gels intended for frozen products packaging.

2 Materials and methods

The research material included lyophilized solidified and modified gelatine gels. The processes used to produce them involved crosslinking, swelling, freezing and sublimation drying. To obtain 5% gelatine solution, demineralized water and gelatine type A (*Gellwe, Poland*) were employed. The chilled sol (5% gelatine) modified by hydrated paper pulp (PP), ground extruded starch (ES) and hydrogel balls (HGB) was casted into cuboid-shaped molds and allowed to remain for 12 hours at the ambient temperature to promote the crosslinking reaction. The appropriate dose rates of the modified sol were to ensure, after the crosslinking process, production of the plates of the same 9 mm thickness. The solidified gel samples were frozen in a chest freezer at natural convection at -33°C temperature. Gel sublimation drying was performed in the Alpha 2-4LD Plus freeze-dryer (*Martin Christ Freeze Dryers, Germany*) at the constant pressure 20 Pa in the chamber, ice condenser chamber temperature -64°C, drying chamber temperature -36°C and drying time of 72 hours.

The obtained material in the form of porous gelatine structures and styrofoam (XPS) as the reference material were analyzed for thermal conductivity (KD2 Pro thermal properties analyzer with KS-1 probe Decagon Devices, USA) and subjected to strength tests (*texture analyzer LFRA 4500; Brookfield, USA*) using the three point bend test (a symmetrical knife of 30-degree blade angle, blade thickness 3 mm, shift velocity 0.5 [$mm \cdot s^{-1}$], sampling time 100 measurements per sec, measurement accuracy ±0.01 N) [9]. Sample density was calculated as the ratio of sample mass to their volume. The mean from 5 replications was assumed as the final result. The obtained porous structures in a form of plates served to

* Corresponding author: katarzyna.kozlowicz@up.lublin.pl

produce packaging for frozen products. To make comparative assessment of the packaging usability, temperature fields recorded at the thawing process of the products placed inside them were studied. The thawing operation was ceased when the product surface temperature reached 18°C. The temperature was measured and recorded by temperature analyzer LB-515P (*LAB-EL, Poland*) coupled with button data loggers for temperature and humidity (type 23). The loggers recorded the surface temperature of a frozen product (carrot) inside the test packs. The measurements were carried out for 22 hours. The obtained data made basis for thawing kinetics analysis. The temperature field profile on the test pack surface during the thawing process was evaluated with TESTO 882 infrared camera (*Testo, UK*). The assessment of temperature distribution was made using the TestoIRSoft software ver. 3.4 assisting the thermal imaging camera.

The research results were analyzed using the Statistica 13 kit and Microsoft Excel program.

3 Results and discussion

Thermal conductivity is a key property of insulating materials. Its interpretation, however, imposes the "converse logic", namely, the lower the thermal conductivity coefficient, the better its thermal insulating properties. The value of this characteristic depicting given material depends on several parameters and factors but predominantly on material composition and temperature [10]. The measured thermal conductivity values of the modified and lyophilized materials and the reference material – styrofoam were tabulated (Table 1). The thermal conductivity of gelatine structures was found within the 0.047 – 0.081 [W·(mK)$^{-1}$] range. The mean thermal conductivity of styrofoam averaged 0.033 [W·(mK)$^{-1}$]. As regards thermal insulation, the obtained gelatine structures, just like styrofoam, can be applied as packaging material for frozen products.

Table 1. Physical properties (mean ± standard deviation) of styrofoam and freeze-dried samples.

Sample	Thermal conductivity [W·(mK)$^{-1}$]	Bending force [N]	Density [kg·m^{-3}]
XPS	0.033 ± 0.002^{a}	3.35 ± 0.17^{a}	36.37 ± 1.15^{a}
PP	0.081 ± 0.008^{b}	11.86 ± 0.27^{b}	90.33 ± 9.07^{b}
ES	0.054 ± 0.004^{c}	4.74 ± 0.41^{c}	159.0 ± 7.94^{c}
HGB	0.047 ± 0.003^{d}	4.17 ± 0.28^{d}	66.33 ± 13.20^{a}

Apart from low thermal conductivity, packaging material must have good mechanical strength. The bend testing has indicated considerably higher mechanical strength of the new material as compared to the styrofoam sample (Table 1). The mean maximum damage parameter established for gelatine material modified by paper mass was 11.86 N, while 4.74 N in case of modification by extruded starch and 4.17 N by hydrogel balls. The lowest damage parameter had the styrofoam sample (3.35 N). Notably, higher force required to damage the sample proves that in this context, gelatine material can be considered a better raw material for packaging production than styrofoam. Hence, it is possible to manufacture packaging of markedly higher strength with, importantly, a similar thermal insulation parameter.

Density is a main parameter in material characteristics as it provides information on the mass per unit volume [11]. The lowest density was recorded for the gelatine structures modified by hydrogel balls (66.33 [kg·m^{-3}]), whereas the highest for those modified by extruded starch (159.0 [kg·m^{-3}]). Yet, the obtained density values were higher than those characterizing the reference material- styrofoam (36.37 [kg·m^{-3}]).

The thawing curve constitutes the major source of the data for determination of the sample temperature at the thawing process [12]. Taking into account the thawing curves established (Fig.1), it was found that the temperature in the styrofoam packaging was below 0°C and it maintained for the longest time, namely over 15.5 h. The temperature of 0°C was recorded in the packaging obtained from gelatine structures after 10 h (modified by paper pulp) and 8 h (modified by extruded starch and hydrogel balls).

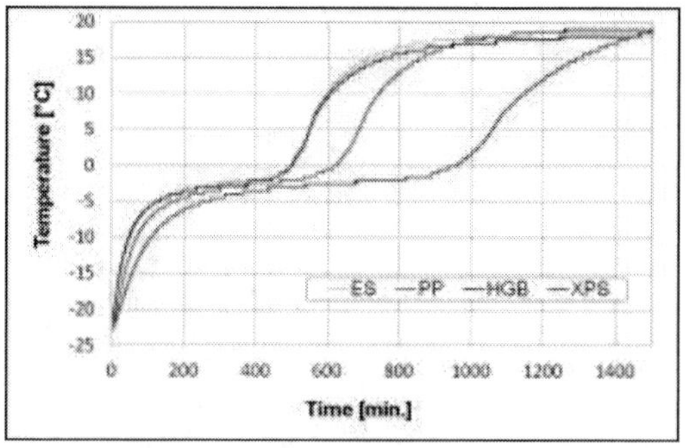

Fig. 1. Thawing curves of frozen product placed in test packs.

The changes in the temperature on the outer surface of test packs with a deep-frozen product were recorded by an infrared camera. Figure 2 presents the infrared image of the surface of packagings made from the studied materials together with analysis of the temperature profile and its extreme values. The point CS1 denotes the minimum temperature, while HS1 – the maximum one. The lowest surface temperature at the CS1 point was established for the packaging obtained from the hydrogel balls-modified material (8.2°C) and the material modified by paper pulp (12.9°C). As for the styrofoam packaging, the minimum surface temperature was 19.5°C. Considering the point HS1, the lowest maximum was determined for the packaging manufactured from the gelatine structure modified by paper pulp (22.1°C), whereas the highest for that extruded starch-modified (23.3°C). The presented temperature distribution profiles showed that the product got frozen (stuck) periodically to the surface of its packaging from the material based on the hydrogel balls-modified gelatine structure.

https://doi.org/10.1051/bioconf/20181002012

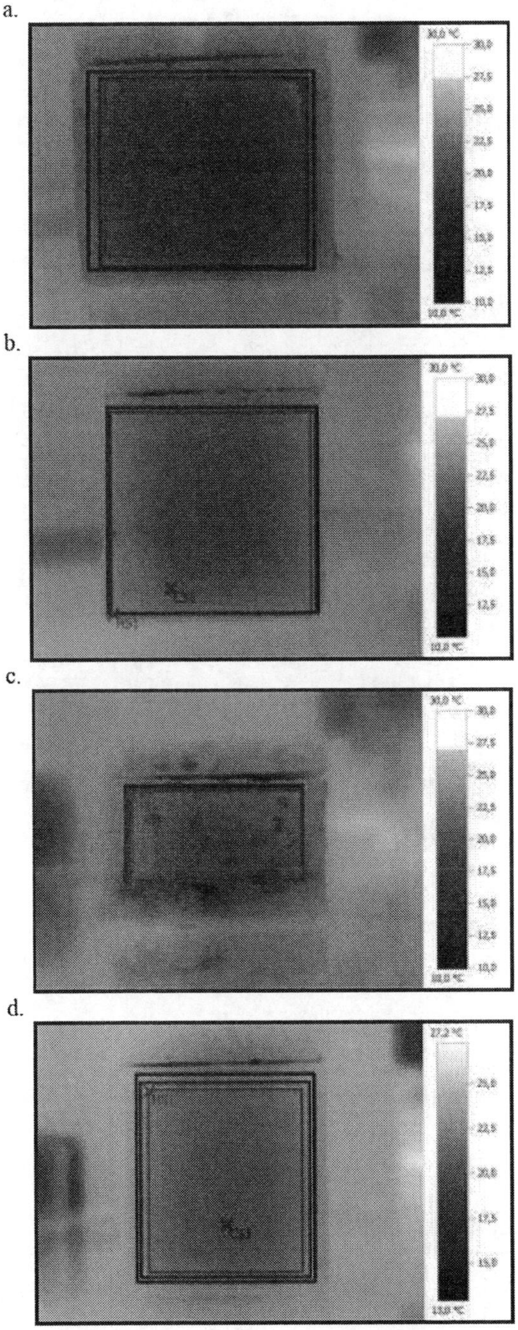

Fig. 2. Infrared image of packaging surface with analysis of temperature profile and its extreme values, a. PP, b. ES, c. HGB, d. XPS

The temperature distribution profiles were analyzed together with the histograms created (Fig. 3). A histogram is a graphical representation of the empirical distribution of characteristics which means that it serves to present the results obtained for some quantitative variables. The histogram for the packaging surface area shows a percentage share of each temperature value denoted as colour pixels on the infrared image.

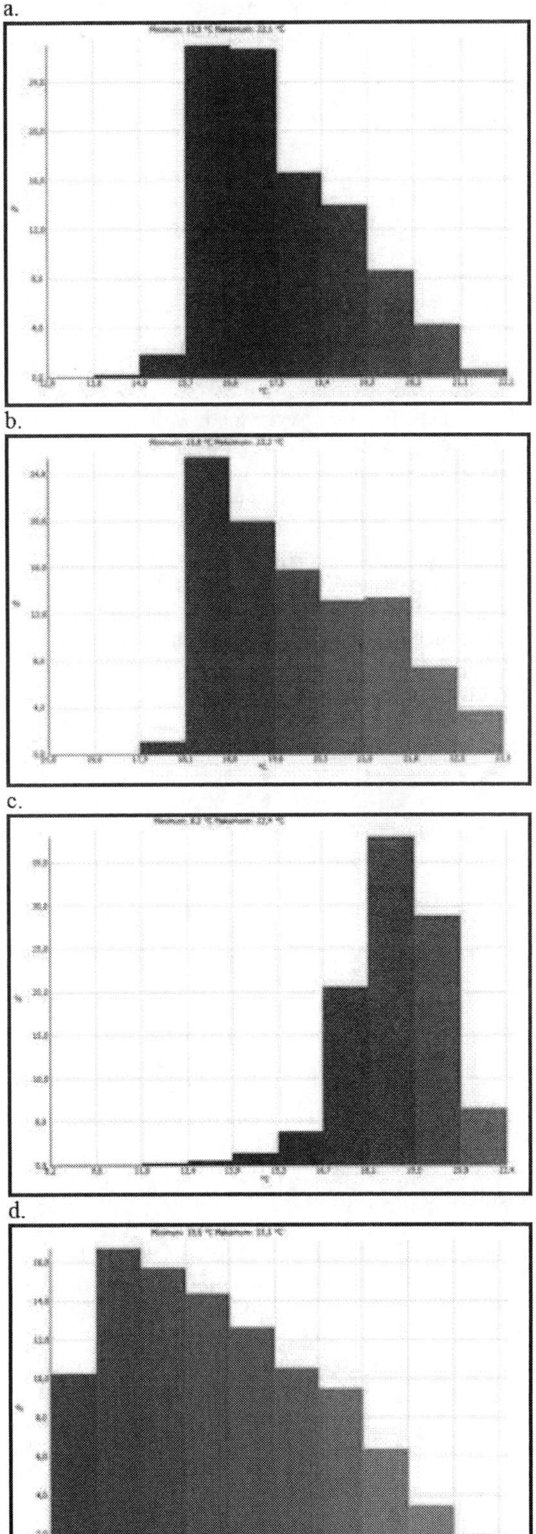

Fig. 3. The histogram of temperature distribution on pack surface: a. PP, b. ES, c. HGB, d. XPS

The highest percentage rate in the entire area for the packaging made of the material modified by paper pulp was determined for the temperatures ranging between approximately 15.7 and 16.6°C, that accounts for circa 27% of the total percentage of the temperatures. As regards the packaging based on extruded starch-modified material, it was determined for the temperatures varying from 18.1 to 18.8°C, which constitutes about 25%. The histogram created for the styrofoam packaging displays that the highest percentage share in the entire area had the temperatures from the 19.9 up to 20.2°C range, that is circa 16% of total percentage of the temperatures.

The research results are consistent with those reported in the earlier studies by Kozłowicz and others [13] who used lyophilized gelatine structures as packaging material for frozen foods with good properties of thermal insulation. The mean thermal conductivity of these structures was 0.054 $[W \cdot (mK)^{-1}]$ and density 71.16 $[kg \cdot m^{-3}]$. Besides, gelatine as a natural-origin polymer is commonly used as a drug delivery carrier, dressing material and tissue engineering scaffolds [14]. The production ease, low density and porosity of lyophilized gelatine structures as well as feasible complete biodegradability of such insulating material make it an efficient replacer of polystyrene in the packaging manufacture process. It is also worth noting that recently many packaging manufacturers have applied natural raw materials more frequently. These are polysaccharides (corn and potato starch, cellulose, gums), animal proteins (casein, collagen, gelatine), plant-based (soybeans, gluten) and lipids (fats oils) [6, 15-17] that allow maintaining a suitable quality of the packaged raw materials.

4 Conclusion

The obtained material that varied in composition displayed the thermal insulation properties similar to these of the reference material – styrofoam. Therefore, the lyophilized gelatine structures are applicable for packaging of agricultural frozen foods. Their mean thermal conductivity ranged from 0.047 up to 0.081 $[W \cdot (mK)^{-1}]$, whereas for the styrofoam tested, it averaged 0.033 $[W \cdot (mK)^{-1}]$. As for packaging material based on lyophilized gels with hydrogel balls and paper mass, frost penetration was observed at only some sites. The investigations have confirmed that the proposed materials made from lyophilized porous gelatine gels with paper mass, hydrogel balls or starch can constitute an alternative to the materials employed presently as packaging for agricultural frozen products. By reason of gelatine properties (a natural polymer), the obtained gelatine structures are fully biodegradable and easily compostable.

References

1. G. Robertson, *Food packaging: principles and practice*, CRC Press, Boca Raton (2013)
2. Regulation (EC) No 1935/2004 of the European Parliament and of the council on materials and articles intended to come into contact with food and repealing Directives 80/590/EEC and 89/109/EEC
3. B. Grabowska, Przem. Spoż. **68**, 9, 16-18 (2014), (in Polish)
4. S. Farris, K.M. Schaich, L. Liu, L. Piergiovanni, K.L. Yam, Trends Food Sci. Technol. **20**, 316-332 (2009)
5. P.A. Pawar, A.H. Purwar, AJER **02**, 05, 151-164 (2013)
6. A. Rejak, A. Wójtowicz, T. Oniszczuk, Przem. Chem. **92**, 11, (2013)
7. V. Siracusa, P. Rocculi, S. Romani, M. Rosa, Trends Food Sci. Technol. **19**, 634-643 (2008)
8. C.J. Weber, V. Haugaard, R. Festersen, G. Bertelsen, Food Additives & Contaminants, **19**(S1), 172-177 (2002)
9. B.M. McKenna, *Texture in food*, CRC Press, Boca Raton (2003)
10. Da-Wen Sun, *Handbook of frozen food processing and packaging*, CRC Press, Boca Raton (2006)
11. P.A. Webb, Volume and density determinations for particle technologists. Micromeritics Instruments Corp., www.micromeritics.com
12. IIR. *Recommendations for the processing and handling of frozen foods*, Int. Institute of Refrigeration, Denmark (2006)
13. K. Kozłowicz, D. Góral, F. Kluza, M. Domin, Z. Kobus, A. Sagan, Ł. Prazner, Przem. Chem., **94**, 10 1742-1747 (2015)
14. A.A. Karim, R. Bhat, Trends Food Sci. Technol. **19**, 644-656 (2008)
15. H.W. Kang, Y. Tabata, Y. Ikada, Biomaterials, **20**, 14, 1339 (1999)
16. X. Wu, Y. Liu, X. Li, P. Wen, Y. Zhang, Y. Long, X. Wang, Y. Guo, F. Xing, J. Gao, Acta Biomater. **6**, 3, 1167 (2010)
17. H. Kim, J. Knowles, H.H. Kim, J. Biomed. Mater. Res. 72, 136 (2005)

Effect of composition and processing conditions on selected characteristics of extruded corn instant gruels enriched with fruits addition

Magdalena Kręcisz[1], *Agnieszka* Wójtowicz[1,*], and *Anna* Oniszczuk[2]

[1]University of Life Sciences, Doświadczalna 44,20-280 Lublin, Poland
[2]Medical University of Lublin, Chodźki 4a, 20-093 Lublin, Poland

Abstract. Cranberries and goji berries were used as natural supplements in extruded instant corn gruels. The effects of additive type and level, as well as extrusion-cooking screw speed on selected properties of extrudates were tested. Corn grit was used as the base raw materials and dry and ground goji berries and cranberries were added at 1, 2, 3, 4 and 5%. Extrusion-cooking of blends was performed with a single screw extruder at temperature ranged 125-135°C using various screw speed during processing. Extrudates were ground below 1 mm for instant gruels. Water absorption, water solubility, as well as colour profile were tested. The results showed that the highest water absorption was evaluated for instant gruels consist the highest amount of goji berries extruded at the highest screw speed during processing, while cranberries addition have no significant effect on water absorption. Increasing amount of dry cranberries and goji berries in the recipe affected on lowering the water solubility of the extrudates. Significant effect of fruits on color coordinates was observed. Increasing amount of dry berries lowered lightness L* and yellowness of instant gruels, especially when goji berries were used in the recipe. Increasing the screw speed during processing decreased intensity b* values.

1 Introduction

The extrusion-cooking technique seems to be a milestone in the production of instant products according to processing simplicity [1]. Full automatization of processing compacted lines and limitation of stuff contact with foods, offering the high hygienic and sanitary requirements extremely important in baby and infant instant food production. The main equipment in processing line is single or twin-screw extruder useful for thermomechanical treatment of cereal raw materials as corn, rice, buckwheat or wheat flour with various additives into extrudates with unique characteristics [2-6]. Equipment needed for processing of instant gruels and baby food is easy to operate, energy saving and more efficient comparing to conventional process with drum drying of cooked cereal slurries. Configuration and operational parameters of extruder make it possible to create products' specific characteristics as viscosity, density, water absorption or instant gruels consistency with water or milk [7]. These features are very important for ready-to-eat (RTE) extrudates as well as for extruded components used in bakery, beverages, chemicals or pharmaceutical products or biodegradable thermoplastic starch [8, 9].

The functional properties of extruded foods plays an important role for their acceptability and properties as water absorption, water solubility, oil absorption indexes, expansion index, bulk density and viscosity of the dough [10, 11]. Raw materials composition, especially carbohydrates and protein content, as well as fibrous fractions of food components, can affect on several physiochemical characteristics of extruded foodstuffs [1-3, 5, 7, 10, 11]. There is a possibility to improve both nutritional and quality characteristics of food by supplementation with some fruits addition [12].

Cranberries and goji berries are nowadays popular additives for food supplementation according to unique composition of these berries.

Cranberry according to its excellent nutritional characteristic is qualified as one of the "superfruits". The chemical composition of cranberry depends on the variety, crop region and storage conditions. Ripped cranberry consists 88% of water, 4.2% of sugar and organic acids, 1.2% of pectin, minerals (potassium, calcium, sodium, selenium), vitamins of A, C, E and B group, flavonoids and anthocyanins, catechins and proanthocyanins [13-15]. Cranberries are valuable diet component known from ancients and used, for example by Indians, in natural medicine for cold, whooping-cough, scurvy, rheumatism, gastrointestinal problems, oral mucositis, angina, urinary tract infection and many others [13, 16]. Cranberries have the potential ability to limit the development and severity of certain cancers and vascular diseases including atherosclerosis, ischemic

* Corresponding author: agnieszka.wojtowicz@up.lublin.pl

stroke, and neurodegenerative diseases of aging [17, 18]. Research results confirm positive effects of cranberries for prevent the bacteria expansion in urinal tract, antibacterial and antimicrobial effect, anti-cancer, anti-inflammatory, protect against food poisoning e.g. due to *E. coli*, inhibit adhesion of Helicobacter pylori, a major cause of gastric cancer, to human gastric mucus, thereby prevent ulcerative disease. NDM (the nondialyzable material) in cranberries, contains about 65% proanthocyanidins, along with a much smaller quantity (0.35%) of anthocyanins, is helpful in treatment of mouth diseases and cold, NDM is limiting growth of bacteria causing gum disease and tooth decay, and also protecting from problems with fungi disorders [15, 19, 20].

More and more popular component in functional and pro-healthy foods are goji berries, growing as fruits of *Lycium barbarum*. According to the high amount of biologically active components also the goji berry is called as "superfruit". Its caloric value is 348 kcal in 100 g of dry berries [21]. Goji berries are used as medicinal in various diseases. The medical power of these fruits was proved by scientists from China since over 2000 years.. The most of the research was based on its antioxidant characteristics because of 15-times higher antioxidant potential than of green tea [22]. Goji berries are characterized red-orange colour formed by carotenoids [23]. One of the most important components of goji berries are polyphenols with high antioxidant activity [24]. The biggest group of components are compexed polysaccharides, and moreover vitamins as ascorbic acid, tiamin, ryboflavin or glucopyranosids. It can be eaten as dry or cooked fruits. In these forms, it could be used for tea, soups, desserts, or other meals [25]. A unique characteristic of goji berry is the effect of the presence of 20 types of amino acids, including 8 of exogenous. These essential amino acids play the important role in human body, i.e. lysine is playing the role in antibodies production, is the component of collagen and supports B_3 vitamin; isoleucine is supporting in haemoglobins production, and methionine causes acidity of urine and destroy bacteria. Goji berries are good source of mineral components, one of the biggest natural source of iron, consist also calcium, chromium, germanium, magnesium, potassium, zinc or selenium [26].

Some results showed the reduction of fibroids (myoma) growing and production of antibodies by spleen in the experiments on mice feeding with polysaccharide complex from goji berries. Additionally, the increase of lymphocytes production and decrease of fat oxidation was observed on mice infected with fibroids [19]. It was proven that components of goji berries inhibit growth or even kill cancer cells. Tang et al. [27] confirmed the results of previous research showed that the consumption of juice from goji berries increased lymphocytes, interleukin-2 and immunoglobulin G levels. Some research confirmed the anti-cancer effect of scopoletin and 2-O-β-D-glucopyranosyl-L-ascorbic acid present in goji berries [28]. The extract from goji berries consist taurine may inhibit progression of eye diabetic retinopathy. High content of bioactive components in goji berries can protect neurons in eye retina in diabetic animals from destroying by oxygen free radicals, excessive amount of glucose and kainic acid [29]. In 2012 it was proved protective effect of goji extracts on human retina nerve cells [30]. Goji berries characterise antioxidant activity thanks to the content of antioxidants, carotenoids, flavonoids and polysaccharide complexes and analogues of vitamin C. Antioxidant capacity of fresh goji berries is comparable with cherries and redcurrant. Goji extract is also affecting on lowering the activity of liver enzymes causing hepatic necrosis when consumed with alcoholic beverages [31].

Presented advantages of cranberry and goji berry suggest that they may be utilized for supplementation of food products enhancing the nutritional characteristics and creating new type of functional foods. So, the aim of this study was the evaluation of the effect of goji berries and cranberries supplementation of selected characteristics of corn-based instant gruels processed with the extrusion-cooking.

2 Materials and Methods

2.1 Sample preparation

Corn grit (purchased from PZZ Lubella Sp. z o. o. Sp. K., Lublin, Poland) was used as the base raw material and dry and ground goji berries and cranberries were added in amount of 1, 2, 3, 4 and 5% of weight. Raw materials were mixed using a ribbon mixer and moistened with a specific volume of tap water to obtain the moisture content before the extrusion-cooking at the level of 14% [32]. The mixing time was set for 15 minutes to obtain a loose structure. The mixtures of raw materials were rested for 1 hour to unify the moisture. The moisture content of raw materials, moistened mixtures and extrudates was checked by the drying method [33] by drying at 130°C for 1 hour using air dryer SWL-53 SDT (Pol-Eko-Aparatura, Wodzisław Śląski, Poland).

2.2 Extrusion-cooking of instant gruels

The prepared mixtures were fed into the extruder's hopper. The extrudates were processed using a single-screw extruder TS-45 (L/D=12:1). The range of the temperatures of the extrusion-cooking process was as follows: 125/130/135°C, respectively in three extruder's sections. The process was carried out by modifying the rotational screw speed between 80, 100 and 120 rpm. A forming die with a single open of 3 mm in diameter was used. After extrusion extrudates were cooled down and ground with a laboratory grinder LMN10 (TestChem, Radlin, Poland) for a granulation below 1 mm.

2.3 Physical properties

2.3.1 Water absorption index (WAI)

Water absorption index (WAI) was determined with centrifugation method [34] for each sample in three replications. In brief, 0.7 g of extruded gruels was mixed with 7 mL of water for 10 min and then centrifuged at rotational speed 15000 rpm for 10 min in T24D type centrifuge (Leipzig, Germany). The supernatant was removed immediately, the remaining gel was weighted and the WAI was calculated as:

$$WAI = w_g/w_s \quad [-] \qquad (1)$$

where: WAI – water absorption index, w_g - weight of gel, [kg], w_s - weight of dry sample, [kg].

2.3.2 Water solubility index (WSI)

Water solubility index (WSI) was determined in triple as solids recovered after total water evaporation at 110°C of supernatant obtained from the WAI analysis [35]. Results were calculated from formula:

$$WSI = w_{ds}/w_s \cdot 100 \ [\%] \qquad (2)$$

where: WSI – water solubility index, w_{ds} - weight of dry solids in supernatant, [kg], w_s - weight of dry sample, [kg].

2.3.3 Colour parameters

Colour measurements were done with the Lovibond CAM-System 500 colorimeter (the Tintometer, UK). CIE-Lab scale was used for evaluation of L^* for lightness, a^* for (+) redness and (-) greenness, and b^* for (+) yellowness and (-) blueness, accordingly. Measurements were performed in 20 replications [9].

2.3.4 Statistical analysis

The obtained results were tested with bidirectional ANOVA analysis of variance with interactions and Tukey's test for comparison of means ($p<0.05$) with the Statistica software (Statistica version 10.0, USA). The first factor was the dry fruits amount (goji berry – G and cranberry - C) and the second was the screw speed (S). RSM (response surface methodology) was used for fitting polynomial models ($Y=b_0 + b_1X_1 + b_2X_2 + b_{11}X_{12} + b_{12}X_1X_2 + b_{22}X_{22}$, where X_1 was the dry fruits amount and X_2 was the screw speed applied) and quadratic equations of tested characteristics depend on variables used in the experiment (Table 1).

3 Results and discussion

The moisture content of instant gruels ranged from 7.10-7.89% for corn-goji berry and 6.85-7.83% for corn-cranberry gruels (detailed data not shown). These levels of moisture content in extruded corn-fruits instant gruels allow to long term storage, stabile shelf life and microbiological safety of products [11, 36, 37].

3.1 Physical properties of corn-goji berry and corn-cranberry instant gruels

3.2.1 Water absorption index (WAI)

The amount of water absorbed by the starch and protein components from corn grits as well as pectin's from goji berries as a gel formed after centrifugation ranged from 5.85 to 6.29. The results showed that the highest WAI was evaluated for instant gruels with the highest amount of goji berries addition extruded at the highest screw speed during processing (Fig. 1).

Table 1. Adequacy of two variables model fitted for tested characteristics of instant corn gruels supplemented with fruits.

Parameter	Fitted model	R^2
	Goji berry	
WAI [-]	$WAI=4.38-0.05G+0.04S-0.01G^2 +0.001GS-0.0002S^2$	0.21
WSI [%]	$WSI=10.21+1.65G+0.07S-0.13G^2 -0.01GS+0.0001S^2$	0.50
L^*	$L^*=73.55-6.13G+0.10S+0.60G^2 +0.003GS-0.0005S^2$	0.95
a^*	$a^*=-9.25+7.56G-0.08S-0.84G^2 +0.003GS+0.0003S^2$	0.96
b^*	$b^*=63.49-2.76G+0.17S+0.16G^2 +0.008GS-0.001S^2$	0.77
	Cranberry	
WAI [-]	$WAI=7.51-0.07C-0.26S-0.03C^2 +0.002CS+0.0001S^2$	0.35
WSI [%]	$WSI=8.55-1.06C+0.18S-0.06C^2 +0.01CS-0.0009S^2$	0.29
L^*	$L^*=80.43-1.89C-0.14S+0.16C^2 +0.0004CS+0.0006S^2$	0.49
a^*	$a^*=-11.17+1.77C+0.07S-0.24C^2 +0.0002CS-0.0003S^2$	0.32
b^*	$b^*=87.05-4.40C-0.34S+0.03C^2 +0.02CS+0.001S^2$	0.79

G – Goji; C - Cranberry; S – screw speed; WAI – water absorption index; WSI water solubility index; L* - lightness, a* - redness/greenness balance, b* - yellowness/blueness balance,

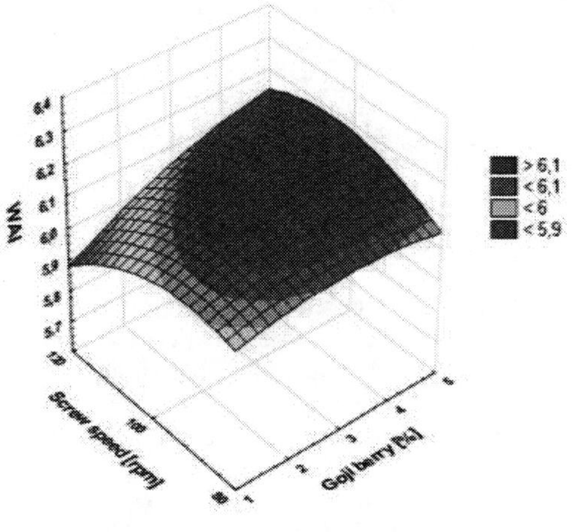

Fig. 1. Water absorption index (WAI) of instant gruels made from corn depending on goji berry content and screw speed applied during processing.

In composition of the recipes of corn grits supplemented with 4-5% of goji berries it was observed the highest water absorption when the screw speed was 100 rpm during the extrusion. The increasing amount of goji berries has shown insignificant effect on WAI values (p_{-value} = 0.686) but the slight effect on water absorption growing was observed with increasing amount of dry fruits in the recipe, especially when 100 and 120 rpm was applied during processing. Statistical analyses confirmed that variable screw speeds during the extrusion-cooking of instant gruels with goji berry addition have insignificant effect on water absorption of tested extrudates (p_{-value} = 0.150).

WAI evaluated for instant gruels processed with the addition of cranberries attained values ranged from 5.86 to 6.25 and were comparable to corn-goji extrudates. The amount of cranberries (p_{-value} = 0.474) and the screw speed applied (p_{-value} = 0.167) have no significant effect on water absorption results. The highest results were observed if 2 and 3% of cranberries were added in the recipe (Fig. 2).

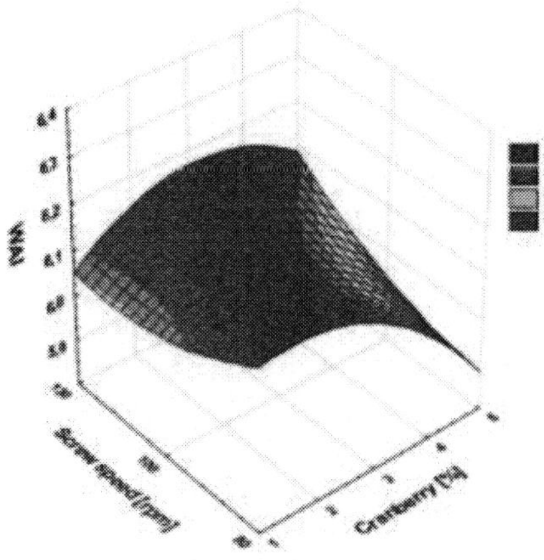

Fig. 2. Water absorption index (WAI) of instant gruels made from corn depending on cranberry content and screw speed applied during processing.

Cranberry addition affected on lower values of water absorption of instant gruels enriched with 4 and 5% of dry fruits content in the recipe when compared with the same amount of goji berry addition. It could be connected with the different composition of fruits used in the experiments.

3.2.2 Water solubility index (WSI)

The values of WSI of instant corn-goji berries gruels made with various screw speed during the extrusion-cooking varied with the wide range from 14.76 to 21.14% (Fig. 3). Addition of dried goji fruits have no significant effect on components solubility in water (p_{-value} = 0.118). Water solubility index was high in values when highest screw speed was applied during processing, but these dependencies were observed only if low amount of goji berries was used as an additive (up to 3%). If higher amount of dried goji was added into the recipe the differences in water solubility were smaller. This could be connected with the application of high screw speed for processing of corn grits and thus intensive shearing treatment of starchy components resulting higher solubility. Increasing amount of pectin-rich additive caused limitation of starch and increased amount of fibrous fractions from goji fruits able to keep the water (as showed by WAI results) and limiting the amount of soluble components passing into the water after centrifugation.

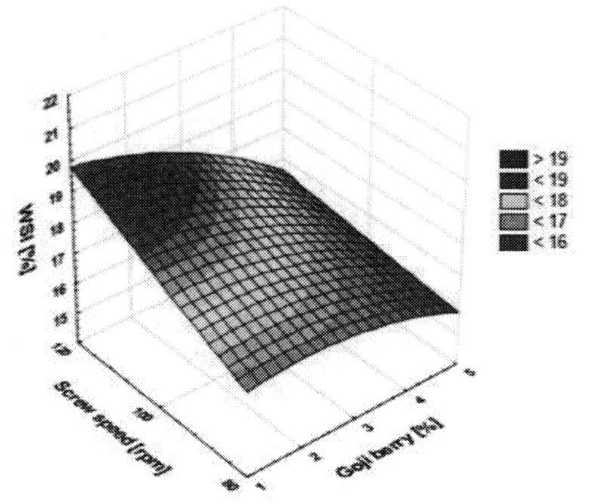

Fig. 3. Water solubility index (WSI) of instant gruels made from corn depending on goji berry content and screw speed applied during processing.

The WSI of instant gruels based on corn with the addition of cranberries was lower and reached values from 15.57 to 18.57%. The cranberry fruits addition has similar effect as it was observed for goji berries. Increasing amount of dry cranberries in the recipe affected on lowering the water solubility of the extrudates.

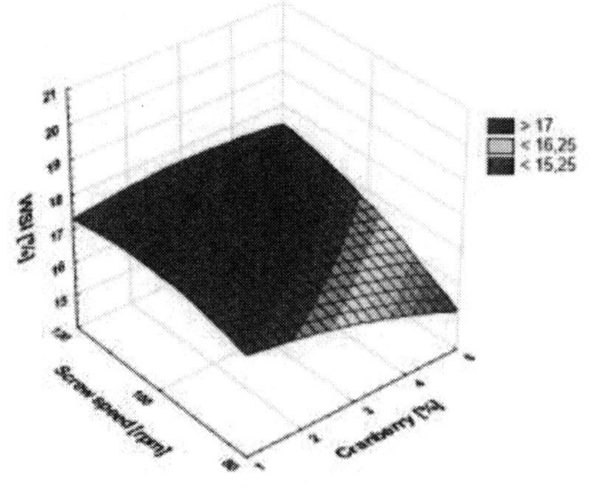

Fig. 4. Water solubility index (WSI) of instant gruels made from corn depending on cranberry content and screw speed applied during processing.

The highest solubility was evaluated for samples extruded at 120 rpm, especially when fruits addition not exceed 3%, what could be the effect of the intensive thermomechanical treatment of ingredients during processing. At 80 and 100 rpm decreasing trend was observed when 4 and 5% of additive was used, but generally cranberry addition has no significant effect on WSI values ($p_{-value} = 0.287$). In general, also in this case the effect of screw speed on WSI was insignificant ($p_{-value} = 0.336$).

Water absorption index (WAI) and water solubility index (WSI) are two indices linked to the estimation of the behaviour of the processed material for application as a binder, a stabilizer, or emulsifier in various food systems [3, 5, 8, 10, 36]. WAI and WSI values extend in a wide range depending on the specific parameters applied during the extrusion-cooking process. Factors affecting the values of these indices can be classified into two main groups: the first group - linked to the raw materials, and the second one - includes parameters related to the extruder equipment and processing conditions. In particular, the first group refers to the initial composition of the raw material, the raw material formulation, the pre-processing treatments, the initial particle size of milled materials, and the milling procedure. The second group is related to the extruder conditions, such as extruder type, last section barrel temperature, or die temperature, feed moisture content, feed rate, screw speed, screw configuration, screw compression ratio, die dimension, and die configuration [36, 37]. Oikonomou and Krokida [11] proposed the mathematical model investigates the effect of die temperature, feed moisture content, and screw speed on the WAI (1.3 to 9.5) and WSI (4.5 to 61.6%) properties for some extruded food products. They found that the model curves do not have a distinct slope as screw speed increases for all the extruded products. Similar trends have been observed in our study, especially when WAI results of enriched gruels were analysed.

3.2.3 Color parameters

The results of color characteristics evaluated for extrudates with fruits addition showed significant differences in CIE-Lab scale coordinates depend on the additive type and amount. The most significant ($p_{-value} = 0.000$) was the effect of goji berry level on $L*$ profile of instant gruels. Red in color goji berry fruits differ strongly from yellow corn grits so significant decrease in lightness was observed in supplemented gruels (Fig. 5). The $L*$ values varied from 64.18 to 73.38. There wasn't observed the effect of screw speed on gruels lightness ($p_{-value} = 0.103$). On the base of the results of $L*$ measurement it can be concluded darker color of final products supplemented with the addition of goji berry than cranberries. Dried fruits were much darker than corn so also the extrudates characterized the lower lightness. $L*$ values noted for instant gruels enriched with cranberries varied from 68.23 to 72.01 (Fig. 6).

The amount of cranberries showed significant effect ($p_{-value} = 0.000$) on extrudates lightness, decreased $L*$ values were observed when the higher amount of dry cranberries was used in the formula but differences were much lower than observed for samples with goji berry addition. In this case the screw speed applied during processing showed the significant effect ($p_{-value} = 0.022$) on samples lightness. It was observed the slight decrease of lightness in extrudates processed with a higher screw speeds, especially when 1 and 2% of cranberries were added to the recipe.

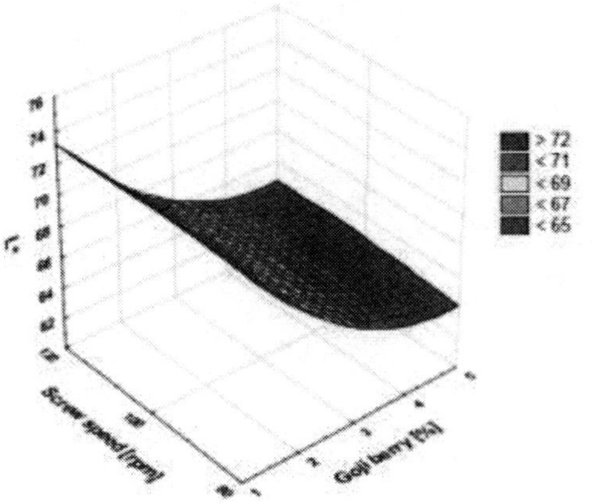

Fig. 5. $L*$ profile of instant gruels made from corn depending on goji berry content and screw speed applied during processing.

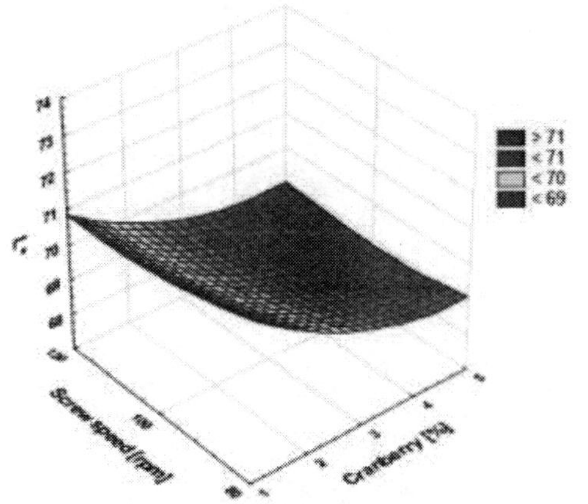

Fig. 6. $L*$ profile of instant gruels made from corn depending on cranberry content and screw speed applied during processing.

On the figure 7 the red-green balance of instant gruels with goji berries is presented. The value of $a*$ profile showed significant effect ($p_{-value} = 0.000$) of fruits addition on increased red tint of extruded gruels. Addition up to 3% of goji berries in the recipe showed green tint presence, the most intensive for samples supplemented with 1% of goji dry fruits. Increasing the amount of additive resulted in formation of red tint with positive index values of a* profile ranged from -7.48 for samples with 1% of additive up to 3.74 for extrudates with 5% of goji berry in formula. This is the result of intensive red-orange color of dry goji berries and the additive increasing amount affected more red tint of instant gruels. The various screw speed applied during processing had no significant effect (p_{-value} = 0.314) on that color coordinate.

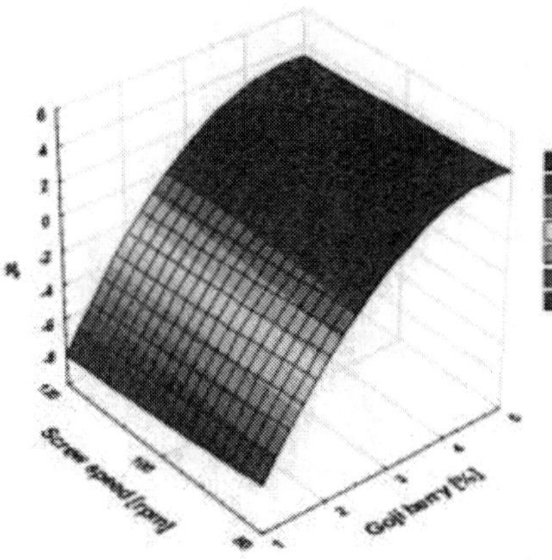

Fig. 7. Profile of *a** balance of instant gruels made from corn depending on goji berry content and screw speed applied during processing.

The values of a* profile of instant gruels supplemented with the addition of cranberries gave negative values of this coordinate showing the green tint of extrudates with dry cranberries in the recipe (Fig. 8). The results varied from -6.62 to -3.19 and the differences were significant (p_{-value} = 0.000). The most intensive greenes was observed for samples with 1% of cranberries and with incresing amount of fruits for 4 and 5% in the recipe lowered greeness intensity. Cranberries characterized red-brown color and extrudates showed lower reddness than observed for gruels supplemented with goji berries. Also for these samples no significant effect of screw speed was observed (p_{-value} = 0.428).

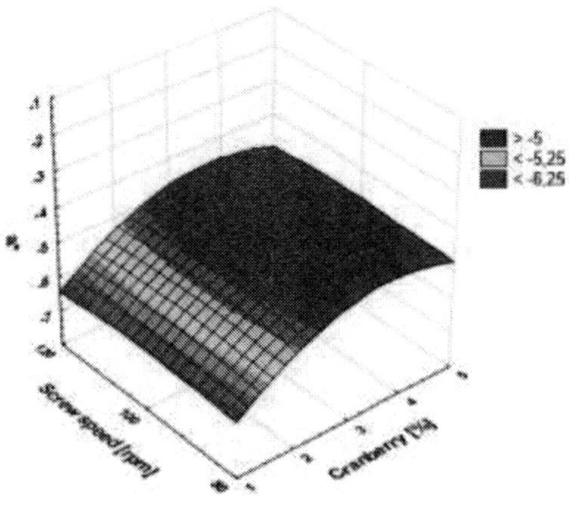

Fig. 8. Profile of *a** balance of instant gruels made from corn depending on cranberry content and screw speed applied during processing.

Figure 9 illustrates the changes in blue-yellow balance of extruded gruels with the addition of goji berries. Values of b* coordinate reach the level of 63.98-68.37 and suggest intensive yellow tint of the extrudates. This color profile is connected with the intensive yellow color of corn grits used in the experiment and red-orange color of dry goji decreased yellowness of tested products. The effect of decreasing the yellowness was significant (p_{-value} = 0.000) with an increase amount of goji fruits in the recipe. Test results showed slight but statistically significant effect (p_{-value} = 0.013) of screw speed on b* values and decrease of yellowness was observed when higher screw speed was applied in the experiment.

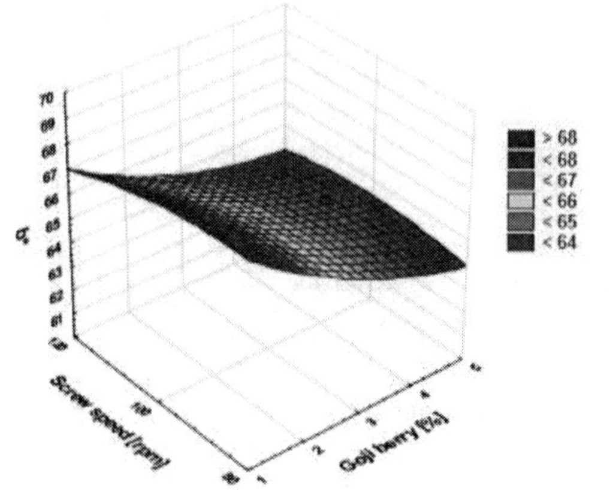

Fig. 9. Profile of *b** balance of instant gruels made from corn depending on goji berry content and screw speed applied during processing.

Similar and more intensive tendency was observed for samples supplemented with cranberries (Fig. 10) but yellowness of tested samples was lower. The addition of dried cranberries showed significant effect (p_{-value} = 0.000) on b* values of extruded gruels. The range of yellowness varied from 64.38 if 1% of cranberries was used in the recipe up to 54.10 for gruels with the highest 5% of additive.

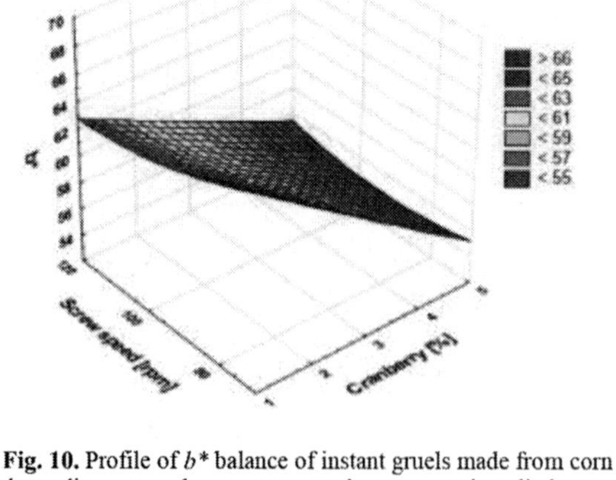

Fig. 10. Profile of *b** balance of instant gruels made from corn depending on cranberry content and screw speed applied during processing.

Yellowness intensity in these samples was lower because of darker color of dry cranberries. The screw speed elevation during processing of samples with 1-4% of dry fruits addition decreased significantly (p_{-value} = 0.022) the yellowness of tested samples and slight increase was observed for products extruded with 5% of cranberries in formula with increasing the screw speed during processing.

4 Conclusions

The addition of dried fruits and screw speed applied during processing of instant corn-based gruels affected on the most of tested characteristics. Increasing amount of fruits and screw speed during the extrusion have no significant effect on WAI and WSI values. Significant effect of higher amount of goji berries and cranberries was observed for color coordinates evaluated with CIE-Lab scale. Increasing amount of dry berries lowered lightness L^* and yellowness of instant gruels, especially when goji berries were used in the recipe. Increasing the screw speed during processing decreased intensity of b^* values.

References

1. P. Mayachiew, C. Charunuch, S. Devahastin, J. Food Sci. **80(12)**, E2782 (2015)
2. A. Altan, K.L. McCarthy, M. Maskan, J. Sci. Food Agric. **88**,1648 (2008)
3. G. Bisharat, I. Katsavou, N. Panagiotou, M. Krokida, Z. Maroulis, Food Sci. Technol. Int. **21(8)**, 613 (2015)
4. E. Fleischman, R. Kowalski, C. Morris, T. Niguyen, C. Li, G. Ganjyal, C. Ross, J. Food Sci. **81(11)**, E2726 (2016)
5. D. Gbenyi, I. Nkama, M.H. Badau, J. Food Res. **4(2)**, 81 (2016)
6. A. Oniszczuk, A. Wójtowicz, T. Oniszczuk, M. Olech, R. Nowak, K. Wojtunik, M. Klimek, W. Krawczyk, M. Hajnos, Open Chem. **13**, 1209 (2015)
7. M. Kręcisz, A. Wójtowicz, A. Oniszczuk Agriculture and Agricultural Science Procedia **7**, 139 (2015)
8. M. Mitrus, L. Mościcki, Chem. Eng. Res. Des. **92,** 778 (2014)
9. A. Wójtowicz, A. Kolasa, L. Mościcki, Pol. J. Food Nutr. Sci. **63(4)** 239 (2013)
10. M.S. Alam, J. Kaur, H. Khaira, K. Gulpa, Crit. Rev. Food Sci. Nutr. **56(3)**, 445 (2016)
11. N.A. Oikonomou, Krokida, M.K., Int. J. Food Prop. **14**, 199 (2011)
12. A. Oniszczuk, M. Olech, T. Oniszczuk, K. Wojtunik-Kulesza, A. Wójtowicz, Arabian J. Chem. (2016), doi:10.1016/j.arabjc.2016.09.003
13. R.P. Feliciano, C.G. Krueger, J.D. Reed, Mol. Nutr. Food Res. **59**, 1292 (2015)
14. A. Stobnicka, M. Gniewosz, A. Mięsuszewska, Bromat. Chem. Toksykol., **XLIV(3),** 650 (2011)
15. M.H. Grace, A. R. Massey, F. Mbeunkui, GG. Yousef, M.A. Lila, J. Food Sci. **77(8)**, H176 (2012)
16. D.L. McKay, J.B. Blumberg, Nutr. Rev. **65(11)**, 490 (2007)
17. C.C. Neto, Mol. Nutr. Food Res. **51**, 652 (2007)
18. A. Basu, M. Rhone, T.J. Lyons, Nutr. Rev. **68(3)**, 168 (2010)
19. L. Gan, H. Zhang, X., Liang Yang, H. Bi Xu, Int. Immunopharmacol. **4,** 563 (2004)
20. C. Bodet, F. Chandad, D. Grenier. J. Dent. Res. **85(3)**, 235 (2006)
21. S. Niro, A. Fratianni, G. Panfili, L. Falasca, L. Cinquanta, M.D. Rizvi Alam, Ital. J. Food Sci. **29** 398 (2017)
22. P. Skenderidis, K. Petrotos, I. Giavasis, C. Hadjichristodoulou, A. Tsakalof, J. Food Process Eng. (2016) doi:10.1111/jfpe.12522
23. C.C. Wang, S.C. Chang, B.S. Inbaraj, B.H. Chen Food Chem., **120(1)** 184 (2010)
24. S.C. Lam, Z. Luo, D.T. Wu, K.L. Cheong, D.J. Hu, Z.M. Xia, S.P. Li., J Food Sci **81(6)**, C1378 (2016)
25. J. Morais Ferreira, B. Azevedo, V. Luccas, H. Andre Bolini, J. Food Sci. **82(3)**, 818 (2017)
26. G-I. Hidalgo, M. Pilar Almajano, Antioxidants, **6**, 7 (2017)
27. W.M. Tang, E. Chan, C.Y. Kwok, Y.K. Lee, J.H. Wu, C.W. Wan, R.Y. Chan, P.H. Yu, S.W. Chan, Int Inflammopharmacol. **20(6)** 307 (2012)
28. B. Kulczyński, A. Gramza-Michałowska, Pol. J. Food Nutr. Sci. **66(2)**, 67 (2016)
29. M.K. Song, N.K. Salman, B.D. Roufogalis, T.H. W. Huang, Biochem. Pharmacol. **82(9)** 1209 (2011)
30. Z. Shen, J. Wang, G. Li, Chinese J. Ophthalmol. **48(9)** 824 (2012)
31. D. Cheng, H. Kong, Molecules, **16(3)** 2542 (2011)
32. A. Wójtowicz, Acta Agroph., **11(2)** 545 (2008)
33. ASAE Standard. ASAE S269.3. Wafers, pellet, and crumbles – definitions and methods for determining density, durability and moisture content (1989).
34. A. Wójtowicz, M. Mościcki, LWT-Food Sci. Technol., **59**, 1175 (2014)
35. N.A. Oikonomou, Krokida M.K., Int. J. Food Prop. **15**, 157 (2012)
36. D.R. Marques, Berwig K.P., Monteiro C.C. F., Oliveira D.M., Monteiro A.R.G., Chem. Eng. Trans. **57**, 1921 (2017)
37. E. Lusac, Rooney L. *Snack Foods Processing* (CRC Press LLC, USA, 2002)

Efficiency of solar radiation conversion in photovoltaic panels

Sławomir Kurpaska, *Jarosław* Knaga, *Hubert* Latała, *Jakub* Sikora and *Wiesław* Tomczyk

[1] Faculty of Production and Power Engineering, University of Agriculture in Krakow, 30-149 Krakow, Poland

Abstract. This paper included analysis the conversion efficiency in photovoltaic panels. The tests were done between February and June at a test stand equipped with three commonly used types of photovoltaic panels: poly- and monocrystalline silicon and with semi-conductive layer made of copper (Cu), indium (In), gallium (Ga) and selenium (Se) (CIGS). Five days of each month were selected for a detailed analysis. They were close to the so-called recommended day for calculations in solar power engineering. Efficiency, calculated as the yield of electrical energy in relation to solar radiation energy reaching the panels was made conditional upon solar radiation intensity and ambient temperature. It was found that as solar radiation intensity and ambient temperature increase, the efficiency of solar radiation conversion into electricity is reduced. Correlation dependence was determined for the test data obtained, describing temperature change of panels depending on climatic conditions. It was found that as panel temperature increases, the conversion efficiency is reduced. Within the tested scope of experiment conditions, the efficiency was reduced in the range between 20.1 and 22.8%. The authors also determined the average efficiency values in individual test months together with average ambient conditions of the environment where the process of solar radiation conversion took place.

1 Introduction

Care of natural environment, real threat of energy crisis as the effect of fossil fuel depletion stimulate the diversification of the present sources of energy and in consequence - the use of renewable energy sources. One of such sources is the energy of solar radiation. It results from the fact that the total amount of solar radiation reaching the surface of the Earth throughout the year is 1,800 times bigger than the current annual world demand for energy [13]. That is why it is becoming increasingly more popular to obtain electricity using the photovoltaic effect. Sun is the strongest source of light reaching the Earth, radiating energy generated as the result of thermonuclear reactions happening within. It emits radiation in all the ranges of electromagnetic waves. The biggest amount of energy – 49% is radiated within visible spectrum (380÷760 nm wavelength) [11]. There are two methods of direct use of solar radiation energy – photothermal conversion (into heat) and photovoltaic conversion, i.e. converting solar radiation energy into electrical energy with the use of the photovoltaic effect. In the past 10 years, there has been a huge development as far as materials and technology used to produce photovoltaic cells and modules is concerned. In Europe, with the programs promoting photovoltaics as a safe and clean source of energy used by some countries (e.g. Germany, Spain), photovoltaic industry is growing extremely fast. Over the decade, we have observed annual increase in PV module production by 45% [9]. Widespread use of solar energy is conditional upon the cost of equipment and the efficiency of solar radiation conversion. The issue of the efficiency of photovoltaic panels has been analysed, for instance by Lu and Yang [3]. The authors determined the amount of generated electricity and the period of return of financial investment in photovoltaic system found on the roof of a building. They also did simulation studies for a diverse spatial set-up of the panels. In their work, Ortega-Izquierdo and Río [6] analysed systems of generating electricity from renewable sources, including photovoltaic panels, commonly implemented in chosen EU countries. They identified benefits, both economical and related to natural environment protection as well as to local labour market growth. Virtuani and Strepparav [12] studied the effects of work of photovoltaic modules made of crystalline silicon, by analysing their operational efficiency as compared to the efficiency determined for standard conditions. The analysis was performed for a south-oriented, horizontal system integrated in a building. New type of modules was analysed by Smyrnakis et al. [10], who, under laboratory conditions, studied the observed effects related to current flow for cells based on silicon nanotubes. They used radiation in visible spectrum as the source of light. Makrides et al. [4] reviewed the most popular photovoltaic systems and provided nominal efficiencies of their conversion. During their studies they compared the efficiency of elements of an identical photovoltaic system mounted in Nicosia and Stuttgart and determined the differences

* Corresponding author: rtkurpas@cyf-kr.edu.pl

in the amount of produced electricity. Hilse and Kapała [1] determined the efficiency of rotational and stationary panels as well as the ability to cover the energy demand of an object. Moharil and Kularni [5] presented their own simulation program concluding that it is useful, and that apart from determining photovoltaic energy results it also allows to determine planned financial expenses. Schallenberg-Rodrı´guez [8] developed a method of estimating the effects of radiation conversion in photovoltaic modules mounted on the roof of residential buildings. This method, being an alternative for many commercial methods, includes the electric energy potential of panels (mono- and polycrystalline) mounted on the roofs of buildings, which depends on the availability of solar energy. It also allows to analyse the costs of such investment. The method was tested in the region of the Canary Islands. Zhai et al. [14] determined the value of harmful substance emissions to the atmosphere with natural gas and coal burning in power plants as well as with the use of photovoltaic panels in selected states of the USA. Based on standard relationships, Latała and Kurpaska [2] calculated theoretical efficiency of silicon panel with different angles found in the region of south Poland.

The review of studies reveals that the issue of producing electricity in photovoltaic panels is analysed in various research centres. Hence, the aim of this paper is to determine the effects of using photovoltaic panels to produce electrical energy, in field tests.

2 Material and method

Studies were performed on a stand built of three panel types (monocrystalline silicon, polycrystalline silicon and CIGS) localised within the Faculty of Production and Power Engineering of University of Agriculture in Krakow. The area of individual panels is: monocrystalline 27.2m^2, polycrstalline 27.2m^{2}and CIGS- 40.1m^2. Figure 1 shows a schematic drawing of the test stand.

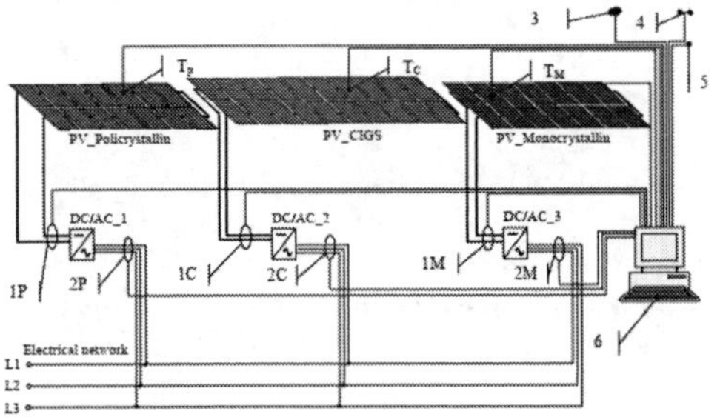

Fig. 1. Test stand schematic drawing.

The test stand is connected to the network by means of three-phase identical inverters (marked in fig. 1: DC/AC_1..3). On primary side, in front of every inverter, a direct-current measurement module (1P, 1C, 1M) was installed, whereas on secondary side there was an alternate-current module (2P, 2C, 2M). During the tests, the authors measured electrical energy yield (LUMEL N27P meters) and ambient parameters: solar radiation intensity (3) - with LP PYRA 02AV pyranometer, wind speed (4) - with MAX 40H cup anemometer, ambient temperature (5) - with PT1000 meters. In addition, the temperature of cells was measured with PT1000 meters fixed directly to the surface of the panels. All the values were monitored with the use of the authors' original measurement computer system (6) with 30s frequency and saved on the computer.

Theoretical analysis

The efficiency of solar radiation conversion into electric energy was calculated as the quotient of available energy and the amount of solar energy reaching the panels' surface, so in differential time $d\tau$ *it is referred to as:*

$$\eta = \frac{\int_0^\tau E_{elektr} d\tau}{F_{PV} \cdot \int_0^\tau R_z d\tau} \tag{1}$$

This equation, with the adopted sampling time (Δτ) can be expressed as:

$$\eta = \frac{\sum E_{elektr} \cdot \Delta\tau}{\sum R_z^{\Delta\tau} \cdot F_{PV} \cdot 10^{-3}} \tag{2}$$

where: E_{elektr} - electrical power on the inverter's secondary side, kW; R_z- solar radiation intensity, W/m^2; F_{PV}- photovoltaic panel area, m^2; Δτ – time (s).

The analysis performed involved finding a functional relation taking account of the influence of factors determining (ambient parameters) the efficiency of the panels' work. The form of the model was chosen on the basis of the highest value of determination factor and this relation was expressed with non-linear estimation with quasi-Newton method with 0.001 coefficient of convergence. The difference between values calculated and measured was determined with average square root error (□).

3 Results and discussion

The tests were done between February and June 2015. Detailed analysis included five days close to the so-called recommended days. Figures 2 and 3 show daily course of measured parameters for selected days: 15/02- fig. 2 and 15/06- fig. 3.

 https://doi.org/10.1051/bioconf/20181002014

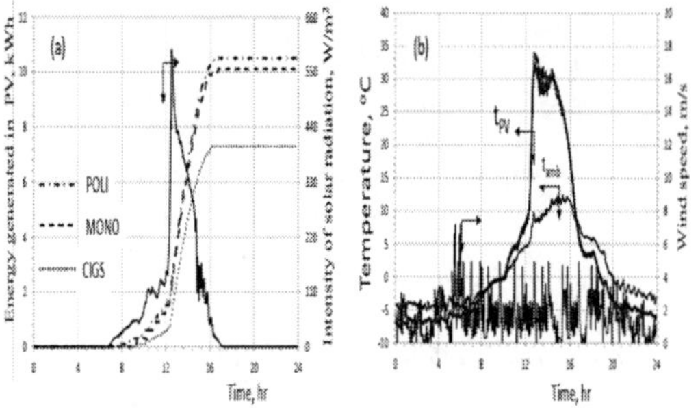

Fig. 2. Daily course of measured parameters (15/02): a) accumulated amount of energy for mono-, polycrystalline and CIGS-type panels as well as solar radiation intensity; b) and the speed of wind, cell temperature (t_{PV}) and ambient temperature (t_{amb}).

It is found that on the discussed day, with the total solar irradiation of nearly 1.5kWh/m^2 and average ambient temperature (during radiation weather) of 4.5°C the amount of solar energy converted into electrical energy falls within the range of 0.05 (CIGS panels) and 0.11kWh/m^2 (polycrystalline silicon panels). During direct solar radiation, the temperature of panels increased from approx. -5 to approx. 33°C.

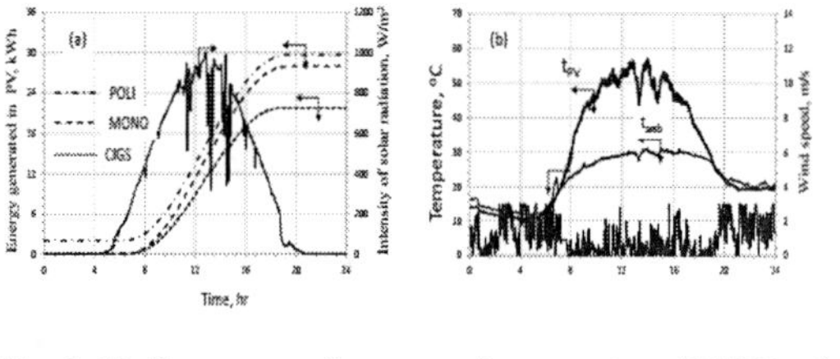

Fig. 3. Daily course of measured parameters (15/06): a) accumulated amount of energy for mono-, policrystalline and CIGS-type panels as well as solar radiation intensity; b) and the speed of wind, cell temperature (t_{PV}) and ambient temperature (t_{amb}).

It is found that on the discussed day, with the total solar irradiation of nearly 7.79kWh/m^2 and average ambient temperature (during radiation weather) of 24.8°C the amount of solar energy converted into electrical energy falls within the range of 0.54 (CIGS panels) and 1.02kWh/m^2 (polycrystalline silicon panels). During solar radiation, the temperature of panels increased from approx. 10 to approx. 57°C.

The attempts to find a functional dependency taking account of the relation between conversion efficiency and ambient parameters failed. The diagram (fig. 4) shows the obtained changes of efficiency in the function of ambient parameters. Wind was not included in further analysis, as its influence on efficiency proved to be statistically negligent.

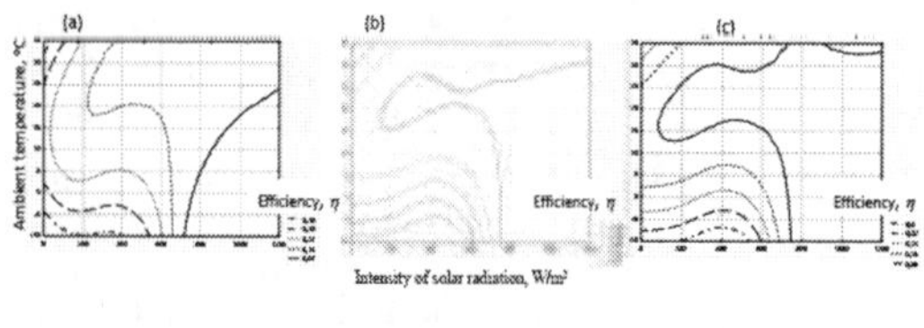

Fig. 4. The influence of ambient parameters on the modules' efficiency: (a) monocrystalline silicon, (b) polycrystalline silicon, (c) module of CIGS.

As can be seen, both the increase of solar radiation intensity and ambient temperature result in decreased efficiency of radiation conversion into electrical energy, because these parameters are partially dependent.

Calculations performed for the given months together with average ambient conditions during radiation conversion are put together in table 1.

Table 1. Average ambient conditions with the efficiency of tested panels in individual months.

Month	Ambient conditions				Panel efficiency (η)		
	MONO, POLY		CIGS				
	$t_{ambient}$, °C	$\Sigma R_{external}$, kWh	$t_{ambient}$, °C	$\Sigma R_{external}$, kWh	MONO	POLY	CIGS
February	4.19	1.67	4.21	1.53	0.123	0.121	0.093
March	6.74	1.68	6.76	1.67	0.089	0.085	0.077
April	11.45	3.31	11.48	3.27	0.107	0.106	0.074
May	14.46	3.25	14.91	3.23	0.137	0.134	0.084
June	20.93	5.43	21.03	4.79	0.096	0.094	0.054

As can be seen, the average efficiency of radiation conversion during the tested period was, for: monocrystalline silicon panels 0.114, polycrystalline silicon panels 0.108, CIGS modules 0.076. Statistical analysis showed that only for CIGS modules (as compared to other silicon modules), the efficiency discrepancies are statistically significant. The analysis shows that both silicon panels performed conversion of solar radiation for identical ambient conditions, whereas in case of CIGS panels, conversion occurred with higher intensity of solar radiation. The performed statistical analysis did not show significant changes in ambient conditions for the periods when radiation was being converted into electrical energy.

The theory describing conversion of solar radiation into electrical energy states that as module temperature increases, the properties of its material structure change, and in consequence power on p–n junction decreases [7]. The change of module temperature is the consequence of ambient temperature and absorption of solar radiation by the module's surface. With the use of quasi-Newton method, the authors found the model describing the temperature of panels in the function of ambient temperature and solar radiation intensity. Among the tested models, the highest value of factor (R^2) was obtained for the relation:

$$t_{og} = a \cdot R_{zew}^{b} + c \cdot t_{otocz} + d$$

Values of parameters a, b, c and d are shown in table 2.

Table 2. Values of parameters of using modules for approximating module temperature.

Specification	Model parameters				R^2	σ°C
	a	*b*	*c*	*d*		
MONO	0.139	0.763	1.05	-2.4	0.93	3.99
POLY	0.165	0.729	1.07	-2.38	0.92	4.18
CIGS	0.105	0.81	0.977	-1.22	0.9	4.58
Application scope: $2 \leq R_{external} \leq 1080$ W/m^2; $-7.2 \leq t_{ambient} \leq 32.5$ °C						

It can be seen that double increase in radiation intensity and ambient temperature causes module temperature (calculated in relation to initial values) to rise from 23.9 to nearly 40%.

When fulfilling the paper's purpose the authors singled out the value of radiation conversion efficiency in relation to module average temperature from the available measurement data. The obtained results are shown in fig. 5.

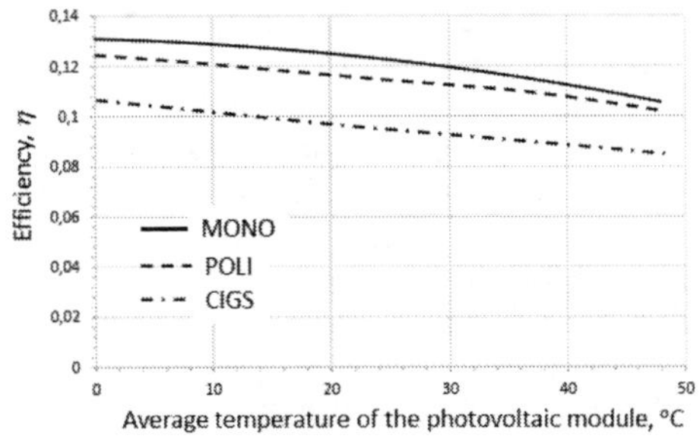

Fig. 5. Average module efficiency in the function of their temperature

It can be seen that as the module temperature increases, their efficiency falls. During the tested period, the increase of temperature from 0 to 50°C, resulted in efficiency decreasing by 20.2 (mono modules), 20.1 (poly modules) and by 22.8% in case of CIGS modules. This phenomenon can be explained with the processes occurring in p–n junction, where, as a result of temperature increase, there is a change in electrical charge carrier separation, the width of band gap decreases, electrokinetic potential increases and there is electrical charge dispersion. As a result, there is a decrease in solar radiation conversion into energy yield.

Conclusion

1. The average efficiency of radiation conversion in the tested ambient conditions was within the range of between 0.076 and 0.114.
2. Double increase in radiation intensity and ambient temperature causes module temperature to rise from 23.9 to nearly 40%.
3. In the tested range, temperature increase from 0 to 50°C, causes efficiency drop by 20.1% for silicon panels and by 22.8% for CIGS panels.

Acknowledgements:
This research was financed by the Ministry of Science and Higher Education of the Republic of Poland

References

1. Hilse D., Kapała J. Inżynieria Ekologiczna, **33**, 41-48, (2013)
2. Latała H., Kurpaska S. Inżynieria Rolnicza **4(129)**, 183- 189, (2011)
3. Lu L., Yang H.X. Applied Energy **87**, 3625–3631 (2010),
4. Makrides G., Zinsser B., Norton M., Georghiou G.E., Schubert M., Werner J.H. Renewable and Sustainable Energy Reviews **14**, 754–762, (2010)
5. Moharil R.M., Kularni P.S. Solar Energy **84**, 691–702, (2010)
6. Ortega-Izquierdo M., Río del P. Renewable and Sustainable Energy Reviews **61,** 372- 383, (2016)
7. Sarniak M. T., *Podstawy fotowoltaiki,* Oficyna Wydawnicza Polit. Warsz., (2008)
8. Schallenberg-Rodrı´guez J. Renewable and Sustainable Energy Reviews **20**, 219–239, (2013)
9. Schiffer H-W (2016). World Energy Resources Solar. https://www.worldenergy.org/ /.../2016/world-energy-resources-20
10. Smyrnakis A., Dimitrakis P., Normand P., Gogolides E. Microelectronic Engineering, Vol. **174,** 74- 79, (2017)
11. Vahrenholt F., Luning S., *Zimne Słońce. Dlaczego katastrofa klimatyczna nie nadchodzi*. Alethia, Warszawa, 2014
12. Virtuani A., Strepparava D. Solar Energy **146,** 113–118, (2017)
13. Zervos A. Renewable Energy Technology Roadmap 20% by 2020, 2009. http://www.erec.org/fileadmin/erec_docs/Documents/Publications/Renewable_ Energy_Technology_Roadmap.pdf
14. Zhai P., Larsen P., Millstein D., Menon S., Masanet E. The potential for avoided emissions from photovoltaic electricity in the United States. Energy 47, (2012), 443- 450

Organizational and technological compatibility of the technological processes of second and third line maintenance of KhTZ-3522 tractors

Roman Kyzminskyj[1], *Ruslan* Barabash[1]

[1]Lviv National Agrarian University, St. Vladimir the Great 1, 80381, Dubliany, Ukraine

Abstract. The work is devoted to systematic substantiation of productivity, production structure and specialization of service stations of KhTZ-3522 tractors. Based on the technical and economic comparative analysis of the results of the technological processes modeling of the second and third line maintenance of KhTZ-3522 tractors, the parametrical series of production structures of service stations both mono- and polytechnological specialization, which can work effectively in a wide range of the general annual program of orders, and in the conditions of seasonal fluctuations of orders for the maintenance of these tractors, are found. The results of calculations of the indicators system for organizational and technological compatibility of technological processes of the second and third line maintenance of KhTZ-3522 tractors in the joint technological flow are presented. The dependencies of these indicators on the total annual program of orders for service stations of different productivity are analyzed. According to the results of the analysis, the polytechnological specialization of such service stations is substantiated.

1 Introduction

Tractors of Kharkiv Tractor Plant (KhTZ) are among the most common in Ukraine. In addition to well-known brands such as KhTZ-150K-09 (drawbar category – 3,0), KhTZ began production of new tractors, in particular KhTZ-3510 (drawbar category – 0,9) and KhTZ-3522 (drawbar category – 1,4), which are intended mainly for small farms.

Efficient use of these new brands of tractors involves timely and qualitative maintenance. First (service interval – 250 hours), second (service interval – 500 hours) and third (service interval – 1000 hours) line maintenance are provided by the manufacturer for these tractors. The list of operations for each type of maintenance is determined by the manufacturer [1, 2]. The operations of each previous line maintenance are included in the following line maintenance. Real first line maintenance are performed by tractor drivers directly at the place of tractors operation. In order to perform operations of second line maintenance of KhTZ-3522 tractors, there is need to use 14 different types of special equipment and tools (r =14), and to perform operations of third line maintenance – 15 different types (r = 15). Therefore, qualified masters at stationary posts of service stations must carry out these complex types of maintenance [3, 4].

However, the technical service base for these new KhTZ tractor brands has not yet been created in Ukraine.

An effective performance of maintenance service on a long-term basis leads to developing loyalty of the client to the company [5]. Technical service is of primary importance in building the client's satisfaction. It should be performed by highly qualified staff and the logistic system of the service should be suited for effective realization of orders placed by users [6].

The substantiation of the program W_A, the performance Q_A, the production structure and the specialization of the technical service enterprises are complex system tasks [7], which involve the use of an iterative procedure for the solution of interdependent parametric and structural optimization tasks [8].

Regarding the technical service processes performed on stationary posts, the task of substantiating their parameters and structure was for the first time solved using the graph theory and schedules for the technological processes of the current repair of tractors [9], but the interdependence between all parameters and performance indicators of the technological process was not taken into account. The application of the iteration procedure has made it possible to eliminate this defect in relation to the tasks of determining the parameters and indicators of efficiency of the technological processes of maintenance of KhTZ tractors [10], as well as the substantiation of the production structure of the service stations of KhTZ tractors [11, 12].

The ground for the determination of specialization of selected production units of the technical service enterprises is organizational and technological

* Corresponding author: rkuzminsky@gmail.com

compatibility (OTC) of the processes performed there [13]. The property of the OTC is determined by the similarity of the construction design of the machines and the technology of the work, the flexibility of the equipment used, and also depends on the overall annual program of work (general program), and on the quantitative proportion of various processes (partial programs) in the general program. If for a certain value of a general program and a certain proportion of partial programs, the cost of implementing different processes in the joint flow is less than the cost of implementing the same processes separately, then it is considered to be OTC of these processes.

In the case of OTC of the different technological processes, a multidisciplinary or polytechnological specialization is appropriate.

For the quantitative assessment of the properties of the OTC, the system of indicators with a certain technical, technological and production content is used [14].

An assumption was made that any ratio of partial programs of orders for different types of technical maintenance in the total program of orders $W_{A\ min}$ is equally probable. Then the coefficient of OTC α_W – is the probability of the occurrence of such ratio of partial programs for which the costs for different types of maintenance in the joint flow on SS of polytechnological specialization are less than the costs of implementing the same processes separately on different SSs of monotechnological specialization. Then also the coefficient level of OTC β_W – is the probability of obtaining for a considerable period of time T such a cost benefit from joining technological processes of different types of maintenance in the joint flow. If only such ratios of partial programs of orders for different types of technical maintenance in the total program of orders $W_{A\ min}$ are taken into account, for which the OTC exists, then the coefficient relative level of OTC γ_W – identifies the ratio between the cost benefit from joining technological processes of different types of maintenance in the joint flow on SS of polytechnological specialization and the total costs for maintenance in the joint flow. The method [15] for calculating these indices has been developed in advance.

The objective of the work is to increase the efficiency of the technical service of the KhTZ-3522 tractors by way of substantiation of the specialization (mono- or polytechnological) of branded service stations of different capacity and production structure from the corresponding parametric series.

2 Methodology

According to the results of modeling of the technological processes of second and third line maintenance of KhTZ-3522 tractors using the theory of graphs and heuristic algorithms of the theory of schedules and analysis of the mutual dependencies between different parameters and performance indicators of these processes [12], the set of possible different variants of production structures of service stations, which provide the same annual Q_A. capacity, has been synthesized.

The feasibility and economic comparison of various possible options was made on the basis of the total expenditure for the entire annual maintenance program ΣZ. The total expenditure ΣZ included the costs determined by the production structure of the service station Z_{SS} (wages of workers, allowance for keeping and depreciation of equipment and industrial premises, payment for power electric etc.), and also the costs which are determined by the technology of performance Z_{lm} (costs for fuel, lubricants and other process materials etc.). The arrangement of the comparison results made it possible to form parametric series of service stations for KhTZ-3522 tractors (Table 1, 2) with different production structure and capacity of monotechnological specialization (for the performance of only second or only third line technical maintenance respectively). The optimal capacity for each maintenance item from the parametric series was determined by the criterion for minimizing specific costs ($C = \Sigma Z/Q_A$) → min.

3 Results and discussion

To the parametric series of service stations for the second line maintenance of KhTZ-3522 tractors (Table 1) five different variants of production structures have been included, that differ in the number of equipment of individual types (high-pressure jet cleaners Karcher K5 Compact K_{r1}, mobile machines Flexbimec-5903 for washing parts K_{r6} and universal devices C-230 for oiling K_{r7}) the number of stationary posts f, and, consequently, the production area. The list of all other types of required equipment for the second line maintenance of KhTZ-3522 tractors is determined by the technology of work [1, 2]. Since all other types of necessary equipment are intended to perform only one operation, the number of equipment of all other types K_r = 1 pcs.

The capacity of each variant of the production structure of the technical service station may vary within a sufficiently wide range, depending on the number of employees engaged, which enables the adaptation of service station performance to seasonal variations in the number of orders (Table 1, 2).

For example, the first variant of the production structure of the service stations for the second line maintenance of KhTZ-3522 tractors can provide maximum annual capacity Q_A = 896 orders per year, if the number of workers is u = 4 persons. If the number of workers is u = 1 person, then annual capacity will be only Q_A = 296 orders per year. For this production structure, optimal performance that meets the minimum specific costs is 799 orders per year, when three workers work (u = 3 persons). The second variant of the production structure of the service station for the second line maintenance of KhTZ-3522 tractors is characterized by an increase in the number of equipment (high-pressure jet washing machines Karcher K5 Compact K_{r1}= 4 pcs, mobile machines Flexbimec-5903 K_{r6} = 2 pcs and universal devices C-230 for filling oil K_{r7} = 2 pcs). This variant can provide maximum annual capacity Q_A =

=1040 orders per year, if the number of workers is $u = 8$ persons. For this production structure, optimal performance is a bit bigger and it is 877 orders per year, when three workers work ($u = 3$ persons).

The next three variants of the production structure of the service stations for the second line maintenance of KhTZ-3522 tractors are characterized by an increase in the number of posts f and working area respectively. Each next variant has a wider range of productivity changes by attracting a different number of workers. The optimal performance of each next variant is also higher than the previous one and it is achieved for an ever-greater number of workers.

To the parametric series of service stations for the third line maintenance of KhTZ-3522 tractors (Table 2) also includes five different variants of production structures. These variants are very similar to the corresponding variants of production structures of service stations for the second line maintenance of KhTZ-3522 tractors and differ only in the use of additional type of equipment ($r = 15$) – the testing bench of tractor electrical equipment group SKIF-1-03. However, it should be noted that for similar variants of the production structure, the maximum and optimal values of annual capacity of service stations for the third line maintenance of KhTZ-3522 tractors are significantly lower than the corresponding values of service stations for the second line maintenance.

Table 1. The parametric series of service stations with different production structure and capacity for the second line maintenance of KhTZ-3522 tractors.

№of SS	Annual capacity Q_A, orders	Amount of equipment and tools*, pcs.			Number of posts f, units	Number of workers u, persons	Economic indicators			
		K_{r1}	K_{r6}	K_{r7}			Z_{SS}, UAH	$Z_{l\text{-}m}$, UAH	ΣZ, UAH	C, UAH
I	296	1	1	1	1	1	117475	1844719	1962194	6629,03
	562					2	165475	3502474	3667949	6526,6
	799^{opt}					3^{opt}	213475	4979496	5192971	6499,34
	896					4	261475	5584015	5845490	6523,98
II	296	4	2	2	1	1	126168	1844719	1970887	6658,4
	593					2	174168	3695671	3869839	6525,87
	877^{opt}					3^{opt}	222168	5465604	5687772	6485,49
	1040					4	270168	6481446	6751614	6491,94
III	296	4	2	2	2	1	138516	1844719	1983236	6700,12
	605					2	186516	3770457	3956973	6540,45
	903					3	234516	5627640	5862157	6491,87
	1217					4	282516	7584539	7867055	6464,3
	1522^{opt}					5^{opt}	330516	9485348	9815864	6449,32
	1739					6	378516	10837726	11216243	6449,82
	1916					7	426516	11940819	12367335	6454,77
	2049					8	474516	12769696	13244212	6463,74
IV	296	4	2	2	3	1	153975	1844719	1998695	6752,35
	614					2	201975	3826546	4028522	6561,11
	915					3	249975	5702426	5952402	6505,36
	1217					4	297975	7584539	7882514	6477
	1510					5	345975	9410562	9756537	6461,28
	1784					6	393975	11118173	11512149	6453
	2049					7	441975	12769696	13211671	6447,86
	2300^{opt}					8^{opt}	489975	14333968	14823943	6445,19
	2493					9	537975	15536775	16074750	6447,95
	2555					10	585975	15923169	16509144	6461,5
V	296	4	2	2	4	1	176542	1844719	2021261	6828,58
	616					2	224542	3839011	4063552	6596,68
	920					3	272542	5733587	6006129	6528,4
	1224					4	320542	7628164	7948706	6494,04
	1522					5	368542	9485348	9853889	6474,3
	1815					6	416542	11311370	11727912	6461,66
	2090					7	464542	13025214	13489756	6454,43
	2379					8	512542	14826309	15338850	6447,6
	2620^{opt}					9^{opt}	560542	16328259	16888801	6446,11
	2797					10	608542	17431352	18039893	6449,73
	3000					11	656542	18696480	19353022	6451,01
	3136					12	704542	19544054	20248596	6456,82
	3234					13	752542	20154805	20907347	6464,86
	3338					14	800542	20802950	21603492	6471,99

*Amount of equipment and tools: $K_{r2} = K_{r3} = K_{r4} = K_{r5} = K_{r8} = K_{r9} = \ldots = K_{r14} = 1$ pcs.

Table 2. The parametric series of service stations with different production structure and capacity for the third line maintenance of KhTZ-3522 tractors

№ of SS	Annual capacity Q_A, orders	Amount of equipment and tools*, pcs.			Number of posts f, units	Number of workers u, persons	Economic indicators			
		K_{r1}	K_{r6}	K_{r7}			Z_{SS}, UAH	$Z_{l\text{-}m}$, UAH	ΣZ, UAH	C, UAH
I	229	1	1	1	1	1	122966	1822650	1945616	8496,14
	440					2	170966	3502035	3673001	8347,73
	589 ^{opt}					3 ^{opt}	218966	4687951	4906917	8330,93
	625					4	266966	4974481	5241448	8386,32
II	229	4	2	2	1	1	131659	1822650	1954309	8534,1
	458					2	179659	3645300	3824959	8351,44
	638 ^{opt}					3 ^{opt}	227659	5077951	5305610	8316
	694					4	275659	5523664	5799323	8356,37
III	229	4	2	2	2	1	144008	1822650	1966658	8588,02
	464					2	192008	3693055	3885063	8372,98
	696					3	240008	5539582	5779590	8304,01
	936					4	288008	7449783	7737791	8266,87
	1118 ^{opt}					5 ^{opt}	336008	8898352	9234360	8259,71
	1254					6	384008	9980799	10364807	8265,4
	1335					7	432008	10625492	11057500	8282,77
	1408					8	480008	11206511	11686519	8300,08
IV	229	4	2	2	3	1	159467	1822650	1982117	8655,53
	471					2	207467	3748769	3956236	8399,65
	704					3	255467	5603256	5858722	8322,05
	932					4	303467	7417946	7721413	8284,78
	1156					5	351467	9200801	9552267	8263,21
	1344					6	399467	10697125	11096591	8256,39
	1533 ^{opt}					7 ^{opt}	447467	12201408	12648874	8251,06
	1656					8	495467	13180386	13675852	8258,36
	1725					9	543467	13729568	14273035	8274,22
	1815					10	591467	14445894	15037360	8285,05
V	229	4	2	2	4	1	182033	1822650	2004683	8754,07
	472					2	230033	3756728	3986761	8446,53
	706					3	278033	5619174	5897207	8352,98
	936					4	326033	7449783	7775816	8307,5
	1169					5	374033	9304270	9678303	8279,13
	1389					6	422033	11055287	11477320	8263,01
	1580					7	470033	12575489	13045522	8256,66
	1739					8	518033	13840997	14359030	8257,06
	1934 ^{opt}					9 ^{opt}	566033	15393035	15959068	8251,84
	2070					10	614033	16475482	17089515	8255,8
	2178					11	662033	17335072	17997105	8263,13
	2250					12	710033	17908133	18618165	8274,74
	2325					13	758033	18505070	19263103	8285,21
	2379					14	806033	18934865	19740898	8297,98

* Amount of equipment and tools: $K_{r2} = K_{r3} = K_{r4} = K_{r5} = K_{r8} = K_{r9} = \ldots = K_{r14} = K_{r15} = 1$ pcs.

For example, the maximum annual capacity of the third variant of the production structure of the service station for the second line maintenance of KhTZ-3522 tractors is Q_A = 2049 orders per year, and the maximum annual capacity of the third variant of the production structure of the service station for the third line maintenance of KhTZ-3522 tractors is only Q_A = 1408 orders per year. This significant difference is explained by the greater labor-intensive characteristic of the technological processes of the third line maintenance.

Because all operations of the second line maintenance of KhTZ-3522 tractors are also performed during the third line maintenance of these tractors, therefore the parametric series of service stations for the third line maintenance corresponds to the parametric series of the polytechnological specialization service stations of KhTZ-3522 tractors, where one can perform these two different types of technical maintenance in the joint technological flow (Table 3).

It should be noted, that the capacity of service stations for maintenance of KhTZ-3522 tractors of the polytechnological specialization depends on the ratio of orders for different types of maintenance: maximum capacity $Q_{A\ max}$ is reached, when all orders will be executed for only second line maintenance, but minimal capacity $Q_{A\ min}$, when all orders will be for the third line maintenance only (Table 3).

Table 3. The results of calculation of indexes of organizational and technological compatibility (OTC) of second and third line maintenance for service stations with different production structure and capacity from parametric series.

№o f SS	Annual capacity Q_A, orders		Amount of equipment and tools*, pcs.			Number of posts f, units	Number of workers u, persons	Indexes of organizational and technological compatibility (OTC)		
	$Q_{A max}$ for second line maintenance	$Q_{A min}$ for thirdline maintenance	K_{r1}	K_{r6}	K_{r7}			α	β	γ
I	296	229	1	1	1	1	1	1	1	0,9553
	562	440					2	1	1	0,5679
	799	589					3	1	1	0,3978
	896	625					4	1	1	0,235
	for first service station in general							**1**	**1**	**0,5865**
II	296	229	4	2	2	1	1	1	1	0,8262
	593	458					2	1	1	0,5022
	877	638					3	1	1	0,3809
	1040	694					4	1	1	0,2612
	for second service station in general							**1**	**1**	**0,5094**
III	296	229	4	2	2	2	1	1	1	0,6696
	605	464					2	1	1	0,4106
	903	696					3	1	1	0,3232
	1217	936					4	1	1	0,2915
	1522	1118					5	1	1	0,2704
	1739	1254					6	1	1	0,2149
	1916	1335					7	1	1	0,1538
	2049	1408					8	0,9957	0,9989	0,084
	for third service station in general							**0,9998**	**0,9999**	**0,2967**
IV	296	229	4	2	2	3	1	1	1	0,5078
	614	471					2	1	1	0,3077
	915	704					3	1	1	0,2507
	1217	932					4	1	1	0,229
	1510	1156					5	1	1	0,2113
	1784	1344					6	1	1	0,1902
	2049	1533					7	1	1	0,1666
	2300	1656					8	1	1	0,1474
	2493	1725					9	1	1	0,1082
	2555	1815					10	0,9183	0,9614	0,046
	for fourth service station in general							**0,9959**	**0,9992**	**0,2065**
V	296	229	4	2	2	4	1	1	1	0,3209
	616	472					2	1	1	0,1798
	920	706					3	1	1	0,1502
	1224	936					4	1	1	0,1465
	1522	1169					5	1	1	0,1422
	1815	1389					6	1	1	0,1373
	2090	1580					7	1	1	0,1307
	2379	1739					8	1	1	0,1235
	2620	1934					9	0,9969	0,9997	0,1067
	2797	2070					10	0,9986	0,9999	0,0916
	3000	2178					11	0,9920	0,9998	0,0728
	3136	2250					12	0,9078	0,9685	0,0445
	3234	2325					13	0,4923	0,5841	0,0228
	3338	2379					14	0,2595	0,283	0,0144
	for fifth service station in general							**0,9636**	**0,9789**	**0,1191**

The system of OTC indicators includes coefficients of OTC α_W, level of OTC β_W and relative level of OTC γ_W, that are calculated for a particular value of the annual program $W_{A\ min}$, provided, that any ratio of partial programs of orders for different types of technical maintenance in the total program of orders $W_{A\ min}$ is equally probable.

Dependencies of OTC indicators of the technological processes of second and third line maintenance of tractors KhTZ-3522 in the joint technological flow on annual program $W_{A\ min}$ (by the number of orders for third line maintenance), received for the fifth variant of the production structure of the service station of the polytechnological specialization (Table 3), are shown in Figures 1-3.

These dependencies are discontinuous functions, intervals of continuity which coincide with the intervals of programs for the fifth variant of the production structure of the service station of the polytechnological specialization, where the number of workers u remains unchanged (Figures 1–3).

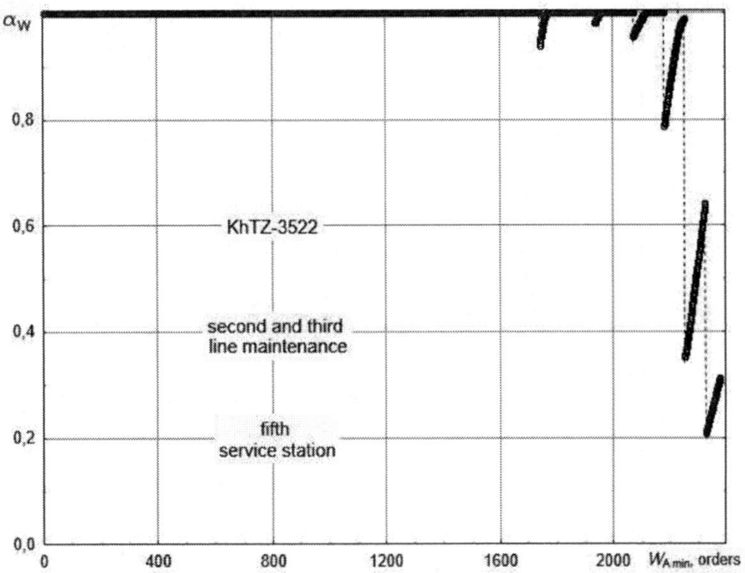

Fig. 1. Dependence of the coefficient of OTC α_W of second and third line maintenance of KhTZ-3522 tractors on the annual number of third line maintenance orders $W_{A\ min}$ for fifth service station.

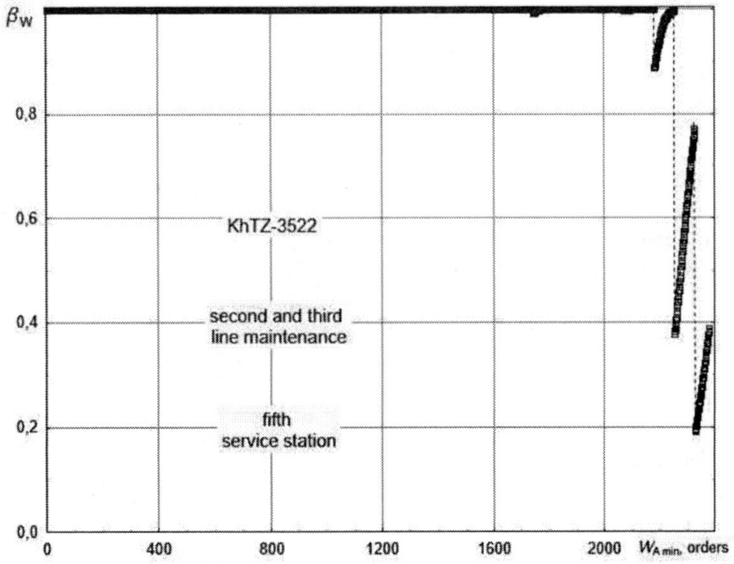

Fig. 2. Dependence of the coefficient of level of OTC β_W of second and third line maintenance of KhTZ-3522 tractors on the annual number of third line maintenance orders $W_{A\ min}$ for fifth service station.

As you can see (Figures 1–3) on the intervals of continuity, with the total annual program $W_{A\min}$ increasing, the values of the coefficient of OTC α_W and the coefficient of the level of OTC β_W remain constant, provided that a complete OTC of technological processes of various types of maintenance in the joint technological flow is in evidence ($\alpha_W = \beta_W = 1$), or increases nonlinearly.

When the program $W_{A\ \min}$ increases, the values of the coefficient of the relative level of OTCγ_W on the intervals of continuity remain constant or non-linearly increase even if a complete OTC of technological processes of various types of maintenance of tractors KhTZ-3522 ($\alpha_W = \beta_W = 1$) in a joint flow is in evidence.

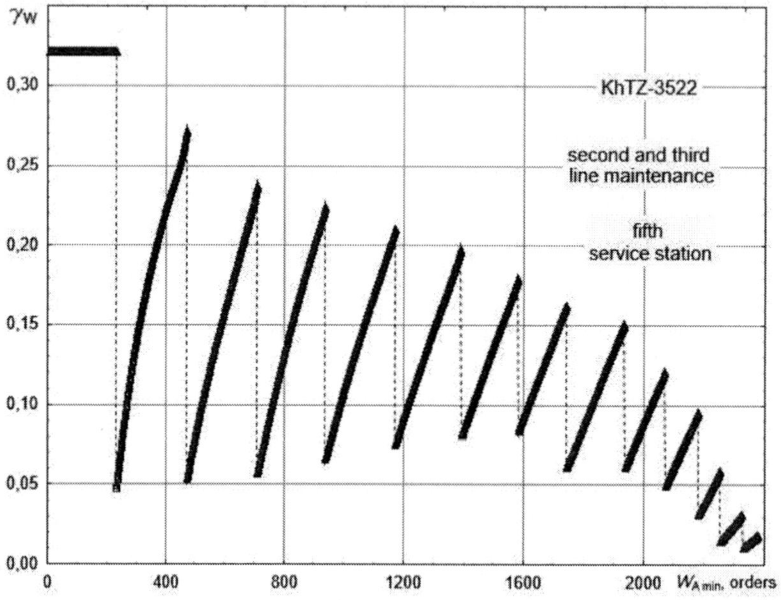

Fig. 3. Dependence of the coefficient of relative level of OTC γ_W of second and third line maintenance of KhTZ-3522 tractors on the annual number of third line maintenance orders $W_{A\ \min}$ for fifth service station.

The analysis of the results of calculating the values of OTC indicators of the technological processes of second and third line maintenance for service stations with different production structure and capacity from parametric series (Table 3), which was conducted under the additional condition that any values of the annual program $W_{A\ \min}$ (by the annual number of third line maintenance orders), that are possible for a constant number of workers u = const, are equally probable, allowed to reveal certain regularities.

Firstly, for each variant of the production structure of the service stations, with an increase in the number of workers, the values of OTC indexes do not increase. For example, for the third variant of the production structure of the service station, where the technological processes of the second and third line maintenance of tractors KhTZ-3522 are executed in the joint technological flow by two workers (u=2 persons) – $\alpha = \beta = 1{,}0$ and $\gamma = 0{,}4106$; and when u = 8 persons – $\alpha = 0{,}9957$, $\beta = 0{,}9989$, $\gamma = 0{,}084$ (Table 3).

Secondly, for service stations with a more complex production structure and higher capacity, the values of the OTC indicators are lower. For example, if for the third variant of the production structure of the service station in general $\alpha = 0{,}9998$, $\beta = 0{,}9999$ and $\gamma = 0{,}2967$, then for fifth variant of the production structure of the service station in general $\alpha = 0{,}9636$, $\beta = 0{,}9789$ and $\gamma = 0{,}1191$ (Table 3).

Thirdly, the most sensitive to the growth of the overall annual program of orders is the coefficient of the relative level of OTCγ.

At the same time, the obtained values of OTC indicators of the technological processes of second and third line maintenance in the joint flow for all five variants of production structures of service stations for tractors KhTZ-3522 designate the expediency of their polytechnological specialization.

4 Conclusions

1. The formed parametric series of production facilities of service stations for maintenance of KhTZ-3522 tractors both mono- and polytechnological specialization (Table 1-3) are the basis for the choice of the effective design solutions with following aspects to be taken into account: firstly, the forecast of annual changes in the total number of maintenance orders due to the changes in the number of these tractors in the service area, and, secondly, the prediction of seasonal fluctuations in the

flow of orders due to the objective variation of use of these tractors in agriculture.

2. The obtained values of OTC of the technological processes of second and third line maintenance in the joint technological flow for all five variants of production structures of the service stations for maintenance of KhTZ-3522 tractors give reasons for the polytechnological specialization of such items.

References

1. *Process flow charts of pre-sales service and maintenance of KhTZ-3512, KhTZ-3522 tractors*, Kharkov Tractor Plant named after S. Ordzhonikidze, p. 80 (2014)
2. *Tractor KhTZ–3512. Operating instructions. 3512.00.001,* Kharkov Tractor Plant named after S. Ordzhonikidze, p. 46 (2014)
3. M.V. Molodyk, A.M. Morgun, L.I. Shapoval at all, *Organizational forms of technical service and forecast of their development in market conditions of management in the agroindustrial complex of Ukraine. Recommendations*, National Scientific Center «Institute of Mechanization and Electrification of Agriculture», p. 170 (2001)
4. O.V. Sydorchuk, S.R. Senchuk, O.V. Kukharuk, *Scientific bases of engineering management of technical service of crop farming*, *Monograph*, p. 172 (2001)
5. A. Bailey, N. Williams, M. Palmer, R. Geering, AGRICULTURAL SYSTEMS, *The farmer as service provider: the demand for agricultural commodities and equine services*, **3** (66), 191-204 (2000)
6. S. Juściński, W. Piekarski, Maintenance and Reliability, *The farm vehicles operation in the aspect of the structure of demand for maintenance inspections,* **1** (45), 59-68 (2010)
7. O.D. Semkovych, R.D. Kuzminskyj, V.E. Chukhrai, M.S. Oliskevych, News of Agrarian Sciences, *Formation and development of the theory of repair and recovery processes,* **Special issue, September**, 90–96 (2001)
8. R.D. Kuzminskyj, Bulletin of Lviv State Agrarian University: Agroengineering research, *Structure, parameters and efficiency of technological processes of repair*, **9**, 50–60 (2005)
9. V.O. Tymochko. Dis. Cand. Tech. Sciences: 05.20.03, *Effective production structure of technological stations of current repairs of tractors of classes 0,9 and 1,4*, p. 241 (1994)
10. R.D. Kuzminskyj, R.I. Barabash, Bulletin of Lviv State Agrarian University: Agroengineering research, *Parameters and performance indicators of technological processes of technical service performed on stationary posts*, **10**, 66 – 73 (2006)
11. O.V. Sydorchuk, R.D. Kuzminskyj, R.I. Barabash, M.A. Mykhajluk, Bulletin of Lviv National Agrarian University: Agroengineering research, *Justification of production structure of service stations for maintenance of KhTZ tractors*, **17**, 54–64 (2013)
12. R.D. Kuzminskyj, R.I. Barabash, Scientific discussion: Issues of technical sciences, *The results of modeling of technological processes of maintenance of KhTZ–3522 tractors*, **4** (25), 97–107 (2015)
13. R.D. Kuzminskyj, Agricultural machines. – Collection of scientific articles, *An organizational and technological compatibility of different objects repair processes in joint technological flow as the basis of formation of thefirm technical service*, **14**, 114–124 (2006)
14. R.D. Kuzminskyj, Theory and practice of development of the agro-industrial complex: Materials of the international scientific and practical forum, *An improvement of the indicators system of organizational and technological compatibility of different objects repair processes in joint technological flow*, **2**, 354–359 (2006).
15. R.D. Kuzminskyj, I.G. Stukalec. Ecological, technological and socio-economic aspects of the use of material and technical base of the agro-industrial complex: Materials of the international scientific and practical forum. *Calculation of the indicators of organizational and technological compatibility of different objects repair processes for single tact value*, 406–409 (2008)

The design of a prototype system which controls the flow of milk in the collector's column of an autonomous milking apparatus

Stanisław Lis[1,*], *Marcin* Tomasik[1], and *Henryk* Juszka[1]

[1]University of Agriculture in Krakow, Faculty of Production and Power Engineering, ul. Balicka 116b Krakow, Poland

Abstract. The research paper presents a prototype system controlling the flow of milk in the collection column of an autonomous milking apparatus. In the Matlab®-Simulink program, the simulation model of such a system was created. Its essential operation is illustrated in the block diagrams. The logic verification of the model's operational accuracy (off-line simulation) did not reveal errors. Therefore a laboratory stand for hardware verification – (on-line simulation) was developed. Within it a virtual feedback loop of the control system (stored in computer memory), with elements of real object control incorporated – in this way a prototype of the device was created. During the hardware verification of the proposed technical solution, operation errors were not observed. In response to certain signals from the measuring elements, the controller correctly calculated control signals for the actuators.

1 Introduction

An observed increase in the milk yield of cows is caused by prolonged milking and consequently an increase in the physical load of teats and udders. Not taking into account individual characteristics of cows in regards to milking can lead to the emergence of disease states of udders including mastitis [2, 9].

One study on individual characteristics of cows concerning the content of milk and pace of its production showed the presence of significant statistical differences between the particular quarters of a cow's udder [3]. It is necessary to undertake research to develop new designs of milking units and better algorithms to control automatic milking. Milking apparatuses which account for the characteristics of cows with regards to milking would contribute to an improvement in the udder health of dairy cows [8, 10]. Analysing the state of techniques with respect to milking equipment in the production market, it can be stated that the current activities of leading companies in this industry are focused on ensuring that farmers expend less effort in the milking process. To a lesser extent, emphasis was placed on fine tuning the parameters of the milking machine to accommodate for the individual characteristics of these animals and to develop a negative pressure control algorithm, which is the primary parameter affecting the course of milking.

It is believed that the optimum solution for a milking apparatus construction/design would be one automatically controlled by negative pressure as a function of milk flow, measured independently for each quarter of the cow's udder [1]. Thus, in the, University of Agriculture in Krakow, research work on designing and constructing an autonomous milking apparatus (AAU) was undertaken. Its task was to individually milk each udder quarter using negative pressure closely corresponding to the intensity of the stream of milk flowing from the teat of the cow. Vacuum separation of the suction and transport has been established [4, 6].

The issues concerning the control of the milking apparatus include two aspects of its functionality – the milk flow and the amount of vacuum suction. The study shows part of the design process relating to one of these aspects in an autontomous milking apparatus, i.e. the system of automatically controlling the flow of milk in the column collector. The purpose of the performed actions was to develop a prototype system which controls the abovementioned process.

2 Purpose and scope of work

The aim of the work was to develop a prototype system controlling the flow of milk in the autonomous milking apparatus (AAU).

The scope of work included:
- determining the design assumptions for the control system of flow in a single column (section) of the AAU collector,
- developing a model for fluid flow control system in AAU using Matlab-Simulink®,
- verification of the model operation in off-line mode,
- construction of research position – a prototype of the device,

* Corresponding author: marcin.tomasik@ur.krakow.pl

- equipment verification of control system (on-line simulation),
- presentation of simulation results.

3 Methodology

The requirements for the designed control system were specified when the simulation model was developed. The correctness of the model was verified by computer simulation (off-line).

A research station was built in which a virtual feedback loop of the control system is stored in computer memory (XPC Target® – toolbox MATLAB-Simulink) including elements of a real object control – in this way the prototype of the device was created. For the purpose of confirming the accuracy of the project's assumptions, a control program was designed and executed which tested the operation and effectiveness of the equipment (online simulation) in the course of which no errors were observed. In response to certain signals of the measuring elements, the controller correctly calculated control signals to the actuators.

Phase assumptions of the project cycle proceed according to the methodology discussed and are presented in block diagram in figure 1 [5, 7, 11, 12].

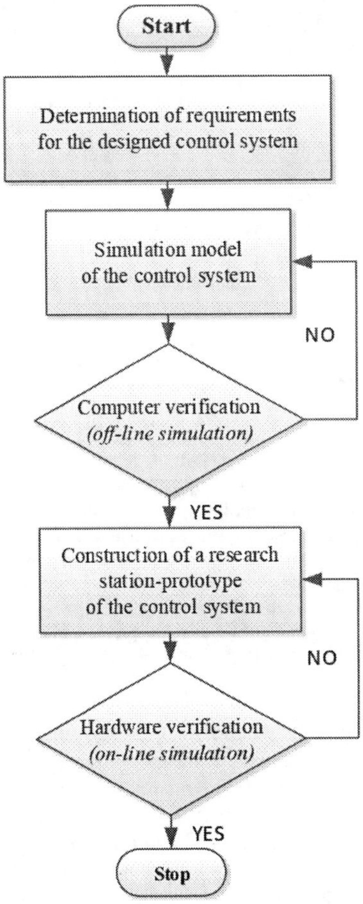

Fig. 1. Methods of application development.

4 The structure of autonomous milking apparatus

According to the adopted methodology (fig. 1), the first stage of the design work was to determine the requirements for the control system. To formulate it, a review of the system of construction for an autonomous milking apparatus was necessary. The requirements are illustrated by the functional diagram shown in figure 2.

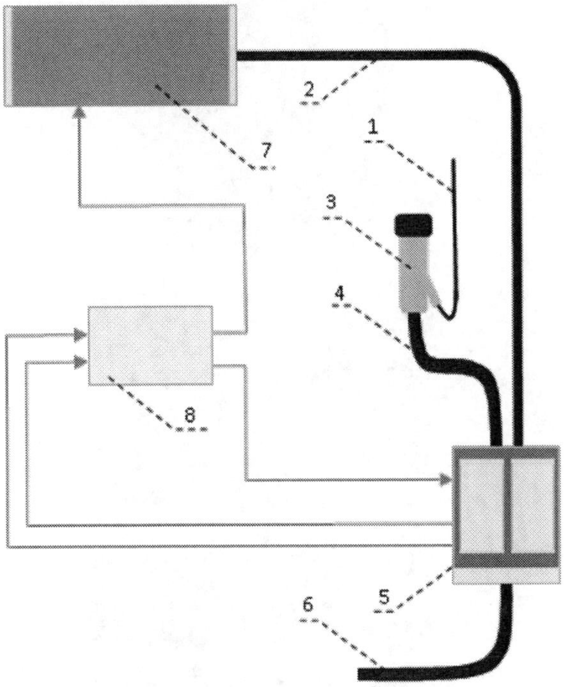

Fig. 2. Schematic diagram of a single column of autonomous milking apparatus, together with the control system: 1– pulsating line, 2 – suction vacuum hose, 3 – teat cup, 4 – short milk tube, 5 – column of autonomous milking apparatus, 6 – milk conductor – vacuum transport, 7 – a system of preparation of vacuum suction, 8 – driver.

A summary of the illustrated construction includes individual milking of each specified quarter of the cow udder, including separation of negative pressure generated by suction from that generated by transport. The collector of the autonomous milking apparatus consists of four independent columns, each of which is comprised of two distinct valve controlled chambers, regulated at both the top and bottom. With this structure, it is possible to separate the vacuum suction of the transport (if one chamber receives milk from a teat, the other empties and vice versa).

This design solution helps to shape the vacuum suction as a function of the effluent stream of milk from the cow teat. Moreover, this solution enables greater use of vacuum suction for conveying milk through to the cooling tank (collecting tank). The design eliminates the mixing of milk from individual quarters of the udder and its movement back under the teats. Figure 3 shows a visualization of the autonomous milking apparatus and a general view of the prototype.

Fig. 3. Visualization of autonomous milking apparatus and overview of the prototype.

In the illustrated milking apparatus, (fig. 3) milking the udder quarter is carried out autonomously by a single column of the collector. A schematic diagram of the column is illustrated in figure 4. Its core elements are two symmetrical chambers (1 and 2) in which, during the milking process, negative pressure is generated by suction or transport in an alternating sequence. The base of the collector (4) has a compact housing structure (3) and includes two nozzles (21 and 22), to which the silicone lines are attached for conducting milk from the chamber (1) and (2) to the milk installation. Pinch valves were installed on these lines (12 and 13), which close the chamber (1 and 2) once it has filled with milk, and open it upon the delivery of negative pressure from transport. The cover of the collector (5) has two nozzles (14 and 15) providing negative pressure to produce suction within the individual chambers. In the cover of the collector, particulate filters are installed (16) in addition to sensors measuring the level of the milk (19 and 20) in each chamber. The level sensors allow for instantaneous volumetric determination of milk flow rate from the teat of the cow based on changes in the level of milk in the chamber with respect to time. The cross-sectional area of chambers (1 and 2) in the measuring range of the level sensor is fixed.

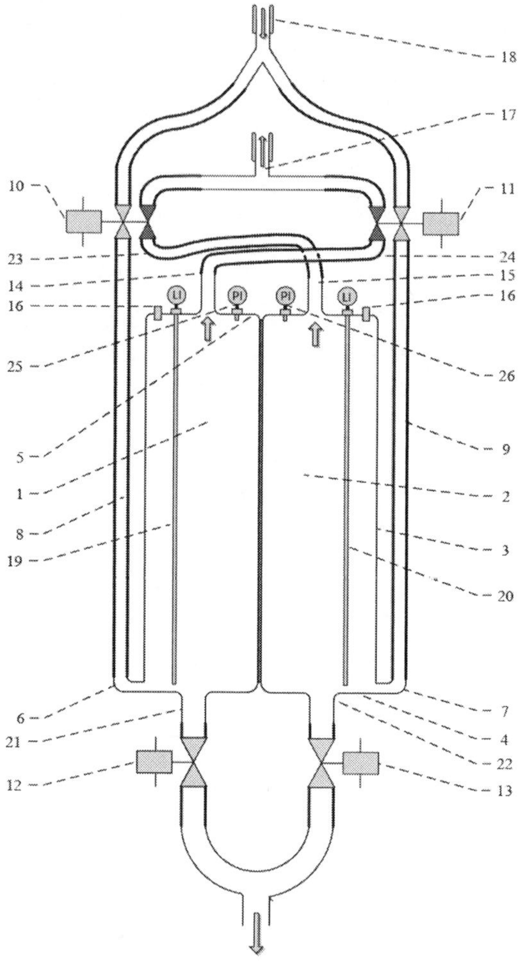

Fig. 4. Column of autonomous milking apparatus: 1 – chamber A; 2 – chamber B; 3 – housing of the collector; 4 – base of the collector; 5 – cover of the collector; 6, 7 – dairy connector pipes; 8, 9 – milk supply lines to the chambers; 10, 11 – double pinch valves Z1 and Z2; 12, 13 – pinch valves ZA and ZB; 14, 15 – vacuum suction connector pipes; 16 – particulate filter; 17 – negative pressure suction inlet; 18 – short milk tube; 19, 20 – CA and CB level sensors; 21, 22 – drain connector to the milk installation 23, 24 – negative pressure supply lines; 25, 26 – vacuum sensors.

Connector pipes (6 and 7) bring the milk from the short milk tube of the teat cup by means of a tee connection (19) to the chambers (1 and 2) of the collector. On the milk supply line (8) into the chamber (1) and on the vacuum supply line that brings milk (23) into the chamber (2), a double pinch valve has been installed (10). This valve shuts off the flow of milk into the chamber (1) and opens the supply of suction to the vacuum chamber (2) and vice versa. Analogously on the milk supply line that brings milk (9) into the chamber (2) and on the vacuum supply line (24) that brings negative pressure into the chamber (1) a double pinch valve has been installed (11). The valve shuts off the flow of milk to the chamber (2) and opens the flow of vacuum suction to the chamber (1) and vice versa.

Construction and functionality of the other three columns of the autonomous milking apparatus collectors are the same.

5 Computer simulation of the control system

The next step in the development of the project was to develop the simulation model of the milk flow system control in the collection column of the autonomous milking apparatus. In essence, the system includes the implementation of the impact object control via actuators in the form of Z1, Z2, ZA and ZB valves. The impact was calculated from the signals from sensors A and C which are installed in the chambers A and B of the column manifold. The control algorithm is illustrated in the diagram in figure 5.

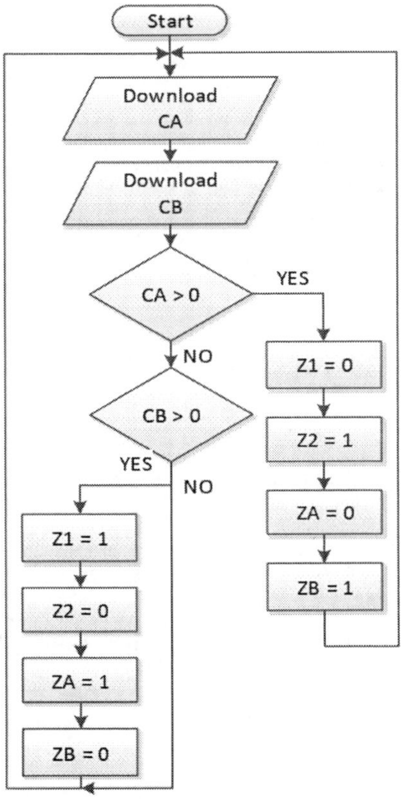

Fig. 5. The control algorithm of autonomous milking apparatus.

Operation of the control program should have the following sequence: If chamber A is full – set the valve terminals Z1=0, Z2=1, ZA=0, ZB=1. This will close the supply of vacuum suction and milk into chamber A. Simultaneously, the exit line from chamber A to the system is opened In the same instant, the vacuum suction supply line and the milk line are opened, the milk supply flows into chamber B. As a result, chamber A will empty and chamber B will be filled. The negative pressure generated by the milk which is discharged from the chamber has a higher value, ensuring faster drainage of the chambers. The system works analogously if the desired level of milk in chamber B is reached. The system of milk flow control discussed above is illustrated in figure 6. The unobstructed lines are shown in green, whereas the closed lines are shown in red.

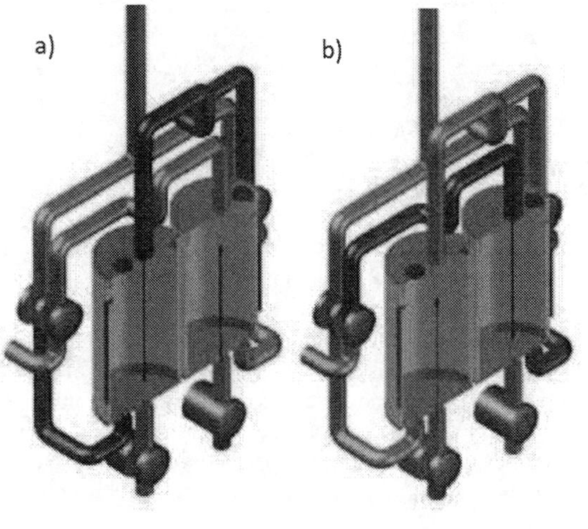

Fig. 6. Operation of the control system: a – if cell A is full, b – if the chamber B is full.

The algorithm of the control system is illustrated in figure 5, entered (in accordance with the standards of the Matlab-Simulink program) in the form of a block diagram illustrated in figure 7.

In this way, a model of the control system was created that allows logical verification of adopted assumptions through a computer simulation (off-line). The results showed correct operation of the modeled control program which properly worked out the impact of feedback on the controlled system. It is, therefore, appropriate to say that the assumptions for the system of milk flow control to the column collector of the autonomous milking apparatus were appropriate for use during on-line hardware simulation.

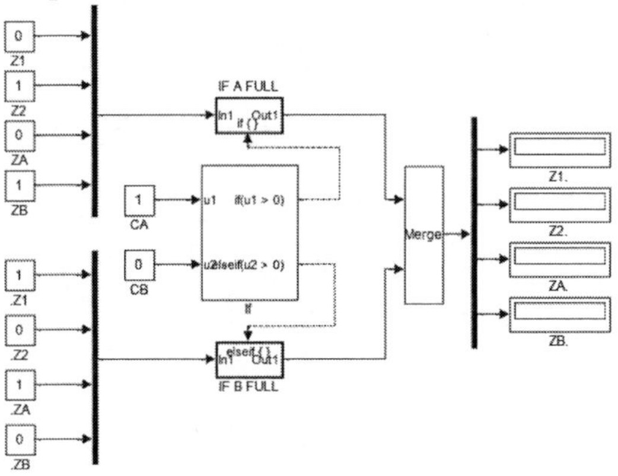

Fig. 7. Model in computer simulation.

6 Hardware simulation of the control system

The hardware simulation provides the ultimate confirmation of the correctness of the design assumptions and the proper operation of the program. In this case it took the form of analysis at the laboratory stand, in which a virtual feedback loop of the control system (stored in Matlab-Simulink) including elements of actual object control – collector column. The result is a prototype of the control system. The research position (test stand) was developed for the necessity of hardware simulation, as illustrated in figure 8.

In order to control the simulation system model illustrated in figure 7 and be able to engage with real elements incorporated into the loop, it has been modified by adding blocks of analog inputs (Analog Input) and digital output (Digital Output) and blocks scaling input and output signals (Fig. 9).

In order to carry out the hardware simulation, for the column collector of an autonomous milking apparatus, a replacement liquid (distilled water) for the milk was brought. It was introduced from a feed tank using a teat cup and a short milk tube to the prototype of the milking apparatus. Each time water flowed through a column of the collector, it was discharged to the second auxiliary tank. Several dozen series of studies were carried out for different flow rates of fluid streaming through the milking apparatus.

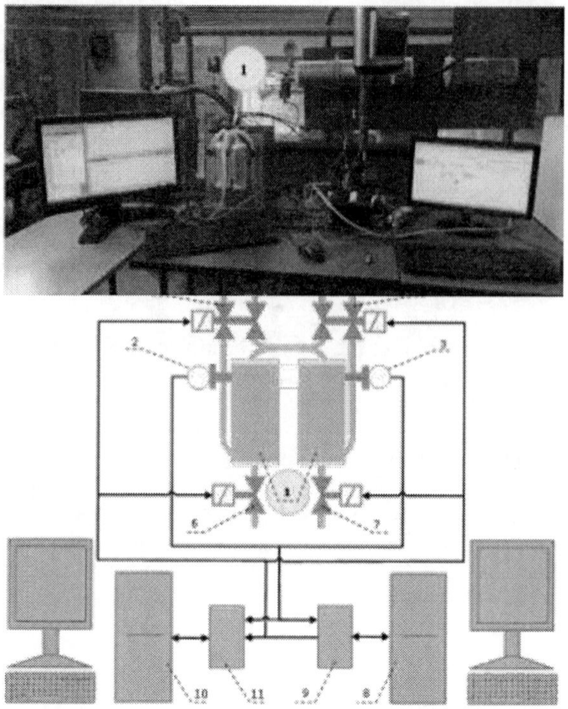

Fig. 8. A general view and schematic diagram of the test stand of the milk flow control system in the column collector of an autonomous milking apparatus: 1 – collector of autonomous milking apparatus; 2 – level transmitter of chamber A; 3 – level transmitter of chamber B; 4, 5 – double pinch valves; 6, 7 – pinch valves; 8 – a control computer; 9 – I/O control computer card; 10 – computer registrant; 11 – I/O recording computer card.

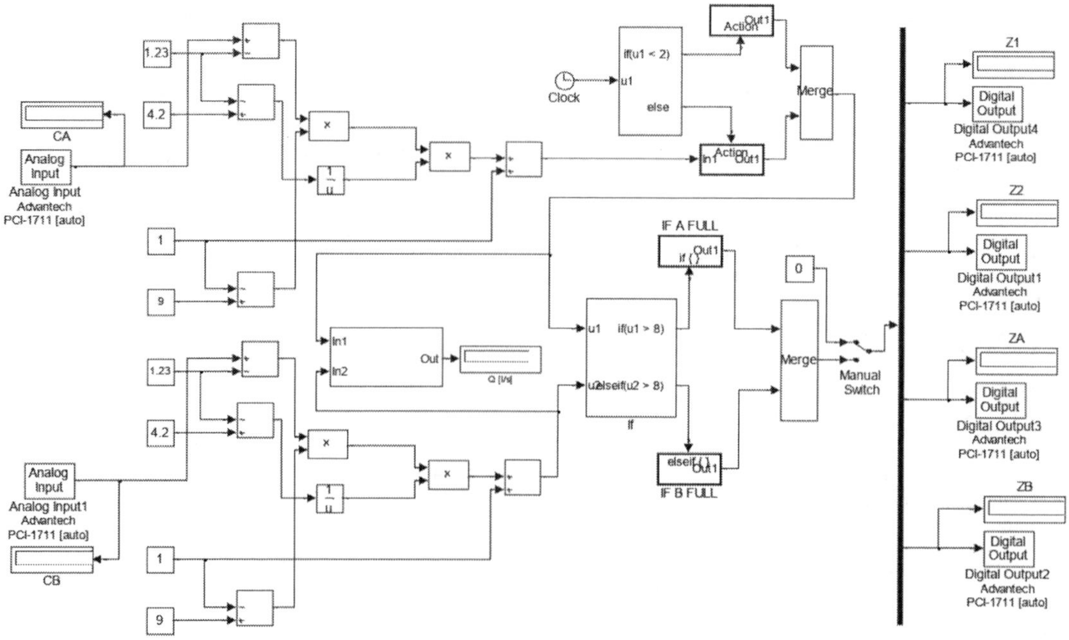

Fig. 9. Model of hardware simulation.

7 Analysis of the results

The registered characteristics during the simulation of the control system are illustrated in figures 10 – 15. The input signals are shown in figures 10, 11, while the output signals (control) are illustrated in figures 12 – 15. In the analysis of waveforms of input signals, illustrated in figures 10 and 11, to the control system (voltage V electric current coming from the liquid level sensors of chambers A and B of the autonomous milking apparatus collector) alternate filling of chambers A and B can be observed. It should also be noted that the chamber empties faster than it fills.

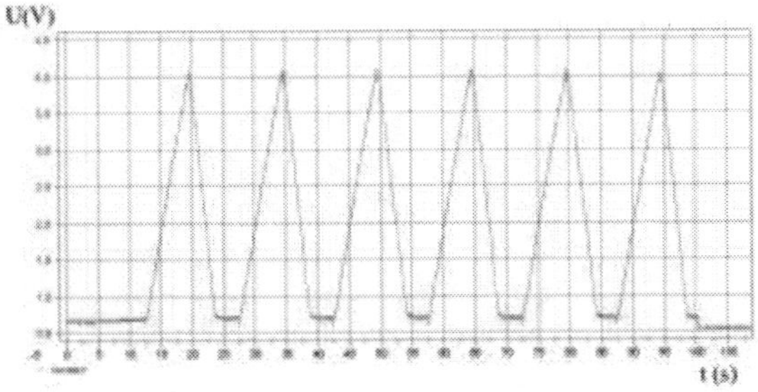

Fig. 10. Characteristics of the voltage (CA) sensor.

Figures 12 and 13 show graphs of output signals of the control system. These are the signals controlling double valves Z1 and Z2. Analyzing these waveforms, the alternating action of the valve terminals can be seen.

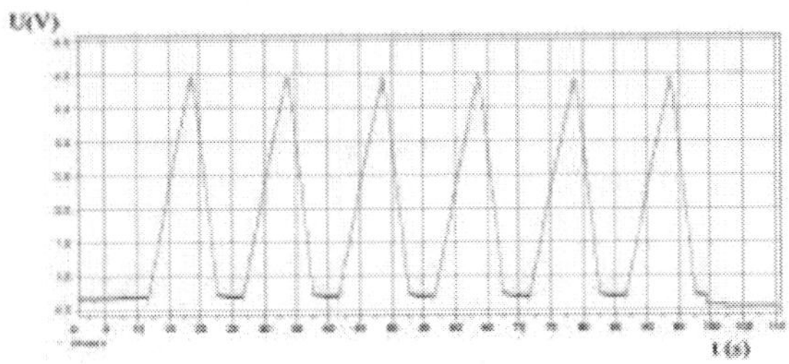

Fig. 11. Characteristics of the voltage (CB) sensor.

When Z1 brings a liquid milk replacement to chamber A, it shuts off the flow of vacuum to chamber B at the same time, however, Z2 releases the vacuum in chamber A, which draws milk replacement into the chamber and closes the milk supply to chamber B. After filling one of the two terminals of the chambers, both valves change their values, and the circuit works in the opposite order.

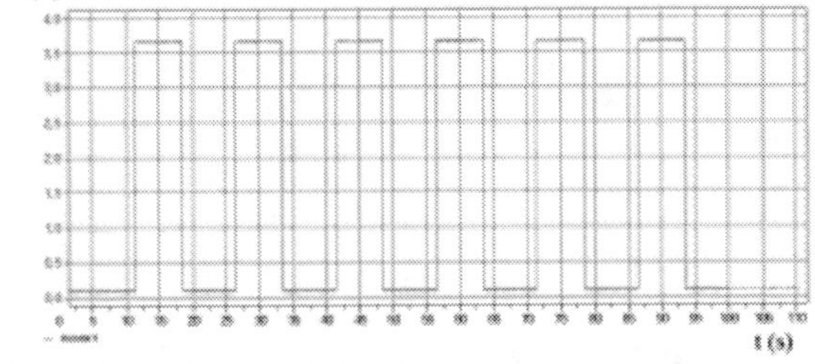

Fig. 12. Z1 valve control signal.

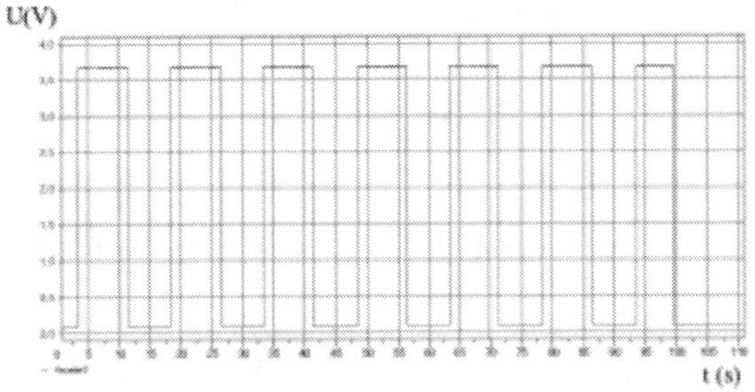

Fig. 13. Z2 valve control signal.

In the following graphs (fig. 14, 15) signals controlling valves were presented with single terminals ZA and ZB. They bring the liquid from chambers A and B to the milking plant (auxiliary tank). You can also note here the alternation of these terminals. Comparing the operation of valve sensor signals in each terminal (fig. 10, 11) it can be stated that the process control is correct.

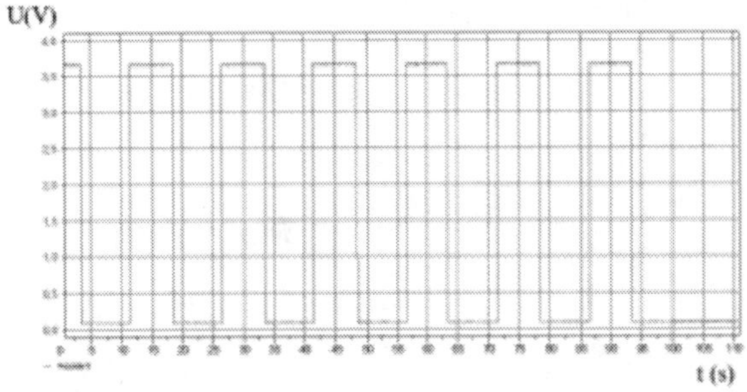

Fig. 14. ZA valve control signal.

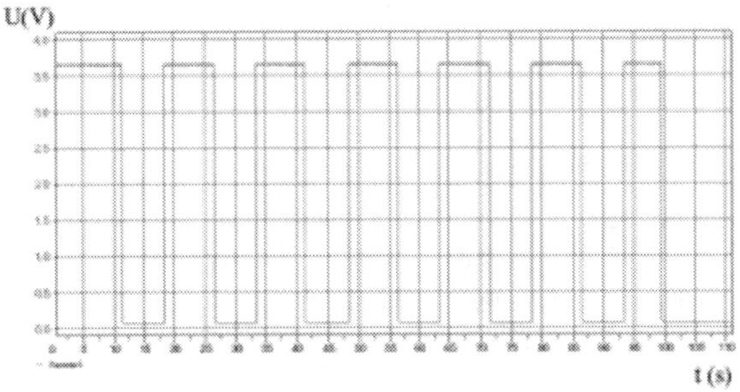

Fig. 15 Control signal of ZB valve.

Analyzing obtained characteristics for the virtual control system with real objects included in the control loop, it should be acknowledged that the elaborated control program is working correctly. It is reasonable to conclude that the design assumptions adopted for the system are correct.

8 Conclusions

1. The logic verification (off-line) of the simulation model of the milk flow control system in the collection column of an autonomous milking apparatus developed in Matlab-Simulink program, demonstrated proper operation. The model calculated the correct value of the

* Corresponding author: marcin.tomasik@ur.krakow.pl

output signal using the simulated values of the input signal.
2. Hardware on-line simulation carried out at the verification stand confirmed the design assumptions. The control system realized adopted assumptions, the measurements carried out for different values of liquid flow has proven its ability to be used for high volume milking in dairy cows.
3. The presented design methodology based on the model enables proper integration and configuration of the control system.

References

1. MA. Davies, E. Maltz, D.J. Reinemann. *Proceedings of the International Symposium Robotic Milking*. Lelystad, The Netherlands, (2000)
2. A.H. Ipema, E. Benders. *Proceedings of the International Symposium on Prospects for Automatic Milking*. Wageningen, Netherlands. EAAP Publication, **65** (1992)
3. A.H. Ipema, P.H. Hogewerf. Computers and electronics in agriculture, **62** (2008)
4. H. Juszka, S. Lis, M. Tomasik. Agricultural Engineering, **8** (133) (2011)
5. I.P. Kurytnik, S. Lis, T. Drożdż, A. Telega. PAK. **5**, (2014)
6. I.P. Kurytnik, M. Tomasik, S. Lis. PAR, **3**, (2015)
7. S. Lis, M. Tomasik, K. Nęcka, T Drożdż, P. Nawara, P. Wrona, M. Oziembłowski. Przeglad Elektrotechniczny, **12** (2015)
8. I. Naumann, R.D. Fahr, G. Lengerken, Archiv für Tierzucht **41** (3) (1998)
9. F. Neijenhuis, J.E. Hillerton, C.O. Paulrud, M.D. Rasmussen, J. Baines. Proceedings of the Annual Meeting National Mastitis Council, **43** (2004)
10. M.D. Rasmussen, E.S. Frimer, E.L. Decker. J. Dairy Sci. **77** (1994)
11. R. Tadeusiewicz, *Biocybernetyka*. Wyd. Naukowe PWN (2014)
12. W. Tarnowski, *Projektowanie układów regulacji automatycznej ciągłych z liniowymi korektorami ze wspomaganiem za pomocą Matlab'a*. Wyd. Pol. Kosz. (2008)

Quality of dried cauliflower according to the methods and drying parameters

Bogusława Łapczyńska-Kordon[1], *Barbara* Krzysztofik[2], *Zygmunt* Sobol[2]

[1]Department of Mechanical Engineering and Agrophysicsof University of Agricultural in Krakow, Poland
[2]Institute of Machinery Management, Ergonomics and Production Processes of University of Agricultural in Krakow, Poland

Abstract. The quality of food products is a complex concept. It can be defined in many ways. The common element of most of these definitions is the condition of meeting the requirements of consumers. Quality determines product compliance with the requirements set by the normalized regulations. The paper attempts to determine the optimal method and parameters of cauliflower drying. In addition, a qualitative assessment of the obtained product was made. The results show that the method and parameters of drying significantly affect the quality of the dried cauliflower. Convection drying guarantees higher drought quality with respect to the color of the sample (higher brightness), taste and odor. Of the drying parameters accepted in the experiment, the most positive effect on the tested parameters was recorded using convection drying at a flow rate of 0.2 ms^{-1} and the least favorable for microwave drying 170 or 210 W.

Introduction

One of the primary ways of food preservation is drying, the reduction of the water content in the raw material. Drying improves product stability during storage and transport [1]. Drying should be carried out so as to prevent adverse changes in the products resulting in deterioration.

Selection of the appropriate drying method significantly influences the properties of the plant material such as: thermal resistance, biological and mechanical-mechanical properties, degree of fineness, and the required quality of dried food products, as well as parameters of the drying agent [2]. The choice of a suitable dryer, pre-treatment and proper process planning, as well as the knowledge of the drying kinetics of a given raw material, are the basis for obtaining a product of good quality [3].

The convection drying method is most commonly used in industry because of the simplicity of construction and easy operation of the equipment. However, the low energy efficiency of such dryers and the low quality of the resulting dried matter are the reasons for the search for methods that produce better products in terms of quality and low cost of production [4], [5]. The disadvantage of this method is the relatively long drying time and the use of high temperatures, which results in qualitative changes in the product. Thus, new technologies are being sought which would be more efficient and enable the receipt of a high quality product at minimal cost and maximum process efficiency.

The alternative method seems to be microwave energy drying, which can improve the quality of the dried product.The use of microwave drying results in better retention of aromatic components, biologically active substances and improved sensory properties of the material compared to a convection drying. Microwave heating causes intense energy release inside the material, as a result has the effect of "inflating" due to rapid evaporation of water [6]. During the process the biological material can be protected from overheating and increase the intensity of evaporation of the water by reducing the pressure inside the dryer. As a result, the drying rate is much higher than with conventional drying methods (eg convection drying). In addition, an important factor influencing the drying kinetics is the selection of the appropriate microwave power [7], [8], [9], [10].

The aim of the study was to analyze the effect of drying method and parameters on dry product quality.

The scope of work included convection and microwave-convection drying of cauliflower particles and evaluation of dry product quality in terms of color, odor, taste and shape (shrinkage).

Material and method

The subject of the study was the quality of dried cauliflower particles obtained from convection and microwave convection drying. The quality assessment included the determination of changes in sensory characteristics: odor, taste and color and shape by drying shrinkage coefficient, water content and density in dry product were also determined.

The research was carried out on a laboratory stand for the convection drying of plant materials, designed and built at the Department of Mechanical Engineering and Agrophysics of the University of Agriculture in Krakow. The stand was equipped with: a drying chamber, in which RADWAG WPE 3000 was placed (± 0.01g), allowing continuous measurement of mass changes during drying, heater with a maximum power of 3.04kW (allowed maximum drying of the drying agent to 50°C), combined with a fan. During drying, the temperature of the drying medium was measured using NiCrNiAl thermocouples (± 0.1° C), at the inlet and outlet of the dryer and inside the drying chamber. Samples were dried at drying agent temperature 50°C.The convection drying took place in a forced air dryer for three speeds: 0.1 (I); 0.2 (II); 0.3 (III) [$m \cdot s^{-1}$]. The airflow velocity was adjusted by varying the diameter of the opening in the convection oven, the diameters being 0.07, 0.12, 0.17 [m].

Microwave-convection was performed in microwave oven Moulinex - Dialog cook Y90, for three microwave power: 170, 210 and 340 [W], at air temperature 50°C.

Cauliflower samples were cauliflower roses with an average weight of 2,2 g. In the driers the material was laid in a single layer. Samples were weighed before and after the electronic weighing of RADWAG WPE 3000 with an accuracy of 0,01 g. During the drying were recorded the temperature and relative humidity of the drying agent. The measurement of these parameters was performed using the multifunction device company ROTRONIC.Three samples of fresh and dried material were taken to determine the density, then weighed on the RADWAG WPS 510 / c / 1 electronic scale in air and in rapeseed oil. A reading accuracy of a mass was 0,001g.

The color of fresh and dried material was determined on the basis of five samples each of which was photographed by a CANON digital camera. Images were analyzed for color, using MULTISCAN.

The taste and aroma were assessed using a five point scale, where 1 was the worst rating, and 5 was the best, using a expert group of evaluators (5 panelists).

The water content of the test material was calculated from the relationship:

$$u = \frac{mp - ms}{ms} \qquad (1)$$

where: u - water content,($kgH_2O(kgs.s)^{-1}$)., m_p - mass of sample, (kg), m_s - mass of dry substance, (kg).

The density of the tested material was determined by a method based on the Archimedes law, using a laboratory scale. The density was calculated from the formula:

$$\rho = \frac{A}{A - B} \cdot \rho_o \qquad (2)$$

where: ρ - sample density ($kg \cdot m^{-3}$), A - mass of sample in air (kg), B - mass ofsample in oil (kg), ρ_o - density of oil ($kg \cdot m^{-3}$).

Drying shrinkage was calculated from the relationship:

$$s_k = \left(\frac{A_p}{A_s} \right)^{\frac{3}{2}} \cdot 100\% \qquad (3)$$

where: s_k - drying shrinkage (%), A_p - sample area (mm^2), A_s - average area of freshmaterial (mm^2).

Each measurement was performed in triplicate.The results of the measurements were subjected to a basic statistical analysis.

Mean and standard deviations were calculated and variance analysis was performed with Duncan's test for dependence between dependent and independent variables. In addition, Pearson's linear correlation analysis was performed for dependent variables. For all statistical calculations the significance level α = 0.05 was assumed.

Results and analysis

The obtained results of studies on density changes, drying shrinkage, taste and odor and water content and color of dried products, depending on the method and conditions of drying, are presented in the form of figures and tables as mean values with standard deviations. The average density of fresh material varied from 0.961 $kg \cdot m^{-3}$ (for microwave-convection drying at 210 W) to 1.101 $kg \cdot m^{-3}$ (for sample III for convection drying at an air flow rate of 0.3 m s^{-1}, for which the highest standard deviation was recorded, ie: 0.097). The remaining deviations from the mean were low, ie: 0.002-0.007 (Figure 1).The highest increase density of simple of the dry material compared to the fresh sample was recorded for convection drying at an air flow rate of 0.3 ms^{-1} (III). The recorded growth was 20.9%. After microwave drying at 210 and 340W and conventional drying at a flow rate of 0,2 ms^{-1} (II), the density of dried material relative to fresh material was reduced.

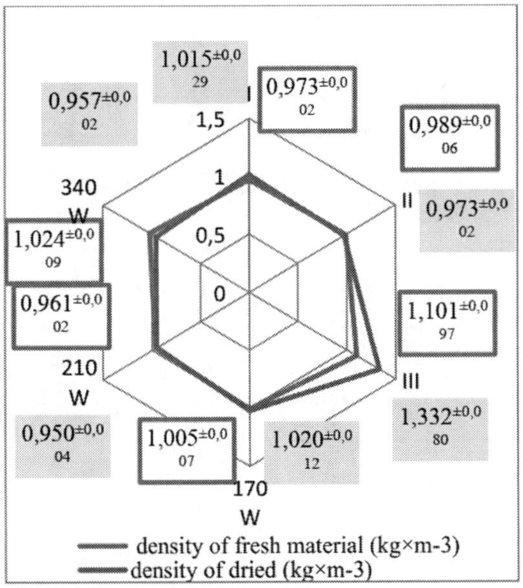

Fig. 1. The density of the samples before and after drying

The greatest decrease in the density of (6.6%) was observed for the material after drying with a microwave power of 340W.

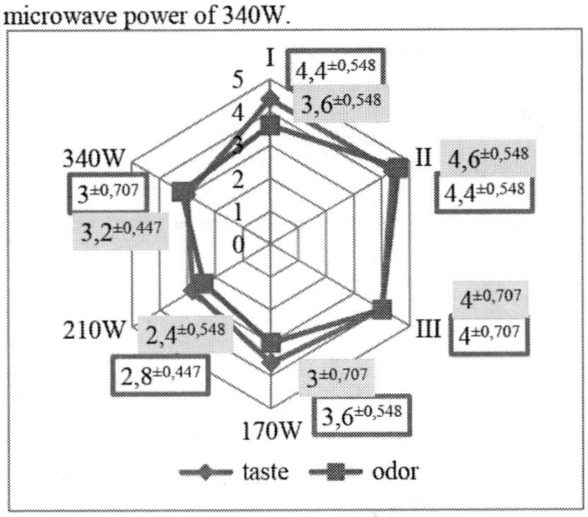

Fig. 2. Results of organoleptic evaluation for taste and odor

Significant statistically differentiation has been found to evaluate the taste and odor of dry products obtained by various drying methods.The higher tasting values on the 5-step scale obtained the drought obtained from the conventional drying method (4-4) (Figure 2).The highest value was obtained with an air flow rate of 0.2 $m{\cdot}s^{-1}$ (II). For microwave- convection drying its value was in the range of 2.8 for drying at 210W to 3.6 for 170W. Similarly, when evaluating the taste, odor evaluation results are formed. Significantly higher values in the sensory evaluation were obtained by conventional methods compared to microwave-convection. Similarly, when evaluating the taste, odor evaluation results are formed. Significantly higher values in the sensory evaluation dried samples obtained by the convection method in comparison with the microwave - convection.

The highest value of drying shrinkage was recorded at air velocity of 0,2 ms^{-1} (II), respectively 4.6 points, lowest at 210W microwave power, 2.4 points respectively. Methods and parameters of a drying affected drying shrinkage and water content significantly (Table 2). The highest values of water content were recorded for dry products obtained from convection drying at air flow velocities of 0,.2 and 0.3 ms^{-1} (II and III), respectively 8.05 and 6.43 kg $H_2O{\cdot}kgs.s.^{-1}$ (Figure 3). The lowest water content was recorded in microwave drying at 210 and 170W microwave power, respectively 1,53 and 3,41 kg $H_2O(kgs.s)^{-1}$.

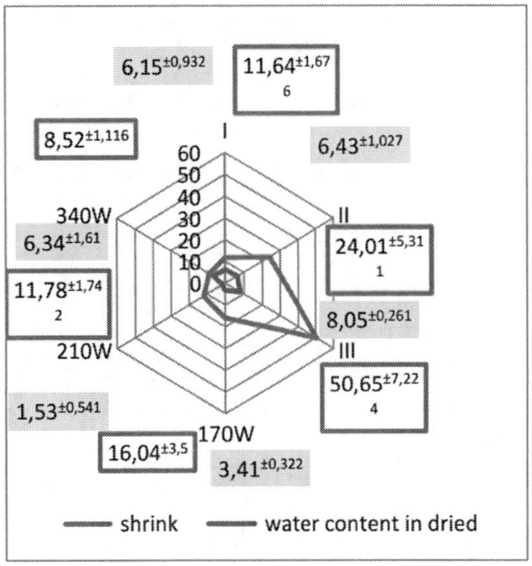

Fig. 3. The water content of dried material (kg H_2Okgss^{-1}), and drying shrinkage (%)

Analysis of drying shrinkage of samples indicates significant differences between drying methods (Table 2). The highest value of the shrinkage was observed for the sample dried convection method using air flow rate of 0.3 and 0.2 $m{\cdot}s^{-1}$, respectively, 50.65 and 24.01%. The highest shrinkage value was recorded for samples with a microwave power of 340W.

Evaluation of color parameters showed that samples before and after drying were significantly different. Compared to the fresh sample, higher sample brightness was observed for convection drying at an air flow rate of 0,2 ms^{-1} (II). On the other hand, it was significantly lower for the dried food obtained from the microwave oven at 170W, for which all other color parameters were also the lowest. Compared to the fresh sample, higher r, b and y parameters were recorded for most drying methods, while the lower values for parameter g (Table 1).

The analysis of the correlation between the dependent variables (Table 3) showed a significant positive correlation between the water content in the dried material and drying shrinkage, density of a sample of fresh and dried. With the density of the fresh sample, the drying shrinkage and the 'g' parameter and the density of the dried were significantly correlated.

 https://doi.org/10.1051/bioconf/20181002017

Drying shrinkage, water content, color g parameter and negative 'r' color are significantly correlated with the dry material density.

Table 1. Parameters of the color of fresh material and dried.

Sample of fresh material	Average rightness	r	b	g	y
	$254.73^{\pm0.20}$	$220.46^{\pm0.82}$	$214.11^{\pm2.94}$	$196.40^{\pm6.23}$	$213.49^{\pm3.25}$
I	$254.08^{\pm0.10}$	$234.71^{\pm0.81}$	$221.90^{\pm3.10}$	$183.85^{\pm7.57}$	$221.07^{\pm2.85}$
II	$254.86^{\pm0.24}$	$228.86^{\pm5.15}$	$221.27^{\pm589}$	$188.51^{\pm6.67}$	$219.40^{\pm5.74}$
III	$254.71^{\pm0.30}$	$223.71^{\pm1.47}$	$220.65^{\pm1.65}$	$196.68^{\pm1.61}$	$218.33^{\pm1.54}$
170W	$252.49^{\pm1.25}$	$220.90^{\pm1.75}$	$211.74^{\pm2.59}$	$171.79^{\pm4.35}$	$209.55^{\pm2.42}$
210W	$254.08^{\pm0.10}$	$234.71^{\pm0.81}$	$221,90^{\pm3.10}$	$183.85^{\pm7.57}$	$221.07^{\pm2.85}$
340W	$254.06^{\pm0.21}$	$235.23^{\pm0.52}$	$223.21^{\pm2.01}$	$186.57^{\pm5.11}$	$222.29^{\pm1.76}$

Table 2. Results of Duncan test - of one-dimensional significance tests for the parameters tested.

Effect	Differences between the test samples				Duncan's test results
	Convection and microwave-convection		I-340W		
	F	p	F	p	for I-340
Drying shrinkage					
Constant parameter	77.709	**0.000**	754.012	**0.000**	I - 170W, 210W, 340W; 210W - 170W, 340W;
Combinations	12.895	**0.001**	74.561	**0.000**	
Moisture content					
Constant parameter	266.192	**0.000**	1022.725	**0.000**	I - II, 340W; II - 340W;
Combinations	22828	**0.000**	34.257	**0.000**	
taste					
Constant parameter	1078.088	**0.000**	1173.429	**0.000**	I – II, III, 170W; II - III, 170W;
Combinations	25.288	**0.000**	6.800	**0.000**	III – 170W; 170W – 210W, 340W; 210W – 340W;
smell					
Constant parameter	796.968	**0.000**	1030.095	**0.000**	I -III, 170W; II – III;
Combinations	23874	**0.000**	8.610	**0.000**	170W – 210W; 340W – I, III, 170W, 210W;
Average brightness					
Constantparameter	3154248.581	**0.000**	6534237.131	**0.000**	I – III, 210W, 340W; II – III;
Combinations	12.436	**0.001**	11.953	**0.000**	III- 210W, 340W; 210W – 340W;
r					
Constantparameter	40404.915	0.000	284639.927	**0.000**	I – 210W, 340W; III – 170W;
Combinations	0.270	0.607	35.031	**0.000**	210W – 340W;
b					
Constantparameter	60986.804	0.000	129485.267	**0.000**	I – II, III, 210W, 340W; II –III, 210W, 340W;
Combinations	1.704	0.202	7.813	**0.000**	III – 210W, 340W; 210W – 340W;
g					
Constantparameter	15423.173	**0.000**	29865.216	**0.000**	I - II, 210W, 340W; II – 210W, 340W;
Combinations	8.989	**0.006**	9.525	**0.000**	210W – 340W;

y					
Constantparameter	53178.646	0.000	142184.619	**0.000**	I – II, III, 210W, 340W; II – III, 210W, 340W; III - 210W, 340W; 210W – 340W;
Combinations	1.071	0.310	10.746	**0.000**	

The bolds were marked by significant relationships at $\alpha = 0.05$

Table 3. Results of linear correlation analysis for dependent variables.

Dependent variables	Drying shrinkage (%)	Moisture content(kg $H_2O \cdot kg_{ss}^{-1}$)	Taste (-)	Odor (-)	Average bright-ness	r	b	g	y	Density of fresh sample ($kg \cdot m^{-3}$)
Moisture content in dried (kg $H_2O \cdot kgss^{-1}$)	**0.620**									
Taste(-)	0.383	**0.635**								
Odor	0.371	**0.566**	**0.541**							
Average brightness	0.377	0.434	0.417	**0.522**						
r	**-0.665**	-0.274	-0.105	-0.149	0.346					
b	-0.138	0.179	0.149	0.167	**0.644**	**0.748**				
g	**0.637**	**0.625**	0.344	**0.491**	**0.674**	-0.049	**0.602**			
y	-0.240	0.100	0.102	0.122	**0.633**	**0.834**	**0.988**	**0.502**		
Density of moist material ($kg \cdot m^{-3}$)	**0.648**	**0.559**	0.121	0.065	0.160	-0.404	0.059	**0.566**	-0.034	
Density of dried material ($kg \cdot m^{-3}$)	**0.876**	**0.647**	0.303	0.111	0.257	**-0.528**	-0.024	**0.637**	-0.118	**0.804**

Conclusions

Analysis of the results allowed to draw the following conclusions:

1. The method and parameters of drying significantly affect the quality of the dried cauliflowers obtained.

2. Convection drying guarantees higher quality of the dried cauliflowers with respect to the color of the sample (higher luminosity), taste and odor.

3. Of the drying parameters accepted in the experiment, the most positive effect on the tested parameters was recorded using convection drying at a flow rate of 0.2 $m \cdot s^{-1}$ and the least favorable for microwave drying 170 or 210W.

4. Significant positive correlation was found between density of dried and moist sample, water content and drying shrinkage.

References

1. B. Krzysztofik, T. Dróżdż, Z. Sobol, P. Nawara, P. Wrona,. *Methods of securing and preserving raw materials and food products*(Monograph. Ed. Agric. Eng. ISBN 978-83-64377-14-3 (2015)
2. B. Bijok, F. Bijok, A. Dąbek,. *Raw materials and food technology, Part II. (Surowce I technologia żywności część II)*. WSiP(2001)
3. E. Biller, A. Wierzbicka, A. Półtorak, Agric.Eng. **9(69)** (2005)
4. D. Witrowa-Rajchert, *Rehydration as an indicator of changes occurring in plant tissue during drying*. (Habilitation dissertation Ed. SGGW in Warszawa)) 29. ISBN 83-87660-95-7 (1999)
5. M. Rząca, D. Witrowa- Rajchert, Acta Agrophys. **9(2),** (2007)
6. J. Oszmiański, *Technology and analysis of fruit and vegetable products. Selected Issues.(*Ed. of Wrocław University of Environmental and Life Sciences (2002)
7. A. Andres, K. Bilbao, P. Fito, J. of Food Eng. **63**. (2004)
8. M. Szarycz, E. Kamiński, K. Jałoszyński, A. Szponarska, Acta Sci. Pol.Technika_Agraria **2(2).** (2003)
9. A. Marzec, M. Zadrożna, ActaAgrophys.**12(2)** (2008)
10. T. M. Lin, T. D. Durance, C. N. Scaman, Food Res. Inter. **31 (2)** (1998)

Modelling fatigue assessment at the vehicle driver's station

Piotr Maksym[1,*], and *Halina* Pawlak[1]

[1]University of Life Sciences in Lublin, Faculty of Production Engineering, Department of Ergonomics, 20-612 Lublin, Głęboka 28. Poland

Abstract. The article presents the principles of fatigue assessment modeling in a driver's station using Bayesian networks. One of the causes of road collisions and accidents is fatigue. The factors determining fatigue are age of driver, psychophysical and health condition, time and length of the route being taken. At present, there is no clear criteria for assessing fatigue among professional drivers, so the objective of assessment is to attempt to design and construct a fatigue assessment model using Bayesian network technology.

1 Introduction

On Polish roads there are many collisions and road accidents every day. The main culprits are drivers of both passenger vehicles and trucks. They carry dangerous consequences. They can cause severe injury, irreversible changes in the human body, and lead to death. Trucks are the most common means of transporting different types of loads, and due to their size in the crash, they can cause more damage than passenger cars.

The profession of a driver, like any other profession, involves a number of dangerous situations that can take place every day. The biggest threat is traffic accidents. Their cause may be a bad technical condition of the vehicle, bad weather conditions or driver fatigue.

The phenomenon that naturally accompanies every job is fatigue. It is a warning signal to the human body from excessive, unfavorable, and in special cases - even health-threatening effort. The natural reaction to feeling fatigued is rest, which allows for regeneration and restore of the body [1].

It is the fatigue of the driver that is influenced by perception, decision-making and motor skills. Driving a car requires careful attention from the driver and a focus on external stimuli. In addition, the organization of the work day, including the length of the route taken, sleep deprivation, breaks, and inappropriate and uncomfortable position while driving and monotony [2-4]. As a result, driver fatigue is a serious safety problem. Such diverse causes are often difficult to identify immediately.

The aim of this paper is to develop a fatigue assessment model at the driver's station using Bayesian technology, with the ability to indicate the accuracy of the probability distribution, which factor has the greatest influence on fatigue.

2 Material and Methods

The process of modelling the fatigue assessment of drivers is related to obtaining a set of empirical data that is used to build and test the Bayesian network. Therefore, a survey was conducted. The author's questionnaire, was based on the Japanese questionnaire of subjective fatigue symptoms, developed by the Research Committee on Industrial Fatigue of Japan Society for Occupational Health [5]. The answers have been derived from subjective feelings of driver fatigue regarding factors affecting fatigue.

The questionnaire included twenty questions, such as participation in traffic collisions and accidents, fatigue symptoms, the length and complexity of the route being taken, time pressure, shift work, exposure to external factors (noise, temperature). The survey was supplemented by expert opinion.

The first step in designing a formal knowledge representation system is to build a conceptual model of the subject area. This is related to the ordering and structuring of acquired knowledge. Non-formal conceptualization is the representation of knowledge in a natural language, containing many ambiguities, best interpreted by the expert. Methodology of designing a formal knowledge representation system assumes ease of transition from the description of the problem in natural language to representation in formal and executable language [6]. The second step is to identify ways to move from the informal conceptualization of the problem of driver's fatigue assessment to the computer model. This approach involves choosing an appropriate system that will be sufficiently expressive to precisely describe the subject matter as fatigue. An additional fact to be considered is the uncertainty with regard to the ratings received. The sources of this uncertainty are both incomplete knowledge about the object being evaluated and the lack of evaluation criteria.

* Corresponding author: piotr.maksym@up.lublin.pl

 https://doi.org/10.1051/bioconf/20181002018

As a measure of uncertainty adopted probability. Attributes of the assessed object and the evaluation of driver fatigue are treated as variables for which values are known to an accuracy of the probability distribution. One of the systems of knowledge representation that fulfills the assumption is a Bayesian-based system.

Bayesian network technology allows expression of cause - effect relationships that exist permanently or temporally, described as cumulative probability distribution. The computational scheme is based on the Bayesian formula and the conditional independence statement, which allows for the cumulative probability distribution as a factorization of conditional probability distributions [7].

The conditional probability of event A provided that B has occurred, denoted $P(A \mid B)$, we call the quotient of the probability of the product $A \cap B$ and probability of event B:

$$P(A \mid B) = \frac{P(A \cap B)}{P(B)}, P(B) \neq 0 \qquad (1)$$

The Bayesian formula is a conditional probability calculation scheme:

$$P(B \mid A) = \frac{P(A \mid B)P(B)}{P(A)} \qquad (2)$$

The Bayesian network is an acyclic directed graph, visualized using an alphabet containing two symbols:

- node

- directed arc

The nodes represent random variables (discrete or pseudo continuous) and arcs connecting the nodes - the relationship between these variables (Fig. 1.).

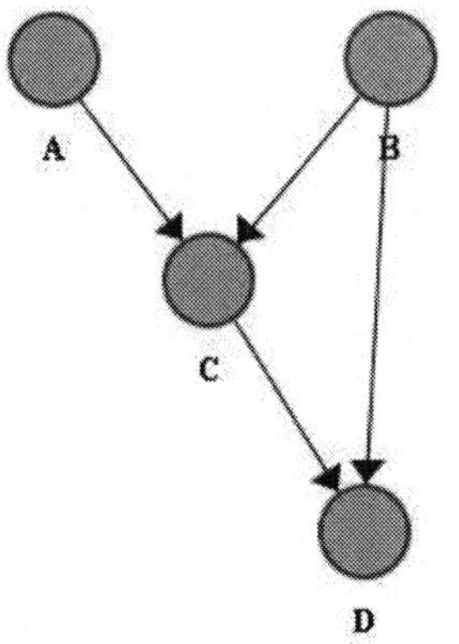

Fig. 1. Example of bayesian network build in BayesiaLab (own work).

Each node is associated with a conditional probability table. Each variable has a strictly defined type, and value domain. Based on such data, a learning process can be performed to determine probability distributions for all network nodes [8-10].

The use of the Bayesian network consists in launching an automatic inference mechanism to determine the probability distribution of variables [11], that applies to:

- prediction – knownig the reason we ask about the consequences,
- explanation – knownig the consequences we ask about the reason.

3 Results

The survey was conducted among fifty professional drivers. Among the respondents were drivers aged between 20 and 46, with a secondary education of 70%. 92% of the respondents answered "Yes" to the question of whether they followed the correct work day and designated breaks. The most common answer to the question about symptoms of fatigue was: back pain, drowsiness, desire to lie down, and other. As part of the "other" answer, respondents were able to list their subjective feelings, for example: variability and a high amount of information, constant attention (Fig. 2.).

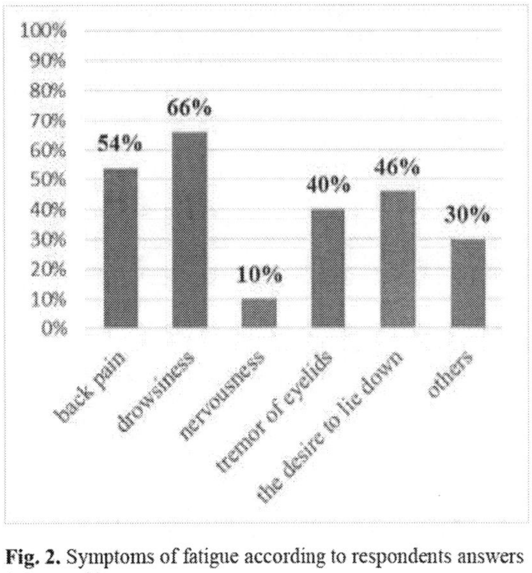

Fig. 2. Symptoms of fatigue according to respondents answers (own work).

According to experts, these statements from drivers, have suggested an increase in mental load, which has been one of the attributes of the Bayesian networks topology. The attribute "fatigue" of drivers is evaluated on the basis of "workload", "stress", "organization of work", "work experience" as a professional driver and "sex" of the driver (Table 1, Fig. 3.).

The obtained results will be subject to an sensitivity analysis, that can be described as a method which define how "sensitive" mathematic model really is when the parameters and the overall structure of the model experience changes.

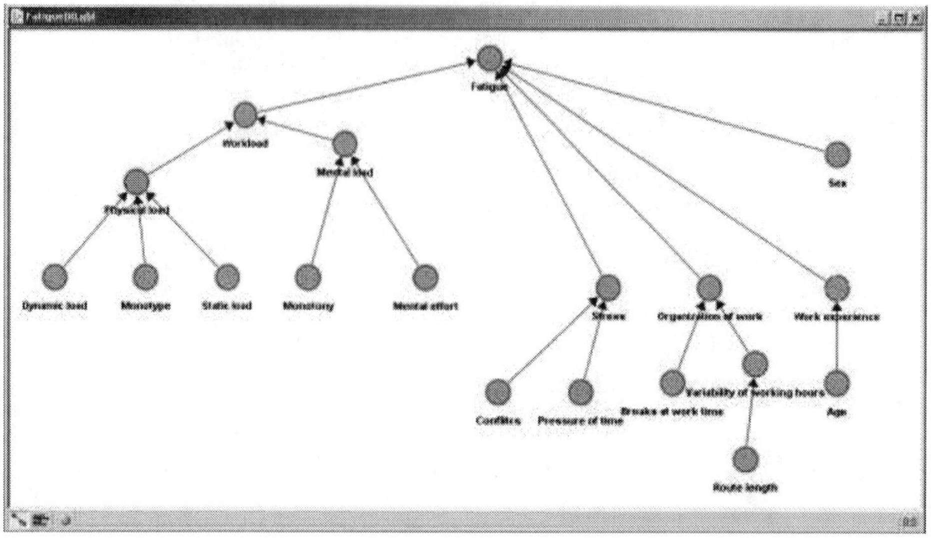

Fig. 3. The network structure of attributes assessments „fatigue" (own work).

Sensitivity analysis allows researcher to find the accuracy needed for an accurate and functional model. If the analysis shows signs of insensitivity of parameter change, then it is probable rough estimation might serve a better information than an actual value. The analysis could points which parameter values are appropriate to use in the model [12, 13].

Table 1. Set of attributes values – "fatigue".

Attributes	Set of values
Workload	Very little Little Average Large Very Large
Stress	Little Average Large
Organization of work	Correct Invalid
Work experience (in years)	0 – 1 1 – 5 5 – 10 Above 10
Sex	Woman Man

4 Conclusions

Taking care of the traffic safety is associated with the awareness of drivers on the phenomenon of fatigue and its effects. At present there is no uniform criteria for fatigue assessment of professional drivers, so an attempt was made to build a fatigue assessment model using Bayesian technology.

The next step of development of the model will be measure of its effectiveness by the sensitive analysis method.

Bayesian networks allow for rapid conversion of expert knowledge and empirical data, expressed verbally (written in natural language), into computer models. The quality of models depends on the conceptualization of the modeled domain. According to the assumptions Bayesian networks should help to quickly assess the degree of fatigue of drivers and explain the reasons for the fatigue.

References

1. A. Łuczak, K. Zużewicz, Bezpieczeństwo Pracy – Nauka i Praktyka. **4**, (2016)
2. H. Häkkänen, H. Summala, Accident Anal. Prev. **33(2)**, 187-196, (2001)
3. A. M. Lasota, K. Hankiewicz, Logistyka **4**, 4422-4428, (2015)
4. P. Thiffault, J Bergeron, Accident Anal. Prev. **35(3)**, 381-391, (2003)
5. H. Yoshitake, Ergonomics **21**, 231-233, (1978)
6. A. Kusz, A. W. Marciniak, Agr. Eng. **12(87)**, 285-294, (2006)
7. F. V. Jensen, T. D. Nielsen, Springer. (2007)
8. Documentation of the program BayesiaLab. http://www.bayesia.com/book
9. J. Y. Halpern, MIT Press, (2005)
10. J. Pearl, Morgan Kaufmann Publishers Inc., (1988)
11. R. E. Neapolitan, Prentice Hall, (2003)
12. L. Breieriova, M. Choudhari, MIT System Dynamics in Education Project, (1996).
13. N.P. Ventikos, G.V. Lykos,, N.A. Vagias Atlantis Computational Intelligence Systems, **6**, 651-676, (2012)

Spectroscopic studies of the quality of WCO (Waste Cooking Oil) fatty acid methyl esters

Arkadiusz Matwijczuk[1,*], *Grzegorz* Zając[2,**], *Dariusz* Karcz[3], *Edyta* Chruściel[1], *Alicja* Matwijczuk[1], *Magdalena* Kachel-Jakubowska[4], *Bogusława* Łapczyńska-Kordon[6], *and Mariusz* Gagoś[4]

[1]University of Life Sciences in Lublin, Department of Physics, Poland
[2]University of Life Sciences in Lublin, Department of Power Engineering and Transportation, Poland
[3]Krakow Technical University, Department of Analytical Chemistry (C1), Poland
[4]University of Life Sciences in Lublin, Department of Machinery Exploitation and Management of Production Processes, Poland
[5]Maria Curie-Skłodowska University, Department of Cell Biology, Poland
[6]Department of Mechanical Engineering and Agrophysics, Faculty of Production and Power Engineering, University of Agriculture in Kraków

Abstract. Different kinds of biodiesel fuels become more and more attractive form of fuel due to their unique characteristics such as: biodegradability, replenishability, and what is more a very low level of toxicity in terms of using them as a fuel. The test on the quality of diesel fuel is becoming a very important issue mainly due to the fact that its high quality may play an important role in the process of commercialization and admitting it on the market. The most popular techniques among the well-known are: molecular spectroscopy and molecular chromatography (especially the spectroscopy of the electron absorption and primarily the infrared spectroscopy (FTIR)).The issue presents a part of the results obtained with the use of spectroscopy of the electron absorption and in majority infrared spectroscopy FTIR selected for testing samples of the acid fats WCO (Waste Cooking Oil) types. The samples were obtained using laboratory methods from sunflower oil and additionally from waste animal fats delivered from slaughterhouses. Acid methyl esters were selected as references to present the samples. In order to facilitate the spectroscopic analysis, free glycerol, methanol, esters and methyl linolenic acid were measured.

1 Introduction

The massive expansion of internal combustion engines is associated from the outset with the use of primarily crude oil as an energy source. In almost one hundred years old history of generally understood automotive industry the only fuel hydrocarbon fractions were extracted from crude oil, which still remains the most important energy source. In this respect, it also has a very large impact on the global

* Corresponding author: arkadiusz.matwijczuk@up.lublin.pl, arekmatwijczuk@gmail.com
** Corresponding author: grzegorz.zajac@up.lublin.pl

political and economic situation. However, the modern world realized that global deposits of oil are largely reduced and the the ever-increasing consumption causes a serious ecological threat such as a greenhouse effect caused by CO_2 gas. This fact has become a powerful driving force for many research programs around the world to search for alternative fuels which would partially or completely replace the conventional fossil fuels and cast little threat to the environment[1-4]. In this context, the fuel constituents derived from renewable sources such as bio fuels seem to play the most important role. Among the various types of bio fuels the biodiesel (most often defined as the fatty acid methyl esters (FAME)) is most attractive. It should be noted that the application of FAME either as fuel for compression engines or ignition or as a gas component considerably changes the properties which have been known so far. Significant differences in chemical structures of hydrocarbons present in diesel and FAME differentiate the behavior of these products and affect their handling, storage conditions as well as the engine compartment. Currently, practically every vegetable or animal fats can be used as a raw material for the production of various types of biodiesel. The most commonly manufactured raw materials are edible vegetable oils and rapeseed oils, soy, sunflower or even palm oil.

Those raw materials, referred to as the first generation, do not resolve major issues such as reduction of CO_2 emission and may even increase problems on the food market [5-6]. Focusing attention on acquisition of raw materials seems to be the solution to this situation. II generation i. e. edible plant oils such as algae and microalgae [7], Jatropha [8] or used vegetable oil can be obtained from households or restaurants.

There are two main factors which prove that waste oil (WCO) is nowadays one of the most promising raw materials for the production of biodiesel [9-12], namely the wide availability of the raw material and the very low production cost. Due to the increasing consumption, the amount of waste oils is growing very quickly. For example, only in the United States about 10 million tons of the WCO is generated annually. This in turn requires environmentally friendly methods for WCO disposal and in this context the use of WCO as a raw material for biodiesel production is considered economically advantageous [13]. It has been estimated that the low cost of biodiesel production from WCO may help in overcoming the economical barriers. Currently the biodiesel production cost is ~1.5 times higher compared to that of conventional diesel, and only this fact speaks much for the opportunity to introduce biodiesel on a larger scale. It is estimated that more than 70% of the cost of biodiesel production is cost-per-acquisition of raw material [14-15]. It should be pointed out that the price of the WCO is 2-3 times lower compared to the price of crude vegetable oils, which can substantially lower the production costs of the same fuel in the form of biodiesel. Also, a significant reduction in the costs of waste disposal is an additional worth-mentioning advantage [16]. Due to the structural transformations taking place in frying oil over time [16], the chemical and physical properties of WCO oils differ from those of fresh oil. As a result of the frying temperature (approx. 190 °C) as well as the presence of O_2 and H_2O from fried products, similar physical and chemical changes occur both in the food, and in oil medium used for frying. These changes can result in the formation of various compounds that often have very complex and unidentified structures, such as free fatty acids (FFA) [17-19]. The presence of solids together with significant quantities of H_2O in the test products (those may come with fried products and are introduced during the washing) may also be problematic. Hence, the fact that the profile of fatty acids composition and the free fatty acids content in raw materials affect physicochemical properties of derived bio fuels is undisputable.

The use of WCO for biodiesel production requires an appropriate technology with advanced pretreatment processes and selection of relevant transesterification reaction parameters. Various contaminants such as water, organic compounds and substances which cannot undergo saponification lower the transesterification yield and overall quality of biodiesel product. The quality of products obtained must be controlled by suitable analytical techniques, one of which is infrared FTIR spectroscopy [20-24].

The main purpose of current work is the application of infrared spectroscopy and FTIR-spectroscopic properties of WCO oils for assessing the suitability of the WCO for the production of biodiesel.

2 Materials and Methods

2.1 Materials

All fatty acid methyl esters tested were obtained from fresh oils using laboratory methods. All frying oils (WCO), waste animal fats and bio-commercial oils used for the experiments meet the requirements of EN 14214:2006 standard.

The following oil materials were used:

1. Refined sunflower oil underwent five cycles of frying potato French fries. Single cycle consisted of heating the oil up to the temperature of 180°C followed by an insertion of vegetable material.

2. Refined sunflower oil underwent five cycles of frying fish fingers. The production cycle was carried out in the same way as in 1.

3. Waste animal fats, were obtained from a slaughterhouse. They were melted down by their slow heating up to the temperature of 60°C (under greatly reduced pressure in order to prevent its degradation). Remains of poured, drained fat were centrifuged and decanted. The raw materials obtained were subjected to esterification in a lab reactor (V = 1000 ml) and adjustable temperature stabilization. The reaction time and mixing speed were based on the available literature [25-27]. In order to begin the trans-esterification reaction the experiments were carried out in temp. of 55 °C. After reaching the desired temperature all tested oil samples were treated by a catalytic mixture consisting of methanol and NaOH. The reaction mixtures were stirred for 2 h using power agitator.

In order to completely separate the Glycerin and esters obtained, the mixture was subjected to 8 h long sedimentation process. The esters were then washed twice with hot water (at 80°C) and dried to remove the remaining alcohol and catalyst used.

Then, each of the selected studies of esters was tested with regard to the peroxide number (LN) [PN-ISO 3960], number of acid (LK) [PN-ISO 660] and composition of higher fatty acids that are marked by gas chromatography. The gas chromatography with nitrogen has been used as the carrier gas, packed column (2.5 m with stationary phase PEGA, which consisted of polyethylene glycol adipate embedded GAS medium-chromium-Q) and a flame ionization detector.

2.2 Methods

Spectroscopic methods. Electronic absorption spectra were recorded at 23°C on double-beam UV-Vis spectrophotometer Cary 300 Bio (Varian, USA) equipped with a thermostated cuvette holder with a 6×6 multicell Peltier block. The temperature was controlled with a thermocouple probe (Cary Series II, Varian, USA) placed directly in the quartz cuvette. The spectra were recorded from 200 to 600 nm.

The measurements of ATR-FTIR background-corrected spectra were carried out in solvents using a HATR Ge trough (45° cut, yielding 10 internal reflections) crystal plate for liquids and were recorded with a 670-IR spectrometer (Varian, USA). Typically, 25 scans were collected, Fourier-transformed, and averaged for each measurement. The IR absorption spectra at a resolution of one data point per 1 cm-1 were obtained in the region between 4000 and 400 cm-1. The instrument was purged with argon for 40 min before and then during the measurements. The Ge crystal was cleaned with ultra-pure organic solvents (Sigma-Aldrich Co.). All experiments were carried out at 20 °C.

3 Results and discussion

Infrared spectrometry (FTIR) is a commonly used tool for analysis of the carbonyl group-containing compounds and for determination of the petroleum products. Its use for determination of the FAME content is based on the fact that the carbonyl C=O vibration, which usually occurs at approximately 1700-1800 cm^{-1} is practically negligible in hydrocarbon fraction, while in case of FAME it is highly intense. It is well-known that esters give highly specific and strong absorption bands caused by vibrations of their C=O and C-O valence bonds.

The absorption band of the carbonyl group C=O in saturated aliphatic esters usually appears at 1735-1750 cm^{-1}. The aryl and α, β-Unsaturated structure may cause slight shift of this band towards a shorter wavenumber values (i. e. 1715-1730 cm^{-1})[28]. The C-O stretching vibration emerges within a much broader range (1100-1300 cm^{-1}), and is also observed in the spectra of other organic compounds such as carboxylic acids, ethers, or alcohols. Methyl esters belonging to the long-chain fatty acids often give rise to absorption maxima at approximately 1250, 1205, and 1175 cm^{-1}[29]. In case of FAME the most characteristic band has its maximum at approximately 1175 cm^{-1}[30].

Other characteristic of a hydrocarbon chain bands which occur in EMKT and ON are:

-methyl ($-CH_3$) and methylene ($-CH_2-$) groups C-H stretching vibrations (both symmetrical and non-symmetrical), present at 2870 and 2965 cm^{-1} (former), and 2925, and 2850 cm^{-1} (latter)[31].

- bending vibrations of the C-H bond: in methyl group present at 1375, 1450 cm^{-1} (and 1438 cm^{-1} in compounds containing carbonyl group); in methylene group at 1465 (scissoring vibration), and 721 cm^{-1} (rocking vibrations)[32].

Interestingly, in the spectra of hydrocarbon fraction of diesel within the range of 1700-1800 cm^{-1} another characteristic band is present, while it is absent in the FAME spectrum. This suggests the presence of various oxidation products (aldehydes, ketones, carboxylic acids) or various oxidation inhibitors, for example -mediated inflammatory diseases. In addition, poor absorption of some in the vibrations of the maximum of approx. 1750 cm^{-1} can be attributed to the presence of small amounts (within 1%) (that is, polymers with carbonyl groups)[33].

Although the formation of such carbonyl compounds is unlikely in high-quality fuels, their presence may cause difficulties in the determination of FAME content. It should be noted, that despite those difficulties the band between 1750-1740 cm^{-1} is considered as optimal for the IR-based identification and quantitative analysis of FAME in medium oil fractions and lubricating oils[34].

Figure 1 shows the selected electron absorption spectra of the methyl esters from sunflower oil after frying French fries (black line) and fish (dashed line) and the spectrum of the selected commercial esters, dissolved in methanol (grey line).

The absorption maxima at approximately 270 nm, for all samples of oils selected for tests (279 nm for the methyl esters of rapeseed oil after frying French fries or fish and 281 nm for commercial esters) clearly indicate the origin of the band as the $n \rightarrow \pi^*$ electronic transition within the carbonyl group (C=O)[35-38].

In case of commercial esters a very broad absorption set of bands is easily noticed (360 - 510 nm), which is not observed in the spectra of sunflower oil esters (after frying French fries and fish).

This region of the spectrum may be characteristic either of various fuel additives e.g. biocides (synthetic or natural) or pollutants. Studies carried out on other esters (not shown) revealed spectral features similar to those presented in Fig. 1.

The FTIR spectra of the investigated esters are given in Fig. 2. The samples were dissolved in appropriate organic solvent and dispersed at the surface of ZnSe crystal and then the solvent evaporated in an atmosphere of inert gas (N_2).

The samples of sunflower oils after frying both French fries and fish have the characteristic maxima at about 1744 cm^{-1} (1743 cm^{-1} for the methyl esters of rapeseed oil) originating from carbonyl C=O for stretch.

The spectrum of methyl esters from oil after frying French fries compared to that of esters from oil after frying fish (Fig. 2, top) shows a remarkable difference in the band shapes within the range of 1770-1670 cm^{-1}. In particular, a clear decrease in the intensity of band at 1744 cm^{-1} is observed, most likely as a result of the formation of strong hydrogen bonding.

The band with the maximum at 1715 cm^{-1} neither is observed in the spectrum of commercial B100 nor in the spectra of the methyl esters of sunflower oil, clearly indicates a lack of ability to create this type of hydrogen bridges.

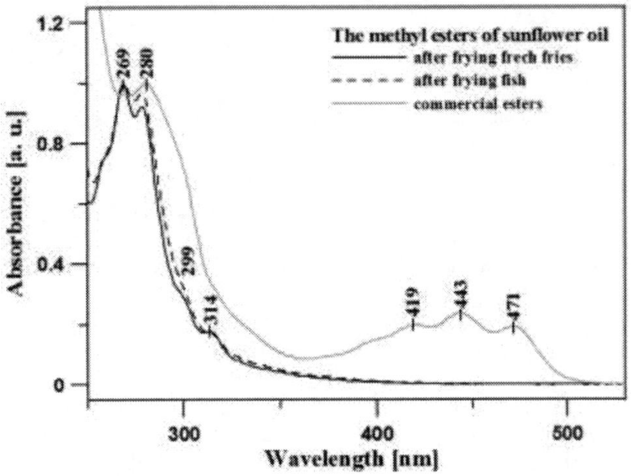

Fig. 1. Electronic absorption spectra of the methyl esters from sunflower oil after frying French fries (black line) and after frying fish (dashed black line) and esters commercial-B100 (grey line), dissolved in methanol (Mt-OH).

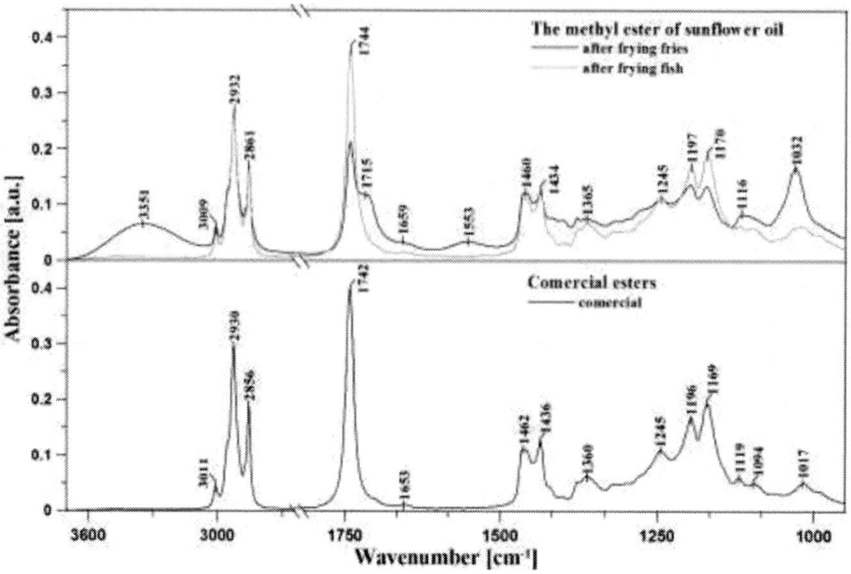

Fig. 2. ATR-FTIR absorption spectra of esters from sunflower oil after frying French fries (solid black line) and after frying fish (dashed black line) (top) and of commercial esters-B100, (bottom). All samples were dissolved in Mt-OH.

With the decrease in intensity of the band at 1744 cm^{-1} and the respective intensity increase of the band at 1715 cm^{-1} a simultaneous increase in intensity of the band at 1032 cm^{-1} associated with the C-O stretching vibrations is observed. This band is also clearly visible in the spectra of methyl esters from oil after frying French fries and fish, while in the spectrum of commercial esters it is least intensive, which is most likely associated with the greatest intensity of carbonyl C=O vibration in this spectrum. Bands in the area of 1100-1300 cm^{-1} also originate from C-O stretching, but only minimal differences are observed in this region between all spectra[34]. These bands together with a decreasing trend of hydrogen bonding between groups of C=O H-O-H show small but noticeable increase in their intensity.

A similar trend is also observed in the spectrum of commercial esters. Interestingly, the area between 3680-3070 cm^{-1}, with a broad band showing the maximum at approximately 3351 cm^{-1} results from the presence of a high content of methanol and a significant amount of H_2O in the tested samples of esters[39]. This effect may also depend on whether the esters were used in frying French fries or fish. The studies have also revealed a glycerol content in the samples tested. the sample of commercial esters contains the lowest amount of glycerol. Analogical observation was made for methanol and methyl ester of linolenic acid content, which presence in the test esters is very well correlated with changes in the intensity of the bands at 3351 cm^{-1}[40-41]. An increased concentration of these compounds results in a significant decrease in intensities of these bands

Table 1. Positioning of FTIR absorption maxima with assignment of appropriate vibrations in bio fuels tested (for the methyl esters of sunflower oil after frying French fries, fish and for commercial esters-B100 recorded within the range of 3700-900 cm^{-1}.)

FTIR Position of bands [cm^{-1}] **Esters of sunflower oil**		**B100**	**Type and origin of vibrations**
3351w	3351st	–	ν (O-H) in H_2O
3009	3009	3011	ν(C-H)
2932	2932	2920	$CH_{2,\ s+as}$, $CH_{3,\ s+as}$ *aliphatic group of triglycerides*
2861	2861	2856	
1744st	1744w	1742s	C=O *group of Ester*
1715	–	–	
1659	1659	1653	ν(C=C)
1553	1553	–	$\nu(C=C)_s$
1460	1460	1462	-C-H (in CH_2) *bending*
1434	1434	1436	=C-H (cis-) *bending*
1417	1417	1418	
1404m	1404m	1403m	ν (=C-H (*cis-*))
1376	1376	1377	-C-H $_s$ (in CH_3)
1365	1365	1360	-C-H $_s$ (CH_3) *bending*
1317	1317	1319	δ(CH)
1277st	1277st	1276m	δ(CH_2)
1245	1245	1245	-C-O *stretching*
1197	1197	1169	or -O-CH_2-C
1170	1170	1194	-C-O
1116	1116	1119	ν(C-C)
1132	1132	1117	
914m	914m	910m	-OH
860	860	859	$\nu(C\text{-}O\text{-}C)_s$
839st	839st	840w	$\nu(C\text{-}O\text{-}C)_s$ or ν(C-C)
721	721	722	-CH_2- *rocking*

ν – stretching vibrations, δ – bending vibrations, s – symmetrical, as – asymmetrical, st – strong, * denotes frequencies that may originate from the solvent used

4 Conclusions

This study presents the application of infrared spectroscopy and FTIR as a tool for analysis of the quality of the selected fatty acid-methyl esters as potential biodiesel components. The experiments primarily focused on the methyl esters obtained from waste oils WCO. The results obtained were compared with the spectra of commercial esters (B100) and comply to commercial standards EN 14214.

When using both spectroscopies: UV spectroscopy as well as FTIR infrared spectroscopy, key areas have been examined. The essence of the examination was to determine suitability of the information of the spectral test on the WCO ester type in order to apply them as a bio fuel.

It has been proved that the analysis of key areas in the UV-Vis and FTIR spectra of the selected waste oils may serve as an inexpensive source of information on the composition of these oils and their suitability as raw materials for biodiesel production.

References

1. [1] Group, U. N. E. P. B. W. Management, UNEP/Earthprint, 2009.
2. [2] X. Wang, M.K. Lim, Y. Ouyang Energ. Econ. (2017)
3. [3] H.V. Lee, J.C. Juan, T.Y. Yun Hin, H.C. Ong Energies **9**, 611 (2016)
4. [4] Z. Bakhtiyari, M. Yazdanpanah, M. Forouzani, N. Kazemi RENEW SUST ENERG REV **69**, 341-349 (2017)
5. [5] A. Guldhe, B. Singh, N. Renuka, P. Singh, R. Misra, F. Bux Springer, 2017, pp. 47-62.
6. [6] D.S. Khang, R.R. Tan, O.M. Uy, M.A.B. Promentilla, P.D. Tuan, N. Abe, L.F. Razon RESOUR CONSERV RECY **119**, 12-23 (2017)
7. [7] A. Klok, P. Lamers, D. Martens, R. Draaisma, R. Wijffels Trends Biotechnol. **32**, 521-528 (2014)
8. [8] H. Ong, A. Silitonga, H. Masjuki, T. Mahlia, W. Chong, M. Boosroh Energy Convers. Manage. **73**, 245-255 (2013)
9. [9] A.N. Phan, T.M. Phan Fuel **87**, 3490-3496 (2008)
10. [10] R. Shan, C.Lv.P. Zhao, H. Yuan, J. Yao Energy Convers. Manage. **127**, 273-283 (2016)
11. [11] Y.M. Isa, E.T. Ganda RENEW SUST ENERG REV **81**, 69-75 (2018)
12. [12] T. Maneerung, S. Kawi, Y. Dai, C.H. Wang Energy Convers. Manage. **123**, 487-497 (2016)
13. [13] S. Jose Springer, 2017, pp. 281-313.
14. [14] J. Sakulsuraekkapong, R. Pairintra, S. Thepa ENERG SOURCE PART B 1-6 (2017)
15. [15] H.H. Mardhiah, H.C. Ong, H. Masjuki, S. Lim, H. Lee RENEW SUST ENERG REV **67**, 1225-1236 (2017)
16. [16] W. Sawaengsak, T. Silalertruksa, A. Bangviwat, S.H. Gheewala ENERGY SUSTAIN DEV **18**, 67-74 (2014)
17. [17] A. Sankoda, N. Harada, K. Iwasaki, S. Yamane, Y. Murata, K. Shibue, Y. Thewjitcharoen, K. Suzuki, T. Harada, Y. Kanemaru Endocrinology **158**, 1172-1180 (2017)
18. [18] J. Sebastian, C. Muraleedharan, A. Santhiagu Cogent Engineering **3**, 1178370 (2016)
19. [19] C.H. Tan, A.A. Ariffin, H.M. Ghazali,C.P. Tan, A. Kuntom, A.C.Y. Choo J Food Sci Technol **54**, 1757-1764 (2017)
20. [20] B. Elzey, D. Pollard, S.O. Fakayode Food Control **68**, 303-309 (2016)
21. [21] N. Cebi, M.T. Yilmaz, O. Sagdic, H. Yuce, E. Yelboga Food Chem. **225**, 188-196 (2017)
22. [22] A. Hirri, M. Bassbasi, S. Platikanov, R. Tauler, A. Oussama Food Anal Methods **9**, 974-981 (2016)
23. [23] F. van de Voort, M. Tavassoli-Kafrani, J. Curtis Anal. Chim. Acta **918**, 1-7 (2016)
24. [24] A. Dominguez-Vidal, J. Pantoja-de la Rosa, L. Cuadros-Rodríguez, M.J. Ayora-Cañada Food Chem. **190**, 122-127 (2016)
25. [25] A. Ramadhas, S. Jayaraj, C. Muraleedharan Renew Energy **29**, 727-742 (2004)
26. [26] C. Sheinbaum, M.V. Balam, G. Robles, S. Lelo de Larrea, R. Mendoza Waste Manag. Res. **33**, 730-739 (2015)
27. [27] M. Szmigielski, W. Piekarski, D. Andrejko, B. Slaska-Grzywna, A.

Maslowski, G. Zajac, A. Sagan, T. Jaskiewicz, I. Rachanczyk Przem Chem **93**, 649-651 (2014)
28. [28] H. Omidvarborna, A. Kumar, D.S. Kim Sci. Total Environ. **544**, 450-459 (2016)
29. [29] N.L.P. Hernandez, Bonon, J.O. Bahú, M.I.R. Barbosa, M.R.W. Maciel, R. Maciel Filho J. Mol. Catal. A: Chem. **426**, 550-556 (2017)
30. [30] Z. Wu, H. Li, D. Tu Anal. Methods **8**, 2581-2587 (2015)
31. [31] H. Omidvarborna, A. Kumar, D.S. Kim Fuel Process. Technol. **140**, 113-118 (2015)
32. [32] S.E. Mahesh, A. Ramanathan, K.M.S. Begum, A. Narayanan Energy Convers. Manage. **91**, 442-450 (2015)
33. [33] S.N. Rabelo, V.P. Ferraz, L.S. Oliveira, A.S. Franca IJESD **6**, 964 (2015)
34. [34] C.L. Cunha, A.S. Luna, R.C. Oliveira, G.M. Xavier, M.L. Paredes, A.R. Torres Fuel **204**, 185-194 (2017)
35. [35] A. Matwijczuk, D. Karcz, R. Walkowiak, A. Niewiadomy, S. Wybraniec, G. Karwasz, M. Gagos Przem Chem **95**, 1894-1898 (2016)
36. [36] Matwijczuk, A., Karcz, D., Walkowiak, R., Furso, J., Gładyszewska, B. e., Wybraniec, S., Niewiadomy, A., Karwasz, G. P., Gagoś, M., J. Phys. Chem. A **121**, 1402-1411 (2017)
37. [37] A. Matwijczuk, A. Górecki, D. Kamiński, B. Myśliwa-Kurdziel, L. Fiedor, A. Niewiadomy, G.P. Karwasz, M. Gagoś, M., J Fluoresc **25**, 1867-1874 (2015)
38. [38] M. Gagoś, A. Matwijczuk, D. Kamiński, A. Niewiadomy, R. Kowalski, D.P. Karwasz J Fluoresc **21**, 1-10 (2011)
39. [39] A.S. Luna, I.C. Lima, C.A. Henriques, L.R. de Araujo, W.F. da Rocha, J.V. da Silva Anal. Methods (2017)
40. [40] J. Zhou, Y. Xiong, Y. Gong, X. Liu Fuel **202**, 23-28 (2017)
41. [41] I. Shancita, H. Masjuki, M. Kalam, S. Reham, S. Shahir Energy Fuels **30**, 4790-4805 (2016)

Sugar beets grown in the strip-tillage system at different soil cultivation depths

Natalia Mioduszewska[1,*], *Mariusz* Adamski[1], *Ewa* Osuch[1] and *Andrzej* Osuch[1]

[1] Poznań University of Life Sciences, Instytute of Biosystems Engineering, Wojska Polskiego 28, 60-637 Poznań, Poland

Abstract. Modifications in soil cultivation affect the quality and quantity of the yield of sugar beets. The crop density, yield and the external quality of sugar beet roots are highly dependent on the preparation of soil. The intensity and precision of fragmentation of the superficial soil layer and the adequate depth of plantation must enable normal development of sugar beets. The aim of the study was to determine the dependence between the depth of soil cultivation in the strip-tillage system used for sugar beet production and the state of the plantation after the emergence of crops as well as the yield quality and quantity. The research encompassed investigations of different depths of strip-tillage systems used for sugar beet production. The systems were evaluated according to the dynamics of crops' emergence, density, yield and external quality of sugar beet roots.

1 Introduction

Apart from the climate and soil conditions, the technique of agrotechnical procedures largely influences the sugar beet yield quality and quantity. It ensures the adequate depth of cultivation, regular sowing depth as well as adequate and timely treatments [1-3]. The agricultural technique is decisive to the yield volume and quality, including the external quality of roots. The soil cultivation technology affects the productivity of crops, because it is simultaneously decisive to the dynamics of plants' growth at an early phase of their development and it provides plants with favourable conditions for growth and development until they are harvested [4-6]. Well-balanced emergence affects crop density and it is decisive to the plantation density and regularity. These factors are essential for the high technological quality of yield and minimal loss during the harvesting of roots [7-8]. Individual agrotechnical procedures are supposed to guarantee yield stability and adequate technological properties of roots for sugar extraction [9].

The traditional cultivation system with soil reversion is the most common. However, in recent years, for economic and environmental reasons, the classic tillage system has been replaced by simplified systems, including the strip-tillage system [10]. Apart from the reduced energy outlay resulting in lower costs of production, simplified technologies (without ploughing) have significant influence on protection of the environment [11]. Lower intensity and frequency of treatments applied to soil protect it from degradation, aid natural biological processes and maintain its productivity [12-14].

The sugar beet production technology with the soil cultivation system, which includes selection of the cultivation depth, affects a wide range of yield characteristics, such as the root processing quality, the size and shape of the root system [7]. Therefore, the aim of this study was to determine the dependence between the depth of soil cultivation in the strip-tillage system used for sugar beet production and the state of the plantation after the emergence of crops as well as the yield quality and quantity.

2 Material and methods

A field experiment was conducted on the farm in Bełcz Mały, Commune of Wąsosz, Lower Silesian Voivodeship, Poland. Experimental plots were located on loamy sand soil, classified as IVa in the soil valuation system.

The experiment included five sugar beet growing technologies which differed in the soil cultivation depth:

1. strip tillage at a depth of 15 cm,
2. strip tillage at a depth of 20 cm,
3. strip tillage at a depth of 25 cm,
4. strip tillage at a depth of 30 cm,
5. strip tillage at a depth of 35 cm.

The soil was tilled and sugar beets were sown on 4 April. Other agrotechnical procedures, such as fertilisation and treatments, were identical in all the systems and they were conducted according to the sugar beet growing guidelines. The plot area was 600

* Corresponding author: natalia.mioduszewska@up.poznan.pl

x 21.6 m. Seeds of the Agronom sugar beet cultivar (Sesvanderhave) were sown in the plots. In theory, the sugar beet seeds were spaced at 18 cm from each other. A Köckerling Master combined cultivator (working width 5.4 m) connected to a 12-row Kverneland Monopill SE seed drill was used in the plots.

The state of the plantation after the emergence of crops was evaluated on the basis of the emergence dynamics, the distribution of plants in rows and the crop density. The emergence was recorded every second day after the appearance of the first sugar beet seedlings at the phase of cotyledons. In each plot all emergences of sugar beets in an area of 10 m^2 were counted in five replicates.

The distribution of plants in rows was evaluated as follows. In each plot 100 consecutive distances between plants in randomly selected rows were measured at an accuracy of 5 mm in five replicates. According to the methodology of the International Sugar Beet Institute in Brussels (IIRB - Institut International de Recherches Betteravières) the measurement results were classified into one of the three categories: double sowing, normal spacing and overspaced sowing [15]. The classification criteria were as follows:

• double sowing – plants spaced up to 0.5 XIST,
• normal spacing – plants spaced from ≤ 0.5 XIST to > 1.5 XIST,
• overspaced sowing – plants spaced more than 1.5 XIST,

where:
XIST – the actual spacing between plants as calculated in measurements of normally spaced plants.

The final density of plants was calculated by counting plants in randomly selected sites of an area of 10 m^2 in five replicates in each plot.

The state of the plantation before harvest was evaluated according to the yield of roots, the height of the aerial part of the roots and the morphological traits of the roots, i.e. their length, maximum diameter and weight. The height of the aerial part of the roots was measured from the leaf base to the surface of the soil. In order to measure the morphological traits the root tops were cut off manually at the upper line of the living shoot buds. The root length was measured from the site where its top was cut off to the end of the tail with a diameter of 10 mm. The diameter was measured at the widest part of the proper root. Before the roots were weighed, their tops had been cut off and cleaned. 100 roots were collected from five randomly selected rows in each plot and the height of their aerial part, their length, maximum diameter and weight were measured.

The yield of the roots from each plot was measured according to the weight of 500 roots and the final density of sugar beets.

The space between plants in a row, the root length and weight were analysed with the Plantator ver. 2 computer program, which calculated the percentage share of the roots in the following groups and intervals [16]:

• root length:
0-10 cm; >10-15 cm; >15-20 cm; >20 cm
• root diameter:
0-4,5 cm; >4,5-7 cm; >7-9 cm; >9-11cm; >11-13 cm; >13-15 cm; >15 cm
• root weight:
0-300 g; >300-600 g; >600-900 g; >900-1,200 g; >1,200-1,800 g; >1,800 g.

3 Results

Dynamics of emergence. The first emergences of sugar beets in the experimental plots began 18 days after the seeds had been sown. This means that there were average conditions in the plantation. It is assumed that under favourable conditions sugar beets start emerging 7-12 days after sowing the seeds, under average conditions – 14-18 days after sowing, whereas under unfavourable conditions – even as late as 30 days after sowing [2]. During the first measurement the smallest number of seedlings was noted in the system where the soil was cultivated at a depth of 25 cm. This technology was characterised by the most dynamic and unequal emergence of the plants. The emergence was the most balanced in the strip-till technology where the soil was cultivated at a depth of 15 cm. This system was characterised by the largest number of emerged plants at all the three terms of measurements. All the five cultivation systems were characterised by similar dynamics of emergence. The number of consecutive emerging plants varied comparably, without considerable deviations, except the strip-till technology at a depth of 25 cm, where the number of plants decreased at the fourth term. The dynamics of emergence is illustrated in Figure 1.

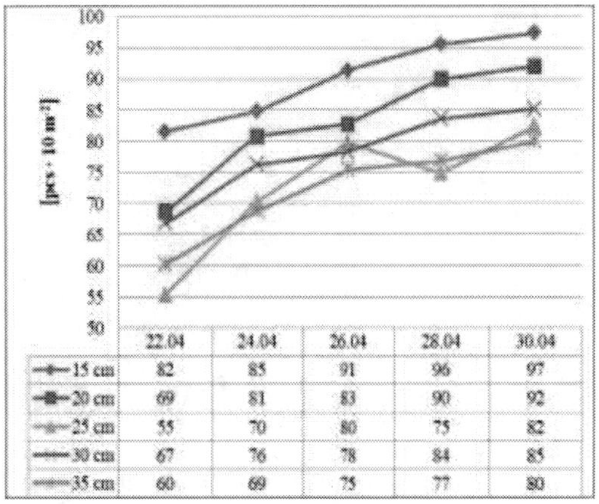

	22.04	24.04	26.04	28.04	30.04
15 cm	82	85	91	96	97
20 cm	69	81	83	90	92
25 cm	55	70	80	75	82
30 cm	67	76	78	84	85
35 cm	60	69	75	77	80

Fig. 1. The dynamics of emergence in the strip-till technology systems.

The field emergence capacity in the strip-till technology systems ranged from 69.4% to 80.2% (Table 1). The technology where the soil was cultivated at a depth of 30 cm was characterised by the lowest emergence capacity. The highest capacity was noted in the technology where the soil was cultivated at a depth of 15 cm.

Table 1. The average field emergence capacity in the strip-till technology systems [%].

Strip-tillage depth	15 cm	20 cm	25 cm	30 cm	35 cm
Field emergence capacity	80.2	72.6	70.3	69.4	70.2

The best field emergence capacity was observed in the technologies where the soil was cultivated at depths of 15 and 20 cm. The field emergence capacity was only satisfactory in the systems where the soil was cultivated at depths of 25, 30 and 35 cm.

Density of plants. The density of plants was diversified in all the strip-till technology systems (Fig. 2). The highest final density of plants, i.e. 99,400 plants per hectare, was noted in the system where the soil was cultivated at a depth of 15 cm. The lowest density, i.e. 86,000 plants per ha, was observed in the strip-till system where the soil was cultivated at a depth of 30 cm. The measurements also showed that in comparison with the standard cultivation depth of 25 cm, when the soil cultivation was 5 cm shallower, the density of plants increased by 2,800 pieces per ha. When the soil cultivation depth was reduced by 10 cm, the density of plants increased by as much as 12,200 pieces per ha. According to Gutmański [17], the optimal density of sugar beets on average soils ranges from 81,000 to 95,000 pcs· ha^{-1}. When the density increases, so does the yield of leaves and sugar, but the yield of roots is usually lower. At higher densities the roots are smaller, their size is more equal and they have better processing value. Therefore, the optimal density of plants was noted in the systems where the soil was cultivated at depths of 20, 25, 30 and 35 cm.

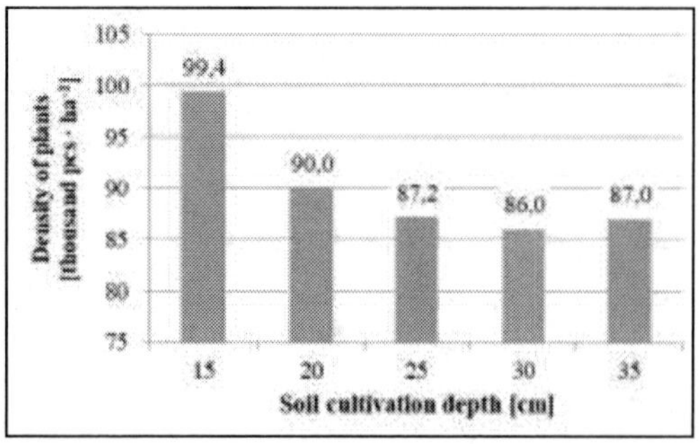

Fig. 2. The final density of sugar beets according to the strip-tillage system

Distribution of plants. Adequate distribution of plants in a row is very important not only for good density of the plantation but also for minimal loss of roots during harvest. From the point of view of the harvesting technique, irregular distribution of plants in a row, irregular heights of aerial parts of roots and considerably diversified heights of individual sugar beets in the field are unfavourable. The analysis of the results showed that the highest percentage of adequately distributed plants in a row resulted from the strip-till technology where the soil was cultivated at a standard depth of 25 cm (83.4%). A lower soil cultivation depth decreased the adequate distribution of seeds by about 2%, whereas a higher depth decreased the adequate distribution of plants by as much as 8.4% (Table 2).

Table 2. The percentage of adequately distributed plants in a row according to the strip-tillage system [%]

Strip-tillage depth	15 cm	20 cm	25 cm	30 cm	35 cm
Percentage of adequately distributed plants	81.4	81.6	83.4	75.2	75.0

The remaining percentage included chiefly single gaps and less frequently – double and triple gaps (Fig. 3). The most gaps were noted in the systems where the soil was cultivated at depths of 30 and 35 cm (17.4%). When the soil cultivation depth increased, so did the number of double and triple gaps. Their share amounted to 6% and 6.8% in the soil cultivated at depths of 30 and 35 cm, respectively. The share of single gaps in the systems where the soil was cultivated at depths of 15, 20 and 25 cm amounted to 14.8%, 15% and 14.6%, respectively. The share of double and triple gaps in these three technologies did not exceed 3.4%. The share of double sown seeds in all the systems was not greater than 0.8%.

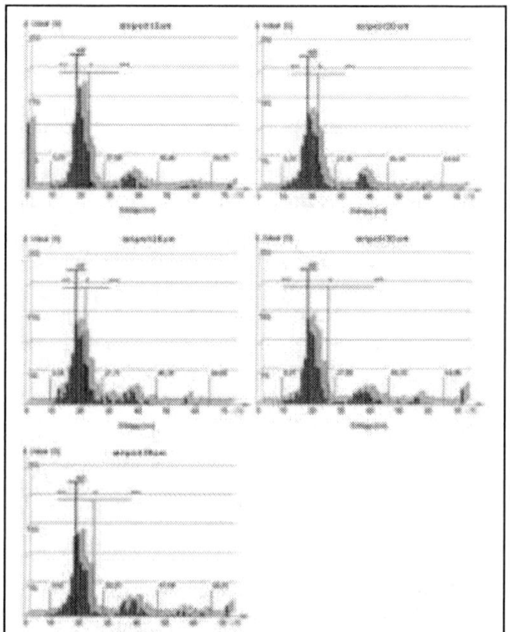

Fig. 3. The share of adequately sown plants, double sown plants and gaps according to the strip-tillage system (Share - %, Spacing - cm)

The strip-tillage systems did not differ significantly in the height of the aerial parts of the roots (Fig. 4). The average heights ranged from 4.81 cm to 5.98 cm.

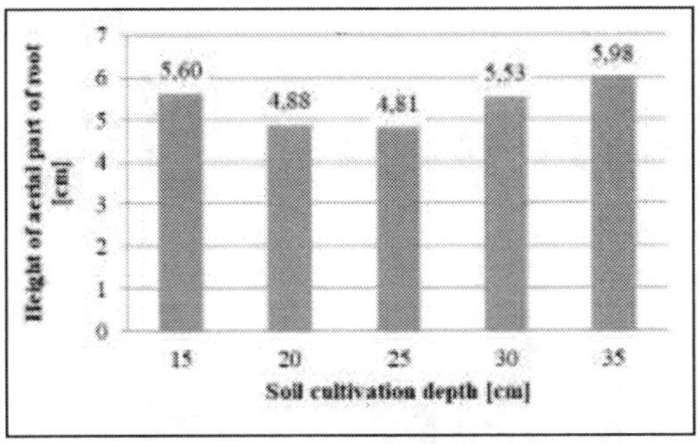

Fig. 4. The average height of the aerial part of the sugar beet roots according to the strip-tillage system

External quality of roots. The biometric traits of sugar beet roots are very important at harvest and they may affect the technological value of raw material and its suitability for processing into sugar. It is difficult to pull out heavy roots from soil. Too long roots may increase the loss during harvesting because their ends may be broken. The size of roots is also decisive to their external quality, because medium roots (500-700 g) are more suitable for processing into sugar [17]. It is noteworthy that roots with a diameter of 12-15 cm are the least resistant to damage, whereas those with a diameter of 7-9 cm are the most resistant [18-19].

The analysis of the results showed that when the soil cultivation depth was 5 and 10 cm greater than the standard depth (25 cm), the root length increased by 2.1 cm and 2.9 cm, respectively (Fig. 5).

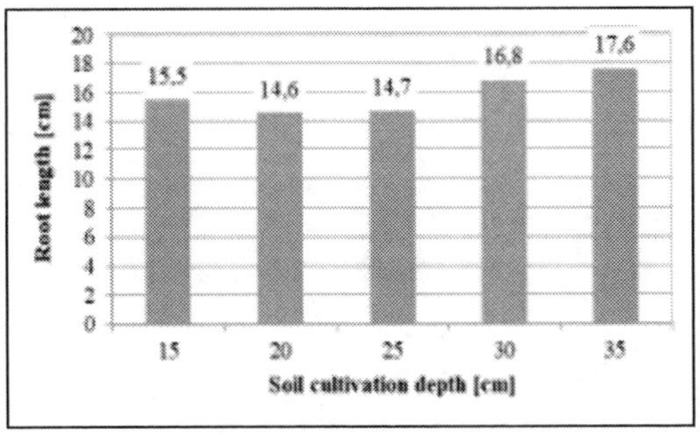

Fig. 5. The average sugar beet root length according to the strip-tillage system

The highest percentage of roots longer than 20 cm was noted in the systems where the soil was cultivated at depths of 30 and 35 cm (Fig. 6). These technologies were also characterised by the lowest percentage of the roots which were shorter than 10 cm. The highest percentage of medium-length roots, i.e. >10-15 cm, was noted in the systems where the soil was cultivated at depths of 20 and 25 cm (51.8% and 47.6%, respectively).

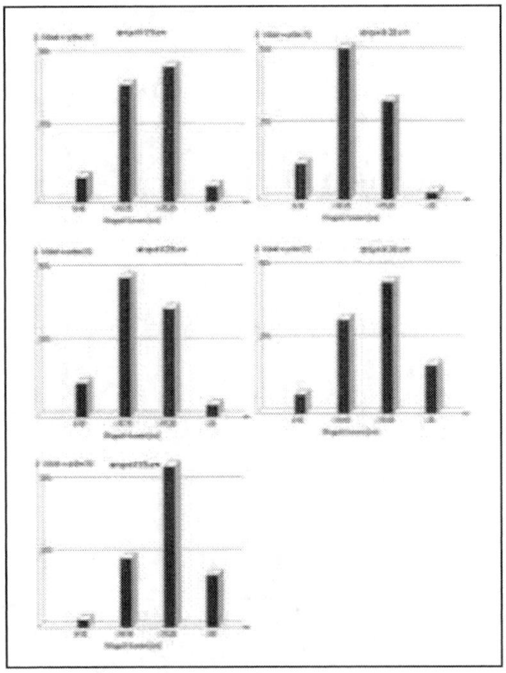

Fig. 6. The share of root lengths of 0-10 cm, >10-15 cm, >15-20 cm and >20 cm according to the strip-tillage system (Share in variant - %, Root length - cm)

The soil cultivation systems did not differ significantly in the diameter of sugar beet roots (Fig. 7). The smallest average maximum diameters of 11.3 cm and 11.5 cm were noted in the systems where the soil was cultivated at depths of 20 and 25 cm. The average maximum root diameter in the other strip-till systems ranged from 12.0 to 12.3 cm.

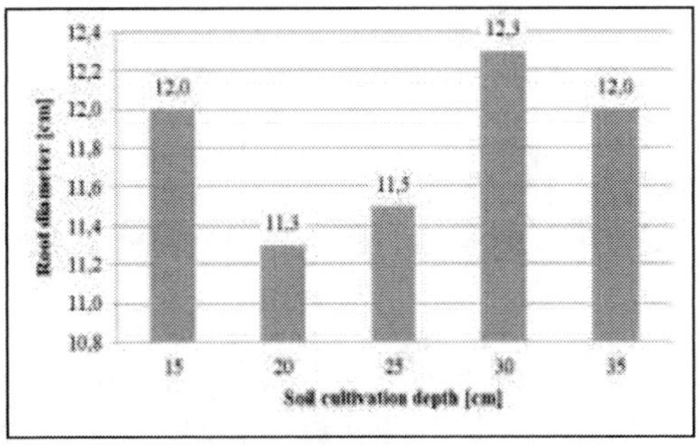

Fig. 7. The average maximum sugar beet root diameter according to the strip-tillage system

The most roots with the maximum diameters larger than 13 cm, which were the most vulnerable to damage [17], were noted in the strip-tillage technology where the soil was cultivated at a depth of 15 cm (Fig. 8).

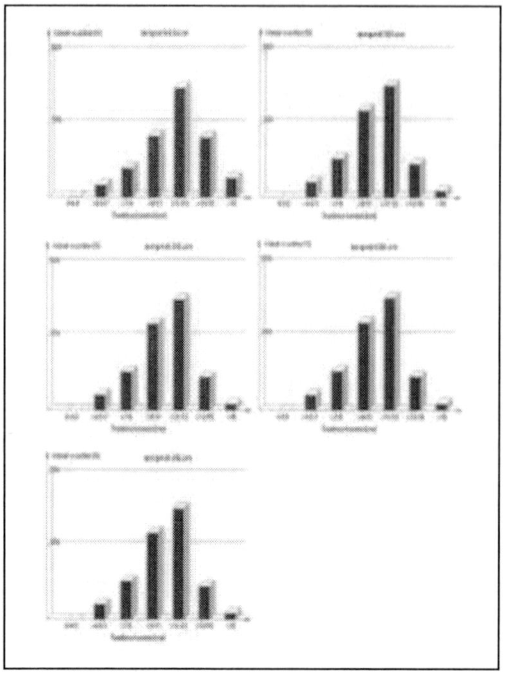

Fig. 8. The share of root diameters of 0-4.5 cm; >4.5-7 cm; >7-9 cm; >9-11cm; >11-13 cm; >13-15 cm and >15 cm diameter according to the strip-tillage system (Share in sample - %, Root diameter - cm)

The smallest average root weights of 873.0 and 904.8 g were noted in the systems where the soil was cultivated at depths of 20 and 25 cm, respectively. The average root weight in the other variants was larger than 1,000 g.

The analysis of the percentage share of roots in individual weight intervals showed that the strip-tillage systems where the soil was cultivated at depths of 20 and 25 cm resulted in the best conditions for the development of medium roots, weighing 600-900 g. The highest percentage of the heaviest roots was noted in the strip-tillage systems where the soil was cultivated at depths of 15, 30 and 35 cm (Fig. 10).

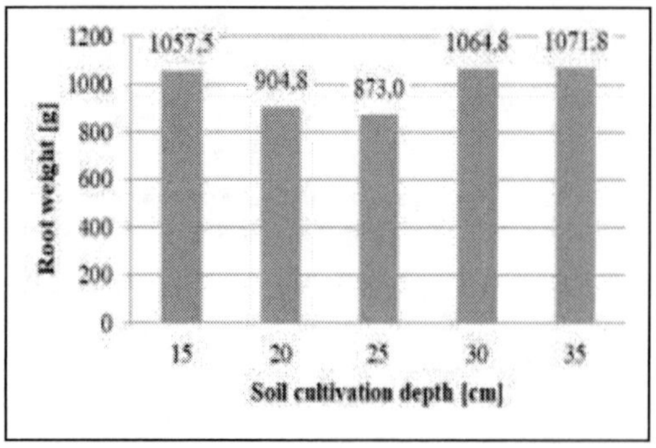

Fig. 9. The average sugar beet root weight according to the strip-tillage system

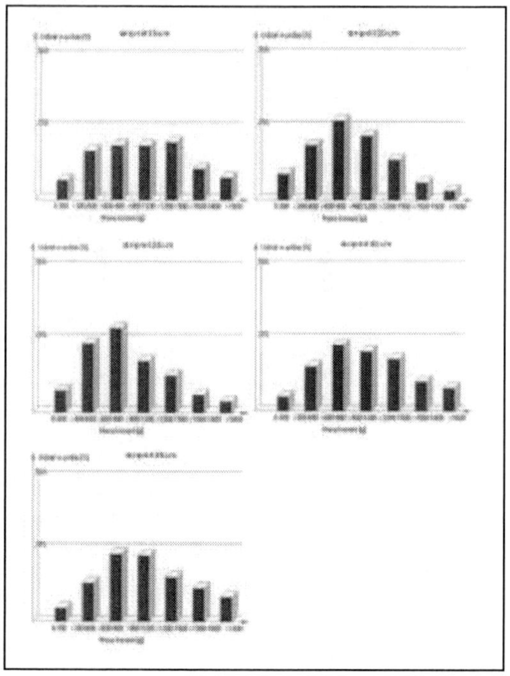

Fig. 10. The percentage of roots in weight intervals of 0-300g > 300-600g > 600-900g > 900-1,200g > 1,200-1,800g > 1,800g according to the strip-tillage system (Share in sample - %, Root weight - cm)

The analysis of the results showed that the yield of roots was the lowest in the system where the soil was cultivated at a standard depth of 25 cm. Both smaller and greater soil cultivation depths resulted in higher yields (Fig. 11).

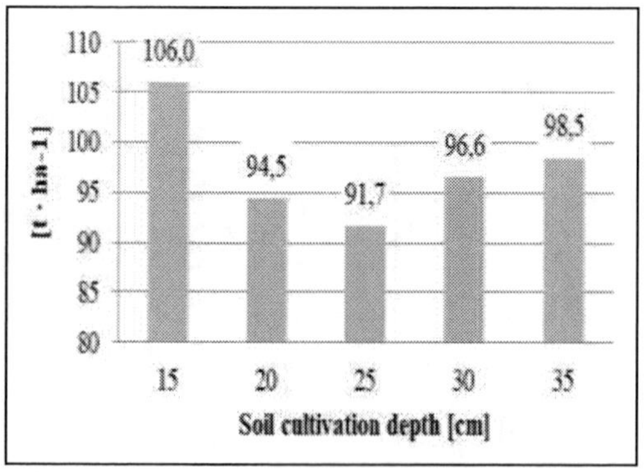

Fig. 11. The yield of roots according to the strip-tillage system

The analysis of the yield of roots showed that when the soil cultivation depth was 10 cm smaller than the standard depth (25 cm), the yield increased by as much as 14.3 $t \cdot ha^{-1}$.

The analysis of the results showed that when the soil cultivation depth was reduced from the standard 25 cm to 15 or 20 cm, neither the state of the plantation after the emergence of plants nor the external quality of the yield became worse. The results also showed that the yield of sugar beets did not decrease whereas the energy outlay could be limited, e.g. by reducing the consumption of fuel.

4 Conclusions

1. The strip-tillage system at a depth of 15 cm resulted in the highest density and the highest yield of sugar beets.
2. The soil cultivation depths of 30 and 35 cm were characterised by the heaviest weights and the greatest diameters and lengths of roots.
3. The reduction of the standard soil cultivation depth of 25 cm did not decrease the yield or the external quality of roots.
4. The greatest soil cultivation depths of 30 and 35 cm in the strip-tillage system caused a decrease in the number of adequately spaced plants and reduced their density.
5. The strip-tillage systems did not differ greatly in the parameters under analysis, but the strip-tillage system where the soil was cultivated at a standard depth of 25 cm tended to reduce the size of roots and decrease the yield.

References

1. J. Mikita, I. Gutmański, Biuletyn IHAR. **222.** 171-183 (2002)
2. N. Mioduszewska, J. Przybył, D. Wojcieszak, *Technologia uprawy buraków cukrowych a powschodowy stan plantacji. Aktualne problemy inżynierii biosystemów,* Wyd. UP Poznań. Monografia pod red. M. Lipińskiego i J. Przybyła. ISBN 978-83-7160-731-8. 33-41. (2014)
3. P. Okoń, J. Rudowicz-Nawrocka, P. Boniecki, K. Koszela, M. Zaborowicz, P. Idziaszek, K. Górna, N. Mioduszewska, *The application supports the process of identifying the diseases that occur on the leaves of sugar beet*, Proc. SPIE 10420, Ninth International Conference on Digital Image Processing (ICDIP 2017), 1042007 (2017/07/21); doi: 10.1117/12.2281724
4. F. Borówczak, *Przyrodnicze i agrotechniczne uwarunkowania plonowania buraków cukrowych. Nowoczesna uprawa buraków cukrowych.* Pr. zbior. pod red. W. Grzebisza. Wyd. AR Poznań. ISBN 83-7160-273-1, 20-28 (2002)
5. R. Gaj, A. Budka, A. Niewiadomska, J. Przybył, N. Mioduszewska, J. Elem., **20(3)**: 571-584. DOI: 10.5601/jelem.2015.20.1.772 (2015)
6. A. Niewiadomska, R. Gaj, J. Przybył, A. Budka, N. Mioduszewska, A. Wolna-Maruwka, Pol. J. Environ. Stud. vol. **25**, no. **5**. ISSN 1230-1485. s. 1803-1812 (2016)
7. B. Michalska-Klimczak, Z. Wyszyński. Zeszyty Problemowe postępów nauk rolniczych. Zeszyt **556**. 181-191. (2010)
8. G. Niedbała, N. Mioduszewska, W. Mueller, P. Boniecki, D. Wojcieszak, K. Koszela, S. Kujawa, R. J. Kozłowski, K. Przybył, *Use of computer image analysis methods to evaluate the quality topping sugar beets with using artificial neural networks*, Proc. SPIE 10033, Eighth International Conference on Digital Image Processing (ICDIP 2016), 100332M (2016/08/29); doi: 10.1117/12.2244875
9. M. Bzowska-Bakalarz, M. Banach, Inżynieria Rolnicza. Nr **3(63),** 95-102 (2005)
10. L. Zimny, A. Zych, R. Wacławowicz, Zeszyty Problemowe Postępów Nauk Rolniczych. Nr **581**. 135–145 (2015)
11. J. Gorzelany, G. Zaguła, M. Zardzewiały. Inżynieria Rolnicza. Nr **8(133)**. 143-151. (2011)
12. C. K. Cavalaris, T. A. Gemtos, CIGR Journal of Scientific Research and Development. Manuscript LW **01 008**. Nr IV. (2002)
13. L. Zimny, P. Kuc, R. Wacławowicz, A. Zych Zesz. Nauk. UP Wroc., Rol. **CVIII, 599**. 75–84 (2014)
14. D. Tarkalson, D. L. Bjorneberg, A. Moore, Journal of Sugar Beet Research. **49(3-4)**. 79-102 (2012)
15. J. P. Vandergeten, J. P. Van der Linden, P. Jarvis, E. Leveque, D. Guiraud de Willot, K. H. Kromer, *Test Procedures for Measuring the Quality in Sugar Beet Production – Seed Drillability, Precision Seeders, Harvesters, Cleaner Loaders*. IIRB. Bruksela (2004)
16. J. Przybył, P. Błażczak, A. Urban, Inżynieria Rolnicza. **2(35)**, 249-258 (2002)
17. I. Gutmański, *Produkcja buraka cukrowego*. Pr. zbior. Państwowe Wyd. Rolnicze i Leśne. Poznań (1991)
18. M. Bzowska-Bakalarz, *Właściwości mechaniczne korzeni buraków cukrowych* Rozprawa habilitacyjna. 166. AR Lublin (1994)
19. M. Rudy, R. Stanisławczyk, E. Głodek. Zeszyty problemowe postępów nauk rolniczych. Z. **558**. 231-238 (2011)

Selected properties of the potato snacks expanded in the microwave radiation

Marcin Mitrus[1,*], *Martyna* Golian[1], *Maciej* Combrzyński[1], and *Arkadiusz* Matysiak[2]

[1]University of Life Sciences in Lublin, Department of Food Process Engineering, Doswiadczalna str. 44, 20-285 Lublin, Poland
[2]POL-FOODS Sp. z o.o., Kolejowa str. 1d, 19-335 Prostki, Poland

Abstract. The results of measurements of the selected properties of the extruded potato pellets and snacks expanded in the microwave field are presented in the paper. The potato pellets with the addition of the baking soda were prepared with a single screw extruder TS-45. The snacks were obtained by pellets expansion in a conventional microwave oven. The expansion index and the hardness of the pellets and the snacks, as well as, the texture properties of the snacks were evaluated during this study. The results showed that baking soda addition reduced the potato pellet expansion during their extrusion. This was an effect of a smaller thickness of the obtained pellets. The addition of baking soda had positive influence on potato snacks expansion in microwave radiation. The higher content of the soda additive resulted in lower hardness of pellets during cutting tests. The opposite effect was observed during texture measurements of the snacks. The addition of baking soda increased hardness of the expanded snacks. Soda addition lowers crispness and fragilityof the potato snacks expanded in the microwave radiation.

1 Introduction

Generally speaking, extrusion-cooking of vegetable raw materials deals with extrusion of ground material at hydro-thermal conditions. With the help of shear energy, exerted by the rotating screw, and additional heating of the barrel, the food material is heated to its melting point or plasticizing point [1, 2]. In this changed rheological status the food is conveyed under high pressure through a die or a series of dies and the product expands to its final shape.

Expansion involves the nucleation and growth of vapour bubbles during intensive heat. Superheated steam pressure inside each bubble is the driven force for bubble growth. Finally, as the vapour quickly evaporates, the extrudate expands. To improve the expansion degree, special additives can be used during extrusion process [3]. Zazueta-Moralez et al. [4] suggested that addition of 0.2% of the calcium hydroxide improves quality characteristic of the extrudates from corn flour. Martinez-Bustos et al. [5] reported that addition of the 0.15% of calcium hydroxide increase radial expansion index but, in the same time, decrease whole expansion index. Lai et al. [6] reported that addition of the sodium bicarbonate increased expansion index, in contrary to Parsons et al. [7] who reported decrease of expansion index caused by sodium bicarbonate.

Extrusion-cooking leads to production of the products with the physical and chemical properties much different in comparison to the raw materials used. The extruders have a capability to perform cooking tasks under high pressure. This aspect may be explained for vulnerable food and feed as an advantageous process since small time span exposures to high temperatures will restrict unwanted denaturation effects on e.g. proteins, amino acids, vitamins, starches and enzymes. Physical technological aspects like heat transfer, mass transfer, and residence time distribution have a strong impact on the food and feed properties during extrusion-cooking and can drastically influence the final product quality. Currently, extrusion-cooking as a method is used for the manufacture of many food and feed products as well as for modification a variety of other materials like thermoplastic starch or cellulose [8-21].

Extrusion-cooked products, often called snacks, belong to so called impulse food. They are growing in popularity because, during the day, consumers prefer a few small portion snacks instead of two or three big meals. In addition these snacks are preferred due to a good taste. Pellets are the extrusion-cooked semi-products for snacks production. They can be kept in dry storage for 12 months due to theirs compact structure and low humidity [1, 2, 17].

Pellets belong to the third generation of snacks. They are produced mainly from cereal and potato starch. In addition to the basic raw materials, other materials (milk, meat and pulses) may be used. However, the total amount of the starch in the recipe should be at least 60% [1, 22].

The snacks are produced from pellets by hot oil frying or toasting. This process leads to final starch gelatinisation and pellets expansion.

*Corresponding author: marcin.mitrus@up.lublin.pl

In comparison to the traditional hot oil frying, the relatively new process of pellets expansion by microwave heating was developed. This process has gained the popularity due to elimination of the fat from the recipe.

The aim of the work was to investigate the effect of addition of the baking soda on potato pellets and snacks expansion and texture properties.

2 Materials and methods

2.1 Materials

Potato starch, potato flakes and potato grits were purchased from Zakłady Przemysłu Ziemniaczanego "LUBLIN" Sp. z o.o. (Lublin, Poland). Baking soda (sodium bicarbonate) and salt were purchased from the local market.

Blends of potato starch, potato flakes, potato grits and salt (50/25/24/1% of d.m.) with addition of 0-2% of baking soda were used for manufacturing of the pellets by extrusion-cooking technique. Just before extrusion water was added to the blends to obtain blends with 32% of moisture content.

2.2 Pellets and snacks preparation

Manufacturing of potato pellets was carried out using a single screw extrusion-cooker TS-45 (Polish design) with L/D=18. A die with one flat opening (30.0x0.3 mm) was used. The temperature zones in the extruder were 85, 95, 65 and 95°C. The extruder screw speed 60 rpm was used.

The extrudate was manufactured in the form of flat ribbons that were cut into square pellets. The pellets were dried to a moisture content 9 ± 0.2% in a shelf drier (40°C) and then stored in sealed polyethylene bags before microwave heating.

The pellets were expanded by using a microwave oven (2450 MHz, 1000W, Philips M914, the Netherlands). Between tests, the oven was open and the chamber was cooled to room temperature. The pellets (10 pieces) were placed on the rotary table of the oven and heated for 45 sec using at the highest power. The heating conditions were based on preliminary experiments and were constant for all samples.

2.3 Expansion index

The expansion index of the pellets was calculated as the pellet cross section area (wide x thickness) to the die opening area ratio.

Expansion index of the snacks was calculated as the thickness of the expanded snack to the thickness of the pellet ratio. The thickness for each sample (pellet and snack) was measured in, marked earlier, 5 points. The average from measurements of 10 samples (50 points total) was taken as a result.

2.4 Texture properties

A Zwick testing machine (BDO-FB0.5TH, Zwick Roell, Germany) was used to measure texture properties for both the potato pellets and expanded snacks.

Hardness of the pellets and snacks was measured during cutting test with Warner-Bratzler test device. The hardness was evaluated as maximum cutting force during straight knife cutting test. The knife speed was 500 mm/min. The measurements were made in 20 repetitions.

The texture properties (hardness, crispness and fragility) of the microwave expanded snacks were evaluated during double compression test with Ottawa cell. The samples, consisted of one layer of 3 snacks each, were tested at a compression speed of 100 mm/min. The samples were compressed to a 1 mm thickness. The measurements were made in 10 repetitions.

3 Results and discussion

3.1 Expansion index

Compared to microwave puffed grains there are very few studies about microwave heated starch pellets and their properties. The expansion index of the pellets without soda addition was relatively high (5.72±0.15). The results showed that addition of baking soda reduced the potato pellet expansion during their extrusion (Fig. 1). This was an effect of a smaller thickness of the obtained pellets. The greater amount of soda caused greater reduction of the pellets expansion index. Thermal decomposition of sodium bicarbonate becomes quite rapid above 120°C [23]. During extrusion in lower temperature a baking soda could bind some water from the bland and act as an emulsifier. As a result, lower expansion index was observed.

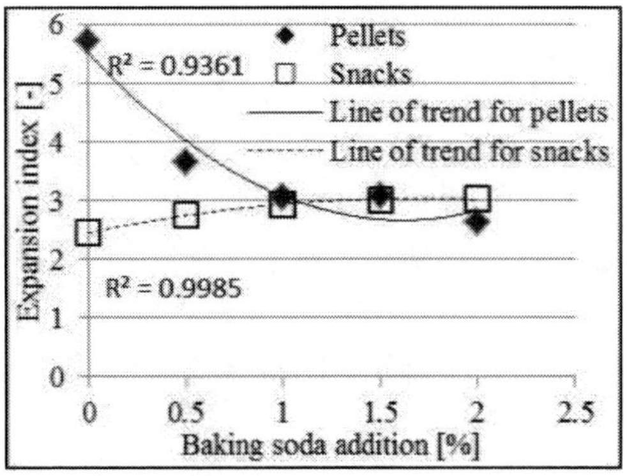

Fig. 1. Changes of expansion index of potato pellets and snacks with addition of baking soda.

Microwave heating is a relatively inexpensive technique that offers advantages such as convenience and fast heating. The formation of the final structure of the snacks is based on the transition of the glassy matrix to the rubbery state during microwave heating. Upon heating, moisture generates the superheated steam necessary for expansion. As the result, starchy matrix undergoes a phase transition to rubbery state and starts to deform under steam high pressure and expansion take place. The final structure sets when moisture evaporates from matrix and without microwave heating matrix cools down [3, 24, 25].

During the research it was found that the addition of baking soda had positive influence on potato snacks expansion in microwave radiation. Expansion index value increased from 2.45±0.25 to 3.02±0.33 with increase of additive level from 0 to 2%. During microwave heating there is a rapid rise in pellet temperature with can lead to thermal decomposition of baking soda and formation of the CO_2 take place. Combined action of CO_2 and steam induces greater snacks expansion during microwave heating. Similar soda effect was observed by other researchers [25]. Additionally, the other products of the thermal decomposition of the baking soda can react with the starch. These reactions can leads to reduction of the glass transition temperature of the starch which can facilitate the expansion of the snacks. Similar interaction between salt and starch was observed by Norton et al. [26] during hot air expansion of the potato pellets.

3.2 Texture properties

The observations for pellets and snacks hardness during cutting tests are presented on Figure 2. Pellets hardness varied from 36.1±0.56 to 14.4±0.45 N and decreased with increase of the baking soda addition. These results can be correlated with the change of the pellets thickness due to their expansion index reduction during the extrusion process.

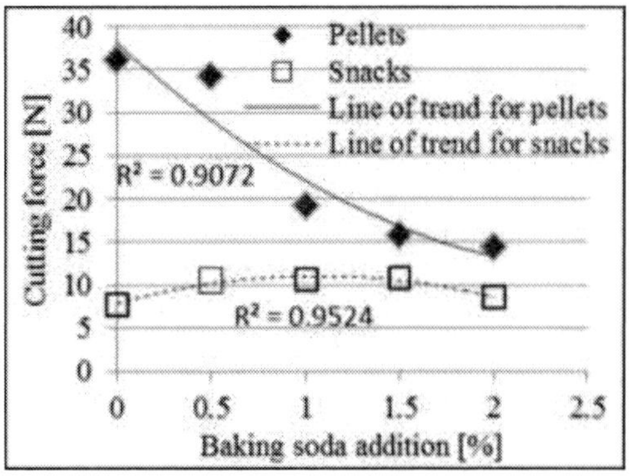

Fig. 2. Changes of hardness of potato pellets and snacks with addition of baking soda.

The opposite effect was observed for snacks expanded during microwave heating (Fig. 2). Maximum cutting force during snacks measurements varied from 7.72±0.1 (for 0% of additive) to 10.7±0.21 (for 1.5% of additive). The addition of baking soda, up to 1.5%, increased hardness of the expanded snacks. Further increase in the amount of the additive, up to 2%,resulted in a slight decrease of the snacks hardness. These results can also be related with snacks thickness changes due to their expansion dependent on sodium bicarbonate content in the blend.

Similar trends in hardness changes were observed for snacks during double compression tests in Ottawa cell. The hardness increased with increase sodium bicarbonate level up to 1.5%. Further increase in addition content resulted in slight potato snacks hardness decrease.

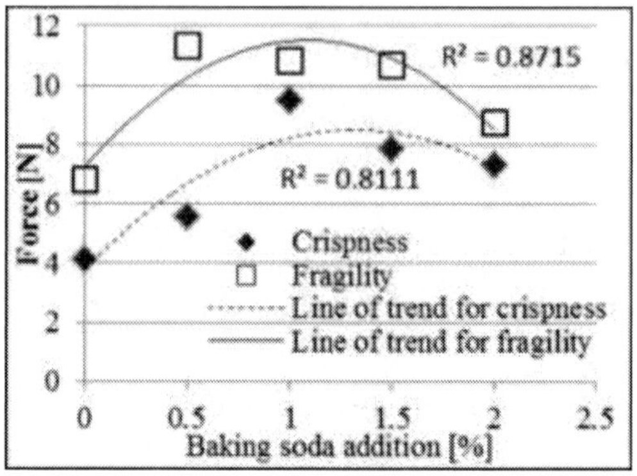

Fig. 3. Crispness and fragility of the microwave expanded snacks with addition of baking soda.

The measurements performed during compression tests revealed the soda bicarbonate influence on snacks textural properties (Fig. 3). Value of force recorded during first crack of the sample was interpreted as crispness. Greater force indicated lower crispness. The baking soda addition had negative effect on snacks crispness. The snacks without soda addition were characterised by the highest crispness. Snacks with sodium bicarbonate had lower crispness. The lowest crispness was observed for snacks with 1% of soda content. Further baking soda content increase caused crispness decrease.

The value of force needed for sample breaking was interpreted as fragility. Higher force indicated lower fragility. The baking soda addition had positive effect on fragility of microwave heated snacks in comparison to snacks without this additive. This additive increased snacks resistance for breaking. Increase in soda additive caused decrease in snacks fragility. In the other hand, this influence of soda on may be received as unfavourable by consumers.

4 Conclusions

1. The investigations showed that, generally, soda bicarbonate had positive influence on potato pellets as well as microwave expanded snacks properties. The addition of baking soda reduced the potato pellet expansion during their extrusion and had positive influence on potato snacks expansion in microwave field.
2. Soda bicarbonate addition decreased potato pellets hardness. In the case of snacks, this additive increased their hardness and fragility but decreased crispness. These changes may be disadvantageous from the consumer point of view.

References

1. Leszek Mościcki, *Extrusion-Cooking Techniques. Applications, Theory and Sustainability* (Wiley-VCH, Weinheim, 2011)
2. C. Mercier, P. Linko, J.M. Harper, *Extrusion cooking* (American Association of Cereal Chemists Inc, St. Paul, 1989)
3. C.I. Moraru, J.L. Kokini. Compr. Rev. Food Sci. Food Saf. **2**, 147 (2003)
4. J.J Zazueta-Moralez, F. Martinez-Bustos, N. Jacobo-Valenzuela, C. Ordorica-Falomir, O. Paredes-Lopez, J. Sci. Food Agric. **81**, 1379 (2001)
5. F. Martinez-Bustos, Y.K. Chang, A.C. Bannwart, P.A. Rodriguez, P.A. Guedes, E.R. Gaiotti, Cereal Chem. **75**, 796 (1998)
6. C.S. Lai, J. Guetzlaff, R.C. Hoseney, Cereal Chem. **66**, 69 (1989)
7. M.H. Parsons, F. Hsieh, H.E. Huff, J. Food Process. Preserv. **20**, 221 (1996)
8. L. Guz, A. Sopinska, T. Oniszczuk, Aquacult. Nutr. **17**, 695 (2011)
9. A. Bouasla, A. Wojtowicz, M. N. Zidoune, LWT Food Sci. Technol. **75**, 569 (2017)
10. T. Oniszczuk, R. Pilawka, Przem. Chem. **92**, 265 (2013)
11. A. Oniszczuk, A. Wójtowicz, T. Oniszczuk, M. Olech, R. Nowak, K. Wojtunik, M. Klimek, W. Krawczyk, M. Hajnos, Open Chem. **13**, 1209 (2015)
12. S. Sharma, N. Singh, B. Singh, Starch/Stärke **67**, 721 (2015)
13. K. Mezreb, A. Goullieux, R. Ralainirina, M. Queneudec, Carbohydr. Polym. **64**, 1 (2006)
14. D. Gumul, M. Krystyjan, K. Buksa, R. Ziobro, T. Zięba, Starch/Stärke **66**, 190 (2014)
15. Y. Ai, K.A. Cichy, J.B. Harte, J.D. Kelly, P.K.W. Ng, Food Chem. **211**, 538 (2016)
16. L.C. Oliveira, C.M. Rosell, C.J. Steel, Int. J. Food Science Technol. **50**, 1504 (2015)
17. X. Tovar-Jimenez, J. Caro-Corrales, C.A. Gomez-Aldapa, J. Zazueta-Moralez, V. Limon-Valenzuela, J. Castro-Rosas, J. Hernandez-Avila, E. Aguilar-Palazuelez, Int. J. Food Science Technol. **52**, 6607 (2015)
18. Q-B. Ding, P. Ainsworth, A. Plunkett, G. Tucker, H. Marson, J. Food Eng. **73**, 142 (2006)
19. M. Wolz, U. Kulozik, J. Food Eng. **209**, 12 (2017)
20. A.A.M. Andersson, R. Andersson, A. Jonsäll, J. Andersson, H. Fredriksson, J. Food Sci. **82**, 1344 (2017)
21. M. Mitrus, L. Moscicki, Chem. Eng. Res. Des. **92**, 778 (2014)
22. A.K. Hardacre, S.M. Clark, S. Riviere, J.A Monro, A.J. Hawkins, J. Texture Stud. **37**, 94 (2006)
23. M. Hartman, K. Svoboda, M. Pohořelý, M. Šyc, Ind. Eng. Chem. Res. **52**, 10619 (2013)
24. C. Boischot, C.I. Moraru, J.L. Kokini, Cereal Chem. **80**, 56 (2003)
25. R.G.M. van der Sman, J.R. Bows, J. Food Eng. **211**, 69 (2017)
26. A.D. Norton, R.W. Greenwood, I. Noble, P.W. Cox, J. Food Eng. **105**, 119 (2011)

Characteristics of electricity quality parameters in agri-food processing plants

Krzysztof Necka[1,]*, *Małgorzata Trojanowska*[1]

[1]Faculty of Production and Power Engineering, University of Agriculture in Krakow, Poland,

Abstract. The paper presents the results of electricity quality measurements conducted in six agri-food processing plants equipped with receivers of different sensitivity to the quality of the supply voltage. In particular, statistical assessment has been conducted for the value of the parameters characterizing the supply voltage, describing the parameters using the basic statistical measures and comparing them with the normative requirements. The paper also presents examples of the hazards to electrical drives and control systems caused by such industrial electromagnetic interferences as voltage changes or deformations of voltage waveform.

1 Introduction

At present, electricity is the basic energy medium necessary for the efficient functioning of most economic entities, especially industrial plants, including agri-food processing plants. Currently, there are approximately a dozen or so thousands of such entities in Poland with different production scales and forms of ownership.

The way in which power is supplied in food processing plants is extremely important. The choice of power supply is mainly determined by the supply reliability. It depends on the amount of losses in the event of a power failure. The amount of losses is estimated at 30-150 times the cost of electricity not supplied [1, 2]. The losses may be even greater in the case of freezing or meat plants.

Additional costs of plants are generated not only due to unplanned downtime associated with interruptions in power supply, but also due to improper operation of equipment resulting from the poor quality of electricity, which causes abnormalities in the manufacturing process and also faster wear and tear or even damagesof electrical equipment [3-7].Electricity consumers are not fully aware that the poor quality of electricity can be the reason for the additional production costs associated with malfunctioning equipment. It can help the plant in gaining competitive advantage by reducing production costs to conduct continuous monitoring or at least periodic energy quality measurements [8-11].

2 Purpose, scope and subject matter of the study

The aim of the study was to analyze the quality of electricity in six selected agri-food processing plants. The

study included two water bottling plants (hereinafter referred to in the paper as plants R1 and R2), three meat plants (plants M1, M2, M3) and one bakery (plant P). Each of the plants is modernly equipped and powered by its own transformer/distribution station 15/0.4 kV. Most of the electrical systems used in them are adjustable electric drives. The paper presents examples of threats caused by electromagnetic disturbances to the electric drives and related control systems.

3 Study method

The quality of electricity is the same as the quality of the supply voltage and includes two groups of disturbances:

- changes, i.e. small deviations from the nominal or desired magnitude, including free voltage changes, voltage fluctuations, asymmetry and higher harmonics of voltages,

- events, i.e. sudden and significant deviations from the nominal or desired magnitude, including rapid voltage changes, voltage dips, voltage increases and transient overvoltages, short power interruptions.

The AS-3 Plus portable parameter analyzer, along with current transformers, was used to test the quality of electrical power in the plants, allowing for both, the detection of changes and the recording of events related to the values of voltage parameters describing its quality. The measurements were made each time at the main switchrooms of the plants, at the points of electricity supply. In particular, the phase and phase-to-phase

*Corresponding author: krzysztof.necka@urk.edu.pl

voltages, voltage asymmetry, the content of higher harmonics and the THD factor for voltages and currents as well asexceedances of the set tolerances of the measured parameters, voltage dips and increases, etc. were continuously monitored.

4 Results

Electricity quality is defined as a set of parameters describing the characteristics of the energy supply process under normal operating conditions [12]. Both the supplier and the customer are responsible for electricity quality. Suppliers, by virtue of the power grid service, are mainly responsible for the supply voltage. Consumers, in turn, are responsible for the quality of the current, affecting its course through the operation of electrical equipment, which can impair the voltage parameters.

The studied agri-food processing plants are equipped with a large number of modern adjustable electric drives. Such electric drive consists of:

- electric motor driving a work machine with adjustable parameters (usually rotational speed or torque),
- a system for converting electrical energy with fixed parameters into electrical energy with variable parameters, resulting from the technological process needs (usually energy-electronic converter),
- a system controlling processing device.

The control is implemented by microprocessor circuits and programmable PLC logic controllers. The whole is complemented by contactors and relays used to connect or disconnect both power and control circuits. All these components are very sensitive to the voltage quality, but not equally sensitive to each of its parameters. The nuisance of the supply voltage disturbances can be ranked in the following order: voltage dips, short power interruptions, voltage values, higher harmonics, overvoltages, asymmetry, fluctuations and voltage frequency [3, 4].

Supply voltage dip is a sudden decrease in the effective voltage to a value in the range of 90% and 5% of the rated voltage [13], followed by a voltage increase to the previous value in a short period of time (usually 10 ms - 1 min). A specific form of the voltage dip is short power interruption. It is recorded when the voltage goes down below 5% of the nominal value. According to the Minister Regulation [12], short power interruptions last longer than 1 second and no longer than 3 minutes.

The regulations in force in Poland do not specify the requirements for voltage dips and short power interruptions. By contrast, according to PN-EN 50160 [13], voltage dips can be recorded from several dozen to one thousand times a year, while the number of short power interruptions ranges from several dozens to several hundred.

In the studied agri-food industry plants, the problem of voltage dips and short power interruptions occurred with varying intensity (Table 1). The largest number of voltage dips, in the number of about 10 thousand annually, was recorded in the bakery, but the number of dips was also too high, as compared to the requirements of the standard, in all of the studied meat plants. The registered dips were, however, characterized by small depths (an average of 10-12% of the rated voltage) and short duration (an average of approx. 1 second). Short voltage supply interruptions were problematic in none of the plants.

Table 1. Characteristics of voltage dips and short power interruptions.

Plant	Annual number of voltage dips	Average duration of a voltage dip [ms]	Mean depth of a voltage dip [%]	Annual number of short interruptions	Average duration of a short interruption [ms]
R1	114	427	19	4	2550
R2	126	355	19	0	0
M1	6811	1124	10	8	4810
M2	7855	1244	10	58	3562
M3	2815	1226	11	4	1600
P	9857	804	12	0	0

Microprocessor circuits are exceptionally sensitive to voltage dips and short power interruptions, and the most common effects include: no data transmission or errors in their transmission.

The inadequate voltage supply that is sustained on the receivers' terminals for a long time is another cause of disturbances in the operation of agro-food industry plants. According to the regulations in force in Poland, 95% of a set of 10-minute average voltage effective values at the point of supply of electricity from a low voltage network shall be within ± 10% of the rated voltage [12]. Exceeding the limit values adversely affects the operation of propulsion engines, resulting in changes in their torque and rotational speed, disrupting their microprocessor systems used to control the motors, as well as the contactors and relays used as connecting elements. The analysis of the voltage measurements in all of the studied plants allowed us to conclude that, in terms of the supply voltage, the normative requirements were fulfilled.

Such a statement could not be tolerated with respect to another, potentially harmful, electromagnetic disturbance which is voltage distortion [14-15]. According to therequirements in force in Poland [12, 13], its evaluation is based on the following criteria:

- the permissible values of individual higher harmonics,
- the permissible value of the Total Harmonic Distortion, THD_U.

The criteria must be fulfilled simultaneously, which in practice means that if one of the higher harmonics approaches the limit value, the remaining ones must be much smaller so that the total THD_U factor value did not exceed the limit value of 8%.

Controlled drives, like most electrical systems, are susceptible to such disturbances as disorted voltage. Higher harmonics in the supply voltage cause, among others, worse operating conditions of the insulation due to increased working temperature, which results in a shorter operating life. With a lot of disturbances of this type, interference can also be recorded in the operation of control and protection devices, caused by the increase of the peak current and voltage.

The analyzes performed (Tables 2 and 3) show that the supply voltage of the studied plants most frequently exceeded the permissible levels of the 5th, 6th, 7th and 15th harmonics, but the facilities differed in the intensity of the exceedances of the limit values for particular harmonics due to different characteristics of the equipment used. However, the number of the events was very low for most of the plants studied.

Table 2. Breakdown of the permissible levels of individual higher harmonics in the supply voltage.

Plant		R1	R2	M1	M2	M3	P
Percentage share of the exceedances of the value permissible for the higher harmonic:	2	18.0	0.0	0.0	0.0	0.0	0.0
	4	9.0	0.0	0.0	0.0	0.0	0.0
	5	0.0	0.0	0.0	0.0	0.7	48.0
	6	27.0	100.0	0.0	0.0	0.0	0.0
	7	0.0	0.0	0.0	0.0	56.7	0.0
	8	9.0	0.0	0.0	0.0	0.0	0.0
	10	9.0	0.0	0.0	0.0	0.0	0.0
	11	0.0	0.0	0.0	0.0	0.4	3.7
	12	9.0	0.0	0.0	0.0	0.1	0.0
	13	0.0	0.0	0.0	0.0	3.5	0.0
	14	9.0	0.0	0.0	0.0	0.1	0.0
	15	9.0	0.0	0.0	0.0	38.5	48.1

The M3 meat plant was the exception, the daily average number of such disturbances was close to 9. During the exceedances of the permissible levels for single harmonics, the THD_U value in plants R1 and P exceeded the permissible level of 8%, and it was close to it for the M3 plant.

Throughout the entire measurement period, the THD_U limit was exceeded in the same plant, but the annual number of the exceedances was at a very different level of 4 to 112 events, respectively. There were also disturbing situations in the M2 and M3 plants, where the THD_U value was close to the limit values.

The current course has a decisive influence on the distortion of the supply voltage waveform. During the assessment of the current course, distortion of higher harmonics, the content of individual harmonics was verified in accordance with the requirements of the *Transmission Network Code*[16].

The analyzes show (Table 4) that exceedances of the permissible distortion levels for the current course were recorded in two of the studied plants and mostly concerned the 3rd, 5th, 11th and 13th harmonics, with a significant number, between 4 and 10 per day.

Table 3. Characteristics of the Total Harmonic Distortion, THD_U.

Plant		R1	R2	M1	M2	M3	P
Annual number of the exceedances of the allowable level of higher harmonics		12	9	0	0	3240	30
THD_U when exceeding the allowable level of higher harmonics [%]	min	2.70	2.10	0.00	0.00	2.70	6.20
	avg	5.43	2.17	0.00	0.00	3.90	7.50
	max	9.80	2.30	0.00	0.00	7.90	9.70
	SD	0.52	0.12	0.00	0.00	0.71	1.13
THD_U hroughout the study period [%]	min	0.50	0.50	0.00	0.00	0.00	0.00
	avg	1.23	1.26	1.82	2.66	2.72	3.45
	max	9.80	2.80	3.06	6.73	7.96	9.99
	SD	0.96	0.24	0.36	0.73	0.82	1.16
Annual number of the exceedances of the allowable level of THD_U		4	0	0	0	0	112

Table 4. Breakdown of the exceedances of the permissible levels of individual higher harmonics in current and the THD_I levels.

Plant		R1	R2	M1	M2	M3	P
Percentage share of the exceedances of the value permissible for the higher harmonic:	3	0.0	84.0	0.0	0.4	0.0	0.0
	5	0.0	72.0	0.0	41.4	0.0	0.0
	7	0.0	0.0	0.0	0.8	0.0	0.0
	9	0.0	0.0	0.0	0.4	0.0	0.0
	11	0.0	0.0	0.0	28.2	0.0	0.0
	13	0.0	0.0	0.0	19.5	0.0	0.0
	15	0.0	0.0	0.0	9.2	0.0	0.0
THD_I [%]	min	0.0	5.0	0.0	7.1	0.0	0.0
	avg	0.0	10.0	0.0	17.2	0.0	0.0
	max	0.0	20.4	0.0	47.7	0.0	0.0
	SD	0.0	3.2	0.0	4.9	0.0	0.0
Annual number of the exceedances of the allowable level of THD_I		0	1389	0	3739	0	0

5 Summary

The results of the research on the quality of electricity in agri-food processing plants were analyzed. Measurements were made at the points of energy supply, as these are the only points for which regulations specify the requirements for the quality parameters of the supply voltage.

Each of the plants studied recorded less or more incidents of unsatisfactory voltage quality, especially with respect to the content of higher harmonics. Four of the studied plants recorded values of the THD_U factor that were higher or close to the value allowed by regulations.

The repercussion of the distortion of the supply voltage waveform, as well as the majority of electromagnetic disturbances analyzed in the work, is among others the incrise of flowing current, with all the consequences.

Improving the quality of electricity is only possible after diagnosing and defining the causes of its deterioration. For this purpose, measurements of voltage quality parameters are required not only in the main switchrooms but also in local switchrooms, directly supplying equipment. The quality of electricity at the point of the connection of the plant to the grid is not the same as the quality of the power supply for the receivers themselves.

Acknowledgements
This research was financed by the Ministry of Science and Higher Education of the Republic of Poland

Bibliography

1. Problemy zaopatrzenia w energię w gospodarstwach rolnych oraz przemyśle rolno- spożywczym. Instytut Energetyki Odnawialnej.http://ieo.pl/en aktualności /1095-problemy-zaopatrzenia-w-energie – w – gospodarstwach–rolnych – oraz – przemysle – rolno-spozywczym
2. Hanzelka, Z., Automatyka, Elektryka, Zakłócenia, Vol. **3**, Nr 7, s. 11-19, (2012)
3. Z. Hanzelka, A. Firlit, IV Konferencja Wytwórców Energii Elektrycznej i Cieplnej. Skawina 25-27.09.2013(2013)
4. Jiansheng H. Zhuhan J. Energy Procedia. Volume **141**, s. 345-351 doi.org/10.1016/j.egypro.2017.11.041 (2017)
5. H. Markiewicz, A. Klajn, Wydawnictwo Polskiego Centrum Promocji Miedzi nr **02/03**, s. 3-20 (2003)
6. K. Strzałka-Gołuszka, J. Strzałka, Biuletyn SEP Kraków, nr **182-183**, s. 39-65 (2010)
7. M. Wardach, A. Kiryluk, P. Cierzniewski, T. Zarębski, VIII Lubuska Konferencja Naukowo-Techniczna – i-MITEL 2014 (2014)
8. Z. Hanzelka, II Ogólnopolska Konferencja „Elektroenergetyka na terenach wiejskich", s. 22-28 (2004)
9. E. Musiał, Konferencja Automatyka, Pomiary, Zakłócenia. INFOTECH 2004, Jurata, 20-22 maja 2004 r., s. 103-122 (2004)
10. A. Pawlęga, Przegląd Elektrotechniczny nr **11**, s. 805-810 (2003)
11. M. Trojanowska, K. Nęcka, Inżynieria Rolnicza nr **7(95),** s. 221-227 (2007)
12. Rozporządzenie Ministra Gospodarki z dnia 4 maja 2007 w sprawie szczegółowych warunków funkcjonowania systemu energetycznego. Dz. U. nr 93 poz. 623, z późniejszymi zmianami.
13. PN-EN 50160. 2010: Parametry napięcia zasilającego w publicznych sieciach rozdzielczych.
14. M. Jasiński, D. Kaczorowska, L. Jasińska, Ł. Jasiński, Zeszyty Naukowe Uczelni Jana

Wyżykowskiego. Studia z Nauk Technicznych, **5**, s. 127–134 (2016)
15. M. Olesz, Zeszyty Naukowe Wydziału Elektrotechniki i Automatyki Politechniki Gdańskiej Nr **50**, s. 71-74, (2016)
16. Instrukcja ruchu i eksploatacji sieci dystrybucyjnej (zatwierdzona przez Prezesa URE decyzją nr DRR-4321-60(5)/2013/KSm). Tauron Dystrybucja (2017)

Impact of size and distribution of installed PV power in E-W direction on the level in which selected consumers' energy needs were met

Krzysztof Necka[1,*], *Hubert* Latala[1], *Sławomir* Kurpaska[1], *Jarosław* Knaga[1], *Anna* Karbowniczak[1]

[1]Faculty of Production and Power Engineering, University of Agriculture in Krakow, Poland,

Abstract. The aim of the study was to analyse the energy needs of selected consumers and to select the PV installed capacity in the east and west panels orientation (E-W). The analysis was carried out for two variants. The first concerned the impact of the installation power of a PV source on the contracted power in its symmetric east and west position. The second variant was the result of previous observations and studies of the authors of this paper. Thus, in this variant, the unbalanced power of PV E-W in relation to the S azimuth has been analyzed, taking into account the characteristics of the recipients' needs and the variation of the PV power installed. The analyses show that an increase in meeting the energy demand for two of the tested plants at symmetrical power distribution to the east and west occurred only when the installed power was increased to the level of approx. 1.4-1.6 of the contracted power. However, the power distribution in the E-W direction has a very strong effect on the energy amount that is generated in the power plant and cannot be used due to the lack of demand in the plant.

1 Introduction

In the recent years we could have observed a dynamic increase of the power installed in the renewable energy sources. At the end of 2016 the capacity of RES installations holding a licence of the Energy Regulatory Office achieved the level of over 8.4 GW and the systems using the wind energy (5.8 GW) and biomass (1.3 GW) had the largest share therein.) [1]. Whereas the highest growth dynamics of the installed power was present in the installations using solar radiation. Within the period of three years the capacity of such installations increased from nearly 2 MW in 2013 to over 99 MW in 2016. Moreover, at the end of 2016 17 thousand of PV systems connected to the network based on declaration were operating with the total capacity of 100 [1, 2] MW. Further growth in the installed power of such systems is very likely due to the decreasing investment expenditures and support system financed with many sources.

Nevertheless, photovoltaic installations constitute an energy source with a highly dynamic change both in the twenty-four-hour, as well as in the annual period. The level of the generated power mostly depends on the incident radiation, the value of which is variable in time. Thus the activities aimed at the increase in the level of adjusting the energy demand of a plant to its availability from the local PV system are targeted [3, 4].

Increase in the simultaneity of occurrence of energy demand and its availability may be implemented through modification of an hourly energy demand profile or the use of a modified solar energy profile depending on orientation of the PV panels [5, 6]. The first of the methods for optimisation of the use of the available energy is limited to a large extent by the necessity of execution of a specific technological process in a plant. In plants that have already been operating it may be very often used to a limited scope or may require large financial expenditures for modernisation of the existing infrastructure. On the other hand, such types of activities bring excellent results if they are implemented as early as at the stage of design of the given facility. In the course of selection of a photovoltaic power station for an existing plant, for which we know the hourly energy demand profile activities, in which the required capacity of photovoltaic panels will be installed with different azimuth, may turn out to be intentional.

Many research teams have dealt with the problem of selection of the optimum orientation and tilt angle of mono- and polycrystalline photovoltaic panels and allowing specific features at the place of location of the installation [7 – 11]. According to such tests the tilt angle of the reception plane of modules depends on the place of location, season and day of the year and meteorological conditions. In all the analysed cases the use of systems for tracking the apparent movement of the sun enabled the growth of the energy yield from a PV installation from several to several dozens percent.

However, micro PV power plants with tracking systems are expensive and may not always be used from

* Corresponding author: krzysztof.necka@ur.krakow.pl

a technical point of view [5, 12]. Therefore, for many applications the only solution which aims at the increase of the level in which the demand is met from a PV system may be construction of a stationary installation with power distribution for different reception plane orientations. In such situation the maximum energy production may be obtained on the condition that the tilt angles value and azimuth for a specific application are optimally set.

2 Materials and methods

Information for implementation of the study regarding the amount of the available solar radiation has been obtained based on the long-term meteorological data available on the websites of the Ministry of Infrastructure and Construction for the place of location of the investigated facilities. The characteristics with regard to the distribution of solar radiation were limited to surfaces inclined under the angle of 30° to the horizontal plane. The analyses considered the availability of energy with the level enabling operation of the CIGS panels. Its permitted limits were specified based on the previous authors' own study which provides that operation of thin-film panels begins when the incident radiation exceeds 50 W·m^{-2} and is present until its value decreases below 40 W·m^{-2}.

The second source of information comprised the results of the authors' own study carried out at the photovoltaic power plant situated at the Faculty of Production and Power Engineering, University of Agriculture in Krakow, Poland. The conducted research concerned the operation of thin-film modules of CIGS type with the power capacity of 230 Wp in a configuration enabling acquisition of information on energy production for panels oriented to the east (1.54 kWp), south (4.4 kWp) and west (1.54 kWp). The energy yield for the specific azimuths of operation was designated in relation to the value of power installed in specific module chains.

The study also used the information on energy consumption of the selected recipients, from which hourly energy demand profiles were developed. Three recipients were selected to the analysis, characterised by a different profile of energy needs, i.e. a dairy cooperative, drinking water intake station and wastewater treatment plant.

The next stage was to specify the way in which the size of the installed power and azimuth of the reception surface of the cells influences the cooperation of the power station with a recipient characterised by the specific energy demand profile. During the study the impact of the change in the power station capacity was analysed within the range of 20 up to 200% of the contracted power, as well as power distribution in the E-W direction. The size designated for a power station with contracted power of a plant and south azimuth was adopted as a point of reference. Due to the amount of the energy available from solar radiation the study was limited to the period with the best solar conditions i.e. the months of June, July and August.

In order to assess the use of the energy generated in a photovoltaic power station a coefficient for meeting the energy demand of a recipient (V_{pz}) was defined, as well as a coefficient for surplus energy acquired from a PV power station (V_{ne}), which specifies the amount of energy produced, but not used by a recipient due to lack of demand in a given period. The use and surplus energy coefficients were defined with the change in the power station capacity for four separately analysed orientations of PV panels arrangement.

$$V_{pz(E,S,W,E-W)} = \frac{E_{pv(E,S,W,E-W)} - E_{n(E,S,W,E-W)}}{E_c} \cdot 100\% \quad (1)$$

$$V_{ne(E,S,W,E-W)} = \frac{E_{n(E,S,W,E-W)}}{E_{pv(E,S,W,E-W)}} \cdot 100\% \quad (2)$$

where:
E_c – energy consumption in the facility,
$E_{PV(E,S,W,\ E-W)}$ – energy produced in the PV power station for orientation accordingly in the east (E), south (S) and west (W), and with the arrangement of panels in the east and west position (E-W),
$E_{n(E,S,W,\ E-W)}$ – surplus energy produced in the PV power station for orientation accordingly in the east (E), south (S) and west (W), and with the arrangement of panels in the east and west position (E-W).

3 Results and discussion

The annual availability of solar radiation enabling the operation of PV installations designated based on typical meteorological years for Krakow depending on the orientation of a PV module changes from 989 kWh·m^{-2} in W direction, through 997 kWh·m^{-2} for E direction up to 1118 kWh·m^{-2} for S direction. The efficiency of operation of a PV power station is influenced not only by the annual availability of solar radiation, but also by its intensity. The conversion efficiency of solar radiation depends to a large extent on the incident radiation. The rated power specified in photovoltaic panels is most often designated in STC or NOCT conditions, in which it is assumed that incident radiation amounts to 1000 and 800 W·m^{-2} accordingly. The conducted research shows that the availability of solar radiation at hourly intervals for the averaged meteorological year at the place of location of a power plant may achieve as much as 1200 Wh·m^{-2}, but the period of its occurrence is very short. In the district of Krakow the most often available energy has a value of its flux up to 200 Wh·m^{-2} and the share of hours for which the available solar radiation exceeds 800 Wh·m^{-2} amounts to less than 3%.

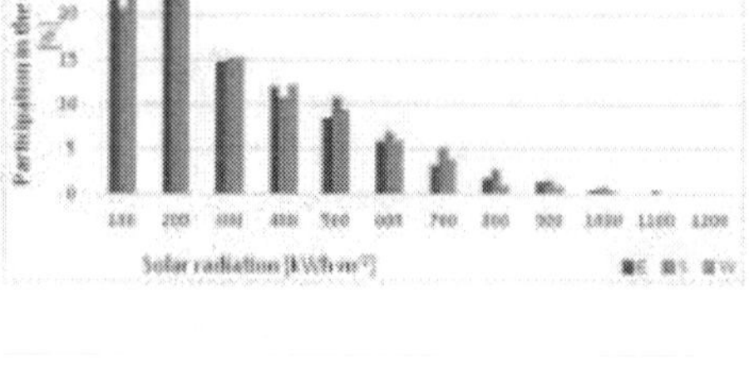

Fig. 1. Histogram of hourly availability of solar radiation depending on panel orientation in annual period

In order to cover the energy demand for a recipient characterised by a specific energy demand profile it is important that solar radiation (Fig. 2) coexists with energy demand (Fig. 3).

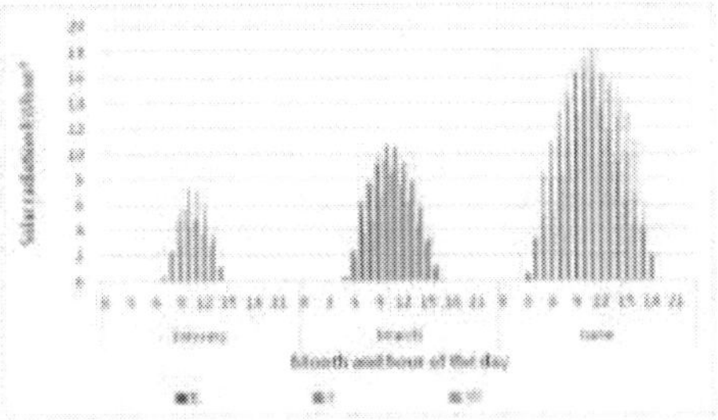

Fig. 2. Averaged distribution of hourly availability of solar radiation in characteristic months with a value allowing for operation of power station depending on orientation of PV module plane

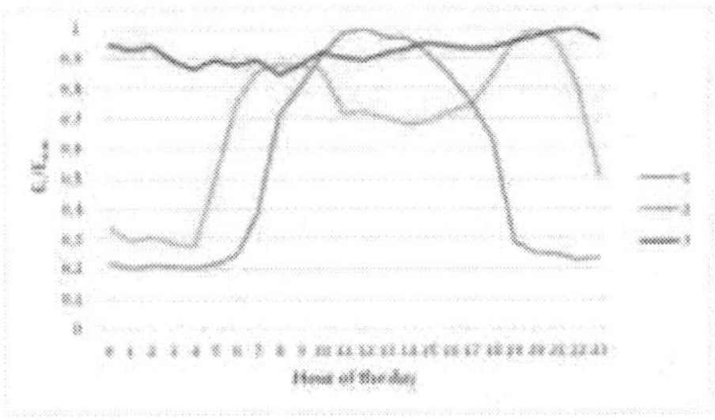

Fig. 3. Reduced energy demand profiles for: 1) District Dairy Cooperative, 2) drinking water intake station, 3) wastewater treatment plant

where:
E_i – energy consumption at a given hour of the day,
E_{max} – maximum hourly energy consumption on a given day

The subsequent part of the study presents the test results for three recipients of electrical energy characterised by different energy demand profiles (Fig. 3). In the course of research the influence of the installed power of the PV power station and the ratio of arrangement of panels in the E-W direction on the level in which energy demand is met, as well as its amount generated the PV power station, but not used by the plant were analysed.

The first part of the research verifies the way in which the change in the installed power of the PV power station in relation to the contracted power of the plant will influence the level in which the demand for electrical energy is met and the amount of the generated surplus energy that the plant did not use. The power of the photovoltaic power station was changed in the range from 20% to 200% of the contracted power. Figures 4 and 5 show the results of the obtained tests.

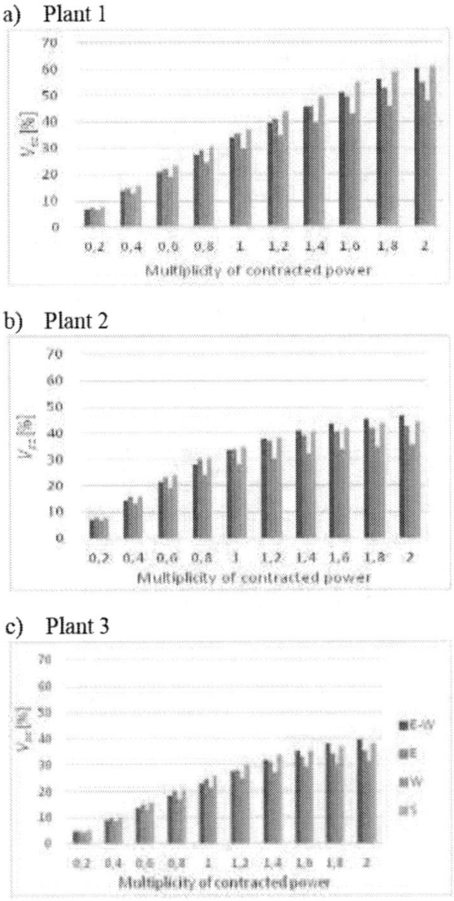

Fig.4. Impact of power of PV power station and panel orientation on the level in which energy demand is met for specific recipients.

The conducted analyses prove that the energy demand will be met to the lowest extent for all the facilities with the W direction panel orientation, regardless the power station capacity. Decrease in the level in which demand is met in relation to the S direction was in the range between 16 and 22%. The change of the orientation in the E direction also did not bring an increase in the level in which the demand was met in relation to the south, and its drop was reduced and fluctuated in the range from 2 to 10%. Only after symmetric distribution of the power in the E-W direction the increase in the level in which the energy demand is met was obtained for plant 2 and 3 (drinking water intake station and wastewater treatment plant), provided that the power station will be over-dimensioned in relation to the contracted power by at least 40-60%. The observed increase in the satisfaction level of the energy demand of a facility from a renewable energy source amounted to 4-5% with doubled efficiency of the installation capacity. However, the analysis of differences between the average twenty-hour satisfaction of the energy demand performed in Statistica program after a change in the azimuth from the south (S) to east-

https://doi.org/10.1051/bioconf/20181002023

west position (E-W) did not indicate statistically significant differences at the significance level of p = 0.05.

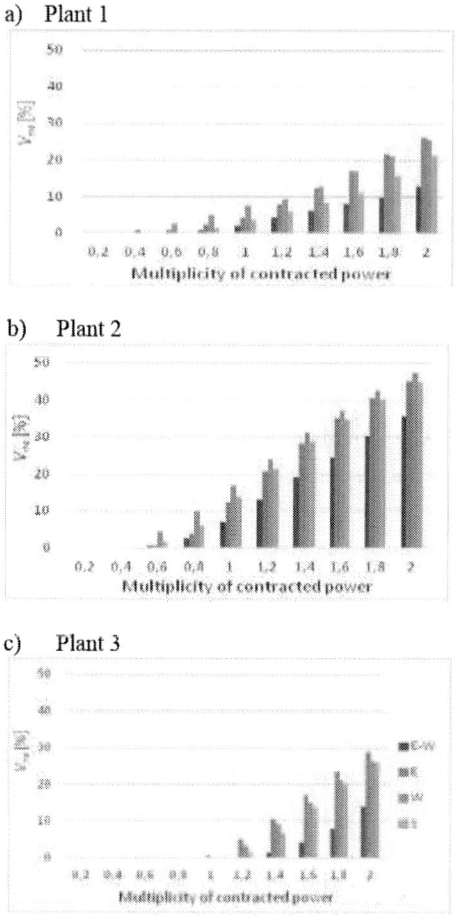

Fig.5. Impact of power of PV power station and panel orientation on the amount of energy that was produced in PV power station, but not used due to lack of demand at the given period for specific recipients

The suggested changes in the capacity of PV installation and the method of installation of panels had a large influence on the level of matching the demand and availability of the generated electrical energy. The performed analyses show that the best solution for the investigated recipients due to minimisation of the surplus of the generated energy is construction of a power station with symmetrically distributed power in the E-W direction. Such a solution enabled the reduction of the share of unused energy by ca. 10%. Alternative solutions consisting in panel orientation only in the E or W direction caused the increase in the amount of unused energy towards the south azimuth. However, the essential changes from the point of view of statistics occurred for the adopted significance level in the amount of surplus energy only for the plant number 3.

In the further part of the study tests results are presented, in which the impact of distribution of the power installed in the specific directions (E-W) on the level in which the energy demand is met and the share of the produced, but unused energy was analysed. The study limited the analysis to two variants of capacity of a PV power station. The first one was equal to the contracted power of the plant, and the second one was 60% higher. Oversizing the power station results from the previous analysis, in which the prevalence of the satisfaction level for a power station with power distributed in the E-W direction in relation to S was observed from this point.

According to the conducted simulations for the first plant where the peak energy demand is present from 10 a.m. to 2 p.m. due to the level in which the energy demand is covered, it is the most favourable to arrange the panels in the S direction, regardless the power station capacity. For the other two plants the most favourable solution of power distribution depends on its amount. In case of the installation power equal to the contracted power in the drinking water intake station, the increase in the level in which the energy demand is covered at the level of 0.7% was observed for power distribution in the E direction 80% and E 70% in relation to the reference level which comprised the level of satisfaction of the demand with orientation in the S direction. In the third plant, the highest level in which the demand was covered was obtained for positioning in the S direction. After the capacity of the PV power station increased by 60% in the second plant, the highest increase of nearly 3% was obtained with the installation of 50-60% of power in E direction. An insignificant growth was observed for the last plant (0.7%) with the share of power in the E direction amounting to 60%. Nevertheless, the differences observed in all the analysed cases were too small to be statistically significant for the adopted significance level.

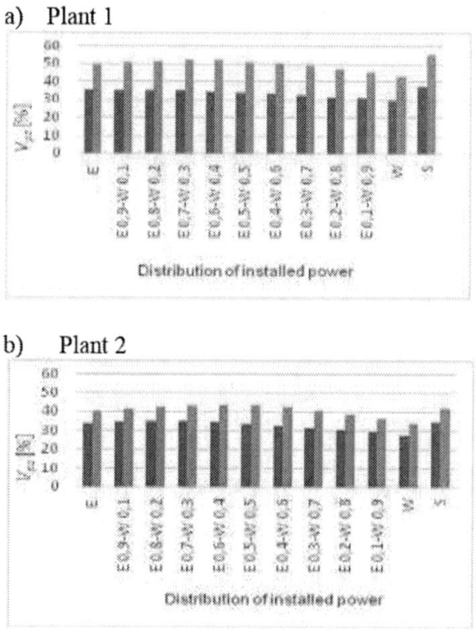

c) Plant 3

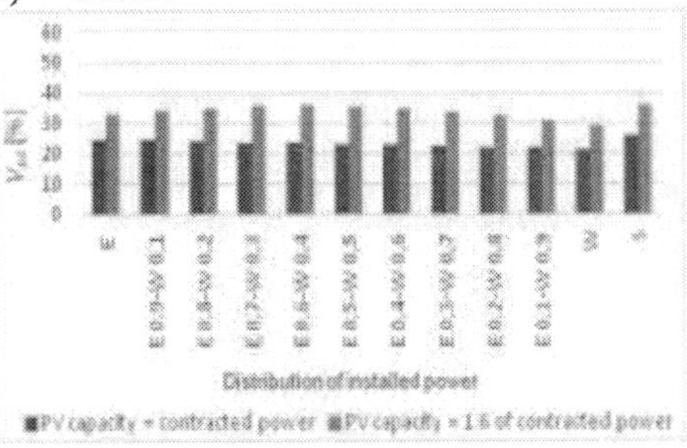

Fig. 6. Impact of power distribution of PV power station in the direction E, S, W on the level in which energy demand was covered for the specific recipients

The share of energy that the tested facilities could not use for all the plants depended on the capacity of a PV power station. In the first facility the surplus energy achieved a low level of nearly 2 and 8% accordingly for the rated power and 60% of oversizing with the power distribution in the E direction amounting to 60%. For the second plant the highest consumption of the produced energy was also registered with 60% of power installed in the E direction and power station capacity equal to contracted power of the plant, and after its oversizing, for symmetric distribution to the specific directions. In case of the last plant the surplus energy occurred only after the increase in the power station capacity and their lowest value was characterised by symmetric power distribution. The differences observed between surplus energy for the south azimuth (S) and after optimal power distribution to the east and west (E-W) for all the analysed cases were sufficiently high so as to assume that they are significantly different from the adopted significance level.

a) Plant 1

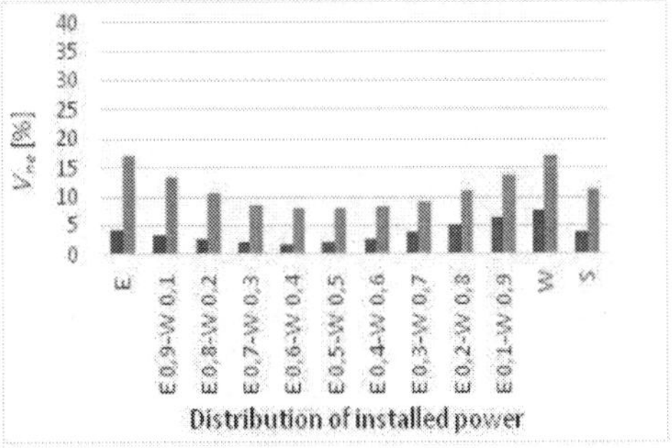

b) Plant 2

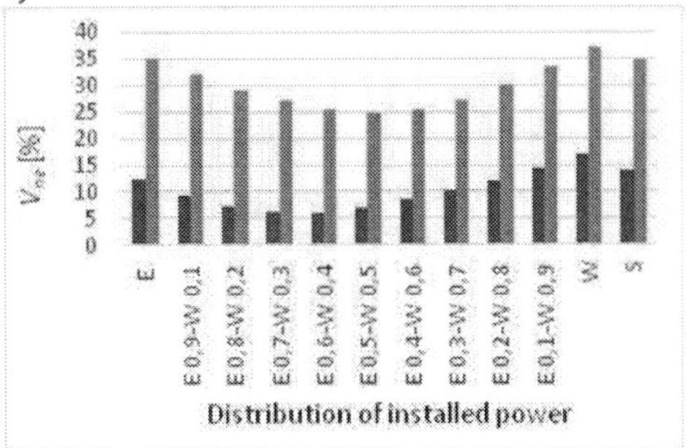

c) Plant 3

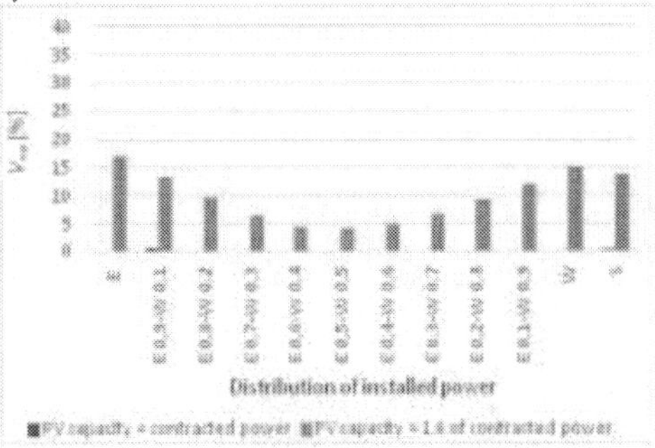

Fig.7. Impact of power distribution of PV power station on the amount of energy produced in PV power station, but unused due to lack of demand at the given time for the specific recipients

4 Conclusions

On the territory of Małopolska Voivodeship 45÷50% of solar radiation has a value of energy flux below 200 $W \cdot m^{-2}$, and nearly 3% of solar radiation approaches the level in which the rated power of PV modules is specified.

The performed analyses show that both the size, as well as distribution of the installed PV power in the E-W direction influences the level in which the energy demand is met and the amount of surplus energy from PV for the selected recipients in a different manner.

Installation of panels only in the E or W direction caused deterioration of cooperation of the power station with the tested facilities. An insignificant increase (4-5%) of the demand satisfaction level was observed for the drinking water intake station and wastewater treatment plant after symmetric power distribution to the E and W and a simultaneous increase of the power station capacity by 40÷60% in relation to the contracted power of the plants. The advantage of such power distribution is the increase in matching the profile of available solar radiation to the profile of energy demand in plants.

The conducted analysis of test results showed that the best solution for the studied recipients due to minimization of the surplus of the produced energy is construction of a power station with power distributed symmetrically in the E-W direction. Such a solution enabled reduction of the share of the unused energy by approx. 10%. On the other hand, alternative solutions consisting in panel orientation in the E or W direction lead to a negative phenomenon of increase in the amount of unused energy with regard to the south azimuth.

The increase in use of the available solar radiation up to 3% was also obtained by changing the ratio of distribution of the installed power in the E and W direction. Such distribution of the installed power of the PV power plant had the greatest impact on the drinking water intake station, the energy demand profile of which is characterised by clear bimodality.

Acknowledgements

This research was financed by the Ministry of Science and Higher Education of the Republic of Poland

References

1. Urząd Regulacji Energetyki. https://www.ure.gov.pl/pl/rynki-energii/energia-elektryczna/odnawialne-zrodla-ener/potencjal-krajowy-oze/5753,Moc-zainstalowana-MW.html. (2017)
2. Rynek fotowoltaiki w Polsce. Instytut Energetyki Odnawialnej (2017)
3. T. Szul, J. Knaga, K. Nęcka, TEKA Komisji Motoryzacji i Energetyki Rolnictwa, vol. **13**, nr 2, 91-96, (2013)
4. J. Knaga, K. Nęcka, Logistyka, **5**, 199-206. (2015)
5. Bugała, G. Frydrychowicz-Jastrzębska, Poznań University of Technology Academic Journals no **79** Electrical Engineering. 47-54. (2014)
6. P. Sawicka-Chudy, M. Cholewa, M. Sibiński, R. Pawełek, Przegląd Elektrotechniczny ISSN 0033-2097, R. 92 NR 9. 58-61. (2016) doi:10.15199/48.2016.09.15
7. H.M.S. Hussein, G.E. Ahmad, .H.H. El-Ghetany, Energy Conversion and Management, Volume **45**, Issues 15–16. 2441-2452 (2004), https://doi.org/10.1016/j.enconman.2003.11.013
8. E. Mehleri, P. Zervas, H. Sarimveis, J. Palyvos, N. Markatos, Renew. Energy, **35**, 2468–2475, (2010)
9. J. Rakovec, D. Kastelec, K. Zakšek, K. Brecl, M. Topic. Orientation and Tilt Dependence of a Fixed PV Array Energy Yield Based on Measurements of Solar Energy and Ground Albedo-a Case Study of Slovenia. INTECH Open Access Publisher (2011)
10. J. Kaldellis, D. Zafirakis, Energy, **38**, 305–314. (2012)
11. A.K. Yadav, S.S. Chandel. Renew. Sustain. Energy Rev., **23**, 503–513. (2013)
12. K. Oprzędkiewicz, J. Teneta, Automatyka. Tom **15**. Zeszyt 2. 381—388, (2011)

https://doi.org/10.1051/bioconf/20181002024

Effect of biomaterials and working pressure of a briquetting machine on physical characteristics and energy consumption of briquette production

Ignacy Niedziółka[1,*], *Maciej* Sprawka[1], *Beata* Zaklika[1], *Artur* Kraszkiewicz[2], and *Artur* Przywara[2]

[1]Univeristy of Life Sciences, Department of Agricultural Machines, Głęboka 28, 20-612 Lublin, Poland
[2]Univeristy of Life Sciences, Department of Machine Operation and Production Process Management, Głęboka 28, 20-612 Lublin, Poland

Abstract. The paper presents an analysis of the influence of biomaterials and working pressure of a briquetting machine on physical characteristics and energy consumption of briquette production. The following types of biomaterials were used in the study: rape, oat and maize straw. Hydraulic piston briquetting machine JUNIOR manufactured by Deta Polska was used for briquetting. During the briquetting process, the working pressures of briquetting machine were 20, 26 and 32 MPa. Depending on the type of biomaterial used and the assumed working pressure of briquetting machine, produced briquettes differed in terms of both their physical characteristics and energy consumption. Based on the analysis of the obtained results, it was found that physical characteristics and energy consumption during briquette production were influenced by such factors as the type of compacted material, its fragmentation as well as granulometric composition and working pressure of the briquetting machine used.

1 Introduction

Biomass is the world's third largest Renewable Energy Source (RES). It is worth using for energy purposes, because biofuels are not harmful to the environment. The amount of carbon dioxide emitted into the atmosphere during its combustion, is balanced by the amount of CO_2 absorbed by plants that reproduce the biomass by means of photosynthesis [1]. In addition, attention should be paid to the declining resources of fossil fuels as well as to the environmental pollution caused by conventional energy using these sources. This contributes to the increased interest in RES. The current trends in the development of power industry are aimed at increasing the use of RES, including biomass [2].

Biomaterials from agricultural and forestry production are the main source of renewable energy in Poland and represent a significant energy potential. These include by-products and waste from agriculture, forestry and agri-food industry [3]. The plant biomass in its unprocessed state, is characterized by low bulk density, making its transport, storage and practical use difficult. For these reasons, there is a need for its agglomeration in the form of pellets or briquettes [4-7]. These biofuels are made of finely fragmented biomass under the influence of suitable pressure and temperature, most often without the addition of any binder. The pressure and temperature occurring during the agglomeration process result in a significant reduction in the volume of compacted biomaterials [8-10]. This also results in a decrease in water content, increased concentration of mass and energy, as well as distribution and use of biofuels produced. Furthermore, the use of biomass in a compact form allows the parametric unification of the fuel, automation of its dosage in the combustion process and obtaining a competitive energy source [11, 12].

Solid biofuels produced from plant biomass arouse interest mainly among distributed energy producers as well as individual consumers. At present, professional energy production is reluctant to use biomass as a fuel in power systems. This is due, among others, lower quality of this kind of fuel and also the instability of its supply. However, it is necessary to look for alternative fuels for the energy sector, hence the aim is to use different types of plant biomass for these purposes [13-15]. New technological solutions using plant fuels to produce electricity and heat, are also sought. The existing technologies for the management of biomass surpluses from agricultural, forestry and orchard production, are also being improved. Growing demand for electricity, as well as a trend towards environmental protection, is driving the search for alternative energy sources [16].

2 Material and methods

Moisture content of biomaterials was determined using the RADWAG MAX 50/1/WH laboratory drier.

*Corresponding author: ignacy.niedziolka@up.lublin.pl

Samples of wet materials were placed in a drying chamber of the weighing-drying device and then dried at 120°C to a constant weight according to the standard PN-EN 15414-3:2011. The calorific value of biomaterials was determined on the basis of the combustion heat determined by means of calorimetric method using an isoperybolic calorimeter of Parr 6400 type. Samples weighing 100 g (±1 g) were taken for the sieve analysis and a laboratory shaker with a set of sieves was applied: 3.15, 2.8, 2.0, 1.4, 1.0, 0.5, 0.25 mm, according to the standard PN-EN 15149-2:2011.

Following biomaterials were used for the production of briquettes: rape, oat and maize straw. These materials were crushed using a H 111/1 type rotary shredder driven by an electric motor of 7.5 kW and equipped with a screen with a diameter of 20 mm. For briquetting of the crushed materials, a hydraulic piston briquetting machine of JUNIOR type manufactured by Deta Polska, was used. During this process, a double feed hopper was applied to the briquetting compaction chamber and three working pressures were applied: 20, 26 and 32 MPa.

Measurements of the physical characteristics of produced briquettes included: length, diameter and weight. Samples of briquettes weighing 1000 g (±10 g) were taken for measurements performed in 5 replicates. Geometric dimensions of briquettes were determined by means of a caliper with a measurement accuracy of ±1 mm while their mass using a laboratory scale with an accuracy of ±0.1 g.

Briquette bulk density was determined on the basis of measurements of their physical characteristics and calculated according to the formula (1):

$$\rho_w = \frac{4 \cdot 10^6 \cdot m}{\pi \cdot d^2 \cdot l} (\mathrm{kg \cdot m^{-3}}) \quad (1)$$

where: ρ_w – briquette bulk density ($\mathrm{kg \cdot m^{-3}}$),
m – mass of briquette (g),
d – outer diameter of briquette (mm),
l – length of briquette (mm).

Measurements of mechanical durability of briquettes were carried out on a test bench in accordance with standard PN-EN 15210-2:2011. The rotational speed of the cylinder was 21 rpm (± 0.1 rpm), a test time of 5 min, and a sample weight of 2000 g (±100 g). After the durability test, the briquettes were screened on a 31.5 mm diameter screen. The mechanical durability of briquettes was calculated according to the formula (2):

$$D_U = \frac{m_A}{m_E} \cdot 100\ (\%) \quad (2)$$

where: D_U mechanical durability of briquettes (%),
m_A – mass of briquettes after durability test (g),
m_E – mass of briquettes before durability test (g).

A specialized registration system was used to measure the energy consumption of the briquetting process of the biomaterials tested. This system uses Lumel N14 electrical network parameters meter, developed with current transformers. With its use, energy consumption was recorded and referred to the resulting briquette mass.

Achieved results of briquettes testing were subjected to statistical analysis using a two-factor analysis of variance and Tukey's test. For this purpose, the SAS Enterprise Guide 5.1 software was used. For statistical analysis, the level of significance α=0.05 was assumed. The obtained analysis results will allow to evaluate the influence of the adopted factors on tested briquette characteristics.

3 Results and discussion

Prior to the briquetting process of biomaterials, their moisture content and calorific value were determined. Based on the achieved results, it was found that the moisture content of materials used for the production of briquettes ranged from 12.8 to 13.5%, while their calorific value was within the range of 16.5-17.3 $\mathrm{MJ \cdot kg^{-1}}$ (Table 1).

Table 1. Moisture content and calorific value of biomaterials used for briquetting.

No.	Kind of biomaterial	Moisture (%)	Calorific value ($\mathrm{MJ \cdot kg^{-1}}$)
1.	Rape straw	12.8	16.5
2.	Oat straw	13.1	16.8
3.	Maize straw	13.5	17.3

Table 2 presents results of granulometric analysis of raw materials used in the production of briquettes. The particle size depended on the nature of raw material and its susceptibility for grinding. The highest particle mass of 3.15 mm was found for maize straw (85.2%), significantly lower for oat straw (46.3%), and lowest for rape straw (31.6%). The smallest mass share of the dusty fraction (<0.5 mm) was recorded for maize straw (2.3%), higher for oat straw (5.1%), while largest for rape straw (9.1%).

Table 2. Granulometric composition of materials used in the production of briquettes (%).

Meshes of sieve (mm) / Kind of biomaterial	3.15	2.8	2.0	1.4	1.0	0.5	0.25	0.0
Rape straw	31.6	1.9	5.7	16.1	21.2	14.4	5.2	3.9
Oat straw	46.3	2.2	6.3	12.5	15.4	12.2	3.2	1.9
Maize straw	85.2	0.8	0.5	1.9	3.6	5.7	1.6	0.7

Figure 1 shows the results of briquette length measurements for the materials tested and the assumed pressure values.

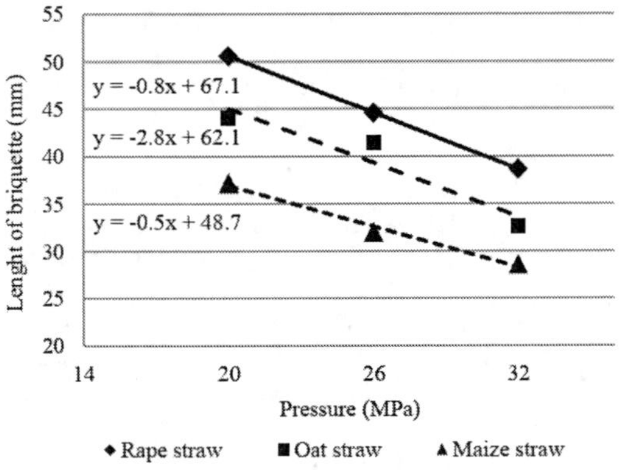

Fig. 1. Average length of briquettes depending on the type of raw material and working pressure of briquetting machine.

For rapeseed straw briquettes, the average length decreased from 50.6 to 38.6 mm, for oat straw from 44.0 to 32.6 mm and for maize straw from 37.2 to 28.6 mm. Analysis of the data revealed that with an increase in working pressure from 20 to 32 MPa, the briquette length decreased by 31% for rape straw, by 35% for oat straw, and by 30% for maize straw. There were statistically significant differences between the briquette length produced from studied raw materials and the working pressure of briquetting machine.

Figure 2 shows the results of mass measurements of briquettes produced from compacted raw materials and assumed pressure values. The lowest mass was obtained for maize straw briquettes (38.6-43.1 g), higher for oat straw (50.4-61.4 g) and highest for rape straw (74.7-81.3 g). The analysis of achieved data suggests that, with increasing working pressure, the weight of briquettes increased by approximately 12, 22 and 9%. There were statistically significant differences between the mass of briquette produced from tested raw materials and the assumed working pressure of the briquetting machine.

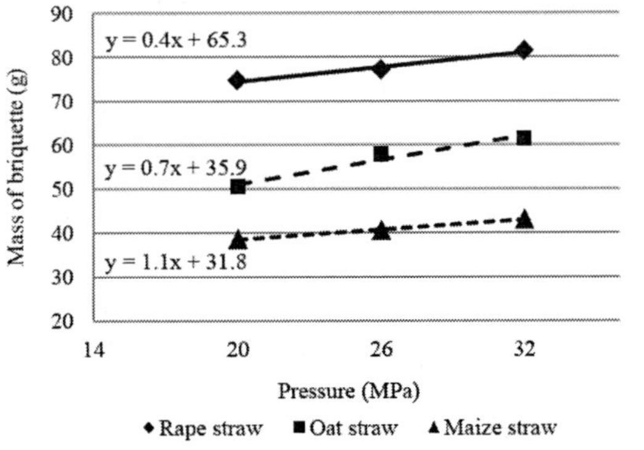

Fig. 2. Average mass of briquettes depending on the type of raw material and working pressure of briquetting machine.

Figure 3 illustrates results of briquette bulk density studies for compacted raw materials and assumed pressure values. The lowest density was recorded for briquettes made of maize straw (530-766 $kg \cdot m^{-3}$), higher for oat straw (589-964 $kg \cdot m^{-3}$), whereas highest for rape straw (753-1075 $kg \cdot m^{-3}$). Analysis of the obtained data shows that, with increasing working pressure, the briquette bulk density increased by approximately 45, 64 and 43%. There were statistically significant differences between the bulk density of briquettes produced from tested raw materials and the assumed working pressure of the briquetting machine.

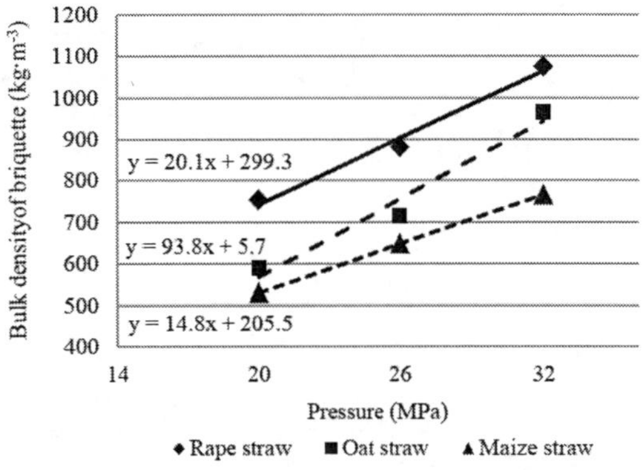

Fig. 3. Average bulk density of briquettes depending on the type of raw material and working pressure of briquetting machine.

Figure 4 shows results of mechanical durability tests of briquettes depending on the type of raw material and working pressure of the briquetting machine. For the assumed working pressure, the lowest mechanical durability was recorded for maize straw briquettes, significantly higher for rape straw, and the highest for briquettes made of oat straw. The analysis of achieved obtained data revealed that the increase in working pressure from 20 to 32 MPa increased the mechanical durability of briquettes from 13 to 16 percentage points.

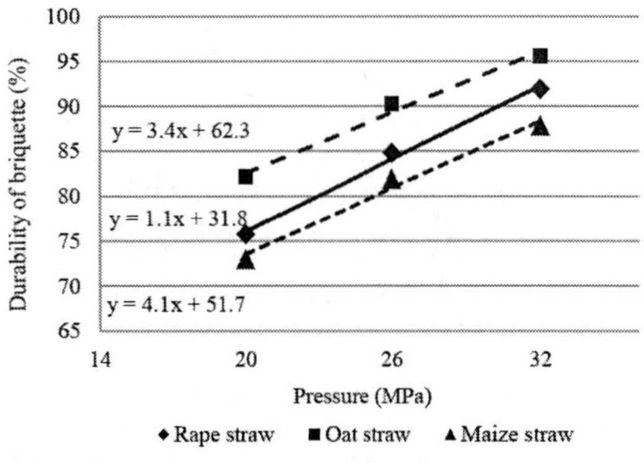

Fig. 4. Average mechanical durability of briquettes depending on the type of raw material and working pressure of briquetting machine.

Figure 5 shows results of the energy consumption tests for briquette production depending on the type of raw material and working pressure of the briquetting machine.

For rape straw briquettes, the average energy consumption of the production was within the range of 109-137 Wh·kg^{-1}, for oat straw 114-129 Wh·kg^{-1} and for maize straw 122-145 Wh·kg^{-1}. Analysis of the data showed that with an increase in working pressure from 20 to 32 MPa, the energy consumption of rape straw briquette production increased by about 26%, for oat straw by 13%, and for maize straw by 19%.

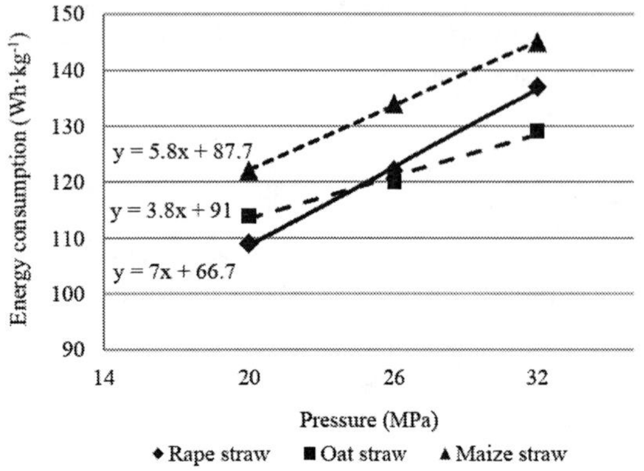

Fig. 5. Average energy consumption of briquettes production depending on the type of raw material and working pressure of briquetting machine.

4 Conclusions

1. Depending on the type of biomaterial used and the assumed working pressure of briquetting presses, resulting briquettes differed significantly both in terms of their physical characteristics and their production energy consumption.
2. It was found that with the increase in the pressure of briquetting machine, the briquette length decreased by 30-35%, while their weight increased by 9% for rape straw, 12% for maize straw, and 22% for oat straw.
3. Increasing the working pressure of briquetting machine from 20 to 32 MPa resulted in an average of 1.5 times the bulk density of produced briquettes. The lowest bulk density was recorded for maize straw compacted at 20 MPa (530 kg·m^{-3}), whereas the highest for rape straw at 32 MPa (1075 kg·m^{-3}).
4. Mechanical durability of briquettes depended both on the type of material to be compacted and working pressure of the briquetting machine. The lowest durability of briquettes was recorded when compacting maize straw (73-88%), while highest for oat straw (82-96%).
5. Analysis of achieved results showed that the physical characteristics and energy consumption of briquette production were influenced by such factors as: type of compacted material, fineness and granulometric composition, as well as working pressure of briquetting machine.

References

1. B.M. Jenkins, L.L. Baxter, T.R. Miles jr., T.R. Miles, Fuel Process. Tech., **54**, 17 (1998)
2. I. Niedziółka (red.), *Technika produkcji brykietów z biomasy roślinnej* (Libropolis Lublin, 2014)
3. A. Grzybek, P. Gradziuk, K. Kowalczyk, *Słoma - energetyczne paliwo* (Wieś Jutra Sp. z o.o., Warszawa, 2001)
4. V. Panwar, B. Prasad, K. Wasewar, J. Energy Eng., **137/2**, 108 (2011)
5. M.K. Sharma, P. Gohil, N. Sharma, Am. J. Eng. Res. (AJER), **04/02**, 44 (2015)
6. S. Mani, L.G. Tabil, S. Sokhansanj, Powder Handl. Process., **15**, 160 (2003)
7. I. Niedziółka, M. Szpryngiel, M. Kachel-Jakubowska, A. Kraszkiewicz, K. Zawiślak, P. Sobczak, R. Nadulski, Renew. Energy, **76**, 312 (2015)
8. R. Hejft, S. Obidziński, J. Res. Appl. Agric. Eng. **60(1)**, 19 (2015)
9. I. Niedziółka, B. Zaklika, M. Kachel-Jakubowska, A. Kraszkiewicz, Ann. Wars. Univ. Life Sci.–SGGW, Agric., (Agric. For. Eng.), **68**, 51 (2016)
10. N. Kaliyan, R.V. Morey, Biomass Bioenerg., **33**, 337 (2009)
11. S. Bielski, B. Dubis, K. Jankowski, Przem. Chem. **94/10**, 1798 (2015)
12. J. Carroll, J. Finnan, Biosys. Eng.,**112**, 151 (2012)
13. P. Mc Kendry, Bioresource Technol., **83**, 37 (2002)
14. I. Niedziółka, M. Szymanek, Teka Komis. Mot. Energ. Rol., **X**, 301 (2010)
15. S. Vassilev, D. Baxter, L. Andersen, C. Vassileva, Fuel, **89**, 913 (2010)
16. P. Gradziuk, *Biopaliwa* (Wieś Jutra Sp. z o.o., Warszawa, 2003)

BIO Web of Conferences **10**, 02025 (2018) https://doi.org/10.1051/bioconf/20181002025
Contemporary Research Trends in Agricultural Engineering

Construction design of apple sorter

Aleksander Nieoczym[1], *Jacek* Caban[2,*], *Andrzej* Marczuk[2], and *František* Brumerčik[3]

[1]Lublin University of Technology, Faculty of Mechanical Engineering, Department of Machine Design and Mechatronics, 36 Nadbystrzycka Street, 20-618 Lublin, Poland
[2]University of Life Sciences in Lublin, Faculty of Production Engineering, Department of Agricultural, Forestry and Transport Machines, 28 Głęboka Street, 20-612 Lublin, Poland
[3]University of Zilina, Faculty of Mechanical Engineering, Department of Design and Machine Elements, 1 Univerzitna Street, 010 26 Žilina, Slovakia

Abstract. When assessing the quality of fruit and packaging process, fruit-producing farms owners decide to evaluate fruit by people or automated sorting lines. The purchase of an automated sorting line generates high costs for the company, but it brings benefits in the form of increased work efficiency, and the better organization of fruit packaging and storage processes. The use of that machinery and equipment is common in agricultural farms as well as in fruit and vegetable processing companies. Despite the widespread use of various types of fruit sorters, the analysis of the operation of the designed device and the study of its technological parameters is still a current research problem. During operation of the devices for sorting fruit there are many technical problems affecting technological processes and quality of fruits. In order to improve the efficiency of sorting fruit, this process should be quickly and accurately. The purpose of this paper is to present the automated apple sorter line construction design, and software for quality controlling fruits. Selected elements of the sorter structure including endurance calculation using the Finite Element Method (FEM) and fruit control system using image analysis were presented.

1 Introduction

The development of technology and the related emergence of mechatronical devices causes that in each industry people are replaced by devices becoming more and more efficient. The human becomes a supervisor for work of the machinery. This supervision is not only mechanical maintenance but also checking on the correctness of the implementation of the programs implemented by the microprocessor. The use of machinery and equipment is also common in agricultural farms as well as in fruit and vegetable processing companies. Market pressure demanding a low-cost, high-quality product means that agricultural companies are forced to replace human work with technically advanced machines [1]. In the case of devices for sorting fruit, optimisation of the operation line technology requires knowledge of the physical characteristics, relevant in the process of separation and sorting fruit so that with a minimum energy it would be possible to receive the final effect of the desired quality [2, 3]. To assess the value of the operating conditions is used a number of approaches, due to the fact that in practice there are different circumstances allowing (or not) the application of certain methods [4]. Designed and built machines are of two types. The first one is a machine that only supports the sorting process, which is not complicated (which makes it cheaper) but does not exclude the use of the human factor. The second type of device is a device with multiple sorting mechanisms equipped with computer-controlled surveillance and vision systems. The most advanced technological lines for sorting fruit have several types of sorting devices, connected to the autonomous system that is capable of "learning" while doing work which translates into quality and significantly reduces the time of its tasks.

There are several types of sorting machines in the fruit processing equipment. Sentinel sorter by Odenberg company is created for two-way pre-sorting. Using this device it is possible to detect faults and foreign bodies in the fruit. This appliance is equipped with colour sensors and sensors type NIR, which are able to scan the surface of apples moving at a high speed. Thanks to it the sorter has a good quality of received image (which translates into the accuracy of the classification), and is maintained high efficiency of segregation [5].

Single-track sorting machine MONO by Sorter company is especially designed for use by small and medium-sized fruit growers. Thanks to their system it is able to sort the fruit by weight, form or colour - depending on its program [6].

Cup sorting machine (CUP) allows the segregation of the fruit in terms of weight, size or colour of colouring. This device is one of the most advanced and the most productive machines of technological sorting line. It is possible to change the number of sorting tracks within

* Corresponding author: jacek.caban@up.lublin.pl

the range from two to twelve. Device software allows the continuous monitoring of the status of the sorting line and its performance [7].

The smaller processors do not decide to purchase the latest sorters. This is due to the high price of sorters and the cost of their servicing. Companies producing these machine in most cases have a monopoly on servicing their own devices due to various forms of patents and security of control programs. The consensus in this situation was to create a little complicated and relatively inexpensive machine that could automatically sort fruit, under the supervision of qualified worker but without more interference in this process.

2 The draft of the sorting line

Already at the stage of preparation of many processing processes in the chemical and food industry, the basic technical conditions of equipment are chosen for the demanded states of quality, form and geometrical dimensions [8, 9]. The apple sorting line design is based on a single, rectilinear separating path where the fruits are sorted and routed to separate trays. The device is equipped with a container of 200kg of fruit. The processing capacity of the entire sorting line is 1.5 tonnes of fruit per hour. The various components of the sorting line are described in Figure 1.

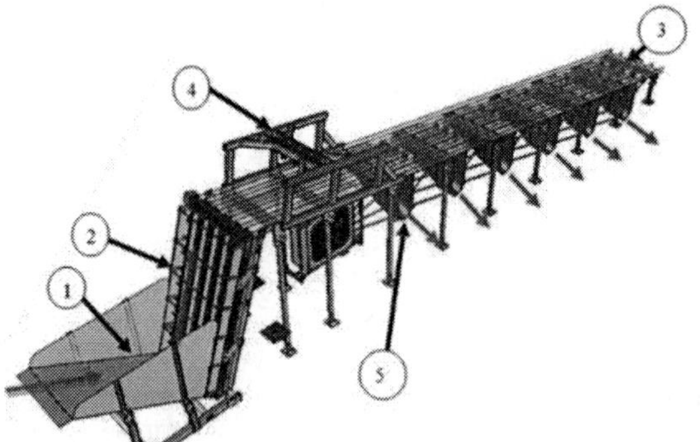

Fig. 1. Solid model of the sorting line for sorting apples (description in the text).

Apples not sorted are poured into the fruit container (1). Thus are transported using the conveyor belt (2), onto the ramp (3). The ramp is at an angle of 20°, so that the fruit may roll over freely. Over the upper part of the ramp are located AVT Oscar F-810C cameras (4), which take a series of photos of the ongoing fruit. Data from cameras are sent to the computer via wireless using the WLAN. The special program analyses the pictures and classifies fruit in accordance with standard [10]. After sorting, the computer sends a signal to the right SSK-B05 driver of stepper motor, which opens the appropriate latch, using 5704X-01engine. When the latch opens, apples classified to the appropriate class fall down through the hole and roll down the gutters (5) to the box below, while the non-standard apples roll to the end of the ramp where they reach the waste container. The line allows to segregate into five selection classes.

2.1 The transport system of non-segregated apples

Model of the fruit container is shown in Figure 2. It consists of a plating made of stainless steel plate mounted on a support structure. The bottom of the plating is the floor (a). It is set at an angle of 30° to the level in the direction of the conveyor belt so that the apples are able to slide towards the belt with a free movement. The floor plates folded at an angle of 90° form the side walls (b). Above the floor there is a shelf (c), which is the back support for the crates, from which the fruit is poured into the container. The shelf is additionally reinforced with a stiffening (d), visible at the I enlargement. The lower part of the floor has notches (e) for the moving line landings. Above the notches, the sides of the container sheet are symmetrically bent at an angle of 90° to the outside (f), and a number of holes (g) are drilled on the resulting surface by which it is possible to fix the container to the outer wall of the conveyor belt.

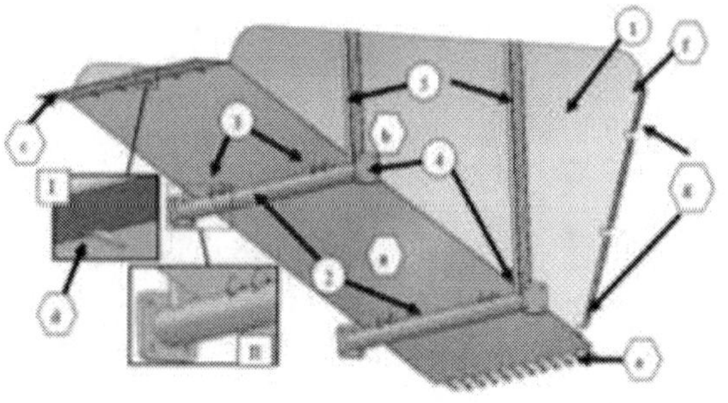

Fig. 2. Structural elements of the fruit container (description in the text).

The plating of the container (1) is based on pipes with a diameter of ø50 mm (2). For a better connection of the flat surface of the sheet with the cylindrical surface of the pipe are welded on to both sides of the pipes network ribs (3), visible on the II enlargement. At the ends of the tubular profiles, are symmetrically welded flat bars (4) with holes for fixing the legs of the container. On the surface of the side walls have been added reinforcements (5) to stiffen the entire profile panel. The solid model showing the conveyor assembly together with the frame and the fruit container is shown in Figure 3.

Fig. 3. The solid model of the conveyor with a frame and a container for fruit.

For the transport of fruit from the container onto the ramp is used the MINITrans H65 belt conveyor [11] with a length of 1500 mm. The original conveyor belt was completely modified (Figure 4). Finger landings have been installed to transport apples at an angle from 0° to 90°.

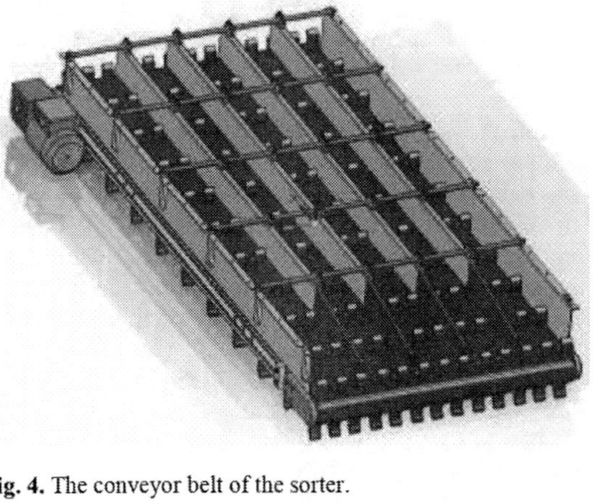

Fig. 4. The conveyor belt of the sorter.

2.2 Segregation

The solid model of the ramp is shown in Figure 5. The welded structure of the ramp (1) is the frame on which the ratchet mechanisms (2) are attached, and the frame (3) on which the cameras (4) are attached is mounted on the upper part. The structure is based on nine pairs of brackets (5), made of C20x40x5 C-profile. Between the supports on the inside of the ramp is a control box (6), which controls the motors of the ratchet mechanism. Complete ratcheting mechanism is shown in Figure 6.

Fig. 5. The solid model of the ramp of the sorting line (description in the text).

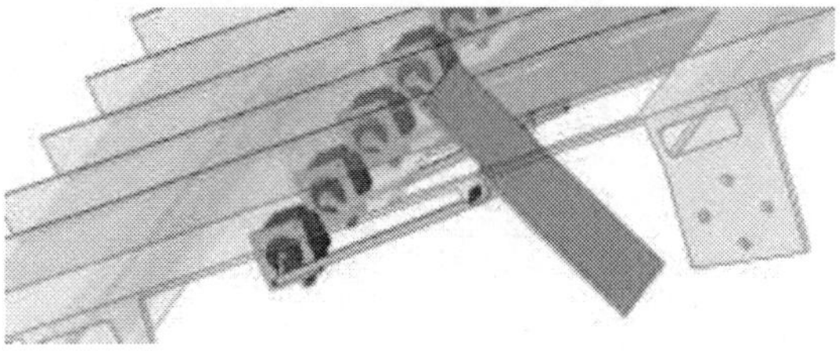

Fig. 6. View of the design of the ratchet.

The process of opening and closing individual latches uses electric motor 5704X-01 by Lin Engineering [12]. This engine is bipolar, powered by 24V DC. SKK-B05 controller was used to control the electric motor 5704X-01 [13]. This device controls the motor via step and direction signals (STEP/DIR). Depending on the frequency of the pulses, the speed of the motor changes and depending on the number of pulses, the angle of rotation of the shaft changes.

As the optical sensor was used the AVT Oscar F 810C camera [14]. It has a high-speed interface (up to 400MB/s) and shutter for recording at 6.7 frames per second (at 1280 x 960 pixels), which significantly reduces the time course of the process of sorting single apples [1]. The camera has also features of automatic image processing. As a result, the images received are free of noise, with better sharpness and colour saturation.

2.3 Strength analysis of selected sorter elements

Endurance calculations were performed using the Finite Element Method (FEM). One of the busiest parts of the sorting line is the fruit container. The container (Fig. 7) is fixed in the holes for the M10 screws (1) and on the face surface of the container sheet (2) (where the structure is attached to the sheet profile of the outer wall of the conveyor belt). For an even distribution of the load on the whole surface of the floor of the container

https://doi.org/10.1051/bioconf/20181002025

(3) was applied pressure of 200 kg of fruit. On the surface of the whole model was generated FEM net, whose characteristics are summarized in Table 1.

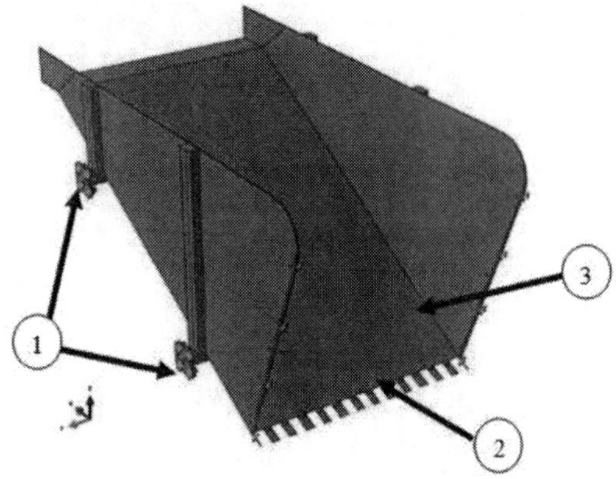

Fig. 7. Model of the container with marked resting places and place of application of pressure on the surface of the fruit container sheet.

Table 1. Setting Word's margins.

Surface	Net type	Number of nodes	Number of components
The side walls of fruit container sheet and mounting sheets	tetragonal of square geometry	75478	35295
The floor of the fruit container	hexagonal with linear geometry	75478	3988

Maximum 50 MPa reduced stresses were obtained on the surfaces of the holes for the fixing screws (item 1, Fig. 7) - Fig. 8.

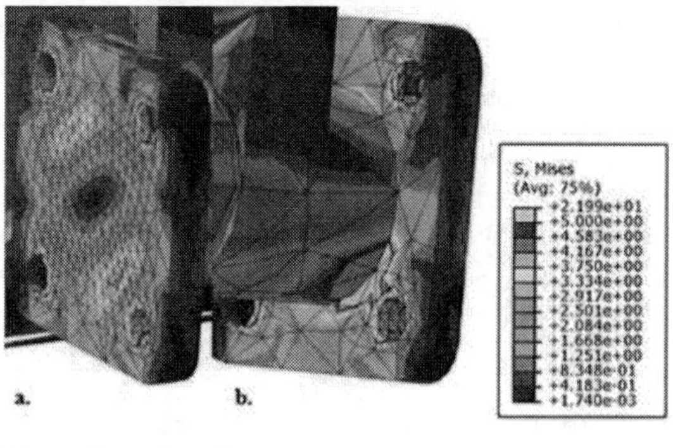

Fig. 8. Place of maximum stress.

Figure 9 shows the stress distribution on the floor surface of the container. Maximal values were obtained in the support areas of transverse tubes (1) and on the surface of the shelf (2).

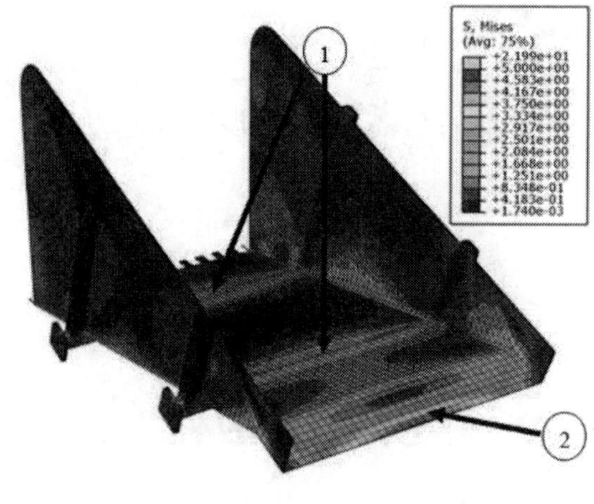

Fig. 9. Reduced stress distribution on the floor of the container.

3 Program controlling the apple sorting process

The control program analyses the photos taken from the camera of a single apple. It distinguishes in them identifying features such as the colour of the fruit, size, pits and fattening on its surface, and then classifies the fruit into the appropriate class. The identification of the fruit ends with sending a signal to the controller to open the corresponding latch. Many research works were devoted to issues of apples quality, including: [2, 15, 16, 17, 18, 19]. However, inspection for damaged, diseased, or contaminated fruit has been hindered by the inability to appropriately orient fruit for imaging and by lack of a method for imaging the whole of the surface of individual fruit [18].

There are many image study techniques. One such option is sub-pixel contour detection of the examined image [16]. Received image is filtered in order to detect the pixels on the border of the abrupt changes on the surface of the apples (pits, fattening). Next, the program, based on the generated pixels, connects them together to form a flat closed curve (Fig.10).

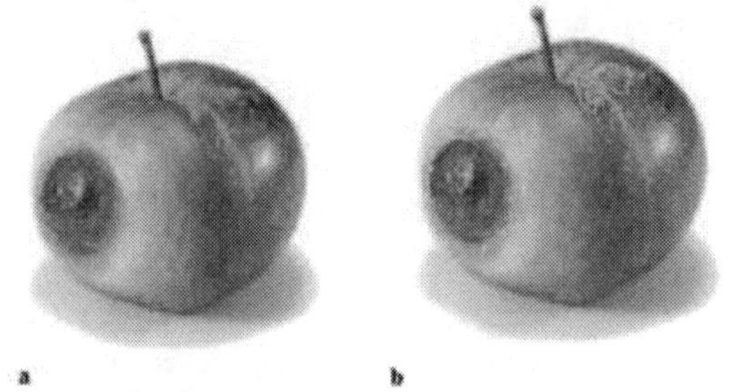

Fig. 10. Apple with marked fattening [16]: (a) visible defects; (b) interpretation of the program.

Another programmatic activity is the addition of the surface areas of the figures obtained and the classification of apples in the range of colour saturation to the corresponding class. High accuracy is achieved with the location of contour lines up to 1/50 of a pixel.

Due to the complexity of calculating the area of closed curves, while limiting the process time, an averaging algorithm can be used in the program to approximate the straight edges of the chaotically located pixels detected by subpixel contour detection.

An example of a program for fruit classification in the processing industry using subpixel contour detection is Ellips EFG 400 software [20]. This program is able to detect defects in fruit but also rate them in terms of the weight and size.

3.1 The selected elements of controlling by EGF400 program

Below are presented some selected tools for controlling the selection process of apples.

1 – The sorting method:

There are several sorting methods by: colour, size, weight, or combinations of these parameters (Fig. 11). It is possible to modify the standard modes and to change their parameters. It is also possible to create your own sorting procedures.

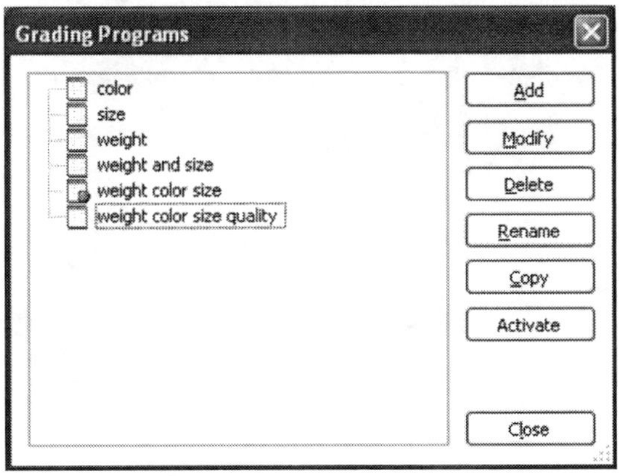

Fig. 11. Sort mode selection window [20].

2 – A quantitative evaluation of the sorting process:

Statistics window (Fig. 12), is equipped with three tabs: Grading Batch, Statistics Batch, and Machine Statistics. A table showing the statistics of each class is displayed and the statistics of the number of apples that do not fall into the standard. The table also shows the total statistics: total weight, percentage of line load, selection speed, grading time. It is possible to generate a chart that may represent the values associated with the work of the sorter.

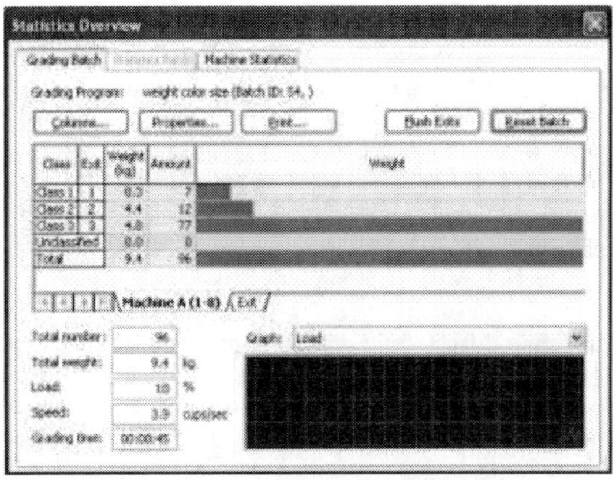

Fig. 12. Statistics window [20].

3 – Report on the measurements:

A window with a measurement table of all apples on a single section (Fig. 13), shows information about each fruit that has gone through the sorting process. The results in the table can be displayed on the basis of the sections or on the basis of selection method. An additional option provides the ability to change the filter measurements and colours – Fig. 14.

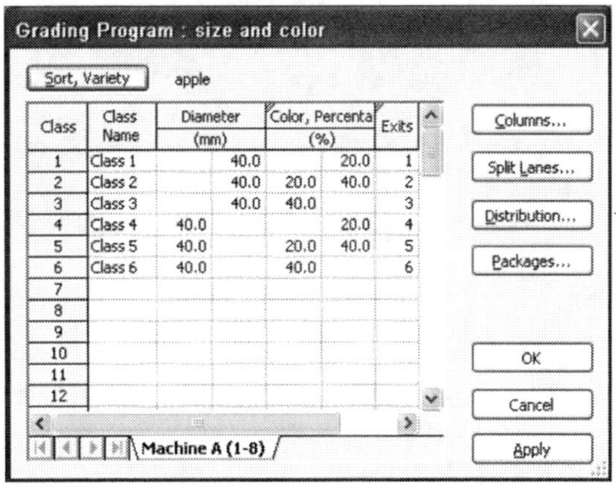

Fig. 13. The results of the sorting in terms of size and colour [20].

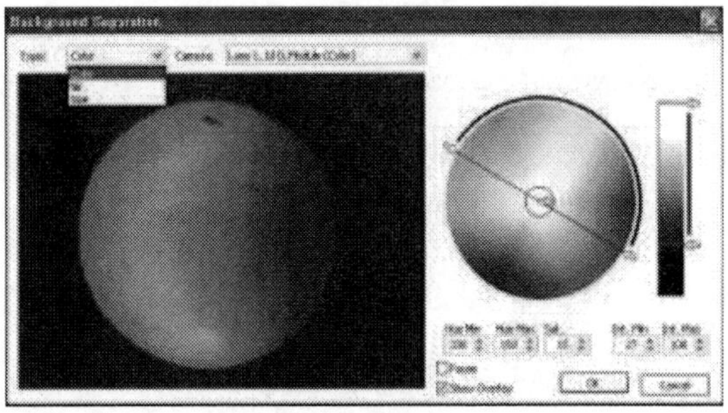

Fig. 14. Advanced base colour settings [20].

Thanks to its intuitive interface and wide range of settings, Ellips EFG 400 is not only suitable for apples but also for segregation of other fruits. Combining and editing selection modes allows you to sort different varieties and fairly accurately detect fruit defects.

4 Closing remarks

The development of mechatronic systems for technological processes in agro-food industries, enforces need to design multifunctional constructions of high-performance. Also a very important factor in the assessment of the quality of the technological process is the condition of sorted fruit. Therefore, fruit inspection and sorting by human and automated imaging systems on the packaging lines are critical while packing premium quality fruits without any defects [17]. In case of apples, defects and disorders include arthropod damage (codling moth, mealybugs, aphid damage), postharvest diseases (blue mold, gray mold, mildew), physiological disorders (bitter pit, superficial scald, core flush), mechanical and field injuries (bruising, frost, hail damage), and fruit finish (sunburns). Damage is generally caused by impacts, vibrations, such processes as harvesting, transport, sorting and storage in bulk [21]. The presented design of the automated apple sorter meets the above mentioned technological and quality criteria.

References

1. A. Marczuk, Eksploatacja Niezawodność, Maintenance Reliability, **4**, 82-90 (2009)
2. J. Jachimski, S. Mikrut, *Próba subpikselowej lokalizacji linii konturowych z wykorzystaniem drugiej pochodnej obrazu cyfrowego,* AGH, Kraków (1998)
3. A.P. Kuczyński, Acta Agrophysica, **2**, 4, 787-802 (2003)
4. A. Marczuk, J. Caban, P. Savinykh, N. Turubanov, D. Zyryanov, Eksploatacja Niezawodność, Maintenance Reliability, **19**, 1, 121-125 (2017)
5. https://www.tomra.com/pl/solutions-and-products/sorting-solutions/
6. http://www.sorter.pl/
7. http://www.sorter.pl/pl/automatyka/
8. J. Flizikowski, J. Sadkiewicz, A. Tomporowski, Przem. Chem. **94**, 1, 69-75 (2015)
9. A. Tomporowski, M. Opielak, Eksploatacja Niezawodnosc, Maintenance Reliability, **14**, 3, 222-227 (2012)
10. Rozporządzenie komisji (WE) nr 1221/2008. http://www.lex.europa.eu
11. http://www.iwb-industrietechnik.de/pl/
12. www.linengineering.com/contents/stepmotors/pdf/5704.pdf
13. http://www.akcesoria.cnc.info.pl/sterowniki_silnikow_krokowych.htm.
14. www.alliedvisiontec.com
15. B.S. Bennedsen, D.L. Peterson, A. Tabb, Comput. Electron. Agr. **48**, 92-102 (2005)
16. A. Groborz, P. Kiełbasa, Inżynieria Rolnicza, **9**, 10, 117-126 (2005)
17. S. Jarolmasjed, C.Z. Espinoza, S. Sankaran, L.R. Khot, Postharvest Biol. Tec. **118**, 35-42 (2016)
18. A.M. Lefcourt, P. Narayanan, U. Tasch, M.S. Kim, D. Reese, R. Rostamian, Y.M. Lo, Biosyst. Eng. **104**, 64-71 (2009)
19. J.A. Throop, D.J. Aneshansley, W.C. Anger, D.L. Peterson, Postharvest Biol. Tec. **36**, 281-290 (2005)
20. www.fruittek.com/fruit-packing.../ellips-vision-sorting.html
21. P. Komarnicki, R. Stopa, M. Młotek, Inżynieria Rolnicza, **2**, 3, 145-157 (2013)

Modelling assessment of farmers workload

Halina Pawlak[1], and *Piotr* Maksym[1,*]

[1]University of Life Sciences in Lublin, Faculty of Production Engineering, Departments of Ergonomics, 20-612 Lublin, Głęboka 28. Poland

Abstract. The article presents the principles of modeling the physical load assessment of farmers using Bayesian Network technology. Despite rapid development of mechanization and automation of work processes in agriculture, many farmers use their own physical strength, such as manual handling of loads. There are periods in which a farmer performs several or even a dozen of variety activities during the 24 hours. It often happens that all these activities do in a hurry, without rest, working after several hours a day, which further strengthens his physical load. The variety of activities often causes problems with the use indicators to assess the physical load gauges, so Bayesian Network technology was used to develop the evaluation model.

1 Introduction

The largest and oldest sector of the economy is agriculture, which employs over 2,2 million people, which accounts for about 15% of the population employed in Poland [1]. Agriculture is characterized by a large variety of work and a wide range of hazards. Farmers, regardless of season and weather conditions, perform their activities every day, and their duration often extends up to 12-14 hours per day. This situation makes the farmer, in spite of access to modern machinery, puts his work under the pressure of time, fearing the quality and yield of the crop, and the variability of the activities and work places [2].

The farmer's workplace is a house, yard, farm buildings, farmland. Often, these are activities requiring manual handling of loads, elevation work, use of multiple hand tools and machines. Animal care is also an important part of the work. With such a range of tasks and responsibilities, it is important to be aware of the type of activities performed as well as the efficiency and physical strength of the farmer [3,4].

The biggest threat to farmers is still and still the physical load. It appears mainly in manual work where human muscle strength is required. Too long time, perform one action could result in the occurrence of musculoskeletal disorders. This is primarily due to improperly adopted body position, use of tools, ambient conditions and many other significant environmental factors.

Despite the visible technical progress and mechanization of agricultural work, farmers are still exposed to physical load.

The aim of the work is to develop methods to assess the conceptualization of the workload of farmers, in the field of physical load, using Bayesian networks technology.

2 Material and Methods

The workload is determined by the knowledge of physical load, mental load and environmental load. Due to the nature of the farmer's work, the physical load was assessed in a particular way. Physical load is understood as a component of the workload, which consists of physical effort, static physical load, and monotype (repeatability activities). Each of these factors, describing physical load, includes some sort of measurement and evaluation uncertainty. The measure of uncertainty is probability, and the description of dependence is the total probability distribution. The calculation scheme is based on the Bayesian formula. Modeling uses Bayesian network technology that enables the expression of cause-and-effect relationships [5-7]. Probabilistic network technology allows for fast, almost automated conversion of expert knowledge, expressed verbally or in natural language, into computer models. Bayesian network is a directed acyclic graph, expressed in the visual with the nodes and directed arcs. Network nodes represent random variables, and arches link nodes to relationships between these variables. Each variable has a strictly defined type and field of values. Each node is associated with a Conditional Probability Table (CPT). Data sources for determining conditional probability distribution may be empirical data or expert judgment. Using this data, you can perform the learning process in order to determine probability distributions for all variables, means network nodes [8-11].

Starting automatic inference mechanisms, the aim of which is to obtain, with an accuracy of probability distribution, the answers to the questions: What grade receives physical load? Which variables affect physical

* Corresponding author: piotr.maksym@up.lublin.pl

load? What should be the values of the physical load components to get the desired rating?

3 Results

According to the demands of the Bayesian Network technology, firstly there are defined area for the individual components (nodes), secondly network topology is created and then is defined code calculation formulas for determining the nodes assessing physical load.

Attribute "physical load" is evaluated on the basis of "energy expenditure", "monotype" and "static load" (Fig. 1.), whose values are given in Table 1.

The method to assess the physical load on the basis of the sum of points obtained in the evaluation of the attributes that determine the load is used to build the network [3, 12, 13].

Table 1. Set of attributes values – „physical load".

Attributes	Set of values		
	verbal	point	kJ/8h
Energy expenditure	Little Average Large Very large	1-25 26-50 51-75 76-100	<3350 3351-6300 6301-8400 >8400
Monotype	Little Average Large Very large	1-25 26-50 51-75 76-100	–
Static load	Little Average Large Very large	1-30 31-60 61-90 91-100	–

The assessment of the physical load was based on the principle of energy expenditure assessment used in the simplified Lehmann method. It consists of summing up the energy expenditure defining the position of the body and the involved muscle groups (type of work) during the eight hour shift [14]. Set of attribute values for the "energy expenditure" attribute are shown in tables 2 and 3.

Another attribute that determines "workload" is "monotype", for which attribute values are given in table 4, and evaluation rules in table 5.

Table 2. Set of attribute values – „body position".

Attribute	Set of values	
	verbal	kJ/min
Body position	Seated On the knees / In Squat Standing Standing leaning Walking	1,25 2,10 2,51 3,35 10,89

Table 3. Set of attributes values – „energy expenditure".

Attributes	Set of values	
	verbal	kJ/min
Work type	Light Average Heavy Very heavy	
Work of the whole body	Light Average Heavy Very heavy	10,5 – 16,8 16,8 – 25,1 25,1 – 35,6 35,6 – 48,2
Work of one arm	Light Average Heavy	2,9 – 5,0 5,0 – 7,1 7,1 – 9,2
Work of fingers and forearm	Light Average Heavy	1,3 – 2,5 2,5 – 3,8 3,8 – 5,0
Work of both arms	Light Average Heavy	6,3 – 8,4 8,4 – 10,5 10,5 – 12,6

Table 4. Set of attributes values – "monotypes".

Attributes	Set of values
Require strength	below 100 N above 100 N
Number of repetitions	below 300 301-800 801-1600 1601-3200 above 3200

Table 5. Evaluation of the nuisances degree of physical load due to monotype of movements [13].

The number of repetitions of stereotyped movements per shift (8h)		Nuisances degree	
strength >100 N	strength <100 N	verbal	point
below 300 301-800 801-1600 above 1600	below 800 801-1600 1601-3200 above 3200	Little Average Large Very large	1-25 26-50 51-75 76-100

The severity of work is influenced by static load, which can be a limiting factor of time and correctness of performed work. If the work consists of activities performed at different body positions, the most stressed position should be taken into account when assessing the static load, provided that the duration of this activity is not less than 3 hours [12, 13].

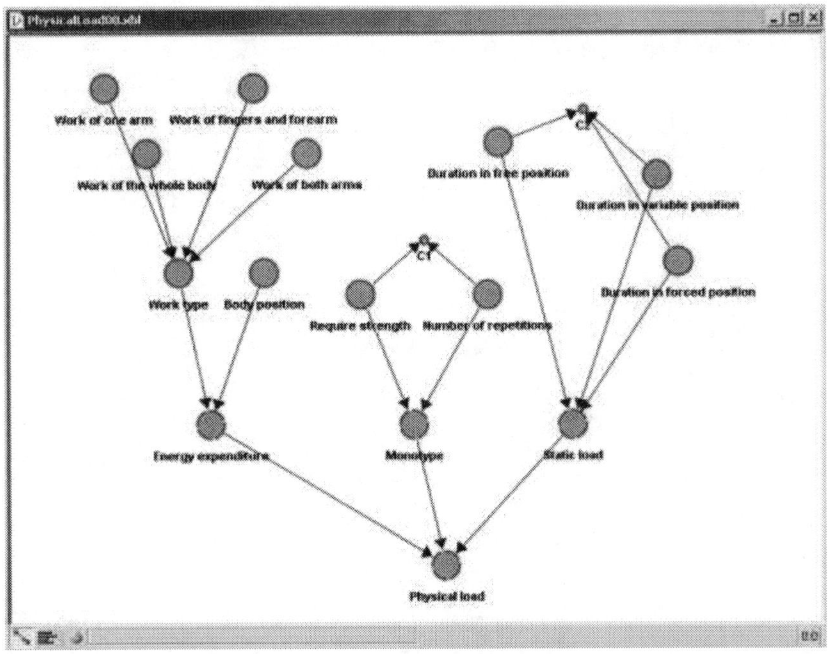

Fig. 1. The network structure of attributes assessment s „physical load" (own work)
where nodes C1 and C2 are constraint (logical) nodes, they allow to introduce dependency constraints on the nodes, for example: between the number of repetitions of movements and the use of force when performing the activity.

The body positions taken into account in the assessment of static load and the way the evaluation are shown in Table 6.

Table 6. Evaluation of static load [12, 13].

The body position at work	Point evaluation	Verbal evaluation
Unforced sitting	1-10	Little
Unforced standing with the possibility of changing position on the sitting	11-20	
Sitting or standing alternating with walking	21-30	
Forced sitting, not bent or slightly bent	31-40	Average
Unforced standing unaffected without the possibility of periodic changing position on sitting	41-50	
Forced standing, not bent with the possibility of changing position on the sitting	51-60	
Force sitting, very bent	61-70	Large
Forced standing, not bent without the possibility of changing position on the sitting	71-80	
Forced standing, bent regardless of ability to change position	81-90	
Kneeling in squat and other positions	91-100	Very large

Attribute „physical load" is evaluated on the basis of the following formula (Fig. 2.).

```
?Physical load?=
If(?Static load?+?Monotype?+?Energy
expenditure?<=30,"Very light",
If(?Static load?+?Monotype?+?Energy
expenditure?<=70,"Light",
If(?Static load?+?Monotype?+?Energy
expenditure?<=120,"Average",
If(?Static load?+?Monotype?+?Energy
expenditure?>120,"Heavy","Very
heavy"))))
```

Fig. 2. Formula code of attribute evaluation of „physical load" (own work).

The attribute "workload" is evaluated for physical and mental load. The nature of work position at the hand-shuffling operations requires force effort, which is associated with considerable physical load. Thus, in the further inference process the attributes describing this load are taken into account (Fig. 3.).

As shown by the histogram a (Fig. 3.), the probability distribution over the values of the attribute "physical load" indicates the value that determines work in this position as heavy. Therefore, we are interested in which of the attributes influencing the physical load has a significant impact on the value of this load.

With a properly organized work process, the physical load should be set at a very light, light or average level. To make the inference, the least favorable value of this range is assumed, ie "average" - histogram a '(Figure 4.). The assumption will be met when energy expenditure and the static load will be reduced (histogram b', d' Fig. 4.).

 https://doi.org/10.1051/bioconf/20181002026

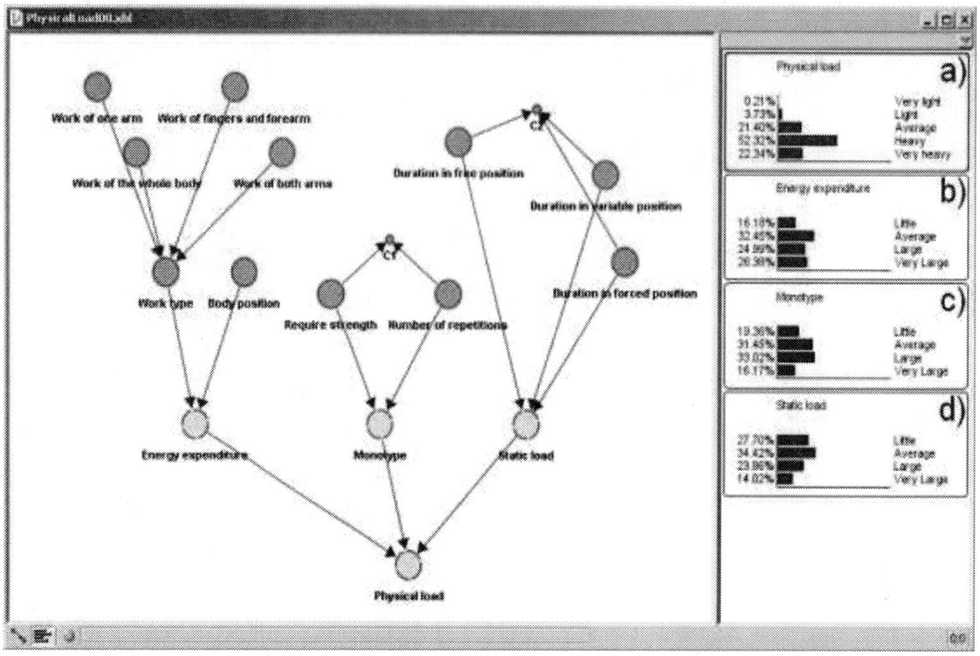

Fig. 3. Evaluation of attribute "physical load" (own work).

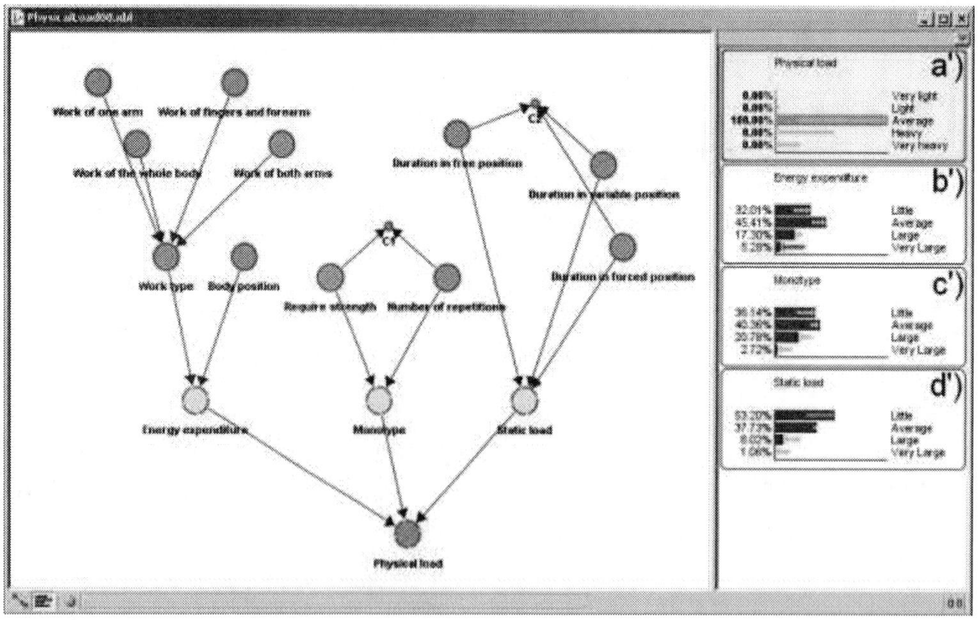

Fig. 4. Evaluation of attribute "physical load" with a priori set value (own work).

The obtained results will be subject to an sensitivity analysis, that can be described as a method which define how "sensitive" mathematic model really is when the parameters and the overall structure of the model experience changes. Sensitivity analysis allows researcher to find the accuracy needed for an accurate and functional model. That kind of analysis identifies a leverage point in the model and a parameter which specific value can significantly influence the behaviour mode of the system [15, 16]. Another method of the effectiveness that will be taken account is ROC curve, which shows the classification quality of the physical load assessment using sample and test data sets [17].

4 Conclusions

The paper presents methods of assessing the physical load of farmers using the formal knowledge representation system.

https://doi.org/10.1051/bioconf/20181002026

Based on empirical data and expert opinions, Bayesian Networks could be used to develop a model for assessing the physical load of farmers. Automatic inference makes it possible to get answers to questions What grade receives physical load? Which variables affect physical load? What should be the values of the physical load components to get the desired rating?

The results are often indicative of the hard work of farmers. For example, cereal harvesting - hand-shuffling operations or animal handling activities. The determining factor of workload is the energy expenditure.

The modular Bayesian Network design allows to expand your network, for example with other criteria for assessing workload. The next step of development of the model will be measure of its effectiveness by the mentioned methods.

References

1. GUS, Zakład Wydawnictw Statystycznych. (2016)
2. H. Pawlak, B. Nowakowicz-Dębek, Agric. Sci. Procedia. **7**, 209-214 (2015)
3. D. Koradecka D. (red.), CIOP, (1997)
4. B. Nowakowicz-Dębek, H. Pawlak, Ł. Wlazło, I. Kuna-Broniowska, H. Bis-Wencel, A. Buczaj, P. Maksym, Ann. Agric. Environ. Med. **21**, 4, 730–734, (2014)
5. Documentation of the program BayesiaLab. http://www.bayesia.com/book
6. J.Y. Halpern,. MIT Press, (2005)
7. J. Pearl, Morgan Kaufmann Publishers Inc. (1988)
8. H. Pawlak, P. Maksym, Agr. Eng. **7(105)**, 173-177, (2008)
9. H. Pawlak, P. Maksym, Agr. Eng. **5(130)**, 231-236, (2011)
10. H. Hołaj, A. Kusz, P. Maksym, A.W. Marciniak, Agr. Eng. **6(131)**, 53–60, (2011)
11. P. Maksym, A. W. Marciniak, A. Kusz, Agr. Eng. **4(129),** 213–220, (2011)
12. J. Kania, PWE. (1980)
13. E. Górska, Oficyna Wydawnicza Politechniki Warszawskiej (2007)
14. D. Koradecka, J. Bugajska, CIOP. (1998)
15. L. Breieriova, M. Choudhari, MIT System Dynamics in Education Project, (1996).
16. N.P. Ventikos, G.V. Lykos,, N.A. Vagias Atlantis Computational Intelligence Systems, **6**, 651-676, (2012)
17. Fawcett T. Pattern Recogn. Lett. **27(8)**, 861-874, (2006)

https://doi.org/10.1051/bioconf/20181002027

Influence of the Harvesting and Ensilage Technology on the Quality of Maize Straw Silage

Jacek Przybył[1,*], *Dawid* Wojcieszak[1], *Ireneusz* Kowalik[1], *Jacek* Dach[1]

[1]Poznan University of Life Sciences, Institute of Biosystems Engineering, WojskaPolskiego 50, 60-625 Poznan, Poland

Abstract. The aim of the study was to assess how the harvesting and ensilage technology influenced the quality of maize straw silages. The research findings showed that it was best to make maize straw silage in a flexible silo, whereas it was least favourable to make it in a film-covered field prism. The choice of the adequate ensilage method results in a high quality silage even if there is high content of dry matter in maize straw.

1 Introduction

At present whole maize plants are the most common substrate for biogas production in Central Europe [1]. However, the use of full-value agricultural products for energy purposes increases their prices and reduces the profitabilityof biogas production [2]. Therefore, it is necessary to develop alternative energy substrates, including agricultural waste and by-products [3]. Maize straw is a by-product available in large amounts, which has not been fully used so far. The production of maize grains leaves crop residue, such as stalks, leaves, cob cores and cob-covering leaves. The crop residue makes 47-50% of the dry weight yield of whole maize plants [4]. There is 0.15 kg of cob cores, 0.22 kg of leaves, 0.14 kg of cob-covering leaves and 0.5 kg of stalks per one kilogram of the dry weight of maize grains [4].

In 2015 maize grain production in Poland amounted to 3.16 million tonnes [5]. Our estimates show that the annual maize straw production in Poland may amount to as much as 4 million tonnes. There is high potential of maize straw as a substrate for biogas production in Poland [6-8]. In Poland maize is harvested for grain in late autumn (October-November), when the air temperature drops and the humidity rises. Under these conditions the moisture of maize grains usually ranges from 30% to 35%. The moisture of maize residue ranges from 47% to 66% [9]. It is impossible to dry maize straw in the field. Ensilage is the only reasonable solution in the storage of maize straw.

The aim of the study was to assess how the harvesting and ensilage technology influenced the quality of maize straw silages.

2 Material and methods

The maize straw ensilage experiment was conducted during production at the Stary Gołębin farm, which belongs to the Top Farms Wielkopolska company. It was abivariate experiment. The first-order factor was thetype of container where maize straw was ensilaged, i.e.

- aheap,
- a flexible silo,
- film-wrapped cylindrical bales.

The second-order factor was the maize straw ensilage method, i.e.

- natural ensilage (CCSN),
- ensilage aided with lactic acid bacterial strains (*Pediococcusacidilactici* 33-11 NCIMB 30085, *P. acidilactici* 33-06 NCIMB 30086, *Lactobacillus plantarum* LSI NCIMB 30083, *L. plantarum* L- 256 NCIMB 30084, *Enterococcus faecium* M74 NCIMB11181)(CCS1),
- ensilage aided with an organic preparation (sodium benzoate, propionic acid and sodium propionate) (CCS2).

Silages were made from the straw left after the maize cultivar PR39A79 (Pioneer company, FAO 300) had been harvested for grain. The dry matter content in the crop residue was 68% and the organic dry matter content was 87.11%.

The yield of maize dry matter consisted of cobs (59%), stalks (19%), leaves (14%) andhusks(8%).

Maize straw was harvested with a New Holland 644 variable chamber baler and a Pöttinger Jumbo 7200 Powermatic loader wagon.

2. 1 Laboratory analyses of silages

Maize straw samples weighing about 2 kg were collected from a flexible silo, prism and film-wrapped cylindrical bales in order to conduct physical and chemical analyses. The analyses of raw silage samples were conducted in the laboratory of the Department of Animal Nutrition and Feed Management, Poznan University of Life Sciences, Poland. The laboratory analyses of the samples included:

- the analysis of dry matter content,

*Corresponding author: jacek.przybyl@up.poznan.pl

- the analysis of the content of volatile fatty acids,
- pH measurement.

The dry matter content was determined by means of the drying and weighing method, according to the Polish Standard PN-ISO 6496:2002. Liquid chromatography was applied to measure the content of volatile fatty acids:lactic acid, acetic acid and butyric acid. The content was measured by means of aWaters 2690 chromatograph with a Waters 2487 absorbance detector. The pH value of silages was determined by measuring the concentration of hydrogen ions with an electrode coupled with a WTW ph 730 InoLab pH-meter.

The results were statistically analyzed. The two-way analysis variance (ANOVA) were performed.

2.2 Assessment of silage quality

The measurements of the content ofvolatile fatty acids: lactic acid, acetic acid and butyric acid were used to assess the quality of silages according to the Flieg-Zimmer scale. This method of assessment of silage quality is based on the percentage of volatile fatty acids in the total weight of silage.

3 Results

3.1 Dry matter content

The highest average content of dry matter was measured in the maize straw ensilaged in the heap, i.e. 76.63%. The lowest content of dry matter was noted in the maize straw ensilaged naturally in film-wrapped bales, i.e. 40.23%. The maize straw ensilaged in a flexible silo contained 48.72% of dry matter (Table 1).

The content of dry matter in the input material was 68.00%. When the maize grain was harvested at the phase of full maturity, maize straw was ensilaged in a flexible silo with the CSS2 preparation. The method resulted in the smallest loss of dry matter, i.e. 5.71%. There was a significantly greater loss of dry matter in the naturally ensilaged straw, i.e. 26.01%. The ensilage in a flexible silo with the CSS1 preparation reduced the content of dry matter by 23.12%.

The lowest content of dry matter was measured in the maize straw ensilaged naturally in film-wrapped bales, i.e. 36.13%. The content of dry matter in the bales with the CSS2 preparation was 39.51%, whereas the content of dry matter in the silage made with the CSS1 preparation was 45.07%.

The content of dry matter in the naturally ensilaged straw was 73.58%, with the CSS2 preparation – 77.42%, with the CSS1 preparation – 78.88%. There were statistically significant differences in the content of dry matter between these silages. The results of the analysis of the content of dry matter show that the process of maize straw ensilage was not initiated in the heap. If there is high content of dry matter in the substrate, it is difficult to condense it. As a result, there are adverse aerobic conditions and the silage becomes overheated, which results in the appearance of moulds and fungi.

3.2 Volatile fatty acids

The ensilage of feeds requires favourable conditions for the development of lactic acid bacteria. A stable silage should contain a large amount of lactic acid (1.5-3.5%) and a medium amount of acetic acid (0.5-1.0%). It might contain trace amounts of butyric acid (0.0-0.1%).

There was diversified content of lactic acid in the silages – from 0.08% to 1.11%. When we analyse the data referring to individual types of containers, we can see that the highest average content of lactic acid, i.e. 1.02% was found in the silages made in the flexible silo. The smallest average content, i.e. 0.18% was found in the silages made in the cylindrical bales. The average content of lactic acid in individual ensilage methods was less diversified and ranged from 0.60% in naturally made silages to 0.65% and 0.66% in silages with preparations CSS2 and CSS1 (Table 2). The difference in the average content of lactic acid between the silages with preparations and in the silage made naturally was statistically significant.

As results from the data, the content of lactic acid in the silages was below the required minimum due to the quality of straw.The high content of dry matter in the ensilaged maize straw may have caused problems in the development of lactic acid bacteria. In consequence, there was low content of lactic acid in the silages.

Table 1. The content of dry matter in the silages, according to the ensilaging preparation and type of silo.

Factors	Type of container (B)			Average	LSD $_{(0.05)}$
Ensilaging preparation (A)	Ensilage heap	Film-wrapped bales	Flexible silo		
	Dry matter [%]				
CSS1	73.58	45.07	41.88	53.51	A 0.734 B 0.734 A*B 1.272
CSS2	77.42	39.51	62.29	59.74	
CSSN	78.88	36.13	41.99	52.33	
Average	76.63	40.23	48.72		

Table 2. The content of lactic acid in silages according to the ensilaging preparation and type of silo

Factors	Type of container (B)			Average	$LSD_{(0.05)}$
Ensilaging preparation (A)	Ensilage heap	Film-wrapped bales	Flexible silo		
	Lactic acid [%]				
CSS1	1.05	0.08	0.87	0.66	A 0.028 B 0.028 A*B 0.050
CSS2	0.59	0.29	1.07	0.65	
CSSN	0.51	0.17	1.11	0.60	
Average	0.72	0.18	1.02		

Table 3. The content of acetic acid in silages according to the ensilaging preparation and type of silo.

Factors	Type of container (B)			Average	LSD (0.05)
Ensilaging preparation (A)	Ensilage heap	Film-wrapped bales	Flexible silo		
	Acetic acid [%]				
CSS1	2.60	0.31	0.23	1.05	A 0.121 B 0.121 A*B 0.209
CSS2	2.47	0.51	0.68	1.22	
CSSN	0.69	0.20	0.67	0.52	
Average	1.92	0.34	0.53		

Table 4. The content of butyric acid in silages according to the ensilaging preparation and type of silo.

Factors	Type of container (B)			Average	$LSD_{(0.05)}$
Ensilaging preparation (A)	Ensilage heap	Film-wrapped bales	Flexible silo		
	Butyric acid [%]				
CSS1	0.14	0.13	0.03	0.10	A 0.044 B 0.044 A*B 0.076
CSS2	0.38	0.42	0.03	0.27	
CSSN	0.06	0.24	0.02	0.11	
Average	0.19	0.26	0.02		

The results of laboratory analyses showed that the highest average content of acetic acid, i.e. 1.22% was found in the silages with the CSS2 preparation. The smallest content, i.e. 0.52% was found in the silages prepared naturally. The average content of acetic acid in maize straw silages with the CSS1 preparation was 1.05% (Table 3).

There was a statistically significant increase in the content of acetic acid in the silages with the ensilaging preparations. This increase may have been caused by the fact that the inoculants containing lactic acid bacteria also contained substances stimulating the production of acetic acid. Acetic acid has positive influence on the aerobicstability of silages, which is important when a silage is collected from a silo.

The silages produced in the experiment also contained butyric acid. The smallest average content of butyric acid, i.e. 0.02% was found in the silage stored in the flexible silo. The highest average content, i.e. 0.26% was found in the silage made from film-wrapped bales. The analysis of variance showed that both factors under study resulted in the production of butyric acid in the maize straw silages (Table 4).

As results from the data, butyric acid was produced in all the silages. This means that there were some irregularities in the ensilaging process.

3.3 The Flieg-Zimmer method

The measurements of the content of volatile fatty acids:lactic acid, acetic acid and butyric acid were used to assess the quality of silages according to the Flieg-Zimmer scale. The silages produced naturally and stored in the flexible silo and the silage inoculated with the CCS1 preparation were characterised by good quality, whereas the quality of the silage with the CCS2 preparation was satisfactory.

Table 5. The ranking of maize straw silages according to the Flieg-Zimmer scale.

Maize straw ensilaged	pH	Volatile fatty acids [%]			Score	Quality according to Flieg-Zimmer scale
		Lactic acid	Acetic acid	Butyric acid		
Naturally in ensilage heap	7.76	0.51	0.69	0.06	23	poor
With CSS1 preparation in ensilage heap	7.35	1.05	2.6	0.14	22	poor
With CSS2 preparation in ensilage heap	7.43	0.59	2.47	0.38	8	low
Naturally in film-wrapped bale	6.06	0.17	0.20	0.24	7	low
With CSS1 preparation in film-wrapped bale	7.96	0.08	0.31	0.13	0	low
With CSS2 preparation infilm-wrapped bale	4.91	0.29	0.51	0.42	-4	low
Naturallyin flexible silo	4.07	1.12	0.67	0.02	72	good
With CSS1 preparation in flexible silo	4.39	0.87	0.23	0.03	76	good
With CSS2 preparation in flexible silo	4.52	1.07	0.68	0.03	52	satisfactory

The silages made in cylindrical bales had low content of lactic acid (0.08-0.29%) and they contained butyric acid – their quality was low. The quality of the silages made in the prism was also low (Table 5).

The ensilaging capacity of vegetatively older green forages was reduced due to the increased content of raw fibre and reduced content of sugar. It was also a problem to displace air from the ensilaged mass. In consequence, sugars were additionally consumed in the process of respiration.

4 Conclusions

The research findings led to the following conclusions:

1. Maize straw with high content of dry matter can be used as a substrate for good quality silage.
2. A flexible silo is the best container for maize straw ensilage.
3. The results of the analysis of the content of dry matter show that the process of maize straw ensilage was not initiated in the ensilage heap.
4. The choice of the adequate ensilage method has significant influence on the quality of maize straw silage.

References

1. J. Dach, W. Czekała, P. Boniecki, A. Lewicki, T. Piechota. AMR 909, 305-10 (2014)doi.10.4028/www.scientific.net/AMR.909.305
2. E.M. White, G. Latta, R.J. Alig, K.E. Skog, D.M. Adams. Energy Policy **58**, 64-7(2013) http://dx.doi.org/10.1016/j.enpol.2013.02.029
3. M. Monch-Tegeder, A. Lammer, H. Oechsner. Energy **73**, 523-530 (2014). http://dx.doi.org/10.1016/j.energy.2014.06.051
4. S. Sokhansanj, S. Mani, S. Tagore, A. F. Biomass Bioenerg **34**, 75-81 (2010). http://dx.doi.org/10.1016/j.biombioe.2009.10.001
5. CSO. Agriculture in 2015. (Warsaw 2016).
6. M. Cieślik, J., Dach, A. Lewicki, A. Smurzyńska, D. Janczak, J. Pawlicka-Kaczorowska, P. Boniecki, P. Cyplik, W. Czekała, K. Jóźwiakowski. Energy **115**, 495-1502 (2016)http://dx.doi.org/10.1016/j.energy.2016.06.070
7. J. Przybył, W. Kot, D. Wojcieszak, N. Mioduszewska, D. Karol. Agricultural Engineering **4(148),** 103-111 (2013)
8. M. Cieślik, J. Dach, P. Boniecki, D. Janczak, W. Czekała, A. Lewicki, P.C.R. Carmona, J. Pawlicka. ICEE Internatonal Conference on Energy &Environmental 300-306 (2015)
9. K. J. Shinners, B. N. Binversie. Biomass Bioenerg **31**, 576-584 (2007)

Influence of fineness level and applied agglomeration pressure of peppermint herb (*Mentha piperita* L.) on the mechanical properties of the obtained product

*Urszula Sadowska**[1], *Andrzej Żabiński*[1], *Krzysztof Mudryk*[1]

[1]*University of Agriculture in Krakow, Faculty of Production and Power Engineering, Balicka 116 B, 30-149 Krakow, Poland*

Abstract. The objective of the conducted study was to evaluate the impact of the pressure agglomeration process of peppermint herb on the mechanical properties of the obtained product. The separated fractions of peppermint with 0.5-2.5 and 2.5-5 mm particles were compacted using a hydraulic press Fritz Heckert EU 20, with pressure 50, 100, 150 and 200 MPa. A closed matrix with the compression chamber diameter of 15.6 mm was used. Every time, a 2-g herb sample (corresponding to the weight of tea used for the production of tea bags) was poured into the matrix. Thus, compacted herb in the form of a straight cylinder was obtained. When producing the agglomerate compaction work was determined. Strength tests of the obtained agglomerate were conducted using the MTS Insight 2 testing machine. The density of the produced agglomerate, its compaction level and strength in the Brazilian test was calculated.

The obtained results indicate that the values of the tested parameters increase with the increase of pressure in the tested range, yet differences occur between the tested herb fractions. Typically, the agglomerate produced from 0.5-2.5 mm fraction is characterized by a greater density, and the higher level of agglomerate compaction is obtained using 2.5-5 mm herb fraction. The highest strength determined using Brazilian test was determined for agglomerate produced from 0.5-5 mm peppermint herb fraction at 200 MPa pressure and 0.5-2.5 mm fraction using 150 and 200 MPa pressure.

1. Introduction

For centuries, humans have used herbaceous plants for different aims, focusing primarily on their medicinal properties. Taste or aroma of herbal teas often constitute one of our first childhood memories.

One of the most commonly used herbaceous plant species distributed around the world for the preparation of herbal teas peppermint [1, 2, 3, 4]. It typically is sold in loose form(leaf), or portioned, in bags. The present study proposes new form of peppermint tea, where the herb is subjected to the pressure agglomeration process. However, only few medicinal substances possess properties enabling their direct compaction. Yaman et al. [5] believe that the compaction pressure and other parameters of the process shall be selected depending on the compacted material. However, selection of the suitable pressure is not an easy task, as its values change depending on the properties of the processed material [6]. The compaction pressure is one of the most important factors determining obtaining of a product (pellet, blocks) with the desired quality [7]. Koutný et al. [8] state that the main parameters indicating the quality of products obtained through pressure agglomeration, is, above all, the density, as well as their mechanical strength.

The objective of the conducted study was to evaluate the impact of the pressure agglomeration process of peppermint herb on the mechanical properties of the obtained product.

2. Materials and methods

2.1. Plant material

The study material was dried peppermint herb (*Mentha piperita* L.) with the moisture of 11%, which is the upper allowable norm for the species in pharmaceutic terms [9]. From the herb, two fractions of peppermint herb (0.5-2.5 mm and 2.5-5 mm) were separated, using LPzE-2e laboratory shaker with the vibration amplitude of 0-2.5 mm and sieves with square mesh diameter of 0.5 mm, 2.5 mm and 5 mm.

* Corresponding author: urszula.sadowska@ur.krakow.pl

2.2. Methods

2.2.1. Pressure compaction process of pepper mint herb

The bulk density for both obtained fractions was established following the NORM PN-ISO 7971-2:1998 using a 1 dm^3 vessel.

The separated herb fractions were compacted using a Fritz Heckert EU 20 hydraulic press with computer recording of with a closed matrix equipped with a compression the compaction process and the pressing mechanism chamber diameter of 15.6 mm and 82 mm long. During the test, a curve, the so-called compaction characteristics (relationship between the compaction force and cylinder displacement) was recorded, which was the basis for determination of the process' parameters. Based on presented data, it was possible to determine the compaction process work, defined as the area under the curve. The head travel speed during compaction was 10 $mm \cdot min^{-1}$. The process was being carried out at ambient temperature.

The following compaction pressures were used: 50, 100, 150 and 200 MPa. Every time, the 2-g weighed herb sample (corresponding to the weight of tea used for the production of tea bags) was manually poured into the matrix. As a result of the process, herb compacted into the form of a straight cylinder was obtained.

2.2.2. Strength tests

The study of mechanical strength of compacted peppermint herb with the shape of straight cylinder included determination of tensile strength using Brazilian test method. In this method, a cylindrical sample is subjected to a compacting load along the diameter, thus the other name for the method is the disk test. Such types of load result in sample destruction when the tensile strength is exceeded, in the perpendicular direction to the plane of symmetry containing the direction of the load [10]. This method is commonly used to assess the mechanical properties of cylinder-shaped brittle blocks [11]. The obtained agglomerates were subjected to the disk test using a MTS Insight 2 test machine. The compaction process was conducted until cracks beginning in the central portion of the cylinder, spreading along the vertical axis became visible to the naked eye (Fig. 1).

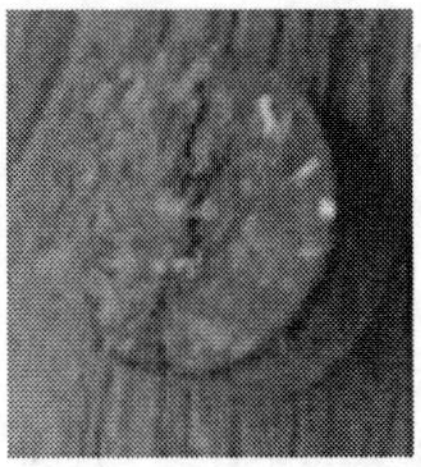

Fig. 1. Example appearance of agglomerate subjected to Brazilian test

Tensile strength during radial compaction was calculated based on the following formula [12, 13].

$$\sigma_n = 2 \cdot Fn \cdot (\pi \cdot d \cdot l)^{-1} \quad (1)$$

Where:
σ_n – tensile strength (MPa)
Fn – agglomerate destructive force (N),
d – compacted sample diameter (m),
l – agglomerate length (m).

Moreover, we measured length, diameter and weight of the obtained agglomerate in order to calculate their density in $g \cdot cm^{-3}$. The measurements were conducted in 20 repetitions for each combination.

The agglomerate compaction level was also determined, as the multiplication factor of volume decrease, following the methodology presented by Skonecki et al. [14] according to the following formula:

$$S_{za} = \rho_{al} \cdot \rho_n^{-1} \quad (2)$$

Where:
S_{za} – agglomerate compaction level
ρ_{al} – agglomerate density after 48 hours of storage ($g \cdot cm^{-3}$)
ρ_n – initial material density in the compression chamber (bulk density) ($g \cdot cm^{-3}$).

After the agglomeration process, the total compaction work was calculated using the registered compaction curves.
Examples of registered courses of the compaction process are presented in Fig. 2.

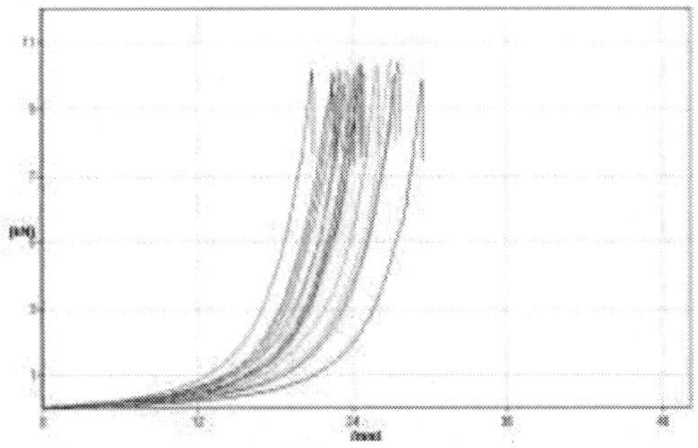

Fig. 2. Selected courses of peppermint herb compaction process

2.3. Statistical analysis

Arithmetic means and standard deviations (SD) were calculated for the obtained results. Statistical analysis was performed using the Statistica 10.0 software (Stat Soft, Inc., Tulsa, Oklahoma USA). Multidirectional variance analysis was used. The significance of differences was evaluated using Duncan test at the significance level of p=0.05, and the results were considerably below the level of

$p<0.05$. Linear regression models were constructed using Microsoft Excel software for the relationship of the compaction work and the applied pressure.

3. Study results

The obtained results of agglomerate strength measurement with the use of the Brazilian test method are presented in Table 1. A significant influence of the applied pressure as well as the separated peppermint herb fraction on its tensile strength was observed. However, a relationship between the applied pressure and separated fraction of peppermint herb was found. According to the Brazilian test, higher strength characterized the agglomerate produced from 0.5-2.5 mm fraction at pressure 50 and 150 MPa as compared to the same pressures for 2.5-5 mm fraction. The phenomenon of higher susceptibility to cracking of agglomerates produced from materials with lower fineness level is reported by Olsson [15].

The highest strength characterized the agglomerate produced at 200 MPa pressure, independently of fraction and 150 MPa only for 0.5-2.5 mm fraction. It can be expected that in the case of the thicker fraction of the herb, empty microspaces could occur between agglomerate particles. The study of Sadowska et al. [16] regarding the grind ability of peppermint agglomerate in a kinetic test also indicate the occurrence of general tendency to improve the mechanical strength of the product when higher pressure is used in the compaction process. However, when subjecting the results of both tests to a comparative analysis, it was observed, that the agglomerate produced from the smaller herb fraction (particularly with lower pressure levels) exhibits better strength parameters in the Brazilian test, whereas in the kinetic test (where its grind ability was tested) it produces inferior results. This phenomenon can be explained by the mechanical wedging-out of the particles, which takes place during compaction of components in the agglomeration process, and then easier separation of the finer particles, due to the limited contact surface of the particles, especially those, which are located on the margins of the product.

Table 1. Values of the Brazilian test of agglomerate produced from different peppermint herb fractions at different pressures

Fraction (mm)	Pressure (MPa)	Brazilian test (MPa)
0,5-2,5	50	0,443 ± 0,078 b*
	100	0,938 ± 0,118 c
	150	1,238 ± 0,052 e
	200	1,191 ± 0,084 e
2,5-5,0	50	0,384 ± 0,066 a
	100	0,921 ± 0,093 c
	150	1,129 ± 0,063 d
	200	1,240 ± 0,070 e

*a, b, c ,d, e, - homogenous groups according to Duncan test, p=0.05

Results of the study regarding the density of the obtained agglomerate are presented in tab. 2. An increase of the agglomerate density was observed along with the increase of pressure applied in the experiment. Similar relationships are provided in Skonecki et al. [17]. Typically higher density characterized agglomerate produced from the finer herb fraction. The relationship for different substrates are confirmed by Mani et al. [18], Mani et al. [7], and Carone et al. [19]. However, in the presented study an influence of the applied pressure and herb fraction and the obtained agglomerate density was observed. The obtained results indicate that at 200 MPa pressure the fineness level of peppermint herb no longer influences the density of the produced agglomerate.

Table 2. Agglomerate density depending on the used pressure and separated herb fraction

Fraction (mm)	Pressure (MPa)	Agglomerate density ($g \cdot cm^{-3}$)
0,5-2,5	50	0,805 ±0,032 B*
	100	0,985 ±0,025 C
	150	1,052 ±0,010 E
	200	1,073 ±0,011 F
2,5-5,0	50	0,769 ±0,024 A
	100	0,973 ±0,023 C
	150	1,037 ±0,008 D
	200	1,066 ±0,010 F

* A, B, C, D, E, F – homogenous groups for agglomerate density according to Duncan test, p=0.05

Changes in the agglomerate density confirm results of its compaction level (Table 3). Analysis of the parameter indicated its increased value along the increase of compaction pressure. The relationship for different substrates are confirmed by Kulig et al. [20]. It also depended on the used material fraction, attaining higher values at lower fineness level of the material and equal pressure for each fraction. This relationship appears to be logical, as the material is characterized by lower density also in the bulk form. Such high level of compaction indirectly explains the possible savings of the storage and transporting spaces which can be generated using the agglomeration process.

Table 3. Agglomerate compaction level depending on the used pressure and separated herb fraction

Fraction (mm)	Pressure (MPa)	Agglomerate compaction level

https://doi.org/10.1051/bioconf/20181002028

0,5-2,5	50	6,242 ± 0,245 a*
	100	7,633 ± 0,193 c
	150	8,156 ± 0,079 d
	200	8,317 ± 0,086 e
2,5-5,0	50	7,124 ± 0,224 b
	100	9,006 ± 0,213 f
	150	9,605 ± 0,074 g
	200	9,869 ± 0,096 h

*a, b, c, d, e, f, g, h – homogenous groups for the compaction level according to Duncan test, p=0.05

Strength parameters of the obtained product shall also be considered in terms of the workload incurred to obtain them. The presented study demonstrated that the level of applied pressure influences the values of compaction work. The total compaction work increased with the increase of pressure value (Table 4). In the majority of cases, no statistically significant differences occurred between the peppermint herb fractions used in the study. Only in the case of the applied pressure of 200 MPa slightly higher values were observed for the finer fraction.

Table 4. Values of compaction work depending on the applied pressure and peppermint herb fraction

Fraction (mm)	Pressure (MPa)	Total compaction work (J)
0,5-2,5	50	41,612 ± 1,788 a*
	100	60,464 ± 2,524 b
	150	78,730 ± 3,522 c
	200	94,432 ± 4,493 e
2,5-5,0	50	41,866 ± 1,960 a
	100	61,908 ± 2,501 b
	150	78,817 ± 1,895 c
	200	92,621 ± 2,597 d

*a, b, c, d, e – homogenous groups according to Duncan test, p=0.05

Regression analysis demonstrated that the relationships of the applied agglomeration pressure and values of compaction work can be presented with linear equations. Results of the conducted analysis, with regression equations explaining these relationships are presented in the following charts (Figs. 3-4).

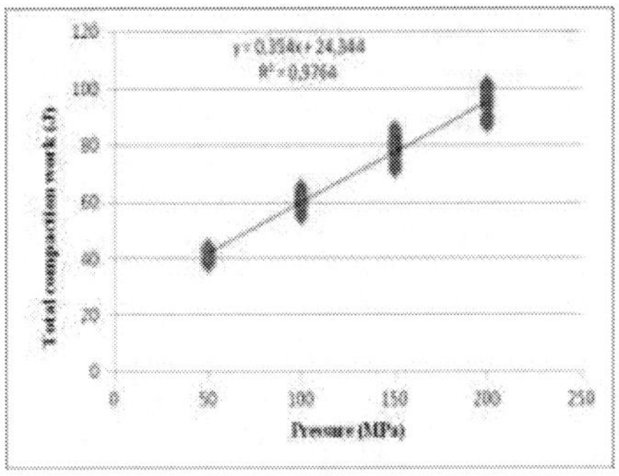

Fig. 3. Relationship between agglomeration pressure and the total compaction work of peppermint herb with fraction 0.5-2.5 mm

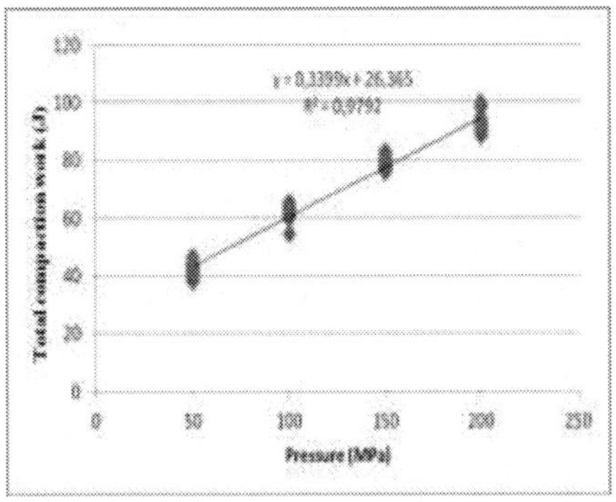

Fig. 4. Relationship between agglomeration pressure and the total compaction work of peppermint herb with fraction 2.5-5.0 mm

High matching index for the linear regression function to the empirical data was demonstrated. In the case of both peppermint herb fractions the coefficient of determination was approx. 0.98.

4. Conclusions

1. It was determined that the mechanical strength of peppermint agglomerates depends on the pressure applied during compaction process as well as the fineness level of peppermint herb. Increase of the applied pressure in the range from 50 to 200 MPa results in increased strength of the agglomerate measured with Brazilian test, produced from 0.5-2.5 mm fraction by approx. 269%, whereas for 2.5-5.0 mm fraction by approx. 323%. In the tested pressure range, an increase in the density of agglomerate from the finer herb fraction occurs by approx. 133%, whereas the thicker fraction about 139%.

2. The work necessary to compact the tested peppermint herb fractions depends on the applied agglomeration pressure. Along with the increased pressure in the range 50-200 MPa, the workload necessary to compact the material to the form of agglomerate are increased twofold. Strict relationships between agglomeration pressure and the work value for individual peppermint herb fractions were found.

3. An analysis of the obtained compaction work values, agglomerate density as well as the results of the Brazilian test allows to conclude that the favorable solution for 0.5-2.5 mm fraction is production of agglomerate at 150 MPa pressure, since as a result it is characterized by high mechanical strength, acceptable density, and the incurred workload is lower than in the case of subsequent pressure used in the experiment. On the other hand, study results regarding 2.5-5.0 mm fraction indicate, that obtaining agglomerates with the highest density and mechanical strength requires the use of 200 MPa pressure.

Acknowledgements
This research was financed by Ministry of Science and Higher Education of the Republic of Poland.

References

1. I. Salamon., P. Labun , Herba Pol, **55** (1), (2009)
2. K. Pytlakowska, A. Kita, P. Janoska, M. Połowniak, V. Kozik , Food Chem. **135,** (2012)
3. A. Szymczycha-Madeja, M. Welna, W. Zyrnicki, J. Braz. Chem. Soc., **24** (5), (2013)
4. A. de Sousa Barros, S.M. de Morais, P. A. T. Ferreira, Í. G. P. Vieira, A. A.Craveiro, R. O. Fontenelle dos Santos, J. E. S. A. de Menezes, F. W. F. da Silva, H. A.de Sousa. IND CROP PROD, **76,** (2015)
5. S. Yaman, M. Sahan, H. Haykiri-Acma, K. Sesen, S. Kucukbayrak, Fuel Process Technol, **68**, (2000)
6. R. Hejft, *Ciśnieniowa aglomeracja materiałów roślinnych* (Biblioteka Problemów Eksploatacji. ITE Radom 2002) ISDN 83-7204-251-9
7. S. Mani, L. G. Tabil , S. Sokhansanj, Biomass Bioenergy, **30** (2006)
8. R. Koutný, B. Čechová, P. Hutla, P. Jevič, Res. Agricul. Eng., **53** (2) (2007)
9. European Pharmacopoeia 6.0, Determination of Essential Oils in Herbal Drugs, 2.8.12, (2008) 251–252
10. J. Mańkowski, P. Ciężkowski, Zeszyty Naukowe Instytutu Pojazdów, **3** (89) (2012)
11. J.R.C. Proveti, G. Michot, INT J FRACTURE, **139**, (2006)
12. Y. Li, D. Wu, J. Zhang, L. Chang, D. Wu, Z. Fang, Y. Shi, Powder Technol. **113**, (2000)
13. G. Ruiz, M. Ortiz, A. Pandolfi, Int. J. Numer. Meth. Engng. **48**, (2000)
14. S. Skonecki, , J. Laskowski, R.Kulig, G. Łysiak, Acta Agroph. **20** (1), (2013)
15. M. Olsson, Wood pellets as low-emitting residential biofuel. Thesis for the degree of licentiate of engineering, Chalmers University of Technology, Gothenburg, Sweden; (2002) http://www.virtualmaze.co.in/sample/Biofuels%20Info/Rhombus%20Power%20-%20biodiesel/5787.pdf dst 25.08.17
16. U. Sadowska, A. Żabinski, A.Szumny, K. Dziadek, IND CROP PROD, Volume **94**, (2016)
17. S. Skonecki, R. Kulig, G. Łysiak, R. Różyło, M. Wójcik, Acta Agroph., **24** (2), (2017)
18. S. Mani, L. G. Tabil, S. Sokhansanj, Evaluation of compaction equations applied to four biomass species. CBE, **46**(3), (2004)
19. M.T. Carone, A. Pantaleo, A.Pellerano, Biomass and Bioenergy, **30**, (2010)
20. R. Kulig, S. Skonecki, S. Gawłowski, A. Zdybel, G. Łysiak, Acta Sci. Pol., Technica Agraria **12**(1-2) (2013)

Analysis of the distribution of the raw material base for biogas production in communes of the małopolskie voivodship

Jakub Sikora[1,*] and *Edyta* Leśniak[1]

[1]University of Agriculture in Krakow, Faculty of Production and Power Engineering, ul. Balicka 116B, 30-149 Krakow, Poland

Abstract. The purpose of the work was to determine the electricity production potential of the available fermentation masses of selected organic fractions in the Małopolskie voivodship, as aggregated at the commune level. The determined potential of biomass production for energy purposes was presented in the form of a spatial distribution, based on data obtained from the queries of the Central Statistical Office. The purpose of the correlation was to facilitate the analysis of the problem of creating agricultural biogas plants in the discussed area. The scope of the study included all communes in the Małopolskie voivodship, as well as public statistics on the sown acreage for cereals, maize, the number of population, the population of pigs and cattle, as well as municipal solid waste. The tool used in the study was the Biogas Calculator of the Masovian Energy Agency and the software for creating spatial databases, ArcView 10.2.1

1 Introduction

In addition to its advantages, the development of civilization, unfortunately, also has disadvantages. One of the issues that involve both economics and ecology is waste management (Ministry of Agriculture and Rural Development, MRIRW). This is an up-to-date topic and it will always be one, because it is associated with everyday activities of every human being. All of us are waste producers, and unfortunately, the scale grows from day to day. The share of the organic fraction is the main criterion for the division of the generated waste [5].

Waste management should consist in reducing the amount of waste, then making maximum use of that, which has already been created, and only then, finally, storage. Unfortunately, it is different in Poland. According to data from the National Waste Management Plan of 2013, over 63% of the generated waste went to landfills. However, there are methods by which waste can find a positive application [3].

The condition of rural areas is affected mostly faeces and urine, i.e. animal faeces, municipal sewage sludge, as well as harmful compounds found in them and their entry into the atmosphere. Therefore, the search for methods allowing the use of these components was launched [7].

One of such methods is the biological processing of the above mentioned masses. An example of this is the anaerobic digestion of organic matter. In addition to the fact that they use different types of substrates depending on the mass available, agricultural biogas plants can also be part of the energy island. This means that they not only produce energy, but can be supplied with a surplus created in another source [4]. In Germany and Denmark, agricultural biogas plants have been operating for 10 - 15 years, and their number is constantly growing [1]. In Poland, the first agricultural biogas plant was built in 2005, and it is still in operation. In 2015, the number of installations included in the Agricultural Biogas Plant Register of the Agricultural Market Agency in Poland was over 50. However, biogas installations in Poland still raise controversy among the public [9, 11].

The input mass for a biogas plant can come from a variety of sources: agriculture, animal husbandry, and industrial or municipal waste. A stable substrate, available for a long time, is crucial in the process of biogas production, which is created during methane fermentation. The fermentation can be interrupted at any time, hence the need to control and properly transmit information regarding, among others, the distribution of the raw material base using information systems.

Against the backdrop of alternative information systems, the SIP spatial information systems are distinguished by the inclusion of spatial information. The information includes both natural and artificial stable objects and social, economic and natural phenomena. Another attribute characteristic for the SIP is the capability of cartographic presentation and analysis in terms of the real world, in relation to the modelled system. Such analysis includes making measurements, determining the proximity of objects or finding points with specific parameters (http://infgeo.strefa.pl).

Local authorities are responsible for shaping the environment and its protection. Improvements implemented in communes involve the use of SIP on various levels, which depend on the nature of the commune and, above all, its size. Communes with larger areas, especially municipalities, use GIS to improve the so-called city management process [11, 12]. In addition to basic information about building plots, objects or

* Corresponding author: Jakub.Sikora@ur.krakow.pl

utilities, the spatial development plan plays a significant role [13, 14].

The purpose of the work was to determine the electricity production potential of the available fermentation masses of selected organic fractions in the Małopolskie voivodship, as aggregated at the commune level. The determined potential of biomass production for energy purposes was presented in the form of a spatial distribution, based on data obtained from the Central Statistical Office's queries. The purpose of the correlation was to facilitate the analysis of the problem of creating agricultural biogas plants in the discussed area.

2 Methodology and subject of study

Data for analysis were obtained from the Local Data Bank of the Statistical Office in Krakow. Data on the sown area and livestock population by type of farm at the municipal level in the Małopolskie voivodship, as well as the population of individual communes, was obtained from the Agricultural Census (PSR) as of Dec 31, 2014. The data were divided into groups:
- cereal mass,
- organic mass,
- animal production waste,
- municipal waste.

The sorted data were collected in the MS Excel spreadsheet, where further calculation activities were performed. Then the mean annual mass of yields in Mg was calculated for individual subgroups. In the case of animal production, the focus was on the annual amount of manure produced. Data referring to the population was used to determine the amount of municipal organic waste in Mg per year.

The obtained data were used to calculate the annual methane yield; the tool used was the Biogas Calculator of the Masovian Energy Agency. The calculator is used to estimate the volume of biogas production, as well as the amount of investment in biogas investments, and its economics.

CH_4yieldfrom vegetable substrates is calculated based on the amount of their yields, animal production – based on the amount of manure produced, and human production - based on the amount of organic municipal waste production, as shown by the following equation:

$$U_{CH4} = P_J * W\,[m^3 * (year)^{-1}] \qquad (1)$$

where:
U_{CH4} - annual methane yield from individual fractions $[m^3*(year)^{-1}]$,
P_J- annual production unit of CH_4(data from the Biogas Calculator) $[m^3* (year * Mg)^{-1}]$,
W - annual amount of input from individual fractions (M_P, or I_O, or I_{OK}) [Mg],

The assumed biomass potential is presented in Table 1.

Table 1. Methane production from selected organic fractions

Type of input	Annual substrate input [Mg $year^{-1}$]	Annual methane production [$m^3 year^{-1}$]
ENERGY CROPS AND AGRICULTURAL WASTE		
Wheat - silage	1	111.52
Maize - silage	1	94.11
Rape - silage	1	167.55
Whole potatoes	1	96.19
Onion	1	43.87
ANIMAL HUSBANDRY WASTE		
Cattle manure	1	45.07
Pig manure	1	43.30
MUNICIPAL WASTE		
Organic municipal waste	1	131.66

The next step was to calculate the nominal power of the generator in the cogeneration system. It was determined from the equation:

$$E_{el} = \frac{E_{tot} * \eta_{el}}{T_p}\,[kW] \qquad (2)$$

where:
E_{tot}- total thermal energy contained in fuel [kWh],
η_{el}- efficiency of electricity conversion [%],
T_p- biogas introduction time [h].

$$E_{tot} = V_{sub} * Q_{bio}\,[kWh] \qquad (3)$$

where:
V_{sub} - flow of biogas volume $[m^3*(d)^{-1}]$,
Q_{bio} - calorific value of biogas $[kWh*(m^3)^{-1}]$ [5].

In order to calculate the total heat energy contained in the fuel, the annual methane yield was divided into 365 days, which resulted in information about the daily yield.

$$V_{sub} = \frac{U_{CH4}}{365}\,[m^3 * (d)^{-1}] \qquad (4)$$

Then, the calorific value of biogas equal to $Q_{bio} = 6$ $kWh*(m^3)^{-1}$ was assumed. The energy thus obtained was multiplied by the efficiency of electricity conversion of $\eta_{el}= 39\%$. Finally, the energy obtained was divided over time (24 hours), which showed the daily power of the system in MW. These calculations were conducted in relation to each of the subgroups.

https://doi.org/10.1051/bioconf/20181002029

Next, data on the methane yield of individual groups and the calculated power of the generator in the cogeneration system were collected in a MS Office Excel worksheet, and columns of data sums from the analyzed subgroups in both categories were added. The next step was to generate spatial databases in the form of maps.

The area of the Małopolskie voivodship was adopted as the subject of the research. The area of agricultural land in Małopolskie was 4.4% of the national total. On the other hand, sowing in Małopolskie accounted for about 3.0% of the national total. The average area of ecological farms in Małopolskie in 2011 was 10 ha. Along with the increase in the area of farms, the proportion of fodder plants increased in relation to other crops [2, 13]. Figure 1a shows the structure of land use in Małopolskie in 2012.

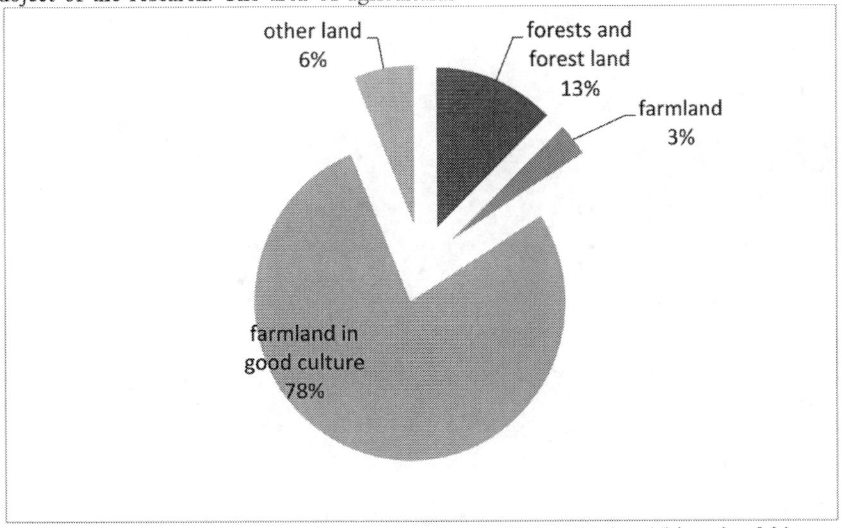

Figure 1 a. Structure of land use in agricultural enterprises in 2012 in Małopolskie voivodship.

The above figure demonstrates that the largest share was occupied by good quality arable land, which constituted as much as 78%. Forests and forest land accounted for 13%. The least acreage was occupied by agricultural land, the soil condition of which was unsatisfactory; it accounted for 3%. The region of Małopolskie voivodship is intensively exploited agriculturally, with the exception of the higher parts of the Carpathians. Agricultural land covers about 58% of the area, of which arable land constitutes approx.. 75% of the area of agricultural land on average.

A spatial database was developed and generated for this area. The Figure below (Figure 1b) demonstrates the study area.

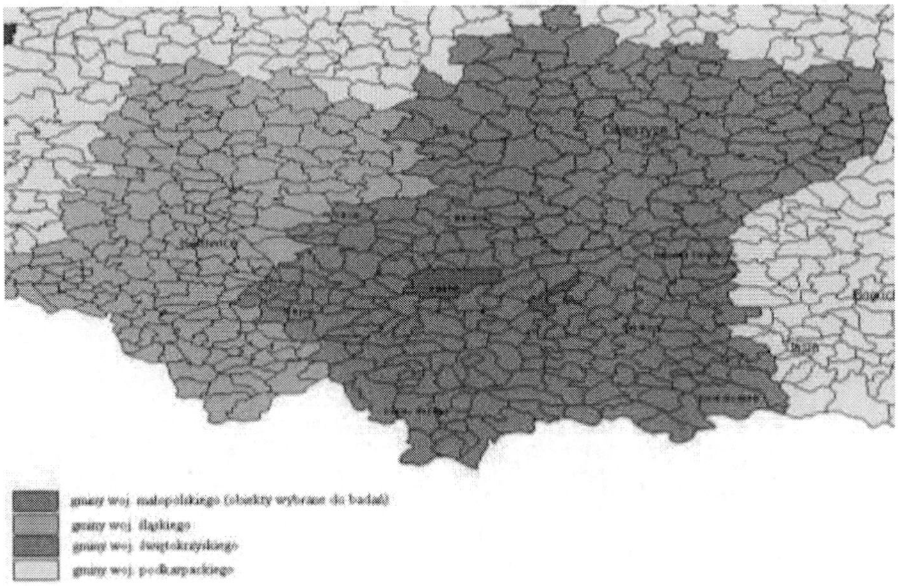

Figure 1b. The study area.

3 Findings

The summary data on the nominal value of Eel shows that in the Małopolskie voivodeship there may be the following maximum number of biogas plants for each group:
- microbiological biogas plants 2
- small biogas plants 6
- large biogas plants 175

Maps were generated based on the obtained data and developed analyses and calculations.

The following drawings (Figure 2 - Figure 7) present the spatial development of the data in question. These are data on the biomethane production potential of individual substrates at the commune level and the electricity generating potential of the obtained biomethane.

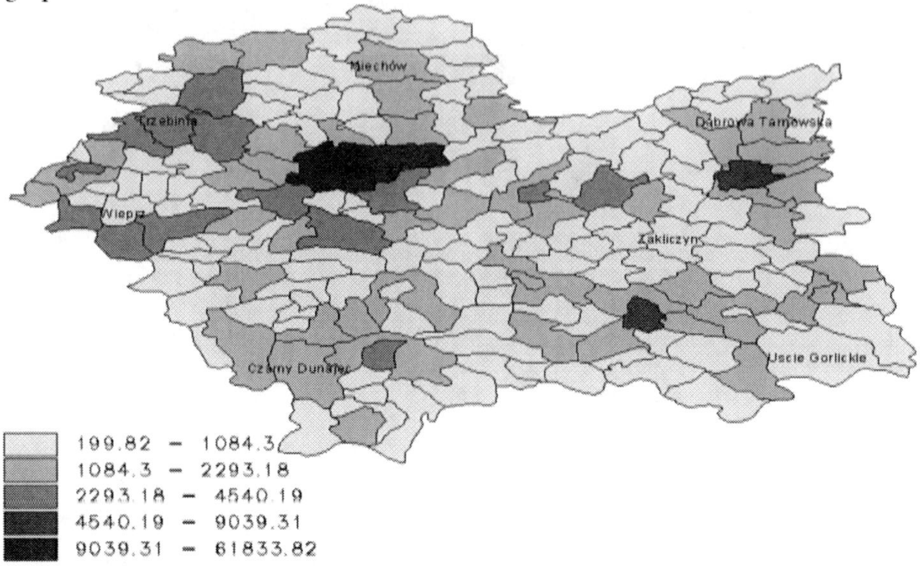

Figure 2. Potential of biomethane production of the organic fraction of municipal waste at the commune level of the Małopolskie voivodship.

The volume of municipal waste production is closely related to the number of inhabitants of a given commune, so the potential of methane production will grow similarly with the increase in the population of the region. Municipalities with the largest CH_4production potential in the Małopolskie voivodship are: the municipality of Krakow with a daily quantity of 6,183,3,82 m^3 of biomethane, the municipality of Tarnow, with over 9,000 m^3 per day, and Nowy Sacz with 6805 m^3per day. The lower range of the scale is occupied by the municipalities of Racławice and Bolesław, with a daily methane yield of approx. 200 m^3. The average value in this category is 1505 m^3CH_4.

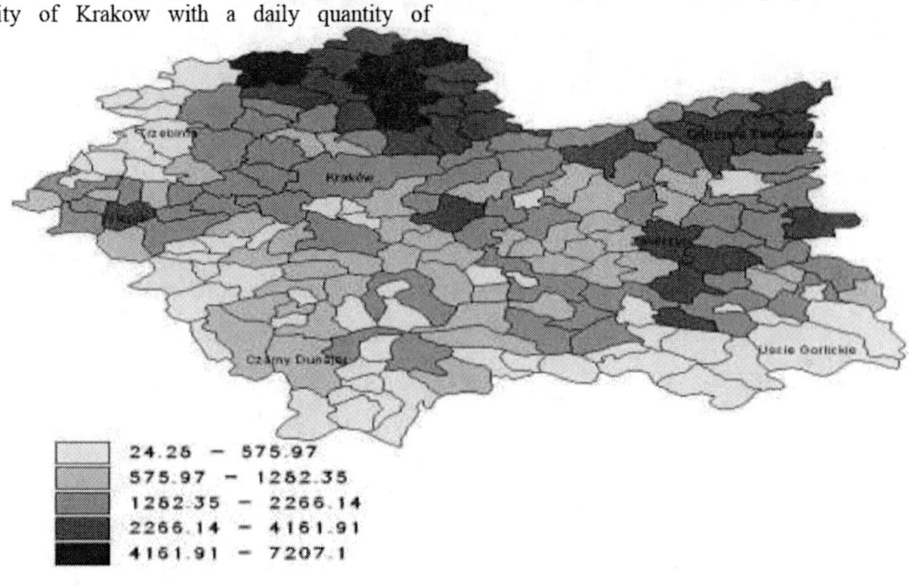

Figure 3. Potential of biomethane production of the production waste mass at the commune level of the Małopolskie voivodship.

https://doi.org/10.1051/bioconf/20181002029

The potential of biomethane production from grain production masses in the communes of Małopolskie voivodship is growing towards the northern part of the area. The minimum value is 24.27 $m^3$for the municipality of Szczawnica, then Bukowno, Zakopane and Sucha Beskidzka, i.e. mountain areas. The maximum value is over 7200 m^3. Such methane yields from cereals are possible in the municipality of Miechów.

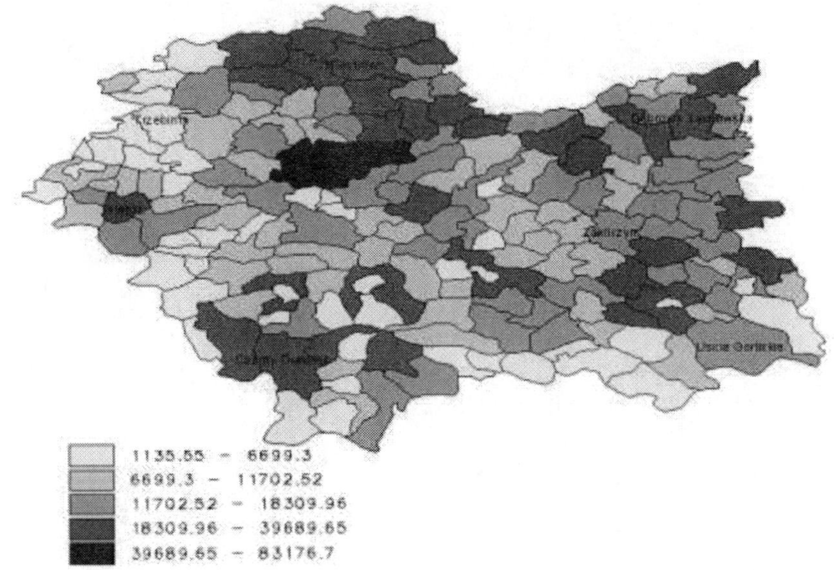

Figure 4. Potential of biomethane production from all studied masses at the commune level of the Małopolskie voivodship.

The potential of biomethane production from all studied masses at the commune level of the Małopolskie voivodship is diversified. A tendency for higher potential to accumulate in the north-western part of the voivodship is noticeable. Municipalities with the highest biogas production potential of the studied fractions, at 100% biogas development, are: the municipality of Krakow, with a daily yield of 83176.7 m^3, then the municipality of Miechów, with over two times lesser potential, and further the communes of Proszowice, Czarny Dunajec and Żabno. The lowest potential in terms of CH_4yield ischaracteristic of the communes Bukowno (1135.5 m^3), Sucha Beskidzka, Szczawnica, Rytro and Jordanów.

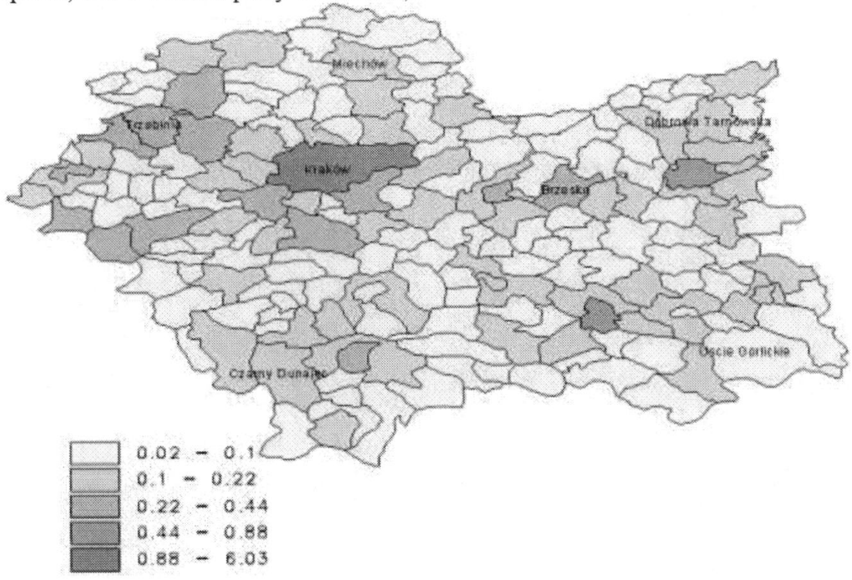

Figure 5. Potential for generating electricity from biomethane obtained from the organic fraction of municipal waste.

In the Małopolskie voivodship, the potential for generating electricity from biomethane produced from municipal waste ranges from 0.019 to 6.028 MW, and as such, it is differentiated. Its mean value is 0.146 MW. The maximum result is assigned to the municipality of Krakow, which is related to the population of the city and its area. Second largest result, only 0.881 MW, is assigned to the municipalities of Tarnow and then Nowy Sacz. Thus, a direct corellation can be observed between the number of residents and the amount of waste, and at the same time the higher energy potential. The lower range of 0.02 MW is taken by the communes. Racławice, Bolesław, Radziemice, Gremboszów and Mędrzechów.

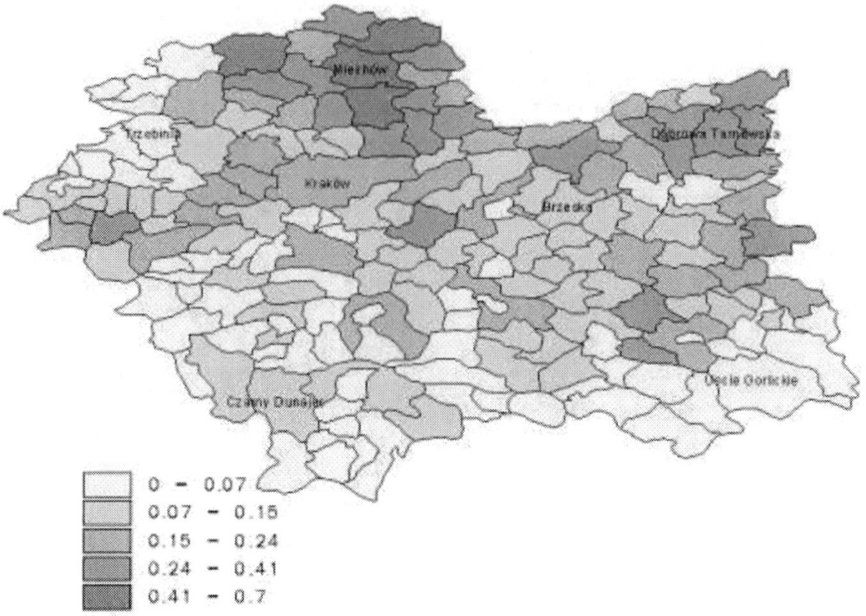

Figure 6. Potential for generating electricity from biomethane obtained from cereal masses.

The energy potential of cereal masses increases towards the north of the voivodship. The spatial distribution of individual plants also demonstrates the soil and climatic conditions of the studied area. It is clear, therefore, that mountain areas are not suitable for growing cereals. The amount of energy possible to obtain from cereals (assuming 100% consumption for biogas purposes) ranges from 0.0023 to 0.702 MW. The average value was 0,129 MW. The values of the lower range are assigned to the communes of Szczawnica, Bukowno, and Zakopane, i.e. mountain communes, while the maximum value of energy potential of biomethane obtained from cereal masses is assigned to the municipality of Miechów in the north of the voivodship.

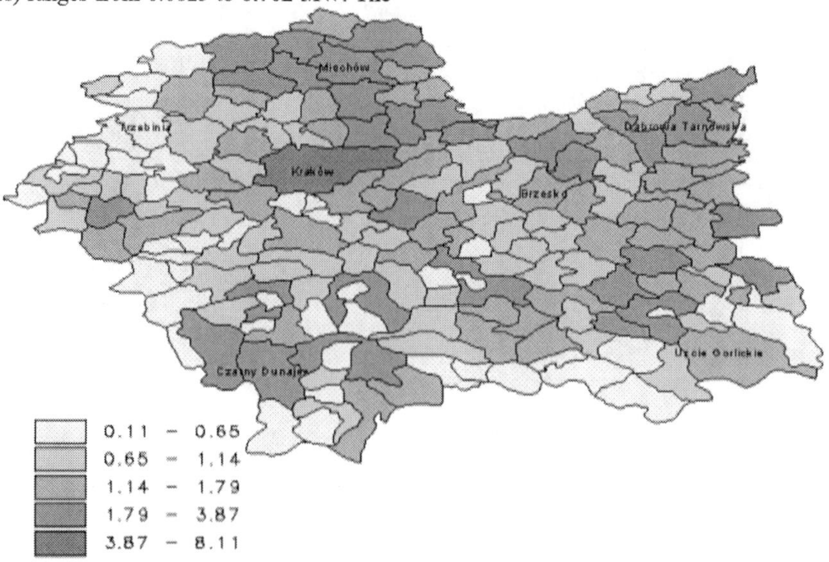

Figure 7. Potential for generating electricity from biomethane obtained from all studied fractions.

When discussing the spatial distribution of all studied fractions, it can be seen that the potential for the generation of electricity from biomethane obtained from all studied fractions in the Małopolskie voivodship is diversified. It ranges from 0.11 MW in the Bukowno commune and 0.13 MW in Sucha Beskidzka to 8,109 MW in Krakow. This is impacted by the amount of municipal waste. The municipality of Krakow with 3.8 MW, is followed by the municipality of Miechow and further by the communes of Proszowice, Czarny Dunajec and Żabno. The average energy potential in the Małopolskie voivodship is 1,216 MW.

4 Conclusion

Implementation of renewable energy sources, including agricultural biogas plants, is one of the ways to protect the environment as well as to save the depleted resources of fossil fuels. The introduction of biogas plants aims at the management of agricultural waste. The waste generated during methanogenesis can be used as a fertilizer rich in organic compounds, or be reused in the next methanogenic cycle. Małopolskie voivodship has the potential to build biogas installations.

In the case of the nominal power of the generator in the cogeneration system, the situation is similar to methane yield. The nominal power of power generators ranges from 0.11 MW to as much as 8,109 MW in Krakow. In Krakow, this power is overestimated by the organic fraction of municipal waste. The best agricultural commune, with the potential of 3.8 MW is the municipality of Miechów. In the medium range were the following communes: Proszowice, Czarny Dunajec and Żabno. The studies included 100% utilization of the tested fractions for biogas purposes.

References

1. Appel F., Ostetmeyer - Wiethaup A., Utilities Policy **41**, (2016)
2. Czarnecka L. et al., Raport o stanie środowiska w województwie małopolskim w 2012 roku – Gospodarowanie odpadami, Voivodeship Inspectorate of Environmental Protection (WIOŚ), (2013)
3. Dec Ł., Sprawozdanie z realizacji Krajowego planu gospodarki odpadami za okres od 1 stycznia 2011 r. do 31 grudnia 2013 r., (2016)
4. Niemiec M, Sikora J., Szeląg-Sikora A., Kuboń M., Olech E., Marczuk A. Przemysł chemiczny 96**(3)**,685-688 (2017)
5. Sikora J., Mruk B., nr III/2/2016, POLISH ACADEMY OF SCIENCES in Krakow, p. 907–917 the Committee for Rural Technical Infrastructure, (2016)
6. Szeląg-Sikora A., Cupiał M., Niemiec M. Agriculture and Agricultural Science Procedia. **7**, 94–98, (2015)
7. Mazovian Engergy Agency (MAE), the Biogas Calculator http://www.mae.com.pl
8. Ministry of Agriculture and Rural Development (MRIRW), Agricultural Biogas Plants in Poland. http://www.minrol.gov.pl/
9. SIP, SIT, GIS, Geomatyka, http://infgeo.strefa.pl/K14/sip_gis_geomatyka.pdf
10. Tytko R., *Urządzenia i systemy energetyki odnawialnej*, (2013)
11. Wróbel, M., Frączek, J., Francik, S., Ślipek, Z., & Mudryk. 12th International Scientific Conference on Engineering for Rural Development - Proceedings ,**12**, 653–657, (2013).
12. Mudryk, K., Frączek, J., Ślipek, Z., Francik, S., & Wróbel, M. 12th International Scientific Conference on Engineering for Rural Development - Proceedings ,**12**, 658–662, (2013).
11. Slovak Association Publishing House, Krakow
12. Valenti F., Porto S. at al, 2017, Use of citrus pulp for biogas production: A GIS analysis of citrus-growing areas and processing industries in South Italy, Land Use Policy 66
13. Kuboń M., Krasnodębski A. Agricultural Economis, **56** , 397-402 (2010)

Drying kinetics of RDF: Experimental investigation and modeling

Karolina Słomka-Polonis[1,*], *Bogusława* Kordon-Łapczyńska[1], and *Jarosław* Frączek[1]

[1] Department of Mechanical Engineering and Agrophysics, Faculty of Production and Power Engineeing University of Agriculture in Kraków, Poland

Abstract. An experimental study was performed to determine the drying characteristics of an oversized fraction of RDF alternative fuel using a laboratory scale hot air dryer at a variety air temperatures and a constant air velocity. For this research the industrial RDF was derived from a Regional Municipal Waste Treatment Facility near the city of Kraków, Poland. The samples of RDF were prepared in two forms: ovesized (unmodified condition) and shreded in a two-drum crusher. In addition, the RDF was sorted into three groups of samples: paper, plastic, textiles. Each form of RDF and each group of samples were dried in hot air dryer at temperatures of 50, 70 i 90 °C and a constant air velocity of 1,5 [$m \cdot s^{-1}$]. The loss of the the samples mass were measured in a continues manner until the equilibrum moisture content was reached. The effective moisture diffusivity [$m^2 \cdot s^{-1}$] and activation energies [$kJ \cdot mol^{-1}$] was amounted. The analysis of the course of moisture content change concludes that that the drying of the RDF alternative fuel occured mainly in the II period of the process during which the transport of water content was carried out by diffusion. And, to a lesser extent, with the surface heat transfer in II period. Based on the calculated data there was a model determined which presented the best possible matching of the course of moisture content change.

1 Introduction

RDF alternative fuel is a combustible fraction of municipal or industrial waste obtained from the separation of the biodegradable fraction. It is typically produced in the Regional Waste Water Treatment Plant (RIPOK). RDF is a heterogeneous fuel whose main components are paper (including cardboard), plastic and textiles [1]. RDF can also be found in admixtures of inert components: glass, wood, organic parts and metals. Their presence is due to an imperfection of the municipal waste screening process. The mass fraction of individual fractions varies, depending on the installation, source of waste: Municipal Solid Waste (MSW), Commercial and Industrial Waste (CIW), Construction and Demolition Waste (CDW) and collection sites: urban and rural [2, 3, 4]. Typically, RDF is dominated by paper: 50-70%, plastic: 30-50%, textiles (<10%) and inert ingredients (<5%). An example of RDF composition according to the source of waste is presented in the Table 1.

The RDF calorific value varies between $MJ \cdot kg^{-1}$ [4]. RDF humidity varies, ranging between 5-23% [3, 4]. This depends on the origin of the waste, the type of screening plant and the time of year for municipal waste collection [6, 7]. The RDF fuel surplus fraction consists of particles greater than 5 cm [8]. It is generated by screening a mechanical waste stream of municipal waste on sieve screens (drum sieves). The particle size depends on the mesh size and typical sieve sizes are 5, 8, 10 cm [9].

RDF can be utilized thermally in combustion, pyrolysis and gasification. Most often it is a product intended for cement kilns, lime plants, waste incineration plants and power plants as well as heating plants [10]. RDF is also subjected to the process of pelleting or briquetting [11]. The moisture content parameter plays an essential role in the design and process optimisation of thermal ultilisation or compaction of the RDF fuel. The presence of moisture has a great impact on the fuel's calorific value and, consequently, on the possibilities of thermal processing. For example, for a cement kiln, humidity cannot exceed 15% [12]. On the other hand, RDF with low moisture content is difficult to handle during transportation within facilities due to dust and blockages on conveyor belts. Moisture also plays a key role in the agglomeration process. Water molecules are involved in the physical-chemical reactions which occur during bonding of fuel elements (adhesion, formation of solid bridges) [13]. However, the water molecules are uncompressible, which at the same time makes the compaction process quite difficult. Too high water content will result in a high relaxation coefficient [14], but too low will increase the energy efficency of the shredding process [15] and increase the risk of spontaneous combustion as well [16]. Determining the required humidity ought to be a compromise between the

* Corresponding author: karolina.slomka-polonis@urk.edu.pl

production process and the disposal of the fuel concerned.

Table 1. Composition of RDF/SRF [3].

Waste component	SRF:		
	C&IW [wt%*]	CDW [wt%]	MSW [wt%]
Paper and	35.6	22.0	30.0
cardboard	24.0	6.8	19.6
Plastic (soft)	16.5	9.2	13.0
Plastic (hard)	8.5	6.0	10.0
Textile	6.4	38.0	8.2
Wood	-	-	0.4
Bio Waste	1.0	2.4	2.2
Rubber	1.2	0.5	2.6
Foam	0.8	0.5	0.5
Metal	0	2.0	0.7
Glass	0	0.6	0
Stone	6.0	11.0	12.8
Fines			

* wt%=wt% percentage by weight (ratio ofmass to totalmass)

Thermal drying to reduce water content has been proved to be an effective method [17]. In many branches of industry, drying has a crucial role: food industry, agriculture, paper industry. There are many methods of drying available: contact, convection, microwave, radiation, dielectric and others [18]. Drying consists by removing moisture from the material through evaporating water, with the share of heat supplied to this process. The most common and simplest method is drying by convection using gas, usually warm air, as a drying medium. In this process, the transport of heat and mass from the interior of the material to its surface is carried out by diffusion, and further on into the surrounding atmosphere by convection. Drying also can be a part of the process, taking part in the preparation of the raw material for the compaction process of solid biofuels [19], and yet is a very energy consuming process. Since RDF is a heterogeneous material composed of many different components, it contains different types of moisture. The paper and textiles would propably be the main water binding materials in alternative fuel [20]. The water contained there would be primarily capillary bounded, while in the plastics the moisture would be collected as a film on the surface. The RDF is a lightweight, porous and bulky material, so convective drying would have a more practical effect due to the ease of flow of the drying medium between the RDF elements. Knowing the drying kinetics has a key role in designing and optimization for the agglomeration or thermochemical processing of raw materials [21]. There are articles concerning drying of paper, paper sludge [i.e. 22, 23, 24], textiles [25] or food waste [26]. These research studies focused on drying kinetics analysis, drying curves analysis, the Effective moisture diffusivity (D_{eff}) and Activation energies (E_a) determinations and calculations the best equation for the drying procces under variable factors (i.e. temperature of drying, air velocity, initial moisture content of the sample). However very little is said in the literature concerning RDF drying. The purpose of this study is to observe the effect of various temperatures at convection drying on the drying kinetics of RDF and it's main fractions (plastic, textiles and paper) in separation.

2 Material and methods

RDF was collected from the RIPOK industrial plant located in Krakow, which was produced from municipal solid waste sorted out of the Krakow area. The analysed fraction was formed on drum sieve with a 10 cm mesh screen installed. Because of that the RDF is mainly composed of large particles (over 3 cm).

Table 2. Composition of presented RDF [27].

Waste component	wt%
plastics	87.44
paper, wood and similar	5.09
textiles	2.60
biomass	1.56
glass	0.64
metal	0.49
hazardous waste	0.00
other, non-combustible	2.17

* wt%=wt% percentage by weight (ratio ofmass to totalmass)

There were five types of samples prepared for the experiment:

- RDF0 – primary fine fraction gained directly from RIPOK
- RDF1 –primary fraction, shredded in a two-drum crusher on the RDF pelletizing line.

In addition, three groups of raw materials were selected from RDF0: paper, plastic and textiles.

Fig. 1. Samples of RDF: A) RDF1, B) RDF0.

There were samples of 12 g (± 5 g) prepared from all raw materials for drying. Samples were dried by convection in the laboratory drier Elkon KBC 65G. The measurement stand diagram is show in Fig. 2. The drying experiments were performed at 50, 70 i 90 °C air temperatures. The drying air velocity was at a constant 1.5 $m \cdot s^{-1}$. All the experiments were replicated five times at each temperature value for each samle type.

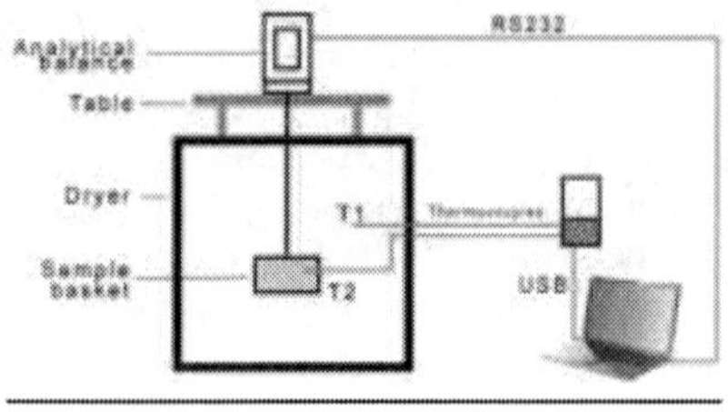

Fig. 2. Measurement stand.

The weight loss was measured by Radwag AS-160.R2 analytical balance (d=0,01 mg). The measurement was recorded continuously of at 5 s interval and saved by the Radwag „Pomiar-Win" software each time. The temperature was measured with two NiCrNiAl thermocouples that were placed inside dryer. With a thermometer that Works in connection with the YC-747UD meter the temperature of the ambient environment- the inside of the dryer (T1) and the temperature of the sample (T2) was measured. The measurements were recorded continuously at 1 s intervals by Temp Monitor_S2 (ver. 1.0.16) software. The drying proces proceeded until the moisture content reached an equilibrum value (until there were no significant variations in the moisture lost). The mean drying time was 25 minutes. A Pol-Eko SLW-53 TOP dryer was used for the samples initial moisture content measurment according to the PN-EN 15414-3:2011 standard [28]. The dry matter calculations were done according to the PN-EN ISO 638:2008 standard [29].

The moisture content (U_τ) was calculated from the following equation:

$$U_\tau = \frac{m_\tau - m_{db}}{m_{db}} \quad (1)$$

Where, m_τ is the mass of the sample at any given time τ, [kg]; m_{db} the mass of the dry basis of the sample, [kg]. The unit of the moisture content is(U_τ) kg water/kg dry basis.

The drying rate (dU/dt) [kg water/·kg d.b.·s^{-1}] was obtained from the following equation [18]:

$$\frac{dU_\tau}{d\tau} = \frac{U_i - U_{i+1}}{\tau_i - \tau_{i+1}} \quad (2)$$

Where U_i is the moisture content at *i*th time [kg water/kg d.b.], U_{i+1} is the moisture conent at the $i+1$-th time [kg water/kg d.b.], τ_i and τ_{i+1} is respectively *i*th and $i+1$-th time of the measurement [s].

The three common models were used to quantify drying kinetics of RDF0 and RDF1. The models are described with simplified equations respectively [18]:

- Henderson-Pabis:

$$MR = (U_\tau - U_e)\cdot(U_i - U_e)^{-1} = a\exp(-k\tau) \quad (3)$$

- Lewis:

$$MR = (U_\tau - U_e)\cdot(U_i - U_e)^{-1} = \exp(-k\tau) \quad (4)$$

- Page:

$$MR = (U_\tau - U_e)\cdot(U_i - U_e)^{-1} = \exp(-k\tau^N) \quad (5)$$

Where, *MR* is the moisture ratio (dimensionless), U_τ is the moisture content at Any given time τ [kg water/kg d.b.], U_i is the initial moisture content [kg water/kg d.b.], U_e is the equilibrium moisture content [kg water/kg d.b.], *k* is the drying rate constant [s^{-1}]. *N* and *a* are the empirical constants.

Taking the natural log of both sides of the equations 3-5 is a way to determine the empirical constants. The form of the normalized equations of the moisture ratio versus time is:

- Henderson-Pabis:

$$\ln(MR) = \ln(a) - k\tau \quad (6)$$

- Lewis:

$$\ln(MR) = -k\tau \quad (7)$$

- Page:

$$\ln[-\ln(MR)] = \ln(k) + N\ln(\tau) \quad (8)$$

The Effective moisture diffusivity (D_{eff}) [m^2·s^{-1}] can be determined from the slope of the normalized plot of the unaccomplished moisture ratio, ln(MR) vs time, using the following equation [31]:

$$D_{eff} = -(4L^2 \cdot \pi^2)\cdot k \quad (9)$$

Where, *L* is the thickness of the sample (slab) of 0,0045m [m], *k* is the drying rate constant [s^{-1}].

The Activation energies (E_a) [kJ·mol^{-1}] was obtained from the reliance of the effective moisture diffusivity D_{eff} on the temperature *T*. The reliance can be described by an Arrhenius type equasion [31]:

$$D_{eff} = D_0\exp\left(-\frac{E_a}{R_C \cdot T}\right) \quad (10)$$

Where the D_{eff} is effective moisture diffusivity [m^2·s^{-1}]; D_0 is the pre-exponential factor of the Arrhenius equation [m^2·s^{-1}], *R* is the universal gas constant (8.3145 kJ·mol^{-1} ·K^{-1}) and *T* is the absolute air temperature [K]. Taking natural logarithm of both sides of Eq. (5) the result can be written as:

$$\ln(D_{eff}) = \ln(D_0) - \frac{E_a}{R_C}\cdot\frac{1}{T} \quad (11)$$

The activation energy (E_a) is determined by plotting the natural logarithm of D_{eff} ($\ln D_{eff}$) versus reciprocal of the absolute temperature *T* [K].

The experimental drying data was tested to find the most suitable model among the above mentioned (3-5). The different constants in the considered models were determined using a linear regression procedure. The adjusted coeficient of determination (R^2), reduced chi-squared (χ^2) and root mean square error (RMSE) were used to inspect the goodness of fit. [18]. The matemathematical model is more suitable for higher values of R^2 and lower values of χ^2, RMSE and %E. The following equations were used to calculate the above mentioned parameters:

$$R^2 = 1 - \left[\frac{\sum(MR_P - \overline{MR}_E)^2}{\sum(MR_E - MR_P)^2}\right] \quad (12)$$

$$\chi^2 = \frac{\sum(MR_E - MR_P)^2}{N - n} \quad (13)$$

$$RMSE = \sqrt{\frac{\sum(MR_P - MR_E)^2}{N}} \quad (14)$$

Where MR_P is the predicted moisture ratio at any time, MR_E is the experimental moisture ratio at any time, $\overline{MR}_E$ is the average experimental moisture ratio, N is the number of observations, n is the number of constants in the drying model.

The calculations and statistical data analysis were performed using R Core Team (2016) software (R Foundation for Statistical Computing, Vienna, Austria. URL https://www.R-project.org/) with the packages: ggpolt, dplyr, stats.

3 Results and discussion

The initial moisture content of all samples were determined using PN-EN 15414-3:2011 standard [29]. As expected the highest average value of moisture content was obtained for paper (~ 41%) and textiles (~25 %). The plastics, that were the main component of the RDF had a moisture content of approximately 10%. There was no significant differnence in the moisture content values between RDF0 and the RDF1 samples. It is clear that the mechanical comminution at the pellet production line had little impact on moisture loss in the material. (Table 3).

Table 3. Moisture content [%] of the specific sapmles:

Sample	Moisture Content [%]
RDF1	21.9057
RDF0	19.1831
Plastic	9.8426
Paper	41.1430
Textiles	24.7662

The shapes of the drying curves versus time for the samples resembled the exponential function (U_τ). The steeper the curve was, the quicker the moisture loss took place. It is cleary evident that these curves were influenced by various temperatures (Fig. 3, 4). The deviations were related to the individual characters of particular samples which could be also noticed in the shapes of the moisture ratio curves (Fig. 3. – bottom plots). There was no I period of drying process in the paper and textiles. In these samples the drying process began in the second period in which the diffusion mechanism is predominant, and which is mostly common for porous matter. The drying rate was going slow at the temperature of 50 °C in the paper, however at the temperature of 70 and 90 °C the rate of changes were faster (Fig. 3A, the bottom plot). The maximum drying rate of the paper was 0.06×10^{-2} kg water/kg d.b. at 90 °C.

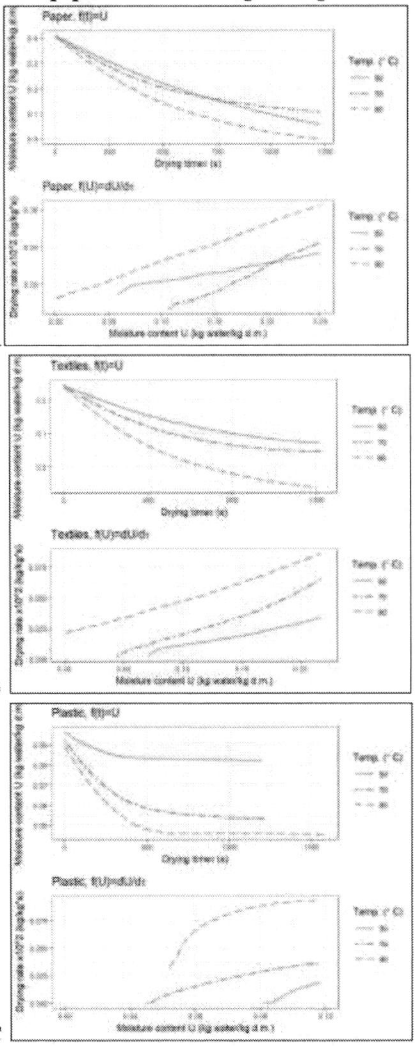

Fig. 3. Effect of drying air temperature and moisture content on the drying rate of A) paper, B) textiles, C) plastic.

Considering the textiles, the drying curves are also noticeably influenced by the various temperatures. With the decreasement of water content, the drying curves are less steep. (Fig. 3B). The paper and textiles were dryed completely at the U = 0 at the temperature of 90 °C after 20 minutes (1250 s). With the decreasment of the temperature the final water content in the both of samples was equal to equilibrium moisture content. The maximum drying rate of the textiles was 0,088 $\times 10^{-2}$ kg water/kg d.b. at 90 °C.

Different kind of curve shapes can be observed in plastisc (Fig. 3C). With the decreasement of water content, the drying curves are less steep, and more flat and paralel to the x axis. There are two periods of drying

A
B

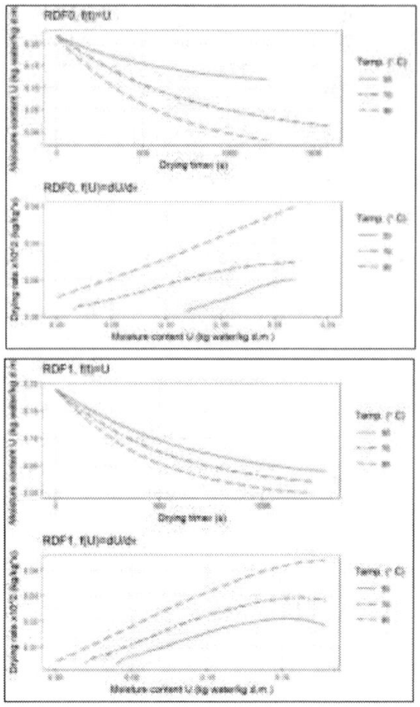

Fig. 4. Effect of drying air temperature and moisture content on the drying rate of A) RDF0, B) RDF1.

process in plastic, where I period is mild and long, and the II one is dynamic and short (especially at the temperature of 90 °C). This is characteristic for porous-capillary and porous materials. The maximum drying rate in plastic is 0,085 x10^{-2} kg water/kg d.b. at 90 °C. The time of drying of 20-25 minutes is too short to remove the water completely from the samples above at 50 °C, in particular for paper and textiles. The plastic and paper exhibit characteristics of hygroscopic material at the lower temperature values, however at 90 °C the samples exhibit the characteristics close to non-hygroscopic material [32].

Based on the shapes of the drying RDF0 and RDF1curves it can be stated that the samples resemble capillary-porous and porous materials. The shape of the curves exhibit the characteristic of paper and textiles drying curves, although the total percentege of them didn't exceed 15 %. However the drying rate was twice more dynamic in RDF0 and RDF1 than in paper and textiles. The first period of drying, related to the constant drying rate occured in both of RDF samples, but it lasted a short time at the temperature 50 and 70 °C. At the 90 °C the first drying period occured slightly in RDF1, but hardy in RDF1 (Fig. 4 – bottom plots of A and B). The first drying period was specific to the plastics olny (Fig. 3C), which is the main component of the RDF (ok. 70 %). The drying rate of RDF was faster in the shredded fraction (RDF1), where the area of of evaporation was larger due to reduction of particie size. At the 90 °C both of RDF samples were dried completely untill the moisture content of U=0 (Fig 4, bottom plots of A and B). At the lower temperatures the RDF resembled more party-hygroscopic material [9], it might contain the areas of a different moisture component. At lower temperatures RDF was dried to an eqiulibrum moisture content, which was 0.110 kg water/kg d.b. at T=50 °C and 0.001 kg water/kg d.b. at T=70 °C for RDF0, and 0.041 kg water/kg d.b. T=50 °C i 0.010 kg water/kg d.b. T=70 °C for RDF1 respectively.

Table 5. Empirical constants of the Henderson-Pabis, Levis and Page equasions of RDF0 and RDF1.

Sample	Temp (°C)	Henderson-Pabis equation			Levis equation		Page equation		
		k [s^{-1}]	a	R^2	k [s^{-1}]	R^2	k [s^{-1}]	N	R^2
RDF0	50	-4.93 ×10^{-3}	0.9238	0.9696	-5.90 × 10^{-3}	0.9239	5.57 ×10^{-3}	0.6617	0.9879
	70	-1.68 ×10^{-3}	1.0982	0.9907	-1.59 × 10^{-3}	0.9937	2.47 ×10^{-3}	0.9254	0.9910
	90	-4.10 ×10^{-3}	1.5709	0.7031	-1.59 × 10^{-3}	0.6598	4.28 ×10^{-3}	0.9312	0.9682
RDF1	50	-1.23 ×10^{-3}	0.0123	0.9950	-1.13 × 10^{-3}	0.9941	3.96 ×10^{-3}	0.8304	0.9934
	70	-1.83 ×10^{-3}	0.9785	0.9997	-1.86 × 10^{-3}	0.9992	5.61 ×10^{-3}	0.8272	0.9890
	90	-3.92 ×10^{-3}	1.5876	0.7152	-3.36 × 10^{-3}	0.9287	3.98 ×10^{-3}	0.9505	0.9808

Table 6. Model prediction evaluation for the RDF0 and RDF1.

Sample	Temp (°C)	Henderson-Pabis model			Levis model			Page model		
		RMSE	χ^2×10^{-3}	%E	RMSE	χ^2×10^{-3}	%E	RMSE	χ^2×10^{-3}	%E
RDF0	50	0.0221	4.921	2.50	0.0349	1.228	4.496	0.0139	0.195	1.8200
	70	0.0253	6.440	5.32	0.0201	4.322	4.197	0.0249	0.621	11.583
	90	0.1523	23.434	47.89	0.1630	26.711	62.73	0.0498	2.493	54.701
RDF1	50	0.0157	0.25	2.49	0.0170	0.291	3.480	0.0181	0.327	4.414
	70	0.0046	0.002	1.18	0.0069	0.005	1.523	0.0259	0.677	9.140
	90	0.1453	0.021	37.01	0.0633	4.008	42.811	0.0378	1.278	47.30

Based on the results of the experiment the three methematical drying models were fit to the experimental data. The calculations are reveled in Table 5 and Table 6. All of the models provided are a very good fit to the experimental data of drying RDF samples at various temperatures. Of all the models tested, the Page model offered the highest value for R^2. However, the Henderson-Pabis and Levis models had a great fit at the temperatures of 50 i 70 °C as well. At the temperatures of 50-70 °C the value of R^2 parameter with the Levis model reached above 0,98 for RDF0 i RDF1 samples (Table 5). The χ^2 test values reflect to the goodness of model fit and are low (less than 0.02), but quite varied depending on samples types and the temperature of process in all three models (Table 6). The lowest χ^2 test values were obtained from the Page model. The highest values of the χ^2 test were related to the lowest values of R^2 parameter and concerned the Henderson-Pabis and Levis models for modeling of drying RDF0 and RDF1 at 90 °C.

The highest values of %E parameter were obtained at the temperature 90 °C, where they reached even 62% (Table 6). The drying ratio curve of RDF1 and RDF0 at 90 °C was more linear (Fig. 4 A,B) and it's deviation from values of lower temperatures propably was caused by describing it with exponential equations. Compared to RDF0, the values of %E were lower in the RDF1 sample. Maybe from the comminuted sample, which RDF1 was, more homogeneous samples were aqcuired.

Table 7. Effective diffusivities for the RDF0 and RDF1.

Sample	Temp [°C]	D_{eff} [$m^2 \cdot s^{-1}$]	E_a [$kJ \cdot mol^{-1}$]
RDF0	50	4.0457×10^{-7}	52.07
	70	13.7892×10^{-7}	
	90	33.6521×10^{-7}	
RDF1	50	10.0707×10^{-7}	28.26
	70	15.0547×10^{-7}	
	90	32.1453×10^{-7}	

The raise in effective moisture diffusivity D_{eff} was related to increasement of the temperature and decreasement of the paticle size of the samples. At 50 °C in RDF1 was almost two times higher than in RDF0 and was about 10.0707×10^{-7} [$m^2 \cdot s^{-1}$] (Table 7). The activation energy (E_a) for diffusion calculated from the slopes of normalized equation of D_{eff} (10) was found to be 52.07 $kJ \cdot mol^{-1}$ for RDF0 and 28.26 $kJ \cdot mol^{-1}$ for RDF1. Comminution of the RDF resulted in a decrease in activation energy.

4 Conclusions

1) The paper and textiles had the highest moisture content in RDF composition of 41% and 22% respectively. Moisture content of the plastic did nor exceed 10 %. The comminution of RDF caused insignificant difference in the moisture content between RDF0 and it's shredded form of RDF1.

2) Based on the analysis of the drying kinetics of the RDF0 and RDF1 it could be stated that RDF has more features of porous material than porous-capillary. The first period of drying occured only in plastic and slightly in RDF0 and RDF1 at lower drying temperatures of 50 and 70 °C. The drying process begun with second period of drying in paper and textiles. The moisture and mass movement was mainly by diffusion and less by capipillary for ces.

3) Of all of the three drying kinetics models of RDF tested, the Levis model had the best fit to the experimental data. The obtained values for R^2 was greater than 0.98 and χ^2 test values were in the range of $0.3 - 2{,}4 \times 10^{-3}$. None of the mathematical models were able to satisfactory describe the performance of drying RDF at the temperature of 90 °C.

4) The Effective moisture diffusivity (D_{eff}) values were affected by increasing temperature. The highest values of 33.6521×10^{-7} $m^2 \cdot s^{-1}$ were obtained at 90 °C, and the lowest of 4.0457×10^{-7} $m^2 \cdot s^{-1}$ were obtained at 50 °C for RDF0. The Activation energy (E_a) varied from 28.26 $kJ \cdot mol^{-1}$ to 52.07 $kJ \cdot mol^{-1}$ for RDF1 and RDF0 respectively. The lower energy is needed to remove water from comminuted RDF.

5) Drying process of the RDF using hot air at the temperature of 90 RDF °C allows for removal of the moisture content completely, until U = 0. At lower temperatures of 50 °C and 70 °C the RDF could be dried untill the equillibrium moisture content.

This work was supported by funds of a specific grant for development of young scientists of the Hugo Kołłątaj University of Agriculture in Kraków (grant number BM-4633/2016) and the Laboratory of Production Technology and Quality Estimation of Biofuels in the Chair of Mechanical Engineering and Agrophysics at the University of Agriculture in Kraków.

Nomenclature

E_a	activation energy, $kJ \cdot mol^{-1}$
D_{eff}	effective moisture diffusivity, $m^2 \cdot s^{-1}$
D_0	pre-exponential factor of the Arrhenius equation, s^{-1}
d.s.	dry solid
k	drying rate constant, s^{-1}
MR	moisture ratio (dimensionless)
N	number of obserwations
U_τ	moisture content at any time, kg water/ kg d.s
U_i	initial moisture content (dry basis), kg water/kg d.s.
U_e	equilibrum moisture kontent, kg water/ kg d.s
T	absolute temperature, K
τ	time, s

References

[1] Schwarzböck T. et al. (2016), Fuel Processing Technology **153,** 101–110 (2016)

[2] E. Iacovidou, J. Hahladakis, I. Deans, C. Velis, P. Purnell, Waste Management, http://dx.doi.org/10.1016/j.wasman.2017.07.001, (2017)

[3] M. Nasrullah, M. Hurme, P. Oinas, J. Hannula, P. Vainikka, Fuel Processing Technology **163** (2017)

[4] M. Malinowski, Infrastructure and Ecology of Rural Areas, **Nr 2/IV/2013**, PAN, Oddział w Krakowie, 125–139, (2013)

[5] Ö. Çepelioğullar, H. Haykırı-Açma, S. Yaman, Waste Management, **48**, 275–284, (2016)

[2] Materazzi et al., Fuel **150**, 473–485 (2016)

[6] C. Bessi, L. Lombardi, R. Meoni, A. Canovai , A. Corti, Waste Management **47** (2016)

[7] M. Malinowski, Episteme **14/2012,** 101-108, ISSN 1895-4421, (2012)

[8] K. Dziedzic, B. Łapczyńska-Kordon, M. Malinowski, M. Niemiec, J. Sikora, Chemical and Process Engineering, **36 (4)**, 381-384, (2015)

[9] C.A. Velis , P.J. Longhurst , G.H. Drew , R. Smith, S.J.T. Pollard, *Production and Quality Assurance of Solid Recovered Fuels Using Mechanical—Biological Treatment (MBT) of Waste: A Comprehensive Assessment, Critical Reviews in Environmental Science and Technology*, **40:12**, 979-1105, DOI: 10.1080/10643380802586980, (2010)

[10] R. Wasielewski, S. Stelmach, A. Sobolewski, Chemik, **65**, **6**, 572-579, (2011)

[11] J. Frączek, K. Słomka-Polonis, K. Mudryk, M. Wróbel, K. Dziedzic, Agricultural Engineering, **Vol. 20**, No.4, 207-217, (2016)

[12] Stowarzyszenie Producentów Cementu (SPC). *Paliwo alternatywne na bazie sortowanych odpadów komunalnych dla przemysłu cementowego*. Kraków, 3. (2008)

[13] N.P.K. Nielsen, D.J. Gardner, T. Poulsen, C. Felby. Wood Fiber Sci, **41:414,** (2009)

[14] H.J.K. Köser, G. Schmalstieg, W. SiemersFuel, **61:791–8**, (1982)

[15] J. Frączek, K. Mudryk, Acta Agrophysica, **14(1)** (2009)

[16] A. Yasuhara, Y. Amanoa, T. Shibamoto, Waste Management **30,** 1161–1164, (2010)

[17] E. Golisz, M. Jaros, M. Kalicka, Technical Sciences **16(4)**, 333–343, (2013)

[18] Y. Tulek, J. Agr. Sci. Tech. Vol. **13**: 655-664, (2011)

[19] C. Strumiłło, *Inżynieria Chemiczna: Podstawy teorii i techniki suszenia*. Wyd. Naukowo-Techniczne, Warszawa, ISBN 83-204-0418-5, (1983)

[32] F. Kneulle, *Suszenie*. Arkady, Wyd. Warszawa, (1970)

[20] J. Frączek J. red., Wyd. PTIR, Kraków 2010, ISBN: 978-83-930818-0-6, (2010)

[21] B.Dobrzański, L. Mieszalski, ISBN-13: 978-83-60489-06-2, (2007)

[22] O.C.Motta Lima, N.C.Pereira and M.A.L.S.Machado, Brazilian Journal of Chemical Engineering, **17(4-7)**, 539-548. https://dx.doi.org/10.1590/S0104-66322000000400018, (2000)

[23] W. Yu, X. Liu, H.S. Su, Y.J. Zhang, Energy and Power Engineering, **9**, 141-148. https://doi.org/10.4236/epe.2017.94B017, (2017)

[24] M.G.A. Vieira, L. Estrella, S.C.S. Rocha, Drying Technology: An International Journal, **25:10**, 1639-1648, DOI: 10.1080/07373930701590806, (2007)

[25] L. Helena, C.D. Sousa, O.C. Motta Lima, N.C. Pereira, Drying Technology, **24: 485–497**, DOI: 10.1080/07373930600611984, (2006)

[26] Y. Choi, B. Jung, N. Sung, Y. Han, J Mater Cycles Waste Manag, DOI 10.1007/s10163-014-0248-8, (2014)

[27] (niepublikowane badania Frączek, Wróbel Modeling of RDF pelletization proces, 2016)

[28] PN-EN ISO 638:2008 - Paper, board and pulps - Determination of dry matter content - Oven-drying method (ISO 638:2008)

[29] PN-EN 15414-3:2011 - Solid recovered fuels - Determination of moisture content using the oven dry method - Part 3: Moisture in general analysis sample

[30] J.S. Roberts, D.R. Kidd, O. Padilla-Zakour, Journal of Food Engineering, **89**, 460–465, (2008)

[31] C.J. Lomauro, A.S. Bakshi, T. P. Labuza, Journal of Food Science, **Volume 50** (1985)

The effect of UV-C stimulation of potato tubers and soaking of potato strips in water on density differences of intermediates for French-fry production

Zygmunt Sobol[1*], *Tomasz* Jakubowski[1] and *Paulina* Wrona[1]

University of Agriculture In Krakow, ul. Balicka 116B, 30-149 Kraków, Poland

Abstract: The paper describes the effect of UV-C stimulation of potato tubers and immersing of potato strips in water on differences in density of intermediate products for French-fry production. The density difference used for description of the experiment was defined as a relationship between the post-treatment density and pre-treatment density. The study was conducted on tubers of the Innovator variety. The studies of density changes induced by absorption of water involved measurements of the mass of potato strips in air and in water. Potato strips measured 10 x 10 mm and 60 mm in length. The strips were cut lengthwise along the longest tuber axis set between the proximal and distal tuber end. Water absorption was investigated by immersing strips in water (1) at a temperature of 20°C for 15 min and (2) at 40°C for 20 min. In addition, the study included the group (3) where strips were blanched at a temperature of 90°C for 2 min and a control group (0) which was not immersed in water. Potato tubers were irradiated by UV-C in the following ways: (1) irradiation on one side for 30 min, (2) irradiation on both sides for 15 min each, (0) control group (no irradiation). The studies were conducted at two dates: (0) after harvest and (1) after 3-month storage. Laboratory and storage experiments were conducted in 2016-2017. The density difference was statistically significantly influenced by storage time, UV-C stimulation and immersion conditions of potato strips. The density difference increased with the increase in storage duration of tubers, immersion duration of strips, water temperature (up to starch gelification temperature) and UV-C stimulation.

1 Introduction

Density of potato tubers is a parameter which affects yield of dried and fried potato products. When density increases, energy required for water evaporation during drying and frying of potato products decreases. Density of potato tubers influences quality attributes of French fries and chips, is decisive for fat absorption during frying and significantly influences their consistency [1,2]. Studies of Sobol [3,4] indicate that tuber density increases during storage. It is mostly caused by tuber transpiration and at the end stage of storage process, also by sprout transpiration. Experiments conducted by Sobol [3,4] demonstrate that peeled potato tubers immersed in water absorb it and their density decreases. Dynamics of tuber density reduction caused by water absorption depends of the storage stage. Those studies suggest that to completely reverse the increase in density of a medium size tuber after an 8-month storage, it should be soaked in water for ca. 3.5 h. Tuber density reduction defined as a difference between density after and before water absorption depends also on tuber size, duration of diffusion process in tubers soaked in water, solvent (water) temperature and potato variety [4,5]. Studies of Sobol [6] also indicate that dynamics of reduction of density of intermediates for French fries and chips production caused by water absorption was many times greater compared with changes occurring in whole tubers. The time needed to balance water losses related to tuber transpiration during 8-month storage, for strips for French-fries production was 5.0 – 7.0% of the time needed for whole tubers and 2.-2.5 % for slices for chip production. According to the opinion of Pilarski et al. [7], the solar spectrum reaching the Earth's atmosphere encompasses a very wide range of wavelengths (from several hundreds to several thousands nm). This radiation crosses the atmosphere and undergoes absorption and scattering due to interaction with air atoms, molecules, and airborne aerosols and particles (natural or of anthropogenic origin). Oxygen, both its molecular and atomic form and ozone absorb UV radiation which changes spectrum of light reaching the Earth's surface. This spectrum is almost completely free of UV-C and UV-B radiation

*Corresponding author: Zygmunt.Sobol@ur.krakow.pl

and contains slight amounts of UV-A. Therefore, in general, UV-C radiation used for agricultural experiments has to be produced by artificial sources. UV radiation as a sterilization method, is currently used for food processing [8,9]. There were also attempts to use this physical method for soil disinfection in tree nurseries [10]. Błażejak et al. [11] performed studies on the use of UV radiation to obtain Gluconacetobacter xylinus mutants which then were applied for biotransformation of glycerol into DHA (the best mutant synthesized 18.00 mg DHA cm^{-3}, i.e. by 32% DHA more than the parent strain). Sławińska et al. [2002] investigated the effect of UV-C radiation on photodegradation of humic acids and demonstrated usefulness of absorption and fluorescence spectroscopy for determination of resistance to ultraviolet radiation-induced degradation. The effect of UV radiation on biological material (soy of the cultivar Augusta) was studied by Szwarc and Skórska [13]. The applied UV irradiation reduced photosynthesis intensity, chlorophyll content and had a negative effect on growth and biomass of above-ground plant parts. It was accompanied by defensive reactions, like increased synthesis of flavonoids and increased activity of antioxidant enzymes (peroxidase and catalase). Allende et al. [14] investigated efficiency of UV-C irradiation in reducing natural microflora inhabiting lettuce leaves. UV-C was also used in procedures aimed to improve quality of onion (Allium cepa L.) and tomatoes (Solanum lycopersicum) during storage [15,16] and asparagus (Asparagus offificinalis L.), especially to reduce pathogen level [17]. As for the potato, Jakubowski and Pytlowski [18] examined the effect of V-C radiation on seed potatoes of the cultivar Jelly and on later ontogenesis of plants. It was shown that irradiation had a significant effect on development of the above-ground parts of potato plants.

The aim of the present studies was to determine the effect of UV-C irradiation of potato tubers and immersion of potato strips in water on differences in density of intermediate products for French fries production at two storage times.

2 Material and Methods

Storage and laboratory experiments were carried out in 2016-2017. The studies were conducted on potato tubers of the variety Innovator. The Innovator potato is one of the most popular varieties used for French-fries production by European, also Polish, companies [19]. It is an early type B variety, with regular tubers round and oval in shape, with shallow eyes, and medium content of starch (14.6 %). This variety is very resistant to darkening of raw flesh and is easily stored after cooking. The studies on density changes induced by water absorption involved measuring the mass of study objects in air and in water. Potato strips measured 10 x 10 mm and 60 mm in length. Strips were cut lengthwise along the longitudinal tuber axis set between proximal and distal tuber end. Absorption process was studied by soaking the potato strips in water (1) at a temperature of 20°C for 15 min and (2) at 40°C for 20 min. In addition, in group (3) strips were blanched at 90°C for 2min. Control strips (0) were not soaked in water. In order to obtain equal measurement conditions, before measurement, the strips were immersed in water and immediately dried (in a two-stage process) using a dry paper towel every time. The same procedure was applied after water absorption was finished and strips were dried in two stages immediately after measurement [3-6]. Calculations of density changes caused by water absorption by potato strips were described in detail in earlier papers by Sobol [3-6]. Difference in density used in description of the experiment was defined as a relationship between post-treatment and pre-treatment density. The studies were conducted at two dates: (0) after harvest and (1) after 3-month storage. A device for UV-C irradiation of biological objects was a cube chamber measuring 0.63 m3 (0.55x0.95x1.2). Its inner walls were made of aluminum (with a high coefficient of reflection). The chamber was equipped with a radiator TUV UV-C NBV 15 W (irradiance of 253.7 nm line at the distance of 1m was 0.42 $W.m^{-2}$, total energy flux of 253.7 nm line was 4.0 W). The radiator was equipped with a reflecting element made of a high quality aluminum with high coefficient of reflection (characteristics similar to mirror). Radiator structure allowed for continuous regulation of the height above the chamber bottom in the range from 0.4 to 1.0 and was equipped with a precise time-lag switch (model AURATON 100). Durability of the UV-C radiator determining stability of parameters of its action was 8000 h. During irradiation, potato tubers were situated on a flat metal bottom with an area of 0.52 m2 (0.55x0.95). Detailed characteristics of the measuring device and method of UV-C irradiation were described in papers by Jakubowski & Wrona [21], Jakubowski & Pytlowski [18]. Potato tubers were irradiated with UV-C in the following ways: (1) irradiation of one side of the tuber for 30 min, (2) irradiation on both sides for 15 min each, and (0) control group which was not irradiated. The stimulation was performed 48 h before cutting potato strips and their immersion in water.

3 Results and Discussion

The three-way analysis of variance demonstrated that all experimental factors, i.e. storage time (1), stimulation (2), immersion conditions of potato strips (3) produced a statistically significant effect on density difference (Tab. 1). The analysis also revealed that two iterations between the experimental factors were statistically significant, namely 1-3 and 2-3.

The effect of storage time (Fig. 1) on density difference was in line with the previous studies by Sobol [3,4,6]. The increase in density difference from |-

0,0034| [g·cm^{-3}] at measurement time (0), i.e. after harvest, to |-0,0061| [g·cm^{-3}] at measurement time (1), i.e. after 3-month storage, was caused by water deficit (increased osmotic pressure) in tubers stored for 3 months vs. measurement after harvest, which developed due to transpiration.

Table 1. Analysis of variance in multiple classification. Effect of storage, stimulation and immersion conditions on the density difference.

Qualitative predictor	Sum of squares	Degrees of freedom	Mean square	F	p
Word free	0,015206	1	0,015206	782,3437	0,000000
{1} Storage term	0,001274	1	0,001274	65,5661	0,000000
{2} Stimulation	0,000163	2	0,000081	4,1914	0,015537
{3} Conditions of immersion of the intermediate	0,011952	3	0,003984	204,9741	0,000000
1*2	0,000111	2	0,000055	2,8526	0,058418
1*3	0,001075	3	0,000358	18,4389	0,000000
2*3	0,000829	6	0,000138	7,1056	0,000000
1*2*3	0,000114	6	0,000019	0,9780	0,439101
Error	0,012595	648	0,000019		

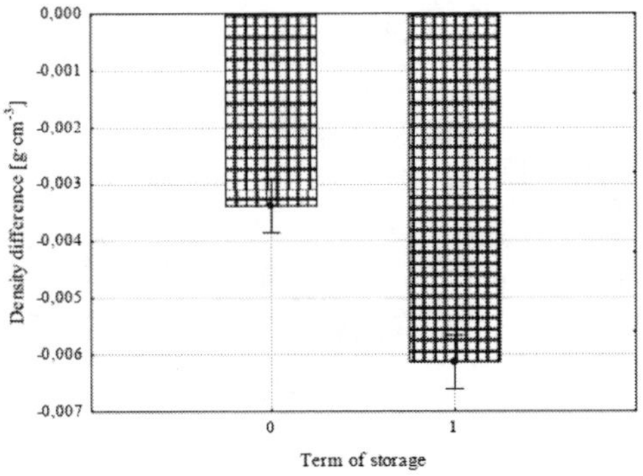

Fig. 1. Effect of storage term on density difference.

Thus, strips cut from tubers stored for 3 months absorbed more water, so a greater density difference was measured. Analysis of the data on the impact of UV-C radiation indicated that irradiation of tubers significantly influenced the density difference of potato strips soaked in water (Tab. 1, Fig. 2).

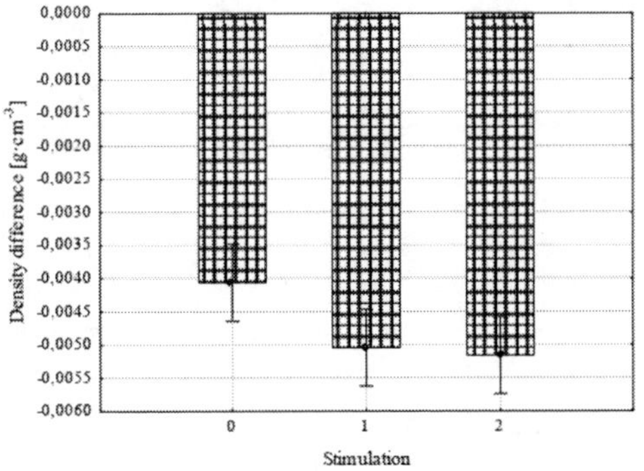

Fig. 2. The effect of the stimulation on the density difference.

The Duncan test (Tab. 2). divided potato strips into two groups depending on the density difference: group 1 comprising strips from stimulated tubers (pooled, irrespectively of stimulation parameters) |-0,0052| [g·cm^{-3}] stimulation 2, |-0,0050| [g·cm^{-3}] stimulation 1, and group 2 containing strips from unstimulated tubers |-0,0041| [g·cm^{-3}].

Table 2. Duncan's test results. Homogeneous groups for stimulation.

Density difference [g·cm^{-3}]			
Stimulation	Mean	1	2
2	-0,0052	****	
1	-0,0050	****	
0	-0,0041		****

Therefore, UV-C stimulation of tubers increased the water absorption effect of strips cut from them. In available literature, the reaction of plants to UV-C is poorly understood. In an attempt to explain the obtained result, namely the UV-C-induced increase in absorption effect of potato strips, it is necessary to analyze possible reactions of biological objects to exposure to ultraviolet radiation because plants developed defensive mechanisms protecting them from its harmful effect. This role is played by flavonoids (flavonols and anthocyanins) able to reduce transmission of UV through epidermis, concomitantly preserving transparency for PAR. Shielding effect protects photosynthetic apparatus and plant parts from DNA damage. Different epidermis formations, like hairs are an additional protective measure because they can scatter up to 70% of incident UV radiation. Besides flavonoids, carotenoids also belong to the dyes able to protect plants from harmful UV effects, due to their antioxidant and reactive oxygen species quenching properties [7,22-24]. UV-C radiation is defined as the

100-280 nm wavelength band. The 100-200 nm wavelength band, called Schumann UV, is not biologically active because it is absorbed by oxygen and water vapor. This limitation causes that biologically active UV-C band is in the 200-280 wavelength range, with the most active 250-270 range (254 nm). This information is important because is allows for focusing on a narrower range in our studies. No information on the effect of absorption (also adsorption) UV-C radiation by potato tubers was found in the available literature. At this stage of research, it can be assumed that the interaction of electromagnetic waves with plant material can be similar to that described in physiotherapy [25]. Following the same reasoning, it can be assumed that the 184.9-290 nm radiation (the range used in phototherapy) penetrates into the potato tuber to the depth of 2 mm (i.e. to a depth of peridermis and directly adjacent flesh). In addition, the irradiated object has to fulfill the Grotthuss-Draper law which states that photochemical transformation in a reactive system can be induced only by radiation absorbed by this system [26-28]. The UV acts on biological material and triggers photochemical reactions (photosynthesis, photolysis, photoisomerization, oxidation, reduction). From the point of view of the present studies, photochemical effect causing photoisomerization seems to be particularly important. This reaction in the potato tuber involves photon-induced transformation of some chemical compounds (organic dyes, like flavonoids) into other isomers [29,30]. Although isomers possess an identical molecular formula as compounds from which they were created, they differ in the sequence and type of atomic bonds, and in stereomeric formula. The above considerations should be treated as hypothetical and as a justification of further studies in this direction. Water immersion conditions had a significant effect on density difference. The greatest difference |-0,011| [g·cm^{-3}] was observed for conditions in which strips were soaked in water at 40°C for 20min. When strips were immersed in water at a temperature of 20°C for 15 min, the density difference was ca. |-0,006| [g·cm^{-3}], while immersion at 90°C for 2min resulted in density difference of |-0,002| [g·cm^{-3}](fig. 3). Such distribution of density difference for soaking strips in water at a lower temperature than starch gelatinization temperature is confirmed by earlier studies by Sobol [5] and result from the following regularities:

- an increase in solvent (water) temperature from 20°C to 40°C caused reduction of its density, thus, the effect of diffusion was more pronounced
- an increase in the time of diffusion (between water and potato strip tissue) elevated density difference.

The lowest difference of strip density was obtained after soaking them in water heated to 90°C for 2 min which resulted from a short duration of diffusion and formation of a layer of gelatinized starch on the strip surface, which formed partly an barrier for water absorption. The Duncan test (Tab. 3) divided the studied strips according to the density difference into three groups, i.e. each combination of soaking conditions statistically significantly influenced the parameter under study.

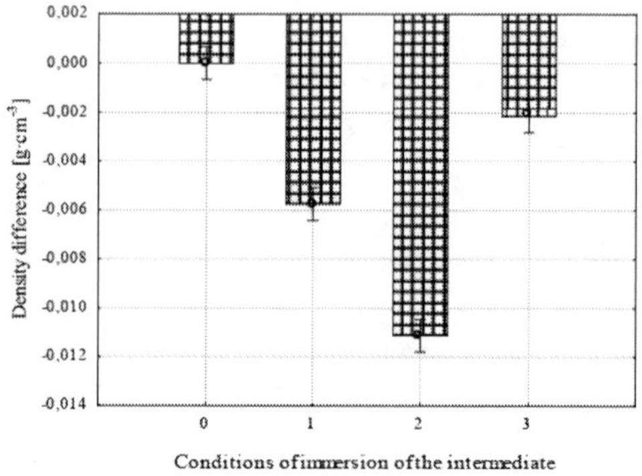

Fig. 3. Influence of immersion conditions on the density differences.

Table 3. Duncan's test results. Groups homogeneous for immersion conditions of intermediates

Conditions of immersion of the intermediate	Mean	1	2	3
2	-0,0111	****		
1	-0,0058		****	
3	-0,0021			****

Detailed analysis of the obtained results (Fig. 4) indicated that the highest density difference |-0,013| [g·cm^{-3}] was obtained for the second combination of soaking conditions, i.e. at 40°C for 20 min.

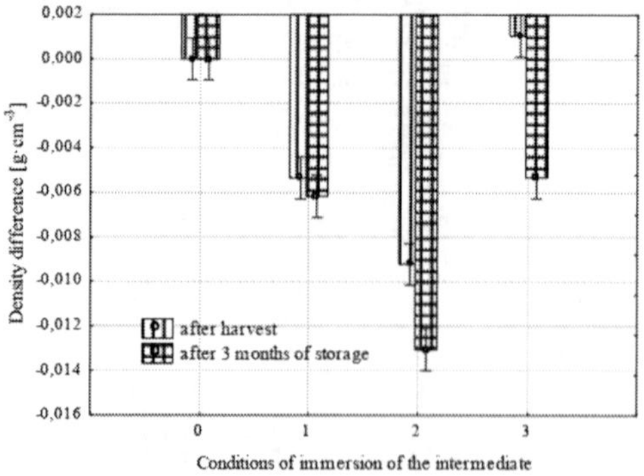

Fig. 4. Influence of immersion conditions of the intermediates, within the storage period, on the density difference.

On the other hand, the highest difference in density change between measurements conducted immediately after harvest and after 3-month storage (ca. |-0,005| [g·cm^{-3}] was obtained for the third combination of soaking conditions, namely at 90°C for 2min. Most probably it was caused both by a high osmotic pressure in cells and low density of water. Analysis of the effect of soaking conditions within groups exposed to different modes of UV stimulation (Fig. 5) on density difference showed that in the second combination of soaking conditions there was a significant difference in this parameter of ca. |-0,005| [g·cm^{-3}] between strips from UV-C-stimulated tubers and unstimulated tubers. No statistically significant diversity in density difference was seen for the remaining combinations of soaking conditions.

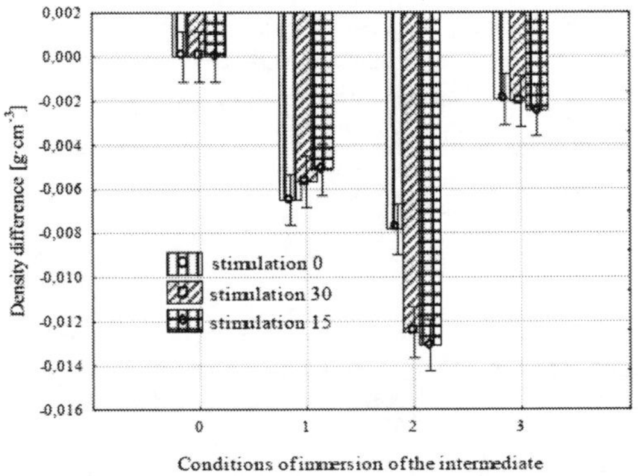

Fig. 5. Influence of immersion conditions of the intermediates, within the applied stimulation, on the density difference.

4 Conclusions

1. Difference in density of potato strips is statistically significantly influenced by storage time, UV-C stimulation and strip soaking conditions.
2. Prolongation of tuber storage increases strip density difference.
3. The used UV-C stimulation of tubers augments difference in strip density.
4. Increase in soaking duration and water temperature (up to starch gelatinization temperature) enhances density difference.

References

1. Sobol Z., Hebda T., Łapczyńska-Kordon B. Technical Sciences, **20(2)** (http://www.uwm.edu.pl/techsci/index.php/no-20-2-2017) (2017)
2. Sobol Z. Agricultural Engineering, Vol. 20, No. **4** , pp.183 -194 (2016)
3. Sobol Z. Acta Agrophysica, **139**, Vol. 8(1), 219-228 (2006a)
4. Sobol Z. Inżynieria Rolnicza, **6(94)**, 231-240 (2007a)
5. Sobol Z. Acta Agrophysica, **142**,Vol. 8 (4), 985-993 (2006 b)
6. Sobol Z. Acta Agrophysica, **151**, Vol. 10(1), 219-228 (2007b)
7. Pilarski; J.; Tokarz; K.; Kocurek; M. Prace Instytutu Elektrotechniki, **256** (2012)
8. Gabriel A. Food Control. 25(2):425–432 (2012)
9. Taze B., Unluturk S., Buzrul S., Alpas H. Journal of Food Science and Technolgy, **52(2)** 1000–1007 (2015)
10. Słowiński, K. Inżynieria Rolnicza, **6(131)**, 207-213 (2011)
11. Błażejak S, Stasiak-Różańska L., Markowski K., Lipińska EActa Sci. Pol., Biotechnologia **10(2)**, 17-24 (2011)
12. Sławińska D., Polewski K. , Rolewski P., Pluciński P., Sławiński J. EJPAU **5(2)**, available online: http://www.ejpau.media.pl/volume5/issue2/environment/art-05.html (2002)
13. Szwarc W., Skórska E. Biuletyn Instytutu Hodowli i Aklimatyzacji Roślin, **245**, 215-221(2007)
14. Allende, J.L. McEvoy, Y. Luo, I. Artes and C.Y. Wang. Food Microbiol. **23** 241-249 (2006)
15. Kasım R., Kasım M. Journal of Agricultural Faculty of Uludag University, **24(1)**, 71-80 (2010)
16. Liu C., Cai L., Lu X., Han X., Ying T. Journal of Integrative Agriculture, **11(1)** 159–165 (2012)
17. Hassenberg, K.; Huyskens-Keil, S.; Herppich, W.B. Journal of Applied Botany and Food Quality, **85(2)**, 174-181 (2012)
18. Jakubowski T., Pytlowski T. Inżynieria Rolnicza. Nr **3** (145). s. 99-107 (2013)
19. Lin Q., Xie Y., Liu W., Zhang J., Cheng S., Xie X., Guan W., Wang Z. J Food Sci Technol. **54(1)** 55–61 (2017)
20. Lisińska G. Zesz. Probl. Post. Nauk Roln. z. **511** część I, 81-94 (2006)
21. Jakubowsk T., Wrona P. Acta Scientiarum Polonorum - Technica Agraria, nr **11** (1-2), 33–41 (2012)
22. Jakubowski, T., Pytlowski, T. Inżynieria Rolnicza, **2(154),** 35-43 (2015)
23. Tevini M In: Tevini M (ed) Lewis, Boca Raton, pp 125–154 (1993)
24. Robin, C.; Hay M.; Newton, P.; Greer, D. Annals of Botany, **74** (1994)
25. Cockeel, C.; Churio, M.; Previtali, C. Biological Reviews **74** (1999)
26. Poon, T.; Barnetson, R.; Halliday, G. Journal of Investigative Dermatology, **125(4)** (2005)
27. Negash, L.; Björn, L., Physiologia Plantarum, **66** (1986)
28. Worrest, R.; Häder, P. Photochemistry and Photobiology, **65(2)** (1997)
29. Reddivari, L.; Hale, A.; Miller, J. Journal of Agricultural Food Chemistry, **55** (2007)
30. Keutgen, A.; Pobereżny, J.; Wszelaczyńska, E.; Murawska, B.; Spychaj-Fabisiak, E. Inżynieria i Aparatura Chemiczna, **53(2)** (2014)

Fractional Maxwell model of viscoelastic biological materials

Anna Stankiewicz[1,*]

[1]Department of Technology Fundamentals. University of Life Sciences in Lublin, Głęboka 28, 20-612 Lublin, Poland

Abstract. This article focuses on fractional Maxwell model of viscoelastic materials, which are a generalization of classic Maxwell model to non-integer order derivatives. To build a fractional Maxwell model when only the noise-corrupted discrete-time measurements of the relaxation modulus are accessible for identification is a basic concern. For fitting the original measurement data an approach is suggested, which is based on approximate Scott Blair fundamental fractional non-integer models, and which means that the data are fitted by solving two dependent but simple linear least-squares problems in two separable time intervals. A complete identification algorithm is presented. The usability of the method to find the fractional Maxwell model of real biological material is shown. The parameters of the fractional Maxwell model of carrot root that approximate the experimental stress relaxation data closely are given.

1 Introduction

Fractional calculus is a branch of mathematical analysis that generalizes the derivative and integral of a function to non-integer order [1]. Application of fractional calculus in classical and modern physics greatly contributed to the analysis and our understanding of physico-chemical and bio-physical complex dynamical systems, since it provides excellent instruments for the description of memory and properties of various materials and processes. During the last two decades fractional calculus has been increasingly applied to mathematical modelling in physics [2,3], engineering [4,5], and especially to rheology [6,7], where fractional calculus constitute a valuable mathematical tool to handle viscoelastic aspects of systems and materials mechanics. Models involving fractional derivatives and operators have been found to better describe some real phenomena than integer-order differential equations [1-5], whence there are many new exciting areas of fractional models and fractional calculus applications, such as the automatic control, the modeling of biological, medical and environmental. A historical review of applications can be found in [8]. The survey on fractional models from biology and biomedicine is presented in [9], see also other papers cited therein. In recent decades fractional derivatives were found quite flexible, especially in the description of viscoelastic polymer materials [10].

Viscoelastic materials present a behaviour that implies dissipation and storage of mechanical energy. Research studies conducted during the past few decades proved that these models are also an important tool for studying the behaviour of biological materials [11]: wood, fruits, vegetables, animals tissues, etc. Viscoelasticity of the materials manifests itself in different ways, such as gradual deformation of a sample of the material under constant stress (creep behaviour), and stress relaxation in the sample when it is subjected to a constant strain. In general, viscoelasticity is a phenomenon associated with time variations in a material's response. In an attempt to describe some of the above effects mathematically several constitutive laws have been proposed which describe the stress–strain relations in terms of quantities like creep compliance, relaxation modulus, the storage and loss moduli and dynamic viscosity. Some of these constitutive laws have been developed with the aid of mechanical models consisting of combinations of springs and viscous dashpots. The classical Maxwell model, which is a viscoelastic body that stores energy like a linearized elastic spring and dissipates energy like a classical fluid dashpot is, perhaps, the most representative example of such models.

For over five decades classical exponential behaviour models have been widely applied to describe the viscoelastic properties of biological materials. Maxwell, Kelvin-Voight and Zener models are used to mathematical modelling of stress relaxation and creep processes [11-13]. For these models the relationship between the stress and deformation of the material is approximated though an ordinary differential or integral equations.

However, relaxation or creep processes deviating from the exponential Debye decay behaviour are often encountered in the dynamics of biological complex materials [6, 13]. For such materials a stretched exponential decay KWW model (Kohlrausch-Williams-Watts) [12], hyperbolic type decay Peleg model [13] or power type behaviour models [14,15] are used to approximate experimentally obtained relaxation modulus or creep compliance data. Recently, the experimental results obtained by the authors show that the behaviour of some viscoelastic biological materials shows good agreements with that of the fractional models [9,10,16]. To this end, fractional

* Corresponding author: anna.stankiewicz@up.lublin.pl

rheological models, originally pioneered by Nutting [17] and Scott Blair [18], have proven to be a concise and elegant framework for predicting the response of complex fluids such as liquid foods using a small number of parameters [16]. By replacing the springs and dashpots of the classical viscoelastic models by the Scott Blair elements, several fractional models, including the fractional Maxwell, fractional Voigt and fractional Kelvin models, have been proposed [6, 19].

In this paper fractional Maxwell model (FMM) model is considered, which relates the stress to the strain in the material by means of using differential fractional equation [3,6,7]. The FMM admit the closed form of analytical solution in terms of the known Mittag-Leffler function [20]. While fitting data to exponential sum models, like classic Maxwell model, is a very old problem, which has been studied for a long time, until now there are only a few papers concerned mainly to finding fractional Maxwell model.

We often determine the parameters in a model by obtaining the „best-possible" fit to experimental data. Common choice of the model quality measure is the mean square approximation error, leading to a least-squares identification problem. Unfortunately, the original nonlinear least-squares task for FMM is very difficult by means of the infinite series representation of the Mittag-Leffler function [20]. Thus, in [7, 21] an approach is considered, which consists on the determination of the FMM parameters by two approximate Scott Blair models identification in two separate time intervals, for small and large times respectively.

The aim of the paper is to develop a complete procedure for FMM identification using relaxation modulus data from ramp stress relaxation test.

2 Material and Method

2.1 Material

We consider a linear viscoelastic material subjected to small deformations for which the uniaxial, nonaging and izotropic stress-strain equation can be represented by a Boltzmann superposition integral [22]:

$$\sigma(t) = \int_{-\infty}^{t} G(t-\lambda)\dot{\varepsilon}(\lambda)d\lambda, \tag{1}$$

where $\sigma(t)$ and $\varepsilon(t)$ denotes the stress and stain, respectively, and $G(t)$ is the linear time-dependent relaxation modulus. The modulus $G(t)$ is the stress, which is induced in the viscoelastic material described by equation (1) when the unit step strain $\varepsilon(t)$ is imposed. By assumption, the exact mathematical description of the relaxation modulus $G(t)$ is completely unknown, but the value of $G(t)$ can be measured with a certain accuracy for any given value of the time t.

2.2 Fractional Maxwell model

Fractional Scott Blair model [7] is described by the fractional differential equation

$$\sigma(t) = E\tau^{\alpha}\frac{d^{\alpha}\varepsilon(t)}{dt^{\alpha}}, \tag{2}$$

where E and τ are the elastic modulus and relaxation time, respectively, and α is non-integer positive order of fractional derivative of the strain $\varepsilon(t)$. Here, $\frac{d^{\beta}}{dt^{\beta}} = D_t^{\beta}$ means the derivative operator in the sense of the Caputo's fractional derivative of a function $f(x)$ of non-integer order γ with respect to variable t and with starting point at $t = 0$, which is defined by [3]

$$D_t^{\gamma} f(t) = \frac{1}{\Gamma(n-\gamma)}\int_0^t (t-1)^{n-\gamma-1}\frac{d^n}{dt^n} f(t)dt,$$

where $\Gamma(n)$ is Euler's gamma function, $n-1 < \gamma < n$.

The Scott Blair model (2) is intermediate model between ideal spring (see Figure 1a) Hooke's model

$$\sigma(t) = E\varepsilon(t)$$

and Newton's model

$$\sigma(t) = \eta\frac{d\varepsilon(t)}{dt}$$

of ideal fluids, represented by means of an ideal dashpot (see Figure 1b). To illustrate the structure of fractional models a fractional element, in addition to the standard purely elastic and purely viscous element, must be introduced – see Figure 1c.

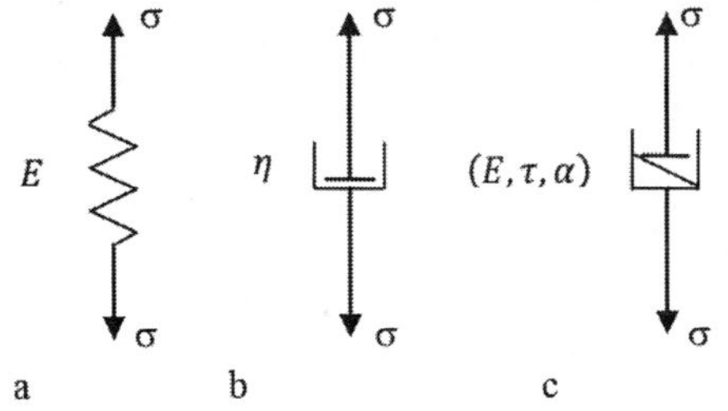

Fig. 1. From left to right: the Hooke (a), Newton (b) and elementary fractional element (c).

Assuming unit-step strain $\varepsilon(t)$ the uniaxial stress response of elementary fractional element (2), i.e. the time-dependent relaxation modulus $G(t)$, is given by [6,7]

$$G(t) = \frac{E}{\Gamma(1-\alpha)}\left(\frac{t}{\tau}\right)^{-\alpha}. \tag{3}$$

Thus, the elementary fractional element (3) is uniquely described by three parameters (E, τ, α), as shown in Figure 1c. This approach makes it possible to include a whole range of dissipation mechanisms in a single (three parameter) rheological element.

Classic viscoelastic Maxwell model is the arrangement of ideal spring in series with a dashpot (see Figure 2a) described by the first order differential equation

$$\frac{d\sigma(t)}{dt} + \frac{E}{\eta}\sigma(t) = E\frac{d\varepsilon(t)}{dt}, \tag{4}$$

which for unit-step strain $\varepsilon(t)$ has exponential type response

$$G(t) = Ee^{-t/\tau}, \tag{5}$$

with the relaxation time $\tau = \eta/E$.

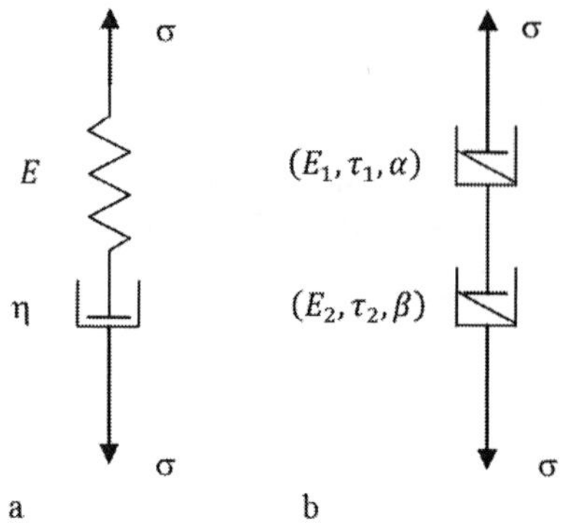

Fig. 2. Classic Maxwell model (a) followed by the fractional Maxwell model (b).

By replacing the spring and dashpot of the classic Maxwell model by two elementary Scott Blair elements (E_1, τ_1, α) and (E_2, τ_2, β) – see Figure 2b – we obtain fractional Maxwell model described by the fractional differential equation [6]. Without loss of generality we assume $\alpha \geq \beta$. Since, in view of (2), for the same stress $\sigma(t)$ the strains for these elements are described by

$$\frac{1}{E_1(\tau_1)^\alpha}\frac{d^{-\alpha}\sigma(t)}{dt^{-\alpha}} = \varepsilon_1(t),$$

$$\frac{1}{E_2(\tau_2)^\beta}\frac{d^{-\beta}\sigma(t)}{dt^{-\beta}} = \varepsilon_2(t),$$

the sum strain $\varepsilon(t)$, which is induced in the series connection of two elementary fractional elements is described by the fractional equation

$$\frac{1}{E_1(\tau_1)^\alpha}\frac{d^{-\alpha}\sigma(t)}{dt^{-\alpha}} + \frac{1}{E_2(\tau_2)^\beta}\frac{d^{-\beta}\sigma(t)}{dt^{-\beta}} = \varepsilon(t),$$

which using the composition rule for fractional differentiations

$$\frac{d^\alpha}{dt^\alpha}\frac{d^\beta f(t)}{dt^\beta} = \frac{d^{\alpha+\beta}f(t)}{dt^{\alpha+\beta}},$$

can be rewritten in compact form as

$$\tau^{\alpha-\beta}\frac{d^{\alpha-\beta}\sigma(t)}{dt^{\alpha-\beta}} + \sigma(t) = E\tau^\alpha\frac{d^\alpha\varepsilon(t)}{dt^\alpha}, \tag{6}$$

where the parameters of FMM (6) are functions of the parameters (E_1, τ_1, α) and (E_2, τ_2, β) of the model components given by

$$\tau = \left[\frac{E_1(\tau_1)^\alpha}{E_2(\tau_2)^\beta}\right]^{\frac{1}{\alpha-\beta}},$$

$$E = \left[\frac{(E_1\tau_1)^{-\beta}(\tau_1)^{\alpha(1-\alpha)}}{[E_2(\tau_2)^\beta]^{-\alpha}}\right]^{\frac{1}{\alpha-\beta}}.$$

For details of the fractional model construction see [6,7]. For the unit-step strain the solution $\sigma(t) = G(t)$ of FMM (6) is known for an arbitrary $1 \geq \alpha \geq \beta \geq 0$ and given by the formula [6,7]:

$$G(t) = E\left(\frac{t}{\tau}\right)^{-\beta} E_{\alpha-\beta,1-\beta}\left(-\left(\frac{t}{\tau}\right)^{\alpha-\beta}\right), \tag{7}$$

where $E_{\kappa,\mu}(x)$ is the generalized two-parameter Mittag-Leffler function defined by series representation, convergent in the whole z-complex plane [3,20]

$$E_{\kappa,\mu}(x) = \sum_{n=0}^{\infty}\frac{x^n}{\Gamma(\kappa n+\mu)}. \tag{8}$$

The above result is obtained in [6] by applying Fourier and Mellin transforms, while in [23] the same result is obtained by an approach involving the Laplace-Mellin transform technique.

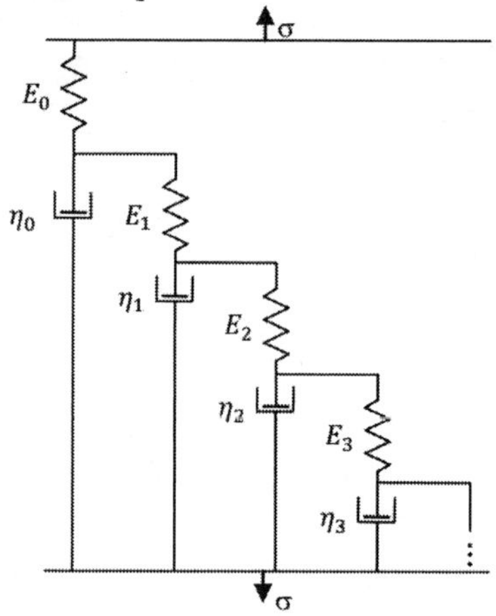

Fig. 3. Sequential ladder realization on the fractional Scott Blair element [6; Fig. 2].

Note, that the fractional Maxwell model (7) is uniquely defined by four parameters (E, τ, α, β), while the classic Maxwell model (4) is defined by means of only two parameters (E, η), or equivalently (E, τ). A deep insight into the complex properties of FMM gives the infinite hierarchical structures of elementary fractional model (2) presented in Figure 3 composed of ideal spring and dashpot elements, which compose the classic Maxwell model.

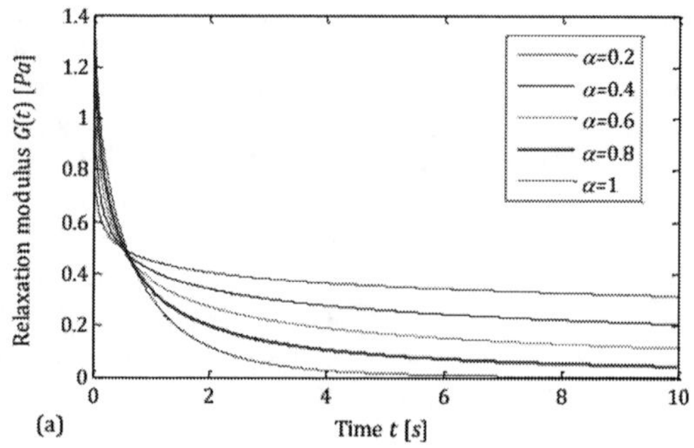

Fig. 4. The FMM relaxation modulus for fixed $\beta = 0.1$.

The course of the relaxation modulus $G(t)$ (7) for fixed β parameter and a few values of the non-integer derivative order α are shown in Figure 4, for fixed α and a few values of β the respective modulus $G(t)$ are plotted in Figure 5. Note, that the curves of FMM relaxation modulus shown in Figures 4 and 5 has the characteristic shape of the relaxation modulus of viscoelastic materials obtained in experiment. The following rule holds: the greater the parameter α is, the shorter the relaxation times are for β being fixed. Similar rule holds for model order β, while α is fixed.

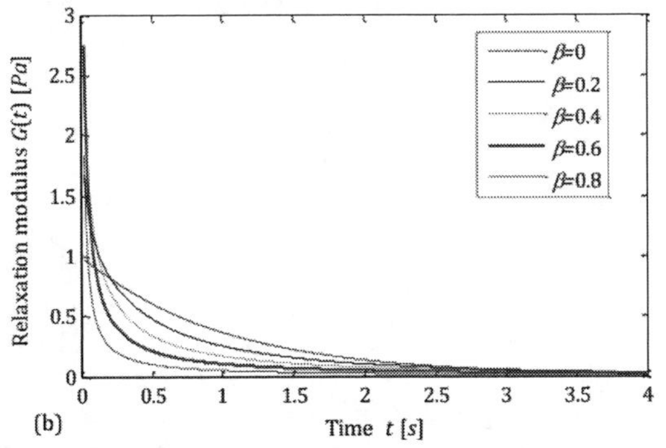

Fig. 5. The FMM relaxation modulus for fixed $\alpha = 1$.

2.3 Approximate fractional models

In the interpretation of the asymptotic behaviour of the FMM a fundamental role plays the asymptotic expansions of the Mittag-Leffler function. Based on (8), the relaxation modulus $G(t)$ (7) can be rewritten as

$$G(t) = E\left(\frac{t}{\tau}\right)^{-\beta}\left[\frac{1}{\Gamma(1-\beta)} + \sum_{n=1}^{\infty}\frac{(-1)^n\left(\frac{t}{\tau}\right)^{(\alpha-\beta)n}}{\Gamma((\alpha-\beta)n+1-\beta)}\right], \quad (9)$$

and since for small t, in particular for $\left(\frac{t}{\tau}\right)^{\alpha-\beta} \ll 1$, the infinite sum of the right hand side of (9) can be truncated, the asymptotic approximation results

$$G(t) \cong \frac{E}{\Gamma(1-\beta)}\left(\frac{t}{\tau}\right)^{-\beta}, \quad (10)$$

here $\cong$ means 'approximately equal'. From the following asymptotic property [24]

$$E_{\varphi,\mu}(-t^{\varphi}) \cong \sum_{n=1}^{\infty}(-1)^{n-1}\frac{t^{-\varphi n}}{\Gamma(\mu-\varphi n)}, \quad (11)$$

which holds for $t \to \infty$, putting $\varphi = \alpha - \beta$ and $\mu = 1 - \beta$ in (11) based on (7) yields

$$G(t) \cong E\left(\frac{t}{\tau}\right)^{-\beta}\left[\frac{\left(\frac{t}{\tau}\right)^{-(\alpha-\beta)}}{\Gamma(1-\beta-(\alpha-\beta))} + \sum_{n=2}^{\infty}\frac{(-1)^{n-1}\left(\frac{t}{\tau}\right)^{-(\alpha-\beta)n}}{\Gamma(1-\beta-(\alpha-\beta)n)}\right],$$

whence for large times, especially for $(t/\tau)^{\alpha-\beta} \gg 1$, we obtain the next asymptotic approximation of FMM in the form

$$G(t) \cong \frac{E}{\Gamma(1-\alpha)}\left(\frac{t}{\tau}\right)^{-\alpha}. \quad (12)$$

The two approximate rules (10), (12) are given in [6] for short times such that $t \ll \tau$ and for large times $t \gg \tau$, respectively, based on power series representations of H-function and fact that the Mitag-Leffler function is expressible through H-function. The same asymptotic approximate models are considered in [7] and [21], but in the last paper stringent applicability conditions, derived above, are imposed. Thus, both for short as well as for large times, relaxation modulus (7) decreases almost according to the time-power elementary fractional element. The approximate models (10), (12) and the exact FMM are summarized in Figure 6, where logarithmic scale is used both for relaxation modulus and time scales. Here, the asymptotic models (10), (12), whose graph is a straight line are linear functions, which slope coefficients uniquely determine the fractional derivatives orders $-\beta$ for small times and α for large times, respectively.

On the basis of this observation in [7] the FMM parameters (E, τ, α, β) of polymers PMMA and PTFE has been determined based on approximate models (10) and (12) in the time intervals defined by $t \ll \tau$ and $t \gg \tau$, respectively. Next, in [21] the PMMA and PTFE fractional models have been corrected by applying more restrictive conditions $(t/\tau)^{-\beta} \ll 1$ and $(t/\tau)^{-\alpha} \gg 1$. In result better fit to experimental data is obtained.

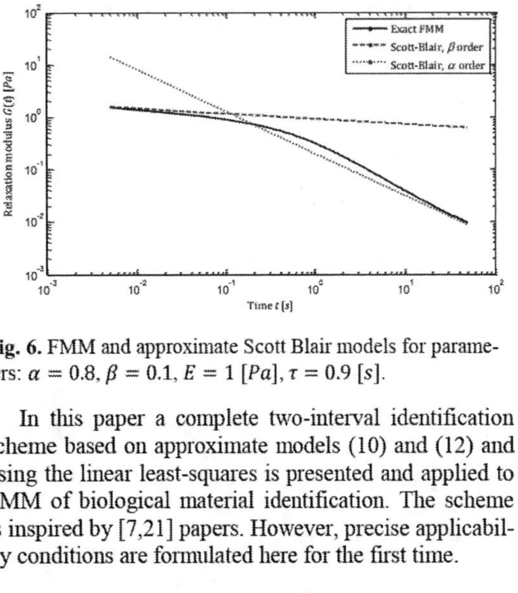

Fig. 6. FMM and approximate Scott Blair models for parameters: $\alpha = 0.8, \beta = 0.1, E = 1\ [Pa], \tau = 0.9\ [s]$.

In this paper a complete two-interval identification scheme based on approximate models (10) and (12) and using the linear least-squares is presented and applied to FMM of biological material identification. The scheme is inspired by [7,21] papers. However, precise applicability conditions are formulated here for the first time.

3 Results and discussion

3.1 Identification

A classical manner of studying viscoelasticity is by two-phase stress relaxation test, where the strain increases during the loading time interval until a predetermined strain ε_0 is reached at given ramp-time, after which that strain ε_0 is maintained constant at that value [11,12,25].

Suppose, a certain stress relaxation test performed on the specimen of the material under investigation resulted in a set of measurements of the relaxation modulus $\bar{G}(t_i) = G(t_i) + z(t_i)$ at the sampling instants $t_i > 0$,

$i = 1, \ldots, N$, where $z(t_i)$ is measurement noise. For computational methods of relaxation modulus determination see, for example [12,25].

In general, identification consists of selecting within the given class of models such a model, which ensures the best fit to the measurement results. Fitting data to the original FMM (7) is a very difficult problem of nonlinear optimization, numerically difficult and often ill-conditioned.

Here, the linear least-squares identification routine will be used to estimate FMM parameters based on the logarithmic transformation of the experimental data and equations (10) and (12) which, respectively, yields

$$log\,G(t) \cong k_1 - \beta log t,$$

$$log\,G(t) \cong k_2 - \alpha\, log t,$$

where

$$k_1 = log\left[\frac{E\tau^{\beta}}{\Gamma(1-\beta)}\right], \quad (13)$$

$$k_2 = log\left[\frac{E\tau^{\alpha}}{\Gamma(1-\alpha)}\right] \quad (14)$$

are introduced for brevity. Denoting $\chi(t) = log\,G(t)$ and introducing the new independent variable $\vartheta = logt$, equivalent linear models are obtained (the sign $\cong$ is neglected for simplicity)

$$\chi(t) = k_1 - \beta\vartheta, \quad (15)$$

$$\chi(t) = k_2 - \alpha\vartheta. \quad (16)$$

Now, in the set of experimental data $\{(\bar{G}(t_i), t_i)\}_{i=1}^{N}$, or equivalently in the set $\{(\chi_i, \vartheta_i)\}_{i=1}^{N}$, where

$$\chi_i = log\,\bar{G}(t_i) \text{ and } \vartheta_i = log\,t_i \quad (17)$$

are log-transformed 'measurement' data, two separable subsets $\{(\chi_i, \vartheta_i)\}_{i=1}^{n_1}$ and $\{(\chi_i, \vartheta_i)\}_{i=1}^{n_2}$ must be determined such that $n_1 < n_2$, probably $n_1 \ll n_2$. The subsets $\mathcal{N}_1 = \{1, \ldots, n_1\}$ and $\mathcal{N}_2 = \{n_2, \ldots, N\}$ of respective indices are chosen during the recurrent identification scheme. It is clear that the union of the sets $\mathcal{N}_1 \cup \mathcal{N}_2$ does not create the set $\mathcal{N} = \{1, \ldots, N\}$. Now, classical linear least squares method can be applied to find optimal approximate models.

As a measure of the model (15) accuracy the mean sum of squares is taken

$$Q_1(k_1, \beta, n_1) = \frac{1}{n_1}\sum_{i=1}^{n_1}[\chi_i - k_1 + \beta\vartheta_i]^2, \quad (18)$$

the respective identification index for the second set of experimental data and log-linearized model (16) is

$$Q_2(k_2, \alpha, n_2) = \frac{1}{N-n_2+1}\sum_{i=n_2}^{N}[\chi_i - k_2 + \alpha\vartheta_i]^2. \quad (19)$$

Therefore, the least-squares identification of the log-linearized models consists of determining the model parameters minimizing the indices (18), (19) by solving the following standard optimization problems

$$Q_1(\hat{k}_{1,n_1}, \hat{\beta}_{n_1}, n_1) = min_{(k_1,\beta)\in R^2}\, Q_1(k_1, \beta,, n_1), \quad (20)$$

$$Q_2(\hat{k}_{2,n_2}, \hat{\alpha}_{n_2}, n_2) = min_{(k_2,\alpha)\in R^2}\, Q_2(k_2, \alpha, n_2). \quad (21)$$

Based on the commonly known results concerning the linear least-squares problem solution, it can be shown that the solutions to (20), (21) tasks there exist and are unique whenever the sampling instants are such that $0 < t_1 < t_2 < \cdots < t_N$. The model parameters optimal in the sense of (21) are given by the known formulas:

$$\hat{\alpha}_{n_2} = \frac{\frac{1}{N-n_2+1}\sum_{i=n_2}^{N}\chi_i\sum_{i=n_2}^{N}\vartheta_i - \sum_{i=n_2}^{N}\chi_i\vartheta_i}{\sum_{i=n_2}^{N}[\vartheta_i]^2 - \frac{1}{N-n_2+1}\left[\sum_{i=n_2}^{N}\vartheta_i\right]^2}, \quad (22)$$

$$\hat{k}_{2,n_2} = \frac{1}{N-n_2+1}\sum_{i=n_2}^{N}\chi_i + \frac{1}{N-n_2+1}\sum_{i=n_2}^{N}\vartheta_i\,\hat{\alpha}_{n_2}. \quad (23)$$

The formulas for optimal $\hat{\beta}_{n_1}$ and $\hat{k}_{1,n_1}$ are analogous. According to definitions (13) and (14) the positivity constraint can be neglected for k_2, k_2 in (20), (21) optimization tasks. However, both α as well as β must be positive. It can be shown, based on the Czebyszev equality, that the optimal parameters $\hat{\beta}_{n_1}$ and $\hat{\alpha}_{n_2}$ are positive, if and only if, the following conditions are satisfied

$$\sum_{i=1}^{n_1}\sum_{j=i+1}^{n_1} log\frac{\bar{G}(t_j)}{\bar{G}(t_i)}\, log\frac{t_i}{t_j} > 0,$$

$$\sum_{i=n_2}^{N}\sum_{j=i+1}^{N} log\frac{\bar{G}(t_j)}{\bar{G}(t_i)}\, log\frac{t_i}{t_j} > 0,$$

for tasks (20) and (21), respectively.

Note, that the models (15) and (16) include two unknown parameters (k_1, β) and (k_2, α). In view of (14) and (13) the pairs of parameters are not independent. Therefore, from (14) and (13) we have

$$\tau = \left[\frac{\Gamma(1-\alpha)}{\Gamma(1-\beta)}10^{k_2-k_1}\right]^{\frac{1}{\alpha-\beta}}, \quad (24)$$

which, in view of (14), next yields

$$E = 10^{k_2}\frac{\Gamma(1-\alpha)}{\tau^{\alpha}}. \quad (25)$$

3.2 Algorithm

Taking into account the above, the calculation of the approximate values of FMM parameters involves the following steps.

1. Perform the stress relaxation test, record and store the relaxation modulus measurements $\bar{G}(t_i)$, $i = 1, \ldots, N$, corresponding to the chosen sampling instants $t_i > 0$.
2. Determine the set $\{(\chi_i, \vartheta_i)\}_{i=1}^{N}$ of log-transform measurement data (17).
3. Based on the course of log-log plot of the measurement data, select in the set of measurements two separable subsets choosing n_1 and n_2, $n_1 < n_2$.
4. Compute the estimates $(\hat{\beta}_{n_1}, \hat{k}_{1,n_1})$ and $(\hat{k}_{2,n_2}, \hat{\alpha}_{n_2})$ according to formulas (22), (23) and, next, compute

the estimates of the relaxation time $\hat{\tau}_{n_1,n_2}$ and elastic modulus $\hat{E}_{n_1,n_2}$ based on (24) and (25).

5. In order to ascertain if the models with parameters $(\hat{\beta}_{n_1}, \hat{k}_{1,n_1})$ and $(\hat{k}_{2,n_2}, \hat{\alpha}_{n_2})$ are a satisfactory approximation of measurement data in time intervals $(0,t_{n_1}]$ and $[t_{n_2}, t_N]$ compute the optimal identification indices and examine if $Q_1(\hat{k}_{1,n_1}, \hat{\beta}_{n_1}, n_1) \leq \delta$ and $Q_2(\hat{k}_{2,n_2}, \hat{\alpha}_{n_2}, n_2) \leq \delta$, for δ, a preselected small positive error. If not, change n_1 and/or n_2 and go to step 4. Otherwise, go to step 6.
6. In order to ascertain if the Scott Blair models (10), (12) with parameters $(\hat{E}_{n_1,n_2}, \hat{\tau}_{n_1,n_2}, \hat{\beta}_{n_1})$ and $(\hat{E}_{n_1,n_2}, \hat{\tau}_{n_1,n_2}, \hat{\alpha}_{n_2})$ are a satisfactory approximation of the original FMM examine if

$$\omega_{n_1,n_2} = \left(\frac{t_{n_1}}{\hat{\tau}_{n_1,n_2}}\right)^{\hat{\alpha}_{n_2}-\hat{\beta}_{n_1}} \ll 1, \qquad (26)$$

$$\Omega_{n_1,n_2} = \left(\frac{t_{n_2}}{\hat{\tau}_{n_1,n_2}}\right)^{\hat{\alpha}_{n_2}-\hat{\beta}_{n_1}} \gg 1. \qquad (27)$$

If both the above conditions are satisfied, stop the procedure taking $(\hat{E}_{n_1,n_2}, \hat{\tau}_{n_1,n_2}, \hat{\alpha}_{n_2}, \hat{\beta}_{n_1}) = (\hat{E}, \hat{\tau}, \hat{\alpha}, \hat{\beta})$ as the FMM parameters. If not, change n_1 and/or n_2 and execute steps 4 and 5.

The stopping rules from Step 5 guarantee the good quality of the log-linearized models in the chosen time intervals and correspond with those commonly used in the optimal identification techniques. The conditions (26), (27) are imposed to guarantee the applicability of Scott Blair models (10), (12) to approximate the original FMM. Both the pairs of conditions must be satisfied simultaneously in order to guarantee good quality of the resulted model. Note also that, in particular, from (27) the obvious necessary applicability condition of the scheme results

$$t_N \gg \hat{\tau},$$

i.e. the experiment time must be significantly larger than the relaxation time of the FMM.

While $\hat{\alpha}_{n_2}$ and $\hat{\beta}_{n_1}$ are determined independently in appropriate time intervals, the relaxation time $\hat{\tau}_{n_1,n_2}$ and the modulus $\hat{E}_{n_1,n_2}$ depend on the identification results in both time intervals, simultaneously.

Note, that (26) and (27) are *a posteriori* conditions, since the applicability of the identification procedure cannot be checked, earlier than after the experiment is performed.

The next example shows, how the identification scheme can be used for identification of FMM of real material.

3.3 Example – FMM of carrot root

A cylindrical sample of 10.9 mm diameter and 8 mm height was obtained from the root of carrot Nantejska variety [14]. During the two-phase stress relaxation test performed by Bohdziewicz [14], in the first loading phase of the time period (0,43.7) seconds the strain rate $0.000033\ [m \cdot s^{-1}]$ is used and the force induced in the specimen was recorded with the mean sampling period $\Delta t_{mean} = 0.11007\ [s]$. Next, during the second phase of the time period (43.7,1187) seconds, at constant strain the corresponding time-varying force was recorded in $N = 341$ measurement points with sampling period $\Delta t_{mean} = 3.34\ s$ (for details see [13,14]). Next, the relaxation modulus measurements $\bar{G}(t_i)$ were computed for the time interval (0,1143.3) seconds using the fast trapezoidal method of approximate relaxation modulus identification presented in [25].

Based on the log-log plot of the experiment results, the time intervals determined by $t_{n_1} = 0.5\ [s]$ and $t_{n_2} = 300\ [s]$ have been chosen and the estimates $(\hat{\beta}_{n_1}, \hat{k}_{1,n_1})$, $(\hat{k}_{2,n_2}, \hat{\alpha}_{n_2})$ of Scott Blair models parameters, and next the estimates of the relaxation time $\hat{\tau}_{n_1,n_2}$ and elastic modulus $\hat{E}_{n_1,n_2}$ have been computed. Freely available [26] Matlab code for evaluating the Mittag-Leffler function has been used. The inaccurate fit of the resulting FMM to the experimental data is illustrated in Figure 7. Next, the recurrent scheme of the algorithm have been realized and successive approximations of FMM parameters have been determined.

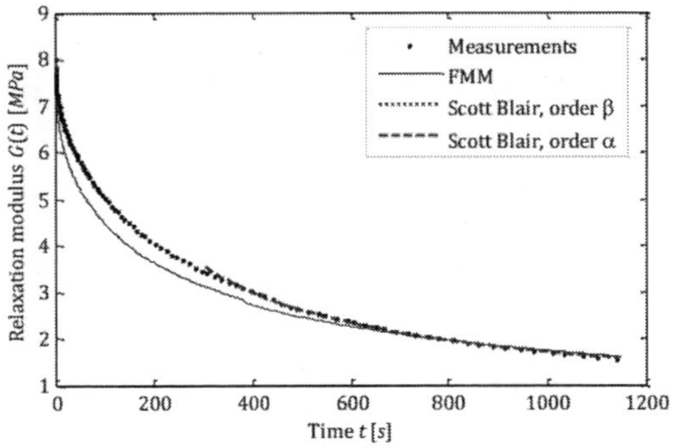

Fig. 7. The inaccurate fit of FMM to experiment data for parameters: $\alpha = 0.6092$, $\beta = 0.0159$, $E = 6.9425\ [MPa]$ and $\tau = 384.7876\ [s]$

The results are summarized in Table 1, where both the model parameters, as well as the factors ω_{n_1,n_2} (26) and $\Omega_{n_1,n_2,mean}$, defined as the mean value of the left-hand side Ω_{n_1,n_2} of (27), are given for a few FMM approximations. Repeated composition of the scheme operations has led to parameters listed in the last row of Table 1. For these model parameters the accurate fit to experiment data is achieved, as shown in the Figure 8. The presented results convincingly prove, that the FMM give satisfactory approximation of the relaxation modulus of the material.

The last column of the table is a value of the empirical mean square model error

$$ERR = \frac{1}{N}\sum_{i=1}^{N}[\bar{G}(t_i) - G(t, \boldsymbol{a})]^2, \qquad (28)$$

where $\boldsymbol{a} = (E, \tau, \alpha, \beta)$ is the vector of FMM parameters. The model errors are not big (see Table 1), but for a few first FMM approximations the accuracy is insufficient, especially in long time region.

Table 1. The results of FMM identification.

$t_{n_1}[s]$	$t_{n_2}[s]$	$\hat{\alpha}_{n_2}$	$\hat{\beta}_{n_1}$	$\hat{\tau}_{n_1,n_2}$	$\hat{E}_{n_1,n_2}$	ω_{n_1,n_2}	$\Omega_{n_1,n_2,mean}$	ERR
0.5	300	0.6092	0.0159	384.7876	6.9425	0.0394	1.7949	0.1379
0.7	300	0.6092	0.0166	388.0466	6.9069	0.0243	1.6841	0.1463
0.7	400	0.6280	0.0166	430.8650	6.8950	0.0203	1.8032	0.0785
0.7	500	0.6363	0.0166	451.4663	6.8896	0.0187	1.7658	0.0563
1	500	0.6363	0.0172	455.1887	6.8537	0.0229	1.7557	0.0620
0.3	500	0.6363	0.0153	443.9948	6.9632	0.0108	1.7863	0.0454
0.3	1000	0.6996	0.0153	639.4535	6.9245	0.0053	1.4764	0.0832
0.2	700	0.6674	0.0149	528.0814	6.9642	0.0069	1.6427	0.0177
0.2	600	0.6538	0.0149	488.5427	6.9723	0.0073	1.7087	0.0185
0.25	500	0.6363	0.0151	443.0339	6.9728	0.0099	1.7890	0.0440
0.25	550	0.6892	0.0139	457.0883	7.0425	0.0063	1.9190	0.0037

Next, the generalized discrete Maxwell model, which presents a relaxation of exponential type given by a finite Dirichlet-Prony series [11,12]

$$G(t) = \sum_{j=1}^{N} E_j\, e^{-v_j t} + E_\infty, \tag{29}$$

where E_j, v_j and E_∞ represent the elastic modulus, relaxation frequencies and equilibrium modulus (long-term modulus), respectively, has been applied to approximate the experiment data. Least-squares approximations are used, and two, three and four-parameter optimal models were determined. The parameters of the classic optimal Maxwell models and the relaxation times $\tau_j = 1/v_j$ are given in Table 2. The differences between the three Maxwell models are characterized by the ERR errors (28) summarized in Table 2, here the vector $\boldsymbol{a} = (E_j, v_j, E_\infty)$. Figure 9 presents experiment data and the optimal Maxwell models.

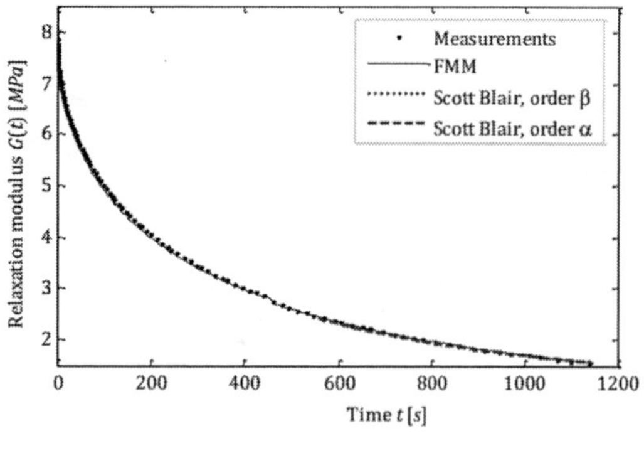

Fig. 8. The time course of FMM and the experimental data, parameters: $\alpha = 0.6892$, $\beta = 0.0139$, $E = 7.0425\ [MPa]$ and $\tau = 457.0883[s]$.

It is easy to observe, that the Maxwell model is inappropriate for description of this relaxation process. Thus, the classic Maxwell models do not fully characterize the true viscoelastic behaviour of the biological material. A better fit to experimental data can be obtained, if the FMM is used. Note, that also the range of the relaxation time estimates is lower for FMM than for classic Maxwell models, even if the conditions (26), (27) of the scheme applicability are not sharply fulfilled – compare Tables 1 and 2.

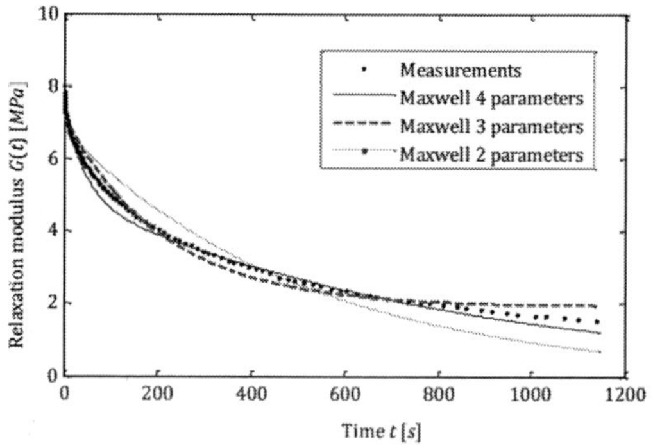

Fig. 9. The optimal Maxwell models fit to the experimental data.

4 Conclusions

In this work, a complete simple procedure is proposed for the fractional Maxwell model identification based on the discrete-time stress relaxation experimental data. The asymptotic approximate log-linearized Scott Blair models, composed of only two parameters, are used in separate time intervals. The linear least squares are applied to find the best Scott Blair models in two separable appropriately chosen time intervals. An applicability conditions of the approximate identification scheme are derived. It is shown that FMM can be used to describe the viscoelastic mechanical properties of biological materials. The effectiveness is compared with the classical two, three and four parameter Maxwell model approximations. While the idea of using the approximate Scott Blair models to find the FMM parameters is known and already appeared in [7,21], the complete identification scheme is presented in this paper for the first time.

Table 2. Maxwell model (29) parameters.

Two-parameter model		**Three-parameter model**	
$E_1\ [MPa]$	6.80761	$E_1\ [MPa]$	5.24942
$v_1\ [s^{-1}]$	0.001975	$v_1\ [s^{-1}]$	0.004694
		$E_\infty\ [MPa]$	1.92311
$\tau_1\ [s]$	506.2194	$\tau_1\ [s]$	213.0323
model error ERR	0.24107	model error ERR	0.06035
Four-parameter model			
$E_1\ [MPa]$	4.95022	$v_1\ [s^{-1}]$	0.00122
$E_2\ [MPa]$	2.64198	$v_2\ [s^{-1}]$	0.02316
		$\tau_1\ [s]$	816.3732
model error ERR	0.03257	$\tau_2\ [s]$	43.1732

The authoress wish to thank Jerzy Bohdziewicz of the Faculty of Life Sciences and Technology, Wrocław University of Environmental and Life Sciences, Wrocław, Poland, for providing measurement results of relaxation tests of carrot roots and helpful suggestions concerning the rheological experiments.

References

1. L. Debnath, Int. J. of Mathematics and Mathematical Sci. **54**, 3413 (2003)
2. R. Hilfer, *Applications Of Fractional Calculus In Physics* (World Scientific, 2000)
3. I. Podlubny, *Fractional Differential Equations* (Academic Press, London, 1999)
4. M.A. Moreles, R. Lainez, Commun. Nonlinear Sci. Numer. Simulat. **46**, 81 (2017)
5. M.D. Ortigueira, IEEE Circuits and Systems Magazine, **8**, 19 (2008)
6. H. Schiessel, R. Metzler, A. Blumen, T.F. Nonnejunacher, J. Phys. A: Math. Gen. **28**, 6567 (1995)
7. A. Hernández-Jiménez, J. Hernández-Santiago, A. Macias-García, J. Sánchez-González, Polymer Testing **21**, 325 (2002)
8. J. Machado, V.K. Tenreiro, F. Mainardi, Commun. Nonlinear Sci. Numer. Simul., **16**, 1140 (2011)
9. T.J. Freeborn, IEEE J. Emerging Select Topics Circuits Syst., **3**, 416 (2013)
10. J. Zhao, L. Zheng, X. Chen, X. Zhang, F. Liu, Applied Mathematical Modelling, **44**, 497 (2017)
11. M.A. Rao, *Rheology of Fluid, Semisolid, and Solid Foods. Principles and Applications* (Springer Science & Business Media, New York, 2014)
12. A. Stankiewicz, *Selected methods and algorithms for the identification of models used in the rheology of biological materials* (Tow. Wyd. Nauk. Libropolis, Lublin 2013)
13. J. Bohdziewicz, *Modelowanie przebiegu odkształceń tkanek parenchymy warzyw w warunkach quasi-statycznych zmian obciążeń* (Wyd. Uniwersytetu Przyrodniczego, Wrocław, 2007)
14. J. Bohdziewicz, Acta Agrophysica, **97**, 499 (2003)
15. J. Bohdziewicz, G. Czachor, Agricultural Engineering, **20**, 15 (2016)
16. C.E. Wagner, A.C. Barbati, J.Engmann, A.S. Burbidge, G.H. McKinley, Food Hydrocolloids, **69**, 242 (2017)
17. P.G. Nutting, Journal of the Franklin Institute, **191**, 581 (1920)
18. G.W. Scott Blair, Periodica Polytechnica Chemical Engineering, **16**, 81 (1972)
19. F. Yang, Ke-Qin Zhu, Acta Mech. Sin., **27**, 866 (2011)
20. R. Gorenflo, A.A. Kilbas, F. Mainardi, S.V. Rogosin, *Mittag-Leffler Functions, Related Topics and Applications. Theory and Applications* (Springer, Heidelberg, New York, Dordrecht, London, 2014)
21. Kai-Xin Hu, Ke-Qin Zhu, Polymer Testing, **30**, 797 (2011)
22. R.M. Christensen, *Theory of Viscoelasticity* (Dover Publications, Mineola, New York, 2013)
23. W.G. Glöckle, T.F. Nonnenmacher, Macromolecules, **24**, 6426 (1991)
24. N.H.T. Lemes, JP.C. dos Santos, J.P. Braga, Applied Mathematical Modelling, **40**, 7971 (2016)
25. A. Stankiewicz, Inżynieria Rolnicza, **16**, 401 (2012)
26. I. Podlubny, Mittag–Leffler function. MATLAB Central File Exchange (2012)

BIO Web of Conferences **10**, 02033 (2018) https://doi.org/10.1051/bioconf/20181002033
Contemporary Research Trends in Agricultural Engineering

Spatial database for division of agricultural plots for the group of vegetableproducers

Anna Szeląg-Sikora[1*], *Joanna* Rorat[1]

[1]University of Agriculture in Krakow, Faculty of Production and Power Engineering, ul. Balicka 116B, 30-149 Krakow, Poland

Abstract. Excessive fragmentation of agricultural enterprises is one of the most important barriers preventing the rational use of land in a large part of Poland. In particular, this problem affects the south-eastern regions of the country. Proper spatial distribution of farmland is one of the most important factors enabling economical effectiveness of agricultural enterprises.
The purpose of the study was to create a spatial database which plotted the agricultural plots of a selected group of producers. The spatial database allowed determining the impact of the plots' location on the organization of field transport between the agricultural enterprise's head office and its plots.
The scope of study was a group of 5 vegetable producerslocated in Goczałkowice-Zdrój, in the commune of Goczałkowice-Zdrój, in the Pszczynapoviat, in the Śląskie Voivodeship.

1 Introduction

In recent years, agricultural production in Poland has developed, however, it still faces a number of problems.Its basic weakness is the unfavourable agrarian structure, which involves, among others, the high level of employment, low level of mechanization and low yield of marketable products of individual enterprises. Prevalent in Poland is small agricultural enterprises, with an area of up to 5 ha, which constitute 53.2% of thetotal. Enterprises ranging from 5 to 10 ha, which constitute 22%, also have a significant share, on decrease in recent years. The largest enterprises, i.e. over 20 ha of agricultural land area, which have the largest development prospects, constitute only 9.4%. The average area of an individual enterprise has slightly increased in recent years; in 2014, it was 10.48 ha [7]. Sill, this seems too small an area for a more effective use of the machine park, increased work efficiency and obtaining higher income. A prospective alternative may be to merge small, individual enterprises into groups, i.e. associations of producers, which allows them to jointly acquire and use machines, thus contributing to the improvement of labor productivity [8, 10]. Agricultural producer groups aim at supporting consortia of agricultural producers, so that, with a stronger position in relation to their customers, they can take over a part of the trade margin, and thus increase their income [9]. Excessive land fragmentation combined with a small area of plots, and also large distances from the enterprise's office often contribute to the abandonment of land use [5, 10].

The group activity of producers allows many benefits related to building a competitive advantage of an agricultural enterprise, not only at the point of selling their products, but also in terms of procurement and processing within the group [1]. Regardless of their size, individual enterprises are too small to be a partner of large entities involved in the agri-food brokerage and trade, and to meet the requirements of competitive agricultural producers associated in corporations. Small agricultural producers are often bound to a single buyer. This causes a lot of uncertainty as to the possibility of selling their products and meeting quality requirements. Joining efforts in a single, representative entity called the "agricultural producers' group" largely reduces this uncertainty, increases competitive power and ensures a more favourable position in relation tolarge commercial buyers and processors. In addition, it provides access to cheaper means of production and necessary services, as well as reduces the risk and uncertainty while maintaining the family character of agriculture [3, 4].

Created by agricultural producers for business purpose, groups adapt production to market demand by developing effective marketing structures, which enable the acquisition of product outlets both at home and abroad [6]. Producers may organize themselves into groupsin order to adapt the production to market conditions, improve management efficiency, plan production with particular emphasis on its quantity and quality, to concentrate supply and organize the sales of agricultural products, as well as to protect the environment [2,11].

2 Purpose and scope of study

The purpose of the study was to create a spatial database, which determined the agricultural plots of a selected group of agricultural producers. This database was created using the spatial information systems technique in the ArcView environment. It contains thematic layers

* Corresponding author: Anna.Szelag-Sikora@ur.krakow.pl

related to the shape and size of agricultural plots and the road network of the studied area.

For the purposes of the analysis, individual enterprises that belong to the studied producer group were coded in letters: A, B, C, D and E, and will be presented in this way. Due to for the different labour intensity of individual crops, an analysis of individual enterprises in terms of crop structure was also carried out.

The studyinvolves an association of 5 vegetable producers.

3 Findings

The studied group of producers jointly manages 620 ha of agricultural land. It specialized in root vegetables, and its rotation crop is cereal, root and oil crops. The listed groups of plants accounted for 41%, 37%, 19% and 3% of the sowing structure respectively (Table 1).

Table 1. Area and structure of crops for the entire group and individual enterprises.

Specification of crops	Group	A [ha]	B [ha]	C [ha]	D [ha]	E [ha]	Structure [%]
beetroot	15	15					2
parsley	20				20		3
spring barley	9					9	2
maize	24	24					4
carrot	198	108	53.5	10		26.5	32
cereal mixture	8	3	4	1			1
spring wheat	33.5	0	0		17	16.5	5
winter wheat	164	47	95		19	3	27
potato	118.5	51	21.5		30	16	19
winter rape	19	0	19				3
oat	9	0	0	9			2
TUZ	2	1	1				0
Total	620	249	194	20	86	71	100

The group's arable fields are located in 15 localities in three poviats of the Silesian Voivodship. The group, i.e. the associated producers own 1 field in 10 localities, these fields are respectively: 37, 12, 12, 14, 12, 13, 14, 12, 43, and 7 km from the head office. Another 2 fields of two member producers are located in 1 village, 13 km from the head office. However, in 4 villages there are 3 fields belonging to 3 member producers, which are separated: 38, 3, 5, 20 km from the headquarters respectively (Table 2).

Table 2. The area of the group's agricultural plots and enterprises A, B, C, D and E in individual localities.

Locality	Distance of fields from the head office [km]	Area of the group's fields [ha]	Areaof fields in individual enterprises [ha]				
			A	B	C	D	E
Bełk	38	190.5		120		30	40.5
Czechowice-Dziedzice	37	37				37	
Ćwiklice	12	14					14
Goczałkowice-Zdrój	3	64	36	18	10		
Kobielice	12	10	10				
Kryry	14	9			9		
Ligota	12	24	24				
Łąka	5	65.5	48		1		16.5
Miedźna	13	19.5		19.5			
Międzyrzecze	14	65	65				
Poręba	12	6	6				
Stanowice	43	2		2			
StaraWieś	7	18	18				
Studzionka	20	61.5	21	21.5		19	
WisłaWielka	13	34	21	13			
Total		620	249	194	20	86	71

As mentioned above, the study covered agricultural enterprises associated in a vegetableproducers group, based in Goczałkowice, in the Pszczynapoviat, in southern Poland,consisting of 5 enterprises (Figure 1).

Figure 1. The head office of the studied agricultural producers group.

A fragment of the spatial distribution of agricultural plots with displayed attributes is shown in Figure 2. The example shows the name of the agricultural plot that is unique in a given production year, displayed as the object label. In the created spatial database, this number is the identifying code (name of the object), based on which other attributes describing the spatial object (agricultural plot) can be combined. In this case, the image is non-editable and shows the surface area of the community in a geographical layout. On the other hand, polygons, which are the thematic layer of the agricultural plots, are fully editable in terms of the added attributes as well as the outline and shape.

Figure 2. The distribution of the registration plots of the studied agricultural producers group was entered in the ArcView program.

The raster coverage of the producer group was prepared in the scale 1:10,000, which covers the area of the spatial layout of the studied enterprises. In the adopted scale, the area is 760 × 320 mm and it would be difficult to use it in an analog format. With the non-scale feature of GIS systems, the analysis of such an area is done on the screen of a traditional computer. The area includes the plots of the enterprises; the image describing its characteristics was entered into the coordinate system. The system to which the raster substrate was adjusted is a commonly used WGS84. The orthophotomaps that producers receive from the Agency for Restructuring and Modernization of Agriculture (ARMiR) are in the 1992 system (this is the most commonly used reference system in Poland). In the ArcView environment, the migration between these two systems is done automatically. The attributes of the prepared polygonal thematic layer showing the distribution of plots can be added and modified in *.dbf format.

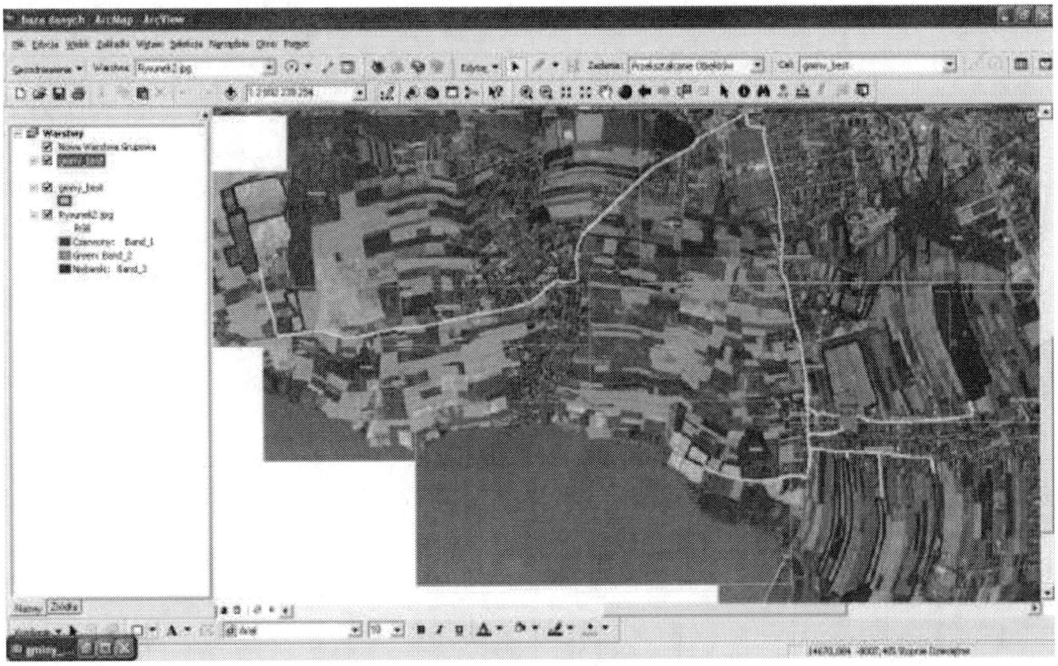

Figure 3. The distribution of the registration plots of the studied agricultural producers group with routes from the group's headquarters to the plot borders.

The prepared spatial database allows automatic exploration of the geographical image of plots with a vector layer depicting the access to the plot. This is shown in Figure 3. The spatial database automatic display of the distance between the head office and the plot. The largest distance of a plot from the head office of the group is 6.44 km. However, the average distance from a plot to the producer group head office is 1.61 km.

The spatial database made using the Spatial Information Systems can be successfully applied to the management of agricultural production as it facilitates creating graphic attachments. The database can be used successfully for archiving and inventory of the of parameters of agricultural production, which has a specific spatial character, easily used in the Spatial Information Systems.

4 Summary and conclusions

Creating an agricultural producers group contributed to a significant improvement in the spatial distribution of agricultural plots. The group manages a total of 620 hectares of arable land. Its area is over 83 times larger than the national average in the Śląskie Voivodship, which was 7.42 ha in 2015. By coming together, the group gained large agricultural plots in 15 localities and the fragmentation of cultivated crops was halved. The group is now more likely to receive assistance and investment funds, and has used this opportunity with the help of a consulting company. The machine park has been supplemented with specialized machines in line with the production direction. As a group,the producers have established their brand on the domestic and foreign market.

The study carried out in the entire agricultural producers group and in individual enterprises A, B, C, D and E made it possible to formulate the following conclusions:

1. Five producers are associated with the agricultural producers group selected for research. Enterprises A and B contributed to the largest percentage of agricultural land in the group, i.e. 40% and 31%,respectively, while the remaining enterprises C, D and E contributed 3%, 14% and 12%respectively.
2. The GIS system environment can be appropriately used for the archiving, storage and visualization of agricultural production data on agricultural plots.
3. The prepared database allows determining the distance from the headquarters of the producer group automatically.
4. The agricultural producers group cultivates 620 ha of farmland, which is 124 ha per 1 producer on average. This area is over 17 times larger than the national average in the Śląskie Voivodship, which was 7.42 ha in 2013.
5. A group of agricultural producers grows 12 species of plants in 15 localities. However, when considering each of the enterprises separately, the number of cultivated plant species would be twice as large and would amount to 24 species grown in 24localities. The founding of an agricultural producers group contributed to the consolidation of agricultural plots. Thus, the area of individual fields increased and the number of crop species decreased by half.
6. Based on the archived attribute data, fertilization, yielding and protection of plants on individual

agricultural plots can be analysed automatically in a tabular form and can serve as the basis for agri-environmental reports requiring the knowledge about crop rotation and fertilization.

7. Together, it is easier for agricultural producers to invest. The group purchased specialized cultivation machines and a modern technological line for washing, polishing and packing vegetables, and built a new storage facility, which would be difficult to carry out individually.

8. The agricultural producers group specializes in the cultivation of root vegetables, especially carrots. In the structure of crops, vegetables occupy 233 ha, which constitutes 38% of the whole cultivation area, including 32% of carrots. Grain crops are grown on an area of 247.5 ha, i.e. 40% of the area, winter rape on 19 ha, i.e. 3% of the area, potatoes on 118.5 ha - 19% of the area; permanent grassland is 2 ha. On the other hand, carrot was grown in 4 individual enterprises, i.e. A, B, C and E.

9. Each of the enterprises declared that the integration with the European Union brought an increase in its area, as well as legalization of lease and improvement in the quality of their products.

10. The agricultural producers group declared that they used a broad range of services regarding consulting, maintenance and repair as well as building renovation.

11. The distribution and sale of agricultural products has been expanded. The sales market was expanded, among others, by export to Eastern Europe. Great emphasis was placed on marketing and advertising.

12. The owners of 4 enterprises are men and 1 is a woman. The average age of the associatedenterprise owners is 40 years.

13. The agricultural producer group uses the full extent of the services of a consulting company, which significantly allows the use of aid funds.

14. In production, great emphasis is placed on the high quality of produced crops.

References

1. Appel F., Ostetmeyer - Wiethaup A., Utilities Policy **41**, (2016)
2. Czapiewska G. Acta Universitatis Lodziensis Folia Geographica Socio-Oeconomica, **13**, 165-178, (2013)
3. Domagalska-Grędys M. Scientific Journal of Warsaw University of Life Science. Series G Agriculture Economics, Vol. 97, No. **4**, Warsaw, 54-59, (2010)
4. Grzywińska-Rąpca M. Scientific Journal of Wrocław University of Economics, 264-268, (2003)
5. Janus J. Infrastruktura i ekologia terenów wiejskich, no. **2**, 71-81, (2011)
6. Kowalski, S., Szeląg-Sikora A., Oleksy-Gębczyk, A. Inżynieria Rolnicza, **6** (131),101-108, (2011)
7. S. Nawar, R. Corstanje, G. Halcro, D. Mulla, A.M. Mouazen, Adv. in Agron. 175-244 (2017)
8. Statistical Yearbook of Agriculture in 2014, Warsaw. Download from: http://stat.gov.pl/obszary-tematyczne/roczniki-statystyczne/roczniki-statystyczne/rocznik-statystyczny-rolnictwa-2015,6,9.html on Oct. 10, 2016.
9. Tabor, S., Cupiał, M. Proceedings XXIX CIOSTA GIGR V Congress, Krakow, Poland, 25-27, 154-157, (2001)
10. Rorat J., Szeląg-Sikora A. Inżynieria Rolnicza, **4** (148),133-142, (2013)
11. Sudduth K.A., Drummond S.T. Agron. J. 99:1471-1482 (2007)

Analysis of heat source selection for residential buildings in rural areas

Tomasz Szul[1,*]

[1]Faculty of Production and Power Engineering, University of Agriculture in Krakow, Poland

Abstract. The research aiming to check whether the output of currently installed boilers matches the use requirements together with estimation of their energy efficiency was carried out on a group of 84 single-family residential buildings located in rural areas. Heating and hot water energy needs were calculated for each building in order to determine the use requirements. This enabled verification whether the currently installed boilers match the actual use requirements in the buildings. Based on the calculations it was determined that the designed average boiler output in the group of buildings subject to analysis is 15.7 kW, whereas the mean rated output capacity of boilers installed therein is 25.4 kW. On average, the output capacity of the installed boilers exceeds the use requirements for the buildings by 60%. To calculate the energy efficiency of boilers, the mean annual boiler output capacity use coefficient was determined. For boilers selected on the basis of standard calculations, the mean coefficient is 0.47. For boilers currently in use it is 0.31, less than the above figure. The above calculations show that if boilers were correctly selected in compliance to the building needs, then the average estimated seasonal efficiency of 65% would be feasible. However, in the current state the achievable efficiency is approx. 55%.

1 Introduction

The Act on the Energy Performance of Buildings [1] implements Directive 2010/31/EU [2] into the Polish law. It introduces two mechanisms designed to reduce energy consumption in buildings: an energy assessment and regular systematic heating system technical-state inspections. The Act explains issues associated with the principles for preparing energy performance certificates for buildings. It prescribes two methods of estimation of energy needs: a method based on standard building use and a method based on actual energy consumption. The legislator also defined the building types for which an energy performance certificate has to be drawn up as well as building types which are exempt from this obligation. The aforementioned Act also sets out the principles for inspection of heating and air conditioning systems in buildings, where by virtue of the provisions the building owner or administrator is obliged to periodically inspect buildings in use, which entails checking the heating system technical-state, including energy efficiency of boilers and ensuring that their output is relevant for the given use requirements. The time period for completing tests and threshold output values for boilers which are to be subjected to the tests are set forth. The lower 20 kW threshold specified in the Energy Performance Act applies to all types of fuels, both renewable and non-renewable. In the past, the Building Law Act [3] set forth the obligation to perform periodic boiler inspections, where the principles for heating system inspections were prescribed together with their frequency, which varied depending on the output. It also specified the types of buildings exempt from the inspection obligation. According to the provisions of the said Act, the obligation to carry out inspections did not apply to owners and administrators of single-family residential houses, structures; homesteads and summer houses. Following amendments to the Building Law Act [4], the provisions in question ,were moved to the Act on the Energy Performance of Buildings [1], however the provision specifying buildings which are excluded from the inspection obligation was not moved, and as such the obligation to inspect boilers prescribed by the Act applies to all buildings, including single-family houses and homesteads which are heated by sources with a rated output of more than 20 kW.

In rural areas the percentage of heat energy used for heating and hot water needs in residential homes constitutes ca. 82% to 85% of the total energy consumption. Most of these buildings (>70%) use solid fuels (coal, wood) for heating purposes and are responsible for emitting ca. 75% and even up to 90% of harmful pollutants into the atmosphere [5-8].

Literature on the subject contains results of research on appropriate selection of heat sources and efficiency of heating systems in households in single-family residential buildings in Belgium and Great Britain, where heating oil and natural gas fired boilers are used [9, 10]. Whereas papers [11, 12] include results of tests on boilers used to heat multi-family residential dwellings

* Corresponding author: t.szul@urk.edu.pl

in Canada as well as office buildings in Great Britain. Authors of the afore-cited papers indicate that heat sources are oversized by between 30% [10, 11] and 300% [9] in relation to the actual demand for heat. Heating systems often operate at low efficiency, in the region of 30% - particularly in buildings with good heat insulation envelopes. At the same time, the authors estimate that correct selection, control and use of heating systems may deliver potential energy savings of ca. 20%, which will also contribute to the reduction of air pollution emissions. In Poland, and particularly in rural areas where single-family residential houses are predominant, research of this type has not been conducted. The objective of the paper was to verify and determine whether heat sources were selected suitably to heating needs (heating, domestic hot water) in single-family residential buildings and also to estimate their energy efficiency.

2 Research area

The research was conducted in Kraków province located in Małopolskie Voivodeship. According to the Polish Central Statistical Office (GUS) data, there are 73.1 thousand single-family residential buildings across the 17 municipalities comprising the research area. 69% of these buildings are heated by solid fuel fired boilers (GUS). This was applied as the basis for determination that 81 buildings are the minimum sample size (for α = 0.95 confidence level and maximum error of 10%) required for the research. A sample of 84 single-family residential buildings heated by solid fuel (coal) fired boilers was used. Within the scope of the research, information was collected pertaining to the heated surface area, building envelope surface where heat loss occurs, heated and ventilated volume, materials used to construct the envelopes, window types, methods for obtaining domestic hot water, type and age of installed boilers and their rated output (as shown on name plates). Information related to any potential thermal-upgrade actions in the buildings was also collected. Figure 1 presents the sample structure in terms of the construction date of buildings according to the Polish Central Statistical Office (GUS) data and buildings for which research was carried out.

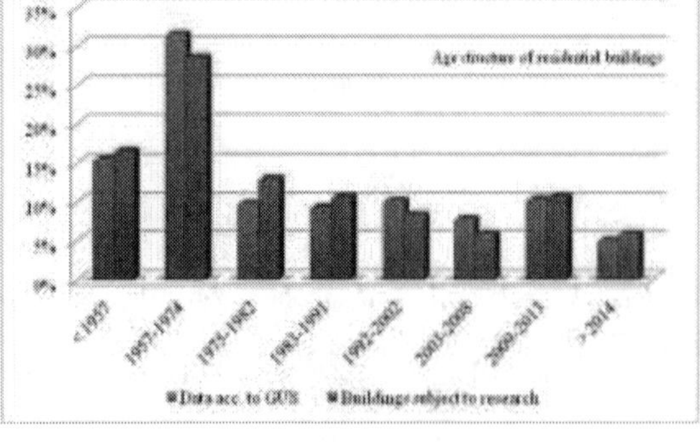

Fig. 1. Age structure of residential buildings in rural areas of Kraków province

Age brackets were determined pursuant to the provisions which are presently in force and prescribe the maximum heat loss coefficient U_{max} for the building envelope during periods when the standards are applicable, as well as pursuant to appropriate industry regulations.

The average floor area to be heated in buildings subject to analysis is 118 m^2 and the average volume is 354 m^3. The analysed buildings, and in particular those built pre-2003 were subjected to thermal-upgrade works - these constitute 39% and in 73% of those, the thermal-upgrades entailed application of additional thermal insulation to the external building walls and the remainder entailed application of additional roof insulation. The average additional external wall insulation in the group of buildings subject to analysis is 7.6 cm, with 9.2 cm for roofs.

Windows were replaced in 76% of the buildings, mainly with PVC single chamber multiple glazed units.

In the group subject to the research, 67% of buildings use a boiler for heating and domestic hot water needs and in the remaining percentage other heating devices are used for domestic hot water needs. Heat sources installed in buildings mainly comprise lower combustion chamber boilers (88%) and retort boilers. The boilers are mainly fired with black coal (different types) and wood. Survey questionnaires indicate that on average wood constitutes 29% of the combusted fuel. On average, the installed boilers are 12 years old. Figure 2 presents a boiler sample structure according to age.

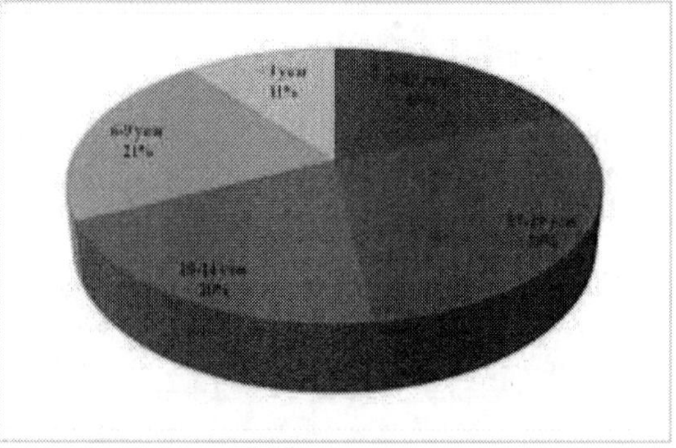

Fig. 2. Boiler sample structure according to age

15 - 20 year old boilers are the largest group with a 29% share, a large percentage of boilers over 20 years old have also been recorded (19%) - these are mostly boilers with cast iron heat exchangers. The new boilers' group (6 - 9 years old) mostly comprises steel chamber boilers. The newest boilers predominantly constitute retort boilers with fuel hoppers.

Boiler sample structure according to output is shown in figure 3.

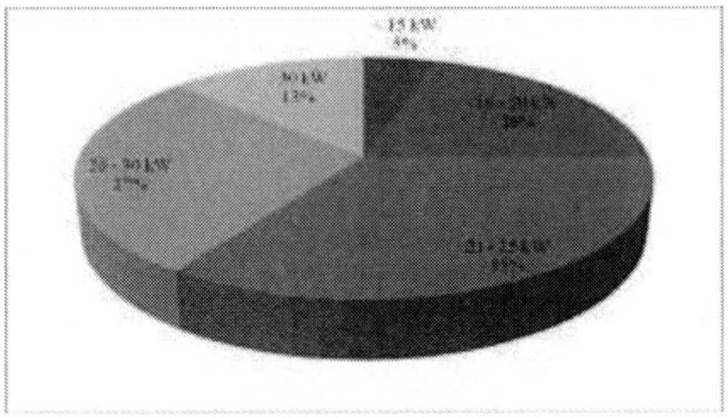

Fig. 3. Boiler sample structure according to rated output

The rated outputs of the boilers are within the 12 to 41 kW brackets, whereas the most populous group constitute boilers between 21 and 25 kW - 35% and the output of approximately 13% of boilers exceeds 30 kW.

3 Methodology

Heat demand for heating buildings was calculated according to PN-EN 12831 standard, temperatures as prescribed by standards were adopted for the calculations, i.e. internal temp. of 20°*C* and external for Climate Zone III, for which the design external temperature is -20°*C*. The heating season length (222 days) and the average external temperature during the heating season (4.04 °*C*) were adopted on the basis of data from the Kraków–Balice meteorological station. The heat source output for domestic hot water needs in a tank based system was calculated on the basis of PN-92/B-01706 standard. Calculations were performed for the actual number of people residing in the given building.

To facilitate comparison between buildings in different age groups, average values for the unitary heating energy need indexes were determined and expressed in W·m^{-2} of the floor area which required heating. Due to the fact that thermal-upgrades were applied to a number of buildings, indexes in given age-groups were also determined. Separate indexes were determined for buildings in their as-is condition (constructed in accordance with the provisions pertaining to the maximum value of the building envelope heat loss coefficient applicable during given time periods) and those subject to thermal-upgrade works. Calculation results are shown in figure 4.

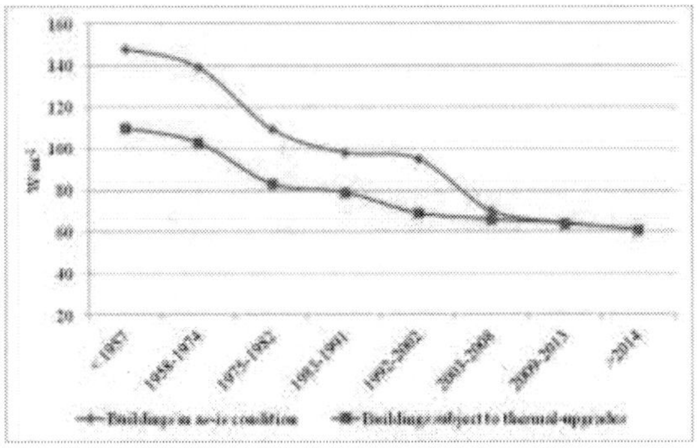

Fig. 4. Unitary heating output needs for buildings

The average energy needs index for the group of buildings subject to analysis was 104 W·m^{-2}. It was 112 W·m^{-2} for buildings in the as-is condition and 78 W·m^{-2} for buildings subjected to thermal-upgrade works. Thermal-upgrade works on buildings constructed pre-2003 reduced building envelope heat loss by approx. 23%.

Heating and domestic hot water energy needs were calculated using the following formula:

$$Q_T = Q_H + Q_W \ \ [kW] \tag{1}$$

where:
Q_H – design building heat load acc. to PN-EN 12831,
Q_W – domestic hot water power needs, based on the hot water tank system, acc. to PN-92/B-01706.

Efficiency of a solid fuel fired boiler as quoted by the manufacturer is a value obtained under test conditions subject to continuous combustion of the boiler's design fuel. According to experience, in actual use conditions, with inferior quality fuel and a contaminated heating surface, the boiler's actual output may be slightly different from that measured in laboratory conditions [13] and as such, in selecting boiler output in practice, the following formula is used:

$$Q_{B\max} = Q_T \cdot 1{,}1 \ \ [kW] \tag{2}$$

Boiler energy efficiency may be estimated by calculating annual average coefficient of boiler capacity use under partial loading [14]. The coefficient stems from the following function:

$$\frac{Q_{B\max}}{Q_{Ave}} = \eta_b \tag{3}$$

Whereas its value can be calculated in accordance with VDI 2067 standard:

$$\eta_b = \frac{1}{\left(\dfrac{b}{b_{Vk}} - 1\right) \cdot 0{,}75 + 1} \tag{4}$$

In order to determine it, an additional ratio has to be estimated, b_{Vk} - hours of full use with reference to $Q_{B\max}$ capacity:

$$b_{Vk} = \frac{b_{Vh} \cdot Q_T}{\eta_d \cdot Q_{B\max}} \ \ [h\cdot(year)^{-1}] \tag{5}$$

where:
Q_{Ave} – seasonal average boiler capacity *kW*,
b – standby time $[h\cdot(year)^{-1}]$,
b_{Vh} – full use hours with reference to Q_H $[h\cdot(year)^{-1}]$,
η_b – boiler output use coefficient under partial load
η_d – average seasonal heat distribution efficiency,
Q_T – design heat demand *kW*,
Q_{Bmax} – installed boiler rated output *kW* (for existing boilers enter Q_K output as quoted on the name plate).

4 Research results review

The conducted research and ensuing calculations made it possible to check and determine whether heat sources were selected appropriately to heating needs in buildings. The average design boiler output in the group of buildings subject to analysis is 15.7 kW, whereas the mean rated output capacity of boilers installed therein is 25.4 kW.

Figure 5 shows a comparison of boiler outputs – design, calculated according to standards and actual for the heat sources in buildings per age group.

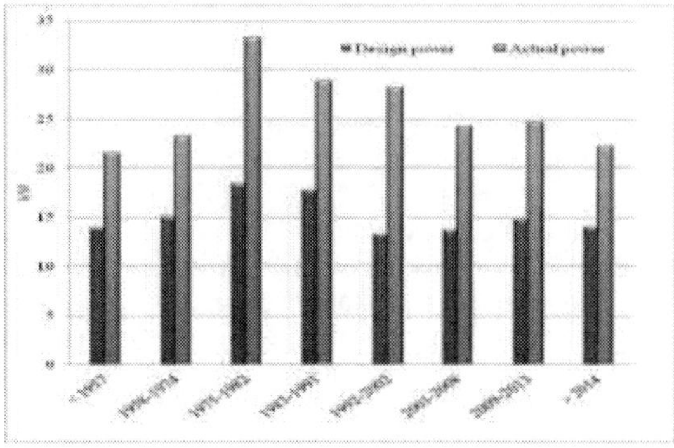

Fig. 5. Comparison between design and actual output of installed boilers

Analysing the data, one may conclude that in every case, rated power of installed boilers exceeds the requirements of the buildings use. Buildings constructed between 1975 - 1982 and 1992 - 2002 exhibit the highest surplus power: the average of approx. 1.8 - 2.1. In other groups, boiler excess power is somewhat lower, within 1.5 - 1.6 range. On average, the output capacity of installed boilers exceeds the buildings' needs by 60%.

The boiler sample structure according to the level to which they match requirements is shown in Figure 6.

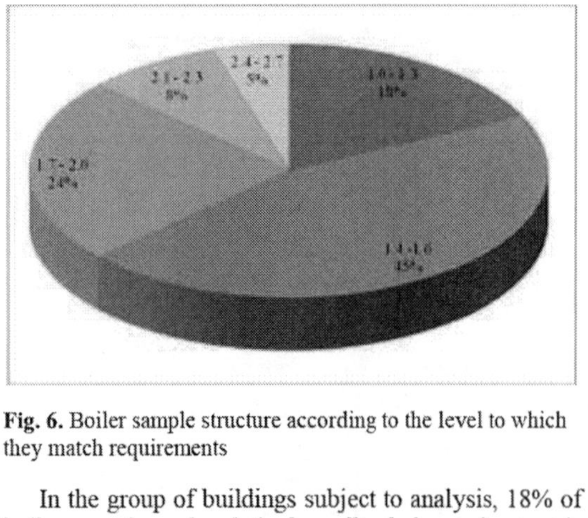

Fig. 6. Boiler sample structure according to the level to which they match requirements

In the group of buildings subject to analysis, 18% of boilers match needs relatively well - their rated output is approx. 1.3 times the actual energy needs of buildings. Most often (45%), rated output of boilers is between 1.4 and 1.6 times bigger. For 13% of the boilers, their output exceeds the demand by more than twofold.

Research has shown, that the amount of boiler excess output depends on whether the boiler is used solely for heating purposes or also for domestic hot water. In this case boilers performing two functions (heating and domestic hot water) better match the needs, with the rated output approx. 50% higher than the designed. If boilers are used solely for heating buildings, their output compared to actual needs is 70% higher (Fig. 7).

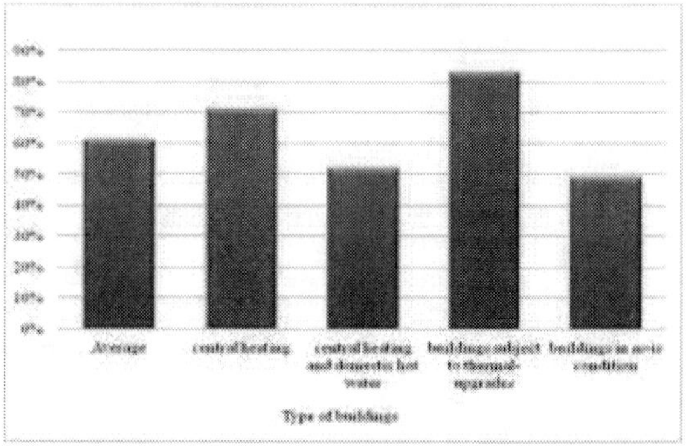

Fig. 7. Degree to which the output currently installed boilers matches the buildings' needs

Even larger differences are seen for boilers installed in buildings whilst still in the as-is condition, and those subsequently subjected to thermal-upgrades. Boilers used as the heat source within buildings in the as-is condition match their needs significantly better - with the average of 48% power surplus. Buildings which were subjected to thermal-upgrades show significantly larger boiler excess power compared to actual needs – here the rated output exceeds the power figures obtained on the basis of design calculations by 80%. In the majority of tested buildings that surplus results from thermal-upgrades, which reduced heat loss through the building's envelope but failed to take into account heating system modernisation.

Estimating boiler energy efficiency entailed calculating η_b – boiler output use coefficient under a partial load. On the basis of calculations, the standby time b was arrived at as well as the hours of full use b_{Vk} with reference to Q_{Bmax} and for the average rated output values Q_K for boilers installed in buildings, for which five variants were identified: rated power of boilers: installed in all buildings, within buildings in the as-is condition, in buildings subject to thermal-upgrades. Furthermore the η_b coefficient was estimated for buildings, where the boiler is solely used for heating as well as for those where it performs both functions, i.e. heating and domestic hot water.

Climate data was used for the calculations from the nearest meteorological station the scope of which extends across Kraków province. Figure 8 shows the coefficient values.

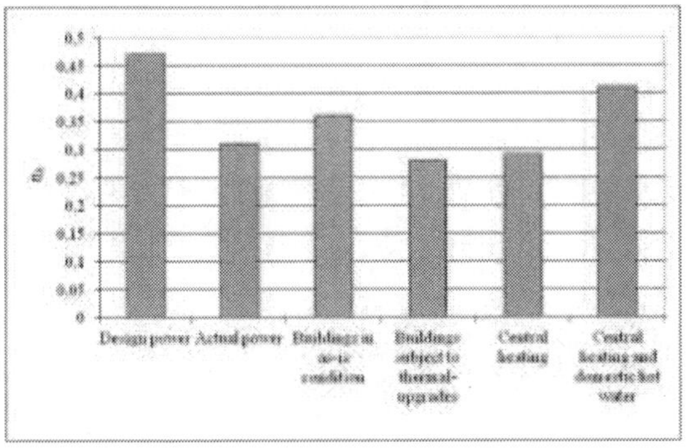

Fig. 8. Boiler output use coefficient under partial load - average annual values

The average annual boiler output use coefficient selected pursuant to calculations based on standards is 0.47, on average. For boilers currently in use it is 0.31, less than the above figure. The lowest value for this parameter was obtained in buildings which were subjected to thermal-upgrade works and where the boiler is used only for heating - 0.28 and 0.29 respectively. It is much better if the boiler is also used for domestic hot water purposes, as that increases the coefficient to 0.41.

The boiler output use coefficient was used to estimate the average seasonal boiler efficiency. Due to the fact that lower combustion hopper chamber boilers constitute nearly 90% of the installed heat sources, calculations were only carried out for this group. Based on information received from the number of boiler manufacturers on the relation between the boiler efficiency and short term output under test conditions, averaged data was plotted to depict the boiler average seasonal efficiency as a function of the boiler load degree (η_b). This made it possible to approximately estimate the efficiency with which heat sources operate. The average seasonal efficiency was determined for the same variants which were analysed for the average annual boiler output use coefficient under partial load η_b. Results are shown in figure 9.

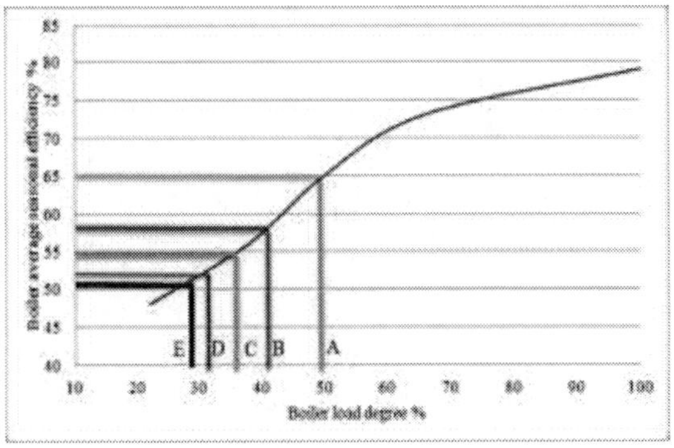

Fig. 9. Boiler average seasonal efficiency as a function of boiler load degree

Boilers correctly selected to match the building needs (A) could exhibit an average estimated seasonal efficiency of 65% which is in accordance with the provisions in the methodology for drawing up energy efficiency certificates in buildings [15], while in the current state (C) the achievable efficiency is approx. 55%. The lowest average seasonal efficiency was reached by boilers in buildings subjected to thermal-upgrades (D) and those which are only used for heating (E). The use of an existing boiler also for domestic hot water needs (B) may improve its efficiency even up to approx. 58%.

5 Conclusion

As a result of the calculations it was determined that the designed average boiler output in the group of buildings subject to analysis is 15.7 kW, whereas the mean rated output capacity of boilers installed therein is 25.4 kW. On average, the output capacity of installed boilers exceeds the building needs by 60%. As shown by the research, the amount boiler excess output depends on whether the boiler is used solely for heating purposes or also for domestic hot water. In this case boilers performing two functions (heating and domestic hot water) better match the needs, with the rated output of approx. 50% higher than the calculated one, whereas if boilers are used solely for heating buildings, their output compared to needs is 70% higher. Even larger differences are seen for boilers installed in buildings whilst still in the as-is condition (constructed in accordance with the provisions pertaining to the maximum value of the building envelope heat loss coefficient) and those subsequently subjected to thermal-upgrades. Boilers used as the heat source within buildings in the as-is condition, significantly better match their needs - with the average of 48% power excess. Buildings which were subjected to thermal-upgrades show significantly larger boiler excess power compared to actual needs - here the rated output exceeds the power figures obtained on the basis of design calculations by 80%. To calculate the energy efficiency of boilers, the mean annual boiler output capacity use coefficient was determined. For boilers selected on the basis of standard calculations, the mean is 0.47. For boilers currently in use it is 0.31, less than the above figure. The lowest value for this parameter was obtained in buildings which were subjected to thermal-upgrade works and where the boiler is used only for heating - 0.28 and 0.29 respectively. The above calculations show that if boilers were correctly selected in accordance with the building needs, then the average estimated seasonal efficiency of 65% would be feasible. However, in the current state the achievable efficiency is approx. 55%. The lowest average seasonal efficiency was reached by boilers in buildings subjected to thermal-upgrades and those which are only used for heating - their efficiency is only approx. 52%.

References

1. Act Dz.U. 2014.1200, (2014)
2. Act Dz. Urz. UE L 153. page. 13, (2010)
3. Act Dz.U. Nr 197. 1373, (2007)

4. Act Dz. U. 2016.290, (2016)
5. T. Szul. J. Res. Appl. Agric. Engng., **53**, 1 (2008)
6. T. Szul. J. Res. Appl. Agric. Engng., **53**, 2 (2008)
7. T.Szul. Inf. Ekol. Ter. Wie., **1** (2010)
8. T. Szul. J. Res. Appl. Agric. Engng., **56**, 1 (2011)
9. L. Peeters, J. Van der Veken, H. Hens, L. Helsen, W. D'haeseleer. En. and Build., **40**, 8 (2008)
10. Z. Liao , AL Dexter, **36**, 3 (2004)
11. Z. Liao, M. Swainson, A. Dexter. En. and Build. **40**, 3 (2005)
12. Z. Yaohong, Z. Liao, RU, **1** (2005)
13. K. Krygier, T. Klinke, J. Sewerynik. Ogrz. Went. Klimat. 1997 (2007)
14. H. Recknagel, E. Sprenger, W. Hönmann, E. Schramek. OKCwCh. **08**, 09 (2008)
15. Act Dz.U. 2015.376 (2015)

Influence of the composition and agglomeration pressure on the compaction level of the fertilizers based on biomass ash and digestate

Marek Wróbel*, *Jarosław* Frączek, *Krzysztof* Mudryk, *Marcin* Jewiarz and *Krzysztof* Dziedzic

University of Agriculture in Krakow, Faculty of Production and Power Engineering 120 Balicka street, 30-149 Krakow, Poland

Abstract. The paper presents the results of research aimed at determining the influence of the composition of the fertilizer mixtures and the compaction pressure on the specific density and density index of fertilizer granules. Investigated mixtures were prepared from fly ash from power plant fuelled by biomass and digestate from biogas plant. The urea, sulfur and phosphorite were also added as enhancing additives. For granule samples made on a strength machine, their specific density was determined on a quasifluid-pycnometer. To determine the effect of agglomerate pressure on the compaction process, the absolute density of the materials was omitted. In such case it was needed to introduce a density index A_f. Such a presentation of the results obtained has made it possible to clearly determine how the density of the test mixture results in the applied agglomeration pressure. The specific density of the resulting granules was in the range of 0.85-1.27 g/cm^3. The determined density index for the given pressure was in the following ranges: 0.44-0.49 g/cm^3 (pressure 100MPa), 0.47-0.51 g/cm^3 (pressure 150MPa) 0.51 - 0.59 g/cm^3 (200MPa pressure). This means that, regardless of the contribution of components to the mixture at the given pressure, a similar degree of compaction was obtained.

1 Introduction

In the anaerobic digestion process which takes place in biogas plants, the biogas and digestate ist obtained which is residue with high fertilizer potential. During this process of the substrate reduction of the organic matter content, decomposition of the compounds, fragmentation of the solid substrate fractions and partial hygienisation, and most importantly cause relatively small losses of the fertilizers occurs [1, 2].

Decomposition deegree usually ranges from 30 to 60% of organic matter. During anaerobic digestion, 80% of the biodegradable substance is converted into biogas [3]. Digestate contains: biomass of fermenting micro-organisms, non-fermented organic compounds and minerals. Because of the nitrogen, phosphorous and potassium content in the mineral form directly accessible to cultivated plants, these substances are very valuable from the fertilizer point of view. In addition, the digestate as organic matter in the soil is mineralized and it is a source of nutrients and positively affects the physicochemical properties of soils. The digestate residue contains nitrogen both in organic and inorganic form [4, 5]. In slurry based digestate mixed with different types biomass, nutrient content falls within the following ranges: general nitrogen - 3.4-4.0; N-NH_4 - 2.4-2.9; P - 0.8-1.2; K - 2.7-3.1 - values are given g/l [6]. The increase in the number of biogas plants on the territory of Poland results in higher amount of digestate to be utilise. The amount of fermentation residue produced by a single biogas plant is difficult to manage locally. According to the European Nitrates Directive [7], the annual nitrogen limit is 170 kg per hectare. On the other hand, research conducted by Dohler and Schliebner [8] shows that it is economically feasible to transport untreated residue up to 10 km from the site.

Therefore, another economically reasonable way of using the fermentation residue for fertilizers need to be found. One of them is the production of organic granulated fertilizers from fermented or organic-mineral fertilizers based on fermentation residues and other mineral substances supplementing nutrients missing from the fermentation. As an additive in such granules may be fly ash from a professional power plant fuelled by biomass. Studies have shown that ash from biomass combustion may find use as a mineral fertilizer [9] or as an ingredient in compound fertilizers with organic substances [10, 11]. These main materials may be supplemented by other components of the fertilizer mixture such as sulfur, urea, phosphate; depending on the required fertilizer composition.

Taking into account the variety of morphological, granulometric and physical properties of these raw materials, the only promising granulation method is pellet production. This form of fertilizer, in the form of stable dimensions and density granules, reduces transportation costs and enables uniform dosing of fertilizer on the field. Irrespective of the ingredients'

* Corresponding author: marek.wrobel@urk.edu.pl

contribution to the mix, the granulation process should allow production a uniform granule. The results of experimental and analytical studies which are designed to investigate the particle trajectory on and off the spreader, the uniformity of the fertilizer distribution depends mainly on the physical properties of the fertilizer. The most often parameters are: particle size, friction coefficient, aerodynamic drag, mechanical strength and density. The specific density of the granules depends not only on the specific density of the raw materials but also on the pressure of the agglomeration.
A comprehensive experimental and analytical study performed to investigate particle trajectories on and off a centrifugal spreader. The performance of fertilizer distributors and hence the evenness of the spread pattern depends to a large extent on the physical properties of the fertilizer.
Five important properties which affect particle motion are reviewed, namely particle size and particle size distribution, coefficient of friction, coefficient of restitution, aerodynamic resistance, and particle strength. As the biomass studies show [12-14], the properties of the raw material, apart from the process parameters, significantly influence the characteristics of the obtained granules. the degree of fragmentation of the raw material has a significant effect on the density of biomass granulates [15]. What is also important, the significant share of dusty fractions in the raw material increases the stability of granules [16, 17]. In the case of fertilizer mixtures based on ash and digestate, the physical parameters of the substrates and their composition have a decisive importance of the granulate characteristics [18, 19, 20, 21].
Therefore, the optimum quality of the raw material, process parameters and the relationships between them need to be establish, so the produced granulate meets the quality requirements.. That is why it is so crucial to know the basics of the agglomeration process of the mineral-digestate mixtures.
According to this information, the main objective of the study was to determine the effect of the composition parameters of the fertilizer mixtures and the agglomeration pressure on the specific density and the degree of compaction of the resulting granules.

2 Material and methodology of research

The research was carried out on fertilizer mixtures which composition was developed by prof. Witold Grzebisz from the University of Life Sciences in Poznan (Member of the Research Consortium). The raw materials for the mixtures production were fly ash from the Połaniec power plant fuelled by biomass and a digestate form Piekoszów biogas plant. Additional components of valorising were sulfur, urea and phosphorite. The bulk composition of the fertilizer mixtures according to which the granules samples were prepared is shown in Table 1. The share of the main constituents (fly ash and digestate) varies from 25 to 70%, each blend contains 5% urea addition. In addition, some blends contain 5% sulfur or 15% phosphate addition.

Table 1. Composition (mass based) of investigated fertilizer mixtures.

Component	B1	B2	B3	B4	B5	B6	B7
Ash	70	65	50	50	45	30	25
Postferment	25	25	25	45	45	45	70
Urea	5	5	5	5	5	5	5
Sulfur	0	5	5	0	5	5	0
Phosphorite	0	0	15	0	0	15	0

In order to minimize the influence of the particle size of the materials included the fertilizer mixtures (especially the digestate which may contain particles larger than 5 mm), all the components of the mixture are uniformly dimensioned by grinding on the IKA A11 BASIC laboratory mill to a grain size of less than 1mm. From such prepared components, specified mixtures were prepared, with moisture content 15%. The moisture level of the mixture was determined on the basis of preliminary studies. Three levels of humidity were tested, 12, 15 and 18%. A sample of 50% ash content and digestate at 200 MPa and humidity of 15% was characterized by the highest density. Depending on the needs, water was added to the mixes to allow for the desired moisture level of the samples. Each of the test mixture variant was prepared in amount about 200g. The weighed ingredients (fertilizers and water) were mixed in a laboratory mixer to produce homogeneous mixtures with a set humidity. The process of agglomeration of fertilizer mixtures was carried out on the EDZ20 type strength machine. Agglomeration pressure levels were set at 100, 150 and 200 MPa. The sample compaction process was carried out in special agglomeration set to produce 8 mm diameter granules (Figure 1).

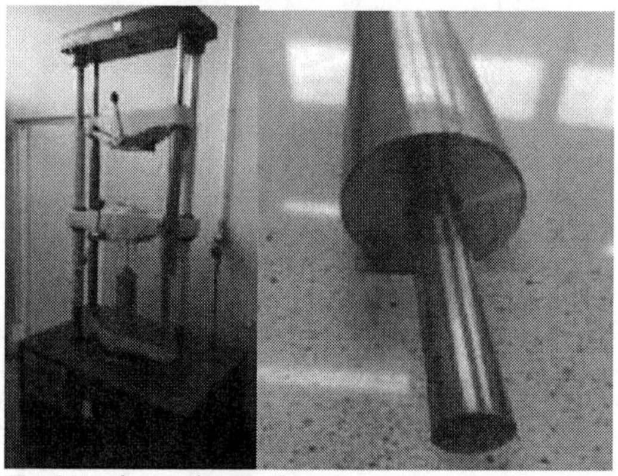

Fig. 1. Agglomeration stand.

The moisture content of component and mixtures (M_{ad}) in as analysed state (air dry) was determined according to the PN-EN ISO 18134-3:2015-11 [19] standard. In this analysis laboratory dryer (SLN 160, POLEKO from Wodzisław Śląski, Poland) was used. The procedure rely on drying of small amount of the material (1 ± 0,05g) in air at 105°C. The moisture content was determined according to difference in sample weight before and after drying [19]. AS 160.R2 analytical balance was used with readability 0,1mg produced by Radwag, Radom, Poland

The absolute density of raw materials (material density without inner pores) was carried out on a AccuPyc 1340 gas pycnometer from Micromeritics, USA. Device measured the volume of the sample by measuring the change in helium pressure within the calibrated chamber volumes. The device uses the direct dependence of Clapeyron on the assumption that the temperature within one measurement does not change (1 and 2).

$$V = \frac{V_{cel} - V_{exp}}{\frac{p_1}{p_2} - 1} \quad (1)$$

$$\rho_{rz} = \frac{m}{V} \quad (2)$$

where: V – volume of the sample
V_{cel} – volume of the sample chamber
V_{exp} – volume of the control chamber
p_1 – pressure before decompression
p_2 – pressure after decompression
m – mass of the sample
ρ_{rz} – absolute density

The specific density measurements of granule were performed with use of The GeoPyc 1360 (Micromeritics Instrument Corp.) based on displacement methods. The sample is placed vessel (inner diameter 50mm) filled with Dry Flo, which is a powder composed of small, rigid spheres with high ability to move (quasi-fluid). The material inside test vessel is agitated and gently compressed (consolidation force is 140N) and all this parameters are controlled by computer. Before measurement apparatus measures the volume of Dry Flo substance in the same way. After measurement is done, based on the diameter of the vessel, displacement (with and without sample), and mass of the sample, density were calculated, by implemented algorithm.

To determine the effect of agglomerate pressure on the compaction process, the absolute density of the materials was omitted. This could be achieved by introducing a density index A_I of material into the analysis. This indicator tells how much of the resulting density of a granule is approaching the common density of the constituents of the granules. Such a presentation of the results make it possible to clearly determine the density of the test mixture resulting in the applied agglomeration pressure. This indicator, bypassing the absolute density factor of the components of the blend, allows comparison of the results obtained for the test mixtures and determine which of them is characterized by high vulnerability for agglomeration Density index can be calculated according to the equation (3):

$$A_I = \rho/\rho_{rz} \quad (3)$$

where: ρ – specific density of the granules
ρ_{rz} – absolute density

Based on the absolute density of raw materials and specific density of granulates, porosity p_Z of granulates was determined.

3 Research results

The results obtained for specific density were ranged from 0.85g/cm^3 (B7 150MPa) to 1.27g/cm^3 (B3 200MPa) (Figure 2). For a given pressure the specific density value is in the range of 0.92-1.05 g/cm^3 (pressure 100MPa), 0.85-1.1 g/cm^3 (pressure 150MPa) 0.93 - 1.27 g/cm^3 (200MPa pressure). For samples B6 and B7 at 100MPa, the specific density of the granules were determined to be higher than at 150MPa. This difference was respectively 0.05 and 0.09 g/cm^3. In other cases, the specific density of the granules increases as the compaction pressure increases. Comparison of the specific density of the obtained granules between the test mixtures is not conclusive. By analysing the obtained results, we can distinguish a group of granules (B1-B3) which specific density is higher in selected pressure groups compared to the others. However, the question arises as to how much is the effect of the applied pressure and how much influence of the density of ingredients constituting the blend? Therefore, the degree of compaction of the obtained granulates was compared on the basis of the A_I index.

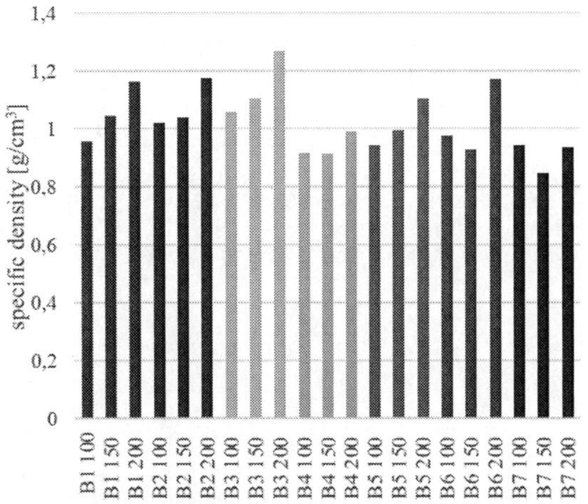

Fig. 2. Specific density of granulates.

On the basis of the absolute density measurement of the individual components of the mixtures, the common absolute density of each of the mixtures tested was calculated. The results of the absolute density determination of the individual components of the fertilizer mixes are presented in Table 2, while the calculated absolute density results of the tested mixtures in Table 3.

Table 2. Absolute density of the fertilizer components

Component	Absolute density [g/cm³]
Ash	2,47
Postferment	1,41
Urea	1,35
Sulfur	2,04
Phosphorite	2,87

The obtained results show that the components that form fertilizer mixtures are characterized by very different values of absolute density ranging from 1.35 g/cm^3 for urea to 2.87 g/cm^3 for phosphate. Depending on the proportion of the component in the mixture, its absolute density influences the absolute density of the mixture (Table 3). By analyzing the results of the absolute density of blends we can divide them into three groups. The first group consists of B1-B3 blends (absolute density above 2.1 g/cm^3), the second group is B4-B6 blends - about 1.9 g/cm^3 and B7 blend with the lowest absolute density of 1.7 g/cm^3.

Table 3. Absolute density of the fertilizer mixturers

B1	B2	B3	B4	B5	B6	B7
2,149	2,127	2,187	1,937	1,915	1,975	1,672

Based on the results obtained, of the specific and absolute density the AI coefficient was detemined and the result is shown in the diagram in Figure 3:

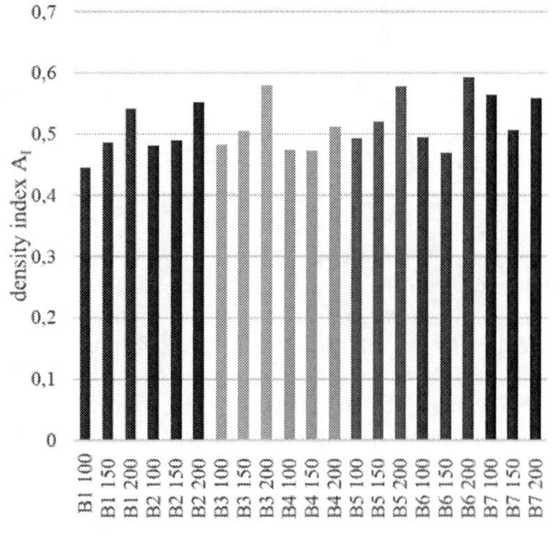

Fig. 3. Density index of granulates.

The level of the compaction factor has been set at an equilibrium level for the given pressure and is in the following ranges: 0.44-0.49 g/cm^3 (pressure 100MPa), 0.47-0.51 g/cm^3 (pressure 150MPa) 0.51 - 0.59 g/cm^3 (200MPa pressure). The obtained results indicate that, regardless of the contribution of components to the mixture at the given pressure, we obtain a similar degree of compaction. This means that the difference in the obtained results of the specific density of the obtained granules results from the different absolute density of the ingredients forming the blend.

Based on the measurements of absolute density of raw materials and specific density of granulates, porosity p_Z of granulates was determined (Fig.4).

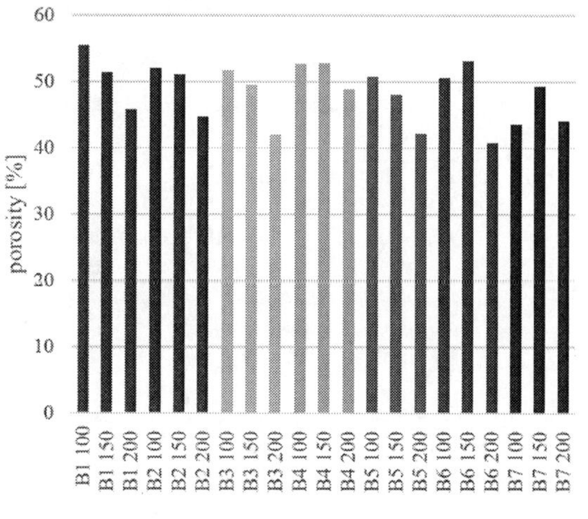

Fig. 4. Porosity of granulates.

The obtained values indicate that the granules made at a given pressure level have a porosity of 41-55%. The presence of pores in the liver at this level is a process-relevant information. This means that it is possible to increase the granule's density by increasing the agglomeration pressure. From a utility point of view, this means that the granules thus produced have at least 40% of the space in which the nutrient-releasing water can penetrate during the decomposition of the granule in the soil.

4 Conclusion

Analysis of the obtained results, confirm that in the majority of cases, the increase of the agglomeration pressure results in an increase in the obtained specific density of the granules, which is a natural phenomenon, as free spaces (pores) are minimized during compression.

Analysis of the density index A_I showed that, regardless of the contribution of components to the mixture at the given pressure, we obtain a similar degree of compaction. On the other hand the specific density of the obtained granules depends mainly on the absolute density of the ingredients forming the blend.

From the practical and technological point of view, the specific density of the granules should be at a comparable level (with the uniform geometry of the granules, the distance on which the granules are spread is determined by specific density) In such case, homogeneous density can be achieved by selecting a compaction pressure to obtain the density 1 g/cm^3. However. That's why further research on variation of agglomeration pressure should be carried out to obtain granules of comparable stability. It is also important to determine if the nutrient release from such granules is similar. The porosity of the granules causes that the granules can be easy penetrated by water what should stimulate fertilizer ingredients release from the granules.

Research was funded by the National Centre for Research and Development and the National Fund for Environment Protection as a part of GEKON program - project no. **GEKON1/05/214543/38/2015** *"Proecology production of organic and mineral fertilizers based on waste: by -porducts of combusiton and biogasification of biomass"*.

References

1. M. Bavutti, L. Guidetti, G. Allesina, A. Libbra, A. Muscio, S. Pedrazzi, Energy Procedia **45**, 1344-1353 (2014). doi:10.1016/j.egypro.2014.01.141
2. G. Grinzi, L. Guidetti, G. Allesina, A. Libbra, P. Martini, A. Muscio, A. Proceedings of 20th European Biomass Conference and Exhibition (2012)
3. A. Kowalczyk-Juśko, M. Szymańska, *Poferment nawozem dla rolnictwa* Fundacja na rzecz Rozwoju Polskiego Rolnictwa, Warszawa, ISBN 978-83-937363-6-2 (2015)
4. Niemiec M, Sikora J., Szeląg-Sikora A., Kuboń M., Olech E., Marczuk A. Przemysł chemiczny 96(**3**),685-688 (2017)
5. R. Vismara, R. Canziani, F. Malpei, S. Piccinini, *Biogas da agrozootecnia e agroindustria* Dario Flaccovio Editore, ISBN 978-88-579-0020-9 (2011)
6. C. Riva, V. Orzi, M. Carozzi, M. Acutis, G. Boccasile, S. Lonati, F. Adani, Sci. Total Environ. **547**, 206-214 (2016)
7. Kuźnia M., Jerzaka W., Łykob P., Sikora J. Journal of Power Technologies. 95(2),158-165, (2015)
8. J. A. Alburquerque, C. Fuente, A. Ferrer-Costa, L. Carrasco, J. Cegarra, M. Abdad, M. P. Bernal, Biomass Bioenergy **40**, 181-189 (2012)
9. Council Directive 91/676/EEC of 12 December 1991 concerning the protection of waters against pollution caused by nitrates from agricultural sources
10. H. Döhler, P. Schliebner, *Verfahren und Wirtschaftlichkeit der Gärrestaufbereitung* Darmstadt: KTBL (2006)
11. X. J. Huang, N. He, M.D. Liu, D. Yang, Y. J. Wang, Appl. Mech. Mater. **535**, 523-527 (2014)
12. L. C. Ram, R. E. Masto, Earth Sci. Rev, **128**, 52-74 (2014)
13. T. Kumar, K. Tedia, D. Devdas, J. Progressive Agric. **6,** 1 78-82 (2015)
14. K. Mudryk, M. Wróbel, *The acquisition and processing of biomass of Virginia mallow for energy purposes.* ed. W. Piekarski, Wieś jutra, Warszawa, ISBN 978-83-62815-10-4 (2011)
15. M. Wróbel, J. Frączek, S. Francik, Z. Ślipek, K. Mudryk, Available at: http://tf.llu.lv/conference/proceedings2013/Papers/121_Wrobel_M.pdf (2013)
16. G. Maj, A. Kuranc, *Wybrane problemy z zakresu ekoenergii i środowiska* ed. W. Piekarski, Towarzystwo Wydawnictw Naukowych LIBROPOLIS, Lublin, ISBN 978-83-63761-46-2 (2014)
17. K. Mudryk, M. Wróbel, Autobusy **13**, 10 166-169 (2012)
18. S. Mani, L. Tabil, S. Sokhansanj, Biomass Bioenergy **30**, 648-654 (2006)
19. A. Karbowniczak, M. Wróbel, J. Rymar, Episteme **26** vol. 1 255-263 (2015)
20. M. Wróbel, J. Frączek, M. Jewiarz, K. Mudryk, K. Dziedzic, Agricultural Engineering 20, 4 207-217 (2017)
21. PN-EN ISO 18134-3:2015-11

BIO Web of Conferences **10**, 02036 (2018) https://doi.org/10.1051/bioconf/20181002036
Contemporary Research Trends in Agricultural Engineering

Influence of extruder screws speed and process temperature on the extrudate shape changes of the maize-spelt blends

Żelaziński Tomasz [1,*], *Ekielski* Adam [1], *Siwek* Adam [2], and *Słoma* Jacek [1]

[1]Department of Production Management and Engineering. Warsaw University of Life Sciences– SGGW, ul. Nowoursynowska 164, 02-787 Warsaw, Poland
[2]Food Technology and Gastronomy Institute, Lomza State University of Applied Sciences, Akademicka 14, 18-400 Łomża, Poland

Abstract. The objective of the study was examination of changes in the shape factors of extruded products, which occur as a result of different settings of the extrusion process variables. Samples analysed included products created by means of the extrusion process from a mixture of spelt flour and cornmeal, with the share of spelt at 70 to 100%. The samples were made with the use of a co-rotating twin screw extruder. Two speeds of extruder screw rotation (300 and 350 rpm) as well as two levels of temperature (120 and 140°C) were set during the investigation. The samples obtained were photographed in a light box, following which they underwent an image analysis with the use of specialist vision software. Four shape-related factors were determined: area, elongation factor, Heywood circularity factor and compactness factor. It was determined that the product shape changed significantly depending on the share of spelt flour in the mixture. Moreover, it was observed that change in the screw rotation speed within the analysed range may cause material changes in the shape of particular extrudates.

1 Introduction

The shape of extruded products is of significant importance not only due to typically marketing aspects, but also from the technical perspective related to further processing thereof. Authors of numerous publications indicate that an evenly porous structure on the products may, for example, improve adhesion of various spices applied at further stages of the production process [1]. Properly selected shape and granulation of products used as toppings may also enable thorough covering of foods [2, 3]. Other shape parameters of products are needed for optimum selection of the grinder or classifier sieves on the production line of, for example, extruded edible coating crumbs [2, 4]. Therefore, the product shape is a parameter which may eventually influence efficiency of subsequent technological processes [5, 6]. Despite numerous advantages of the extrusion process, production of extruded products characterised with regular and repeatable shapes is not an easy process and is still the subject of many studies [7]. In fact, the final product shape depends on numerous parameters of the extrusion process, including composition and humidity of the mixture, process temperature, screw geometry and rotational speed or die nozzle geometry [8, 9, 10, 11]. If a cutter is used, setting of the knife rotational speed is also important. Thus, to ensure a regular shape of manufactured products, constant process monitoring is necessary, e.g. with the use of vision systems, and a quick response to extrusion process disruptions. Consequently, the application of quick vision techniques opens new possibilities of extrusion process control, involving simultaneous product quality control and provision of feedback allowing process adjustment and control.

The introduction of quick vision based analysis with the use of specialist software is, consequently, the subject of numerous studies related to the production process of not only extruded products, but also many other foods [12, 13, 14]. Not only do studies in this area cover investigations into the external shape of products, but also into the shape of pores making up the internal product structure. More detailed analyses include research into the internal 3D product structure [15, 16].

Thus, the objective of the study was to investigate into changes in selected shape parameters of the whole extruded products, which occur as a result of different extrusion process variables settings.

2 Materials and Methods

2.1 Materials and extrusion – cooking

The material used in the investigation was spelt flour and cornmeal (granulation: 0.16 – 0.60 mm). The products were purchased on the local market, Poland.

* Corresponding author: tomasz_zelazinski@sggw.pl

Table 1. Characteristics of raw materials used in the experiment.

Raw material	Corn groats [% d.m.]	Spelt [% d.m.]
Moisture	13,4 ±0.34	13,2 ±0.33
Starch	74 ±1.85	56,5 ±1.425
Proteins	8.3 ±0.21	11,7 ±0.29
Total fat	0.7 ±0.02	2,6 ±0.17

Mixture compositions were prepared using the above ingredients, with spelt flour as the main component at 70, 80, 90 and 100%. The mixtures were prepared in a ribbon mixer where the ingredients were additionally dampened to the moisture level of 14%. Before commencement of the extrusion process, moisture level control of the dampened material was performed with the use of a WPS 210 moisture balance with the accuracy of 0.001g (prod.: Radwag). Prepared as above, the material was thermally processed in an Evolum 25 twin screw extruder (prod.: CLEXTRAL, France) with the screw length to diameter ratio of L:D 27:1. The extruder matrix contained one round outlet opening with the diameter of 2.5 mm. In the tests, the plasticising section temperature was set at 120 and 140 °C. The distribution of temperatures in the extruder cylinder (from the plasticising section to the feeding section, respectively) was 120, 120, 110, 90, 80, 60 and 140, 120, 110, 90, 80, 60 °C. Extruder screw rotation was changed during the tests, at 300 and 350 Rpm. Material flow rate: 5.5 kg.h-1. The extruder head was equipped with a three-legged cutter, which enabled creation of products with the length of 7.5 mm ±0.5 mm. After cooling, extrudates were stored in plastic bags.

2.1 Extrudate properties

The subsequent stage of the investigation involved photographing of randomly selected whole extrudate samples. This was performed with the use of a light boxed fitted with Philips daylight fluorescent lamps with the colour temperature of D65 and a colour Basler A2500 – 14uc (5Mpx) camera (producer: Basler, Ahrensburg, Germany). Before the test, the samples were placed on black paper so that they did not touch each other's edges. Particular photographs were taken in 5 repetitions (a single set consisted of 50 samples). Acquisition of the photographs was performed with the camera software (pylon Viewer). Subsequently, the photographs taken were subjected to a detailed 2D vision analysis with the use of Ni Vision Assistant 2015 (Producer: National Instruments, USA), with 4 identified fators characterising the features of extruded product shape: Area, Elongation Factor, Heywood Circularity Factor and Compactness Factor (Equations 1, 2, 3).

$$\text{Elongation Factor} = \frac{F}{RF_b} \quad (1)$$

$$\text{Compactness Factor} = \frac{A}{W \cdot H} \quad (2)$$

$$\text{Heywood Circularity Factor} = \frac{P}{2\sqrt{\pi A}} \quad (3)$$

Where:

A – Area
W - Bounding Rect Width
H - Bounding Rect Height
F - Max Feret Diameter
RF_b - Equivalent Rect Short Side (Feret)
P – Perimeter

Calculation of particular factors was carried out pursuant to the methods outlined in the IMAQ Vision Concepts Manual 2005. Image analysis covered three levels of image transformation, which were used for analysing subsequent tested samples (Fig. 1). The figures obtained were subjected to Annova statistical analysis and correlation factors were identified, with the data presented in a box-plot. Data analysis was performed with the StatSoft, Inc. (2013), STATISTICA software.

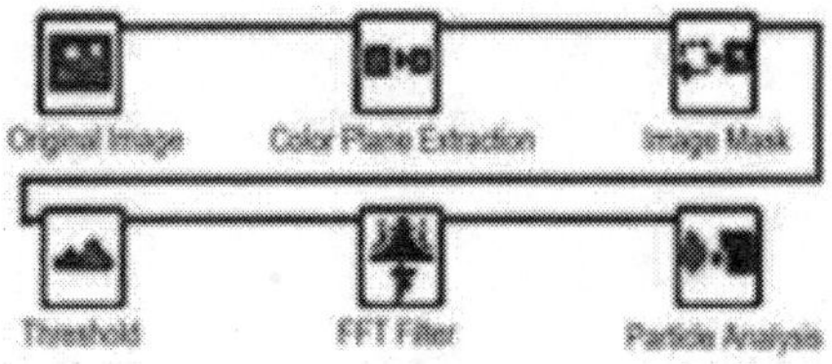

Fig. 1. Tools and settings applied in the Ni Vision Assistant image analysis. Color Plane extraction (RBG Red Plane), Threshold (Manual Threshold Range – 52), FFT Filter (Truancation frequency - 15%, Mode - Low pass).

3 Results and discussion

Analysis of the investigation results led to an observation that the size and shape of extruded products changed along with changes in the share of spelt in the mixture and extrusion process parameters. Significant changes were observed in all analysed samples. In the case of Area analysis of photographed extrudate samples, it was stated that the Area of particular samples, fig. 2a and 2b, increased along with a decreasing share of spelt in the mixture (irrespective of changes in the temperature and rotation of the extruder screw). This can be justified as increase in the share of cornmeal in the mixture (typical structure building ingredient) contributes to growth in expansion parameters (e.g. radial expansion) of many extruded cereal ingredients, including wheat [10, 17]. Therefore, based on the analysis of a single factor – namely, the area of photographed products – one can state whether change in process parameters involves any change in external "visual" quality parameters of the extrudates

Table 1. Correlation coefficients for spelt share (70, 80, 90 and 100%) in the mixture according to various shape coefficients.

Variable	Extrudate produced at temp. 120 °C, screws speed 300 Rpm	Extrudate produced at temp. 140 °C, screws speed 300 Rpm	Extrudate produced at temp. 120 °C, screws speed 350 Rpm	Extrudate produced at temp. 140 °C, screws speed 350 Rpm
Area [mm²]	0.931*	0.926*	0.863*	0.904*
Elongation Factor	-0.949*	-0.902*	0.783*	0.785*
Heywood Circularity Factor	-0.921*	-0.898*	0.876*	0.709*
Compactness Factor	0.928*	0.939*	-0.709*	-0.731*

Correlation coefficients are significant for $p < 0.05$. * - for any spelt share in the compound measure were been 5 repeated.

As variability in particular items is very high (correlation factor from 0.931 to 0.863 for different process settings), this may be at the same time a very sensitive quality indicator (table 2). Analysis of the area of photographed samples is, however, not a very precise parameter and, therefore, other factors defining the extrudate shape were included in subsequent tests.

a)

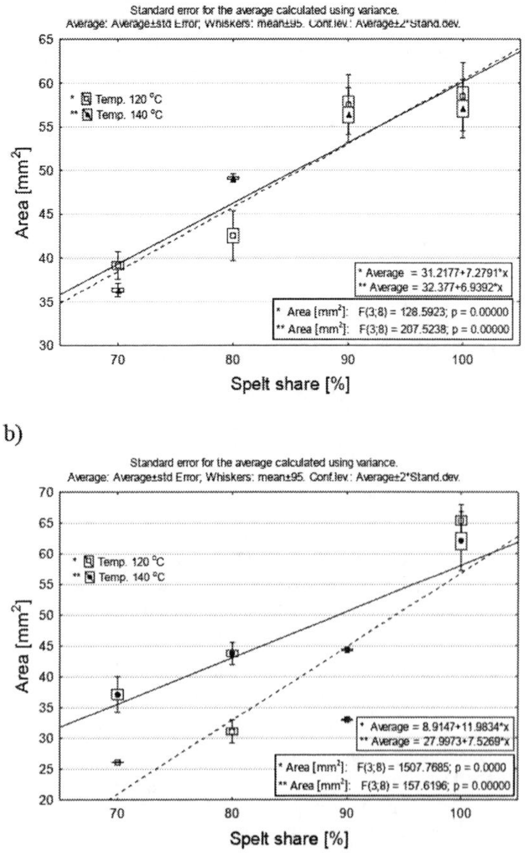

Fig. 2. Effect of spelt flour in extruded blend on the "Area" coefficient extrudates: a) extruder screw rotation 300 Rpm. b) extruder screw rotation 350 Rpm. Test F results and values (ANOVA) are presented for each temp. process.

Figures 3a and 3b illustrate changes in the extrudate Eongation Factor. It was determined that at extruder screw rotation speed of 300 Rpm, the Elongation Factor fell along with increase in the share of spelt in the mixture, whereas at extruder screw rotation speed of 350 Rpm, values of the factor increased. Thus, assuming that other process parameters will not be changed, increase in screw rotation speed itself allows obtaining a product characterised with different elongation degrees. It was moreover determined that the elongation factor changed as a result of temperature increase from 120 to 140 °C, yet the changes were not as intensive as in case of the other parameters. However, an opposite influence of temperature onto the elongation factor was found for samples produced at the screw rotation speed of 350 Rpm. The correlation factor for samples obtained at the screw rotation speed of 300 Rpm was high at -0.949 and -0.902, whereas at the screw rotation speed of 350 Rpm the correlations were 0.783 and 0.782. Elongation factor tests and other shape factors on food products were also performed [18]. The equivalent for the sample elongation visual tests is the extrudate lengthwise expansion factor. Investigations of this kind are commonly described in the literature and they cover changes in lengthwise expansion as a result changes in the mixture parameters, screw rotations, nozzle geometry, process temperature or mixture moisture [19, 20].

a)

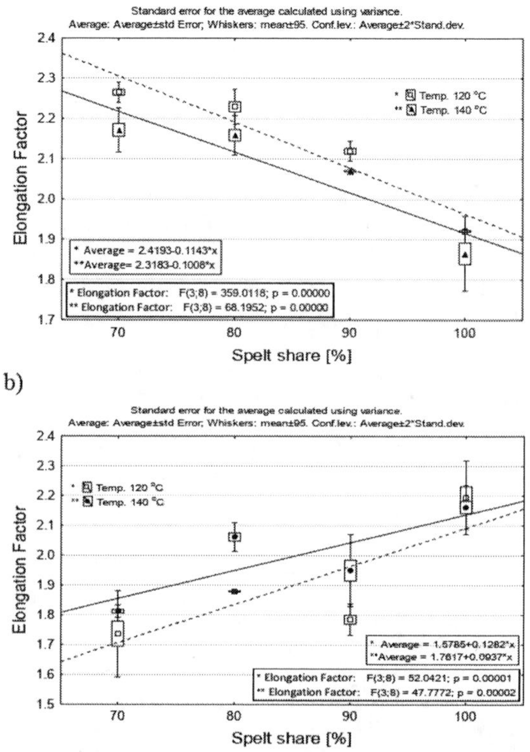

Fig. 3. Effect of spelt flour in extruded blend on the Elongation Factor extrudates: a) extruder screw rotation 300 Rpm. b) extruder screw rotation 350 Rpm. Test F results and values (ANOVA) are presented for each temp. process.

Figures 4a and 4b illustrate changes in the circularity factor for particular extruded products. In these investigations, the parameter defines the possibility to achieve a round shape of extruded products with the use of a round nozzle. It was determined that, along with an increasing share of spelt in the mixture, the circularity factor at the pre-set screw rotation speed of 300 Rpm was decreasing towards the value of 1 and, therefore, products with a higher share of spelt are more round. Significance of the changes is confirmed by the high correlation factor at -0.921 and -0.898, respectively. On the other hand, increase in the rotation speed to 350 Rpm has an opposite effect (increase in the value of this parameter along with growing share of spelt); thus, the product will have a less round shape (correlation factor of 0.876 and 0.709). It was also found that temperature change from 120 to 140°C (350 rpm) in the case of spelt share at 70 and 80% involves an increase in the circularity factor value. This phenomenon is also opposite as compared with the result of test performed at the rotation speed of 300 rpm and proves that, in this case, temperature change in combination with changes in rotation may also have significant impact onto the shape of extruded products. The circularity coefficient studies are common in extruded products research in both the food and pharmaceutical industries [21].

a)

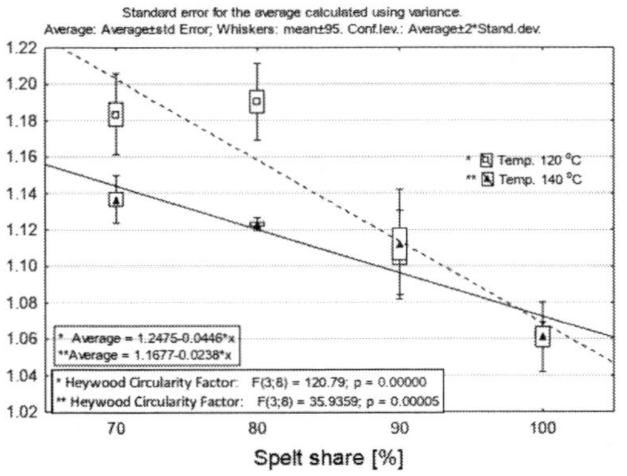

b)

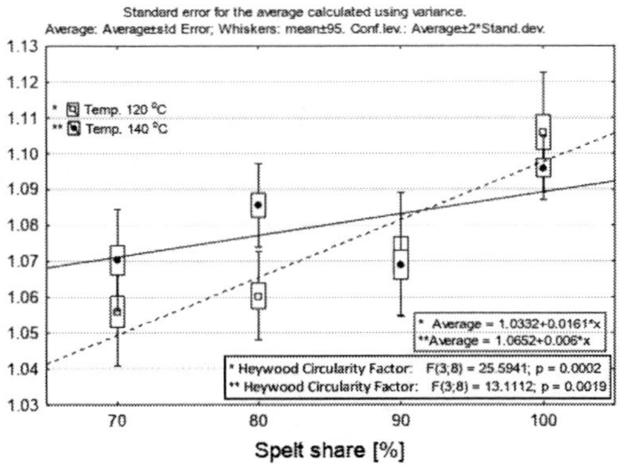

Fig. 4. Effect of spelt flour in extruded blend on the Heywood Circularity Factor extrudates: a) extruder screw rotation 300 Rpm. b) extruder screw rotation 350 Rpm. Test F results and values (ANOVA) are presented for each temp. process.

Figures 5a and 5b illustrate the results of compactness factor analyses. The investigations demonstrated clearly that change in the extruder screw rotation speed had clear influence onto that parameter, which is indicated by high differences in the course of charts for the rotation speeds of 300 Rpm (growing values) and 350 Rpm (falling values). In this case also, variability of results was characterised by high correlation factors at, respectively, 0.928 and 0.939. Another noticeable element was the influence of temperature which, at increased screw rotation speed, resulted in an increased value of the factor (Heywood circularity factor). Thus, also in this case, a change in the range of rotation speed and temperature dramatically changes the shape of extruded products.

a)

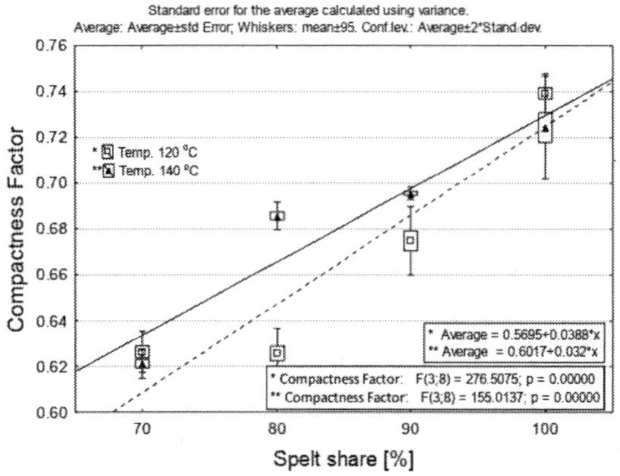

b)

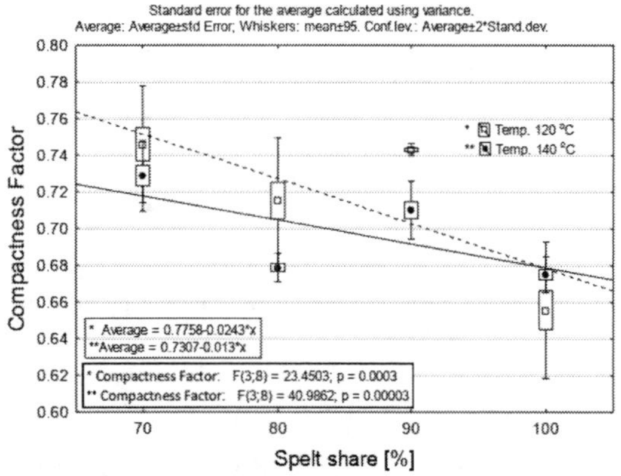

Fig. 5. Effect of spelt flour in extruded blend on the Compactness Factor extrudates: a) extruder screw rotation 300 Rpm. b) extruder screw rotation 350 Rpm. Test F results and values (ANOVA) are presented for each temp. process.

The compactness factor defines the shape of product which allows best adaptation of particular extrudates to one another and, therefore, the higher the value of the factor, the higher number of similar products will fit within a given area. Determination of the value of this factor has a particular useful meaning, e.g. in case of application of such products as topping or semi-products in layered products.

4 Conclusions

The extrusion process parameters which determine significant changes in the shape factors are mainly the mixture composition (spelt/corn) and the extruder screw rotational speed. It was determined that change in the screw rotational speed from 300 to 350 Rpm may have an opposite influence onto changes of extruded product shape factors. Thus, changing the screw rotational speed alone may involve a significant change of the extruded product shape, e.g. from round to elongated. Along with changes in the rotational speed of the screw, opposite influence of temperature onto the final product shape is observed. Vision analysis considering the sample area and elongation factor, Heywood circularity factor and compactness factor provides important information regarding the influence of changes in the extrusion process parameters onto shaping of extruded products.

References

1. S.W. Rosenthal, U.S. Patent No. 4,943,438. Washington, DC: U.S. Patent and Trademark Office, (1990).
2. C.K. Kamalakanth, J. Bindu, K. Bijith, C.N. Ravishankar, T.K.S. Gopal, Development and comparison of extruded breading with conventional bread crumbs for coating fish balls. ICAR, (2013).
3. L. Strauta, S. Muižniece-Brasava, I. Gedrovica, Chemical Technology **66,** 1, 34-39 (2015).
4. S. Muley, T Nandgude, S. Poddar, Asian Journal of Pharmaceutical Sciences **11**, 6, 684-699. (2016)
5. A.M. Bouwman, J.C. Bosma, P. Vonk, J.H.A. Wesselingh, H. W. Frijlink. Powder Technology **146,** 1, 66-72 (2004)
6. A. Ekielski, T. Żelaziński, P. Zdanowska, I. Florczak, INMATEH-Agricultural Engineering, **43**, 2 (2014)
7. J. Fan, J.R. Mitchell, J. Blanshard, International journal of food science & technology **31**, 1, 55-65 (1996)
8. L. Alvarez-Martinez, K.P. Kondury J.M. Harper, Journal of Food Science **53**, 2, 609-615 (1988)
9. C.I. Moraru, J.L. Kokini, Comprehensive Reviews in Food Science and Food Safety **2**, 4, 147-165 (2003)
10. Q.B. Ding, P. Ainsworth, A. Plunkett, G. Tucker, H. Marson, Journal of Food Engineering **73**, 2, 142-148. (2006)
11. H. Cheng, A. Friis, Food and Bioproducts Processing **88**, 2, 188-194 (2010)
12. T. Brosnan, D.W. Sun, Journal of food engineering **61**, 1, 3-16 (2004).
13. C.J. Du, D.W. Sun, Journal of food engineering **72**, 1, 39-55 (2006)
14. C.L.S Lau, Q. Yu, V.Y. Lister ,S.L. Rough, D.I. Wilson, M. Zhang, Chemical Engineering Research and Design **92**, 11, 2413-2424 (2014)
15. P. Babin, G. Della Valle, R. Dendievel, D. Lourdin, L. Salvo, Carbohydrate polymers **68**, 2, 329-340 (2007)
16. P. M. Falcone, A. Baiano, F. Zanini, L., Mancini, G. Tromba, D. Dreossi, M.A.D. Journal of Food Science **70** 4 (2005)
17. A. Arhaliass, J. Legrand, P. Vauchel, F. Fodil-Pacha, T. Lamer, J.M. Bouvier, Food and Bioprocess Technology **2**, 2, 186-193 (2009)
18. C. Costa, F. Antonucci, F. Pallottino, J. Aguzzi, D.W. Sun, P. Menesatti, Food and Bioprocess Technology **4**, 5, 673-692 (2011)
19. G. Shirani, R. Ganesharanee, Journal of food engineering **90**,1, 44-52 (2009)
20. T. Żelaziński, Annals of Warsaw University of Life Sciences-SGGW, **57**, 49-55 (2011)
21. S. Almeida-Prieto, J. Blanco-Méndez F.J. Otero-Espinar, Journal of pharmaceutical sciences, **93**, 3, 621-634. (2004)

AUTHOR INDEX

AUTHOR INDEX